Frontiers in Civil Engineering

(Volume 3)

Numerical Methods and Implementation in Geotechnical Engineering – Part 1

Authored by

Y.M. Cheng

School of Civil Engineering, Qingdao University of Technology, Qingdao, China

Department of Civil and Environmental Engineering, Hong Kong, Polytechnic University, Hong Kong

J. H. Wang

National Base of International Science and Technology Cooperation, for Intelligent Manufacturing Service, Chongqing Technology and Business University, Chongqing, China

Chongqing South-to-Thais Environmental Protection Technology, Research Institute Co. Ltd., China

L. Liang

School of Civil Engineering, Qingdao University of Technology, Qingdao, China

W. H. Fung Ivan

Department of Architecture and Civil Engineering, City University of Hong Kong, Hong Kong

Frontiers in Civil Engineering

Volume # 3

Numerical Methods and Implementation in Geotechnical Engineering – Part 1

Authors: Y.M. Cheng, J. H. Wang, L. Liang and W. H. Fung Ivan

ISSN (Online): 2468-4708

ISSN (Print): 2468-4694

ISBN (Online): 978-981-14-3739-7

ISBN (Hardback/Hardcover): 978-981-14-3737-3

ISBN (Paperback/Softcover): 978-981-14-3738-0

Bentham Science Publishers Pte. Ltd.
80 Robinson Road #02-00
Singapore 068898
Singapore
Email: subscriptions@benthamscience.net

BENTHAM SCIENCE

CONTENTS

PREFACE

For most of the geotechnical problems, particularly those related to real life problems, analytical solutions are usually not available. For both research and practical applications, numerical methods and computer programs are required for many cases. In the recent forty years, many numerical methods have evolved for various kinds of engineering problems. Engineers are now well adapted to the uses of different computer programs for the solution of engineering problems. There is however a major drawback in the current engineering practice in that most of the engineers are not familiar with the basics of the numerical methods, the methods of implementations and the limitations of the numerical methods/programs. In fact, to a certain extent, the methods of implementations and the limitations of the numerical methods are related. In many internal studies using different commercial numerical programs, the authors sometimes found noticeable or even completely different results with different programs or the same program with different default setting for a given problem, and this situation is not uncommon. For a problem with unknown solution, how an engineer assess the acceptability of the computer results is a difficult issue that needs serious attention. In several technical meetings in the Hong King Institution of Engineers, the authors have discussed with some engineers about the appreciation of the limitations of the daily-used engineering programs. If two computer programs can produce significantly different results, how an engineer determine the acceptability of the results actually require deeper knowledge about the basics of the numerical methods and implementations. Interestingly, the authors like to ask the students a question "Different answers can be obtained from different commercial programs. Which results should be accepted, and why should those results be accepted?". In general, the authors challenge the students (undergraduate and graduate students) every year for this question, and virtually this question is never answered properly. The problems in the assessment of the numerical results will also be discussed in this book, which is seldom addressed in other books or research papers.

The authors have participated in different types of geotechnical research and consultancy works in different countries, and has written a book *Frontier in Civil Engineering, Vol.1, Stability Analysis of Geotechnical Structures*, which is well-favored by many students, engineers and researchers. Most of the books on numerical methods seldom address the actual procedures in numerical implementations, but many postgraduates actually need to develop computer programs to consider special constitutive models, loadings, numerical methods, boundary conditions and other effects. In view of the limitations of most of the books at present, the authors would like to write a new book on numerical methods and the implementations based on their previous works, and this new book should be useful for senior undergraduates, postgraduates, engineers as well as researchers.

In this book, finite element method, optimization method, plasticity based slip line method, limit analysis method, distinct element method, Smoothed-Particle Hydrodynamics Method, Spectral Element Method and Material Point Method will be introduced. The present book will not cover dynamic problems which is a big topic, and hopefully this will be covered later by the authors in another book. The authors will also try to explain the methods of implementation for some of these methods through sample computer programs. Sample programs are given and discussed to assist students in developing programs for their own uses. These programs are not meant to be efficient or up-to-date, but will help the students in learning about the implementation of some numerical methods. This book should not be taken as a classical textbook, as the authors do not intend it to be. There are many new contributions to numerical methods in geotechnical engineering over the last 30 years, and many topics can be covered by individual books for detailed discussion. There is also no way for the authors to cover all numerical methods in details in this book. This book is a basic introduction to some more commonly used numerical methods in geotechnical engineering which have been used by the authors for teaching and research, with the discussion of some common commercial program problems, programming techniques and applications.

ACKNOWLEDGEMENTS

The authors would like to acknowledge the support from the Hong Kong Research Grant Council through the project PolyU 5128/13E, the National Natural Science Foundation of China (Grant No. 51778313), the Cooperative Innovation Center of Engineering Construction and Safety in Shangdong Blue Economic Zone. The present book is also partly supported by CityU Strategic Research Grant for unfunded GRF/ECS (SRG-Fd): Enhancement of Building Information Modelling (BIM) in Construction Safety by GIS: 4D Modelling, Geo-spatial Analysis and Topography Modelling (Preliminary Study) (CityU Project No. 7004631).

CONSENT FOR PUBLICATION

Not applicable.

CONFLICT OF INTEREST

The authors confirm that this chapter contents have no conflict of interest.

Y.M. Cheng
School of Civil Engineering, Qingdao University of Technology,
Qingdao,
China
&
Department of Civil and Environmental Engineering, Hong Kong,
Polytechnic University,
Hong Kong

CHAPTER 1

Introduction

Abstract: This chapter is an introduction to the requirement and the various problems that will be encountered during numerical modelling in geotechnical engineering. A large scale tunneling work in Australia will be used to illustrate the necessity to use numerical methods in some real life engineering problems. After that, the authors will introduce a series of numerical problems that may be encountered during the use of commercial engineering programs. Such problem cases may arise from various sources, and engineers are strongly advised to understand the basic principle of each commercial program and to assess the program output with care before accepting the results of analysis. Finally, some of the more important governing differential equations for geotechnical problems are discussed.

Keywords: Errors, Finite element, Governing differential equations, Modelling, Numerical methods, Slope stability, Tunneling.

1.1. INTRODUCTION

For most of the geotechnical problems, particularly those related to real life problems, analytical solutions are usually not available. The authors have carried out many research works and large scale practical projects, and in general, most of the works are complicated in both geometry, applications of loadings, construction sequences, material behavior, ground water conditions as well as other factors. As a good illustration, the construction of the Airport Link project in Brisbane at Australia is a good example (Cheng *et al.* 2019). The project is located beneath the railway embankment of the North Coast Railway line adjacent to Kalinga Park, and the site comprises a thick layer of soft clay. The Airport Link, which is one of the most complex roads and tunnel engineering feats in Queensland's history, will be the first major motorway linking Brisbane city to the northern suburbs and airport precinct. The Link is a 6.7km toll road, mainly underground, connecting the Clem 7 Tunnel, Inner City Bypass and local road network at Bowen Hills, to the northern arterials of Gympie Road and Stafford Road at Kedron, Sandgate Road and the East West Arterial leading to the airport. At one of the project sites, the tunnel section under the QR railway embankment at Toombul is constructed by box jacking technique. The significant size of the

Y.M. Cheng, J. H. Wang, L. Liang, W. H. Fung Ivan (Eds.)

launch box requires 85,000m^3 of spoil to be excavated under the railway embankment. Headwalls, canopy tubes and sidewall nails are constructed to retain the railway embankment for the excavation of the jacking shafts. The challenging ground conditions and requirements for the present project require the combinations of innovative ground support, construction methods and detailed and realistic analysis for the proper execution of the works. In this project, the site is mostly composed of soft clays which are susceptible to ground settlement problem during construction, and a typical section is shown in Fig. (**1.1**). The SPT value for the soft clay is less than ten, whereas the CPT friction ratio for soft clay ranges between 2% and 4% with a mean pore pressure of approximately 0.12 MPa (see also Table **1.1**). The SPT value for the firm clay is approximately 20, whereas the friction ratio for firm clay ranges between 4-8% with a mean pore pressure of approximately 0.38 MPa. The railway has to be maintained in operation during the whole construction to ensure the transportation, and the settlement of the soft clay must be maintained at a low level with minimal disturbance to the railway track. This is technically a very difficult problem, and the original construction proposal is to inject large amount of grout into the ground to stabilize it prior to excavation. However, the cost of the original scheme is extremely high so that a more economical alternative is considered. Ground improvement works underneath the QR railway embankment are hence required for the stability consideration during box jacking stages. A trapezoidal jet grout block constructed immediately behind the headwall is used as a gravity type retaining wall to reduce the earth pressures on the piled headwall. A smaller jet grout block is provided at the north west of the final jacked box location and is used as an anchorage to the northern sidewall nails. A low strength grout wall is installed west of the railway to provide a water cut-off for the TBM launch box. The grout wall is also used in the jacking scheme design to provide adequate anchorage to the geonails at the receiving pit side, eliminating an approximate 10m length of nail with significant time and cost savings. The resulting 'nail anchored' western grout wall can then be used to maintain slope stability, enabling initial excavations in the cut and cover receiving pit to commence early.

Table 1.1. Average properties of ground soil (Young's modulus determined from dilatometer, vane shear and CPT tests).

Soil	Undrained Shear Strength. (kPa)	Young's Modulus. (MPa)	Water Content. (%)	Plasticity Index
Soft clay	20	6	57	25
Firm clay	37	20	46	45

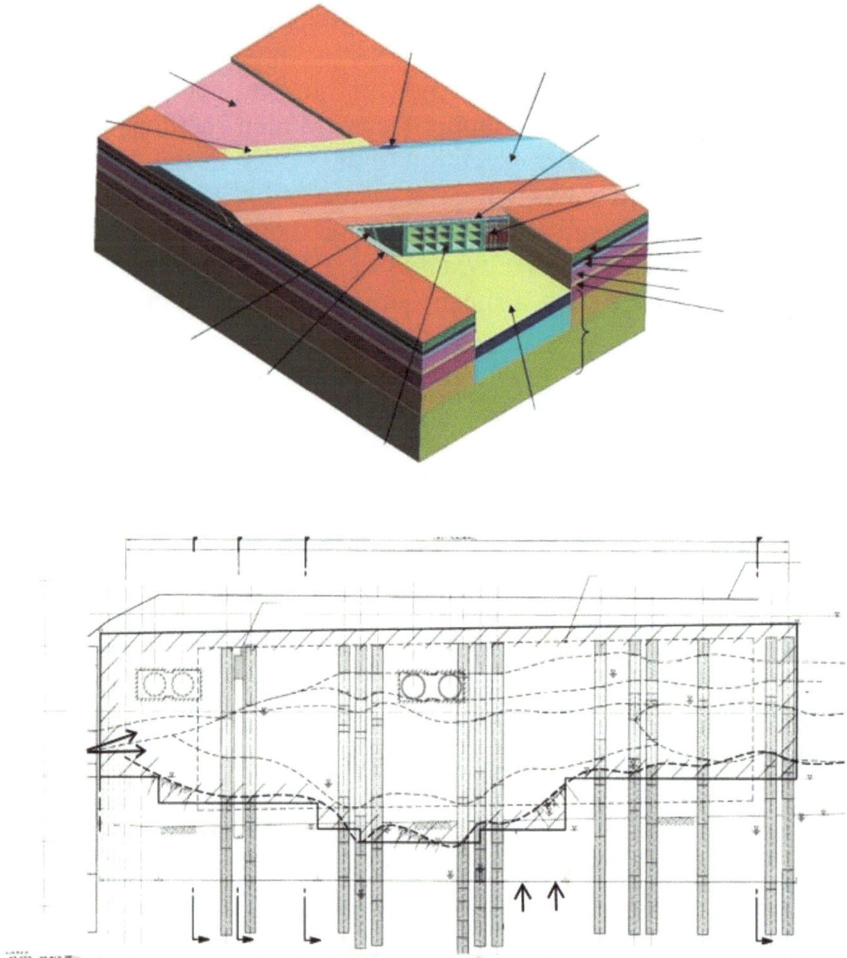

Fig. (1.1). (**a**) Geological condition for the tunnel project in soft clay, (**b**) a typical section of the tunnel work.

In order to optimize the ground improvement design, a combined fracture grouting and GFRP soil nails ground improvement scheme is proposed by the authors as the alternative solution (Cheng *et al.* 2013), and the cost of the alternative scheme is critically reduced to 50% of the original scheme. In the past, fracture grouting was mainly adopted for compensation grouting, and the combined use of fracture grouting and composite GFRP soil nails scheme for soil improvement which is adopted in this project is a pioneer work. The present project has received the Fleming Award in 2011 and Ground Engineering Award in Technical Excellent in 2012 in Australia for the satisfactory performance under such a difficult condition.

The proposed scheme is to utilise approximately 1.0m grid size geonails and face shield to stabilise the soft clay faced slope in advance of the box jacking operation. With the ground improvement from the geonails and the protection from the canopy tubes, a 60 degree cut slope is formed 0.2m in advance of the box, with the box advanced gradually behind the excavation face. The basic principle of the combined geonail and face shield scheme is to obtain some face support from a mining shield and therefore, to reduce the number of geonails required for face stability. Sufficient nails should be installed to provide suitable size shield working compartments, which are considered to be approximately $3m^2$ to $4m^2$.

For the present project, the authors have adopted three-dimensional analysis by program Flac3D, and the number of elements used in the analysis exceeds one million (Cheng *et al.* 2019) which is generated by the use of Patran. The construction of the jacked box is modelled in a realistic manner, and the sequences of computer modelling are shown in Table (**1.2**). For the analysis of the tunnelling process, a three-dimensional analysis was performed as shown in Fig. (**1.2**). The partial factors of safety adopted in the analysis are based on the Australian code AS5100:3. The soil model for different types of clay adopted is the elasto-plastic Cam-Clay model (with a partial factor of safety 0.65), while Mohr-Coulomb model is adopted for sand, fill and rock. Canopy tubes and headwall piles are modelled as elasto-plastic material (with a partial factor of safety 0.6) with a maximum prescribed bending capacity. The geonail is modelled in a way similar to that by Wei and Cheng (2009) (partial factor of safety 0.65 to bond strength and 0.8 to tensile strength). The numerical model is setup based on the jacking drawing series. The computer model needed to consider the complicated ground conditions, the installation and removal of the geonails (due to cutting of nails during excavation), the excavation with a slope surface at the tunnel face, the formation of yield zones (with revision of the stabilisation measures and re-analysis), and the installation of structural supports. The soil nails are modelled in a way similar to that by Wei and Cheng (2009).

Table 1.2. Modelling of Construction Process.

Phase No.	Description	Remarks
1	Initial condition (soil only)	Using 'gravity load' approach
2	Initial condition (railway surcharge)	Activate railway surcharge
3	Construct piling platform	Filling to piling platform levels
4	Apply working load	Activate 10kPa working load
5	Install Trapezoidal Grout Masses and Grout Blocks at receiving side	Change soil to grout properties
6	Install Headwall Piles and excavate to existing ground level	Activate Headwall piles and excavate to Approximately +5mAHD
7	Construct permanent roof slabs of CC421 and restraint beams at the west side	Activate the spring support at west side
8	Install canopy tube from east side and connect with the restrain beam at west	Activate canopy tube elements
9	Install geonails to stabilise west side grout block and commence with excavation in CC421	Activate the soil nail element(cable element) at west side
10	Excavated to -4.5m ADH in CC422 adjacent to south west grout block	Excavate to -4.5m ADH
11	Install fan-shaped sidewall nails from western embankment slope.	Activate fan shaped pile elements at south side
12	Install anchors and walers to CC422 side walls as excavation proceeds	
13	Fill canopy tubes and connect them to the restrain beam	Change properties of canopy tube
14	West: Dig to canopy soffit. East: Excavate to +4.8mAHD.	Excavation initiated at east side
15	Connect Canopies to east headwall piles and Excavate at East to +3.3mAHD.	Canopy tube is activated but NOT connected to the headwall pile at this stage
16	Install 1st geonail and then Dewater and Excavate to +1.8mAHD	The canopy tube is connected to the headwall pile at this stage.
17	Install Geonail to +2.8, Dewater and Excavate to –1.2	Activate geonails within +5.3mAHD and +2.8mAHD.and change the properties of the soil layer to the improved soil mass
18	Install Geonail down to –0.2, dewater and excavate to –4.2	Activate geonails within +2.8mAHD and +0.2mAHD and change the properties of the soil layer to the improved soil mass

(Table 1.2) cont.....

Phase No.	Description	Remarks
19	Install Geonail down to –3.2, dewater and excavate to –6.0	Activate geonails within +0.2mAHD and –3.2mAHD and change the properties of the soil layer to the improved soil mass
20	Install Geonail to –5.2, dig and dewater to –8.2mAHD	Activate geonails within –3.2mAHD and –5.2mAHD and change the properties of the soil layer to the improved soil mass
21	Install Geonail to –7.17 (1m above FEL), Dig and dewater to –8.7 (500mm over-excavation)	Activate geonails within –5.2mAHD and –8.7mAHD and change the properties of the soil layer to the improved soil mass. This stage for ultimate checking, not for movement prediction.
22	Cast Jacking Slabs	
23	JB2 jacked in 1/4	Box modelled with slope crest at about ¼ of the jacked box alignment, carry out the over excavation south end.
24	JB2 jacked in 1/2	Box modelled with slope crest at about at ½ of the jacked box alignment, carry out the over excavation middle of the jacking zone.
25	JB1 jacked in 3/4	Box modelled with slope crest at about at ¾ of the jacked box alignment, carry out the over excavation middle of the jacking zone.
26	JB1 jacked to the end	Box modelled with slope crest at about at the end of the jacked box alignment carry out the over excavation north side of the jacking zone.
23	JB1 jacked in 1/4	Box modelled with slope crest at about ¼ of the jacked box alignment, excavate maximum 200mm in advance of cutting shield and advance box in stages until in final position.
24	JB1 jacked in 1/2	Box modelled with slope crest at about at ½ of the jacked box alignment, excavate maximum 200mm in advance of cutting shield and advance box in stages until in final position.
25	JB1 jacked in 3/4	Box modelled with slope crest at about at ¾ of the jacked box alignment, excavate maximum 200mm in advance of cutting shield and advance box in stages until in final position.
26	JB1 jacked to the end	Box modelled with slope crest at about at the end of the jacked box alignment Excavate maximum 200mm in advance of cutting shield and advance box in stages until in final position.

The numerical models are established based on the arrangement of canopy tubes, sidewall nails and the front face geo-nails. Compartments are formed by temporary concrete blade walls at about 3.5m c/c and two intermediate working platforms with hydraulic steel tables in front. Removable breasting plates are fixed to the steel tables, which can be pushed against the slope face to provide positive face pressure. The compartment sizes have been determined mainly from the practical construction point of view, which is about 3.5m x 3.5m for the present project. During the box jacking excavation, arching of the clay within each compartment is required to maintain stability of the clay face within the compartment, therefore, bearing and frictional supporting force are induced to maintain the face stability. Based on this numerical model, the two-way arching effect of the soils between compartment walls can be modelled.

In order to obtain the individual effect of bulk excavation and jacking operation on the canopy tube and railway, the displacements are reset in the analyses before the jacking operation. During the simulation of jacking process, the displacements are reset in each excavation step to obtain the displacement change in each of the jacking advancement, and the procedures for the numerical modelling are given in Table (**1.2**).

The whole numerical simulation of each box jacking operation will be divided into four stages in each box jacking operation. The shield embedment and 200mm face excavation shall be simulation during the box jacking operation. When the box hits the culverts or timber piles, the obstruction shall be removed before the advancement of the shield. 1700mm overcut shall be allowed in the excavation process. The numerical analyses are divided into two jacking processes (JB1 and JB2). Each jacking process is subdivided into 5 stages. Stage 1 to 5 represent the JB2 jacking process and stage 6 to 10 represent the JB1 jacking process. The five stages excavations are located at 0m, 12.5m, 25m, 37m away from the headwall and 1m away from the west grout block. Each stage is subdivided into five steps. As the area required for over excavation in JB1 is much less than that in JB2, the effect of over excavation at JB1 will not be simulated. As the rapid loading and excavation sequence is demonstrated in the jacking processes, water seepage into the jacking is considered to be insignificant to the design. Therefore, undrained analyses are more suitable to reflect the behavior during the jacking operation.

Based on the three-dimensional finite difference analysis that involved approximately 1 million elements (Fig. **1.2**) to model the entire construction sequences, the stresses, ground settlement, loadings in canopy tubes settlement derived from the canopy tube installation, the bulk excavation in the shaft and the ground improvement study (*i.e.,* Geonail). Based on the results from numerical modelling, the stabilization measures are revised several times in order to maintain a balance between cost and stability of construction. Since there are vast amounts of computational results from the numerical analysis, only some selected results are shown in Fig. (**1.3**) for illustration.

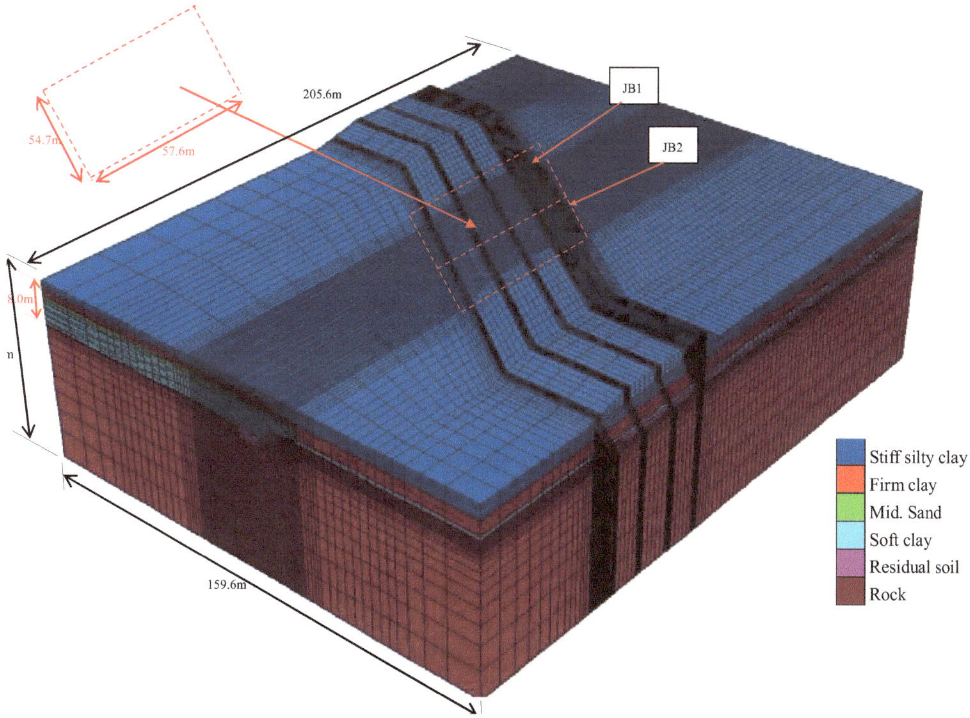

(a) Side view of finite difference mesh.

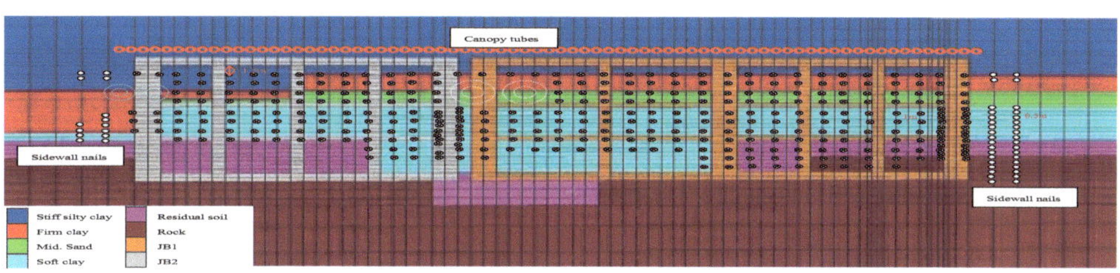

(b) Finite difference mesh for a normal section; the reinforcement is shown.

Fig. 1.2 cont.....

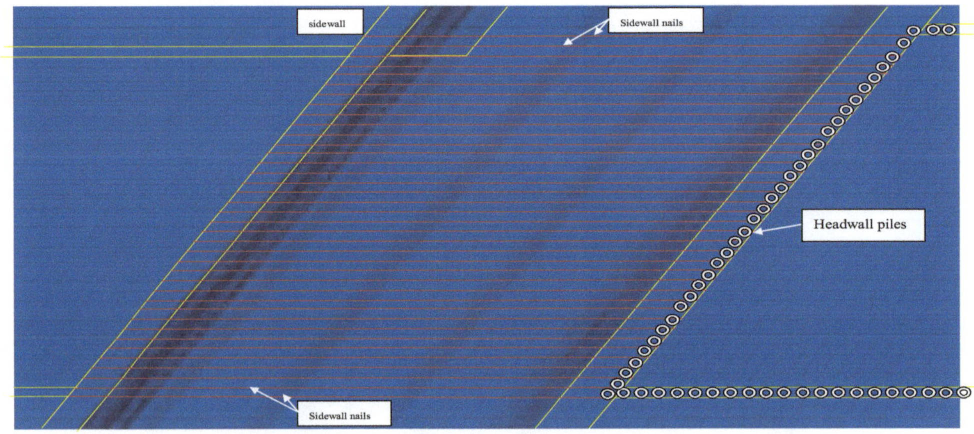

(c) Finite difference mesh; the headwall piles and side nails are shown.

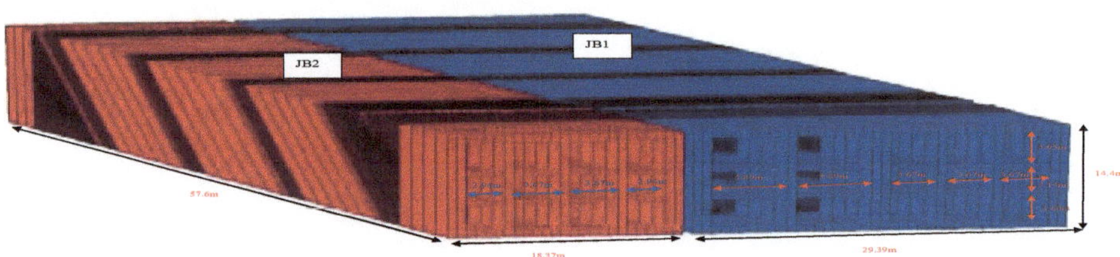

(d) Finite difference mesh during construction.

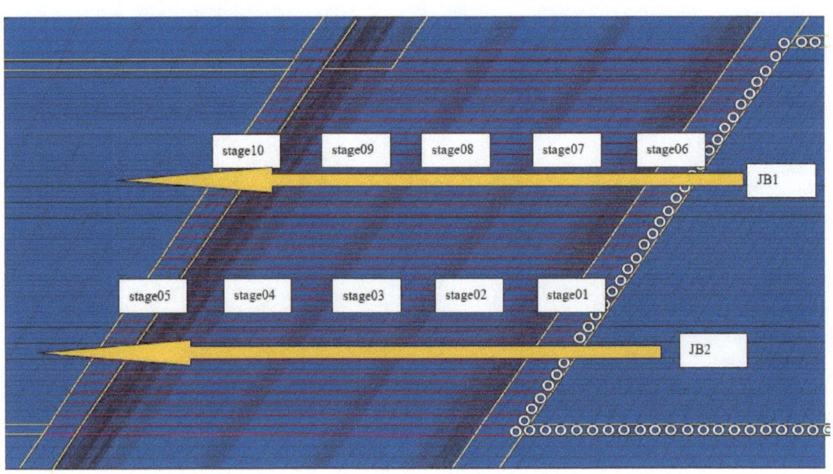

(e) Box jacking sequence in numerical modelling.

Fig. 1.2 cont.....

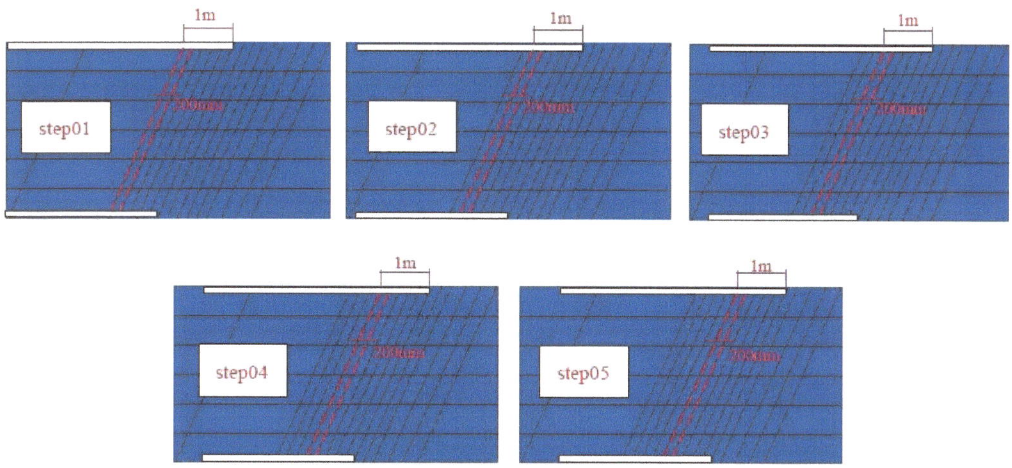

(f) Numerical simulations step in each jacking stage.

Fig. (1.2). Three-dimensional finite difference mesh in this study.

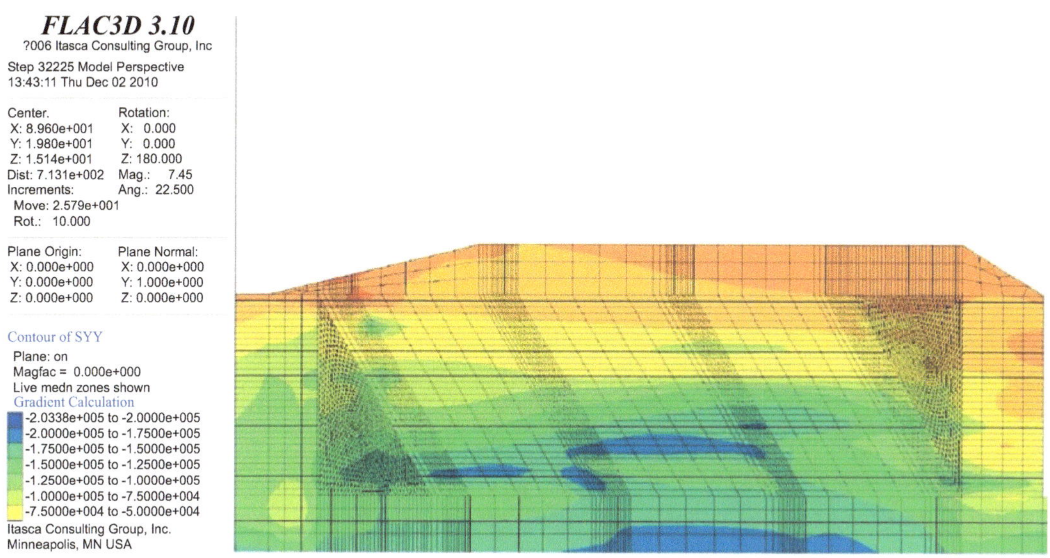

(a) σ_{yy} at sidewall position for stage 6, step 3.

Fig. 1.3 cont.....

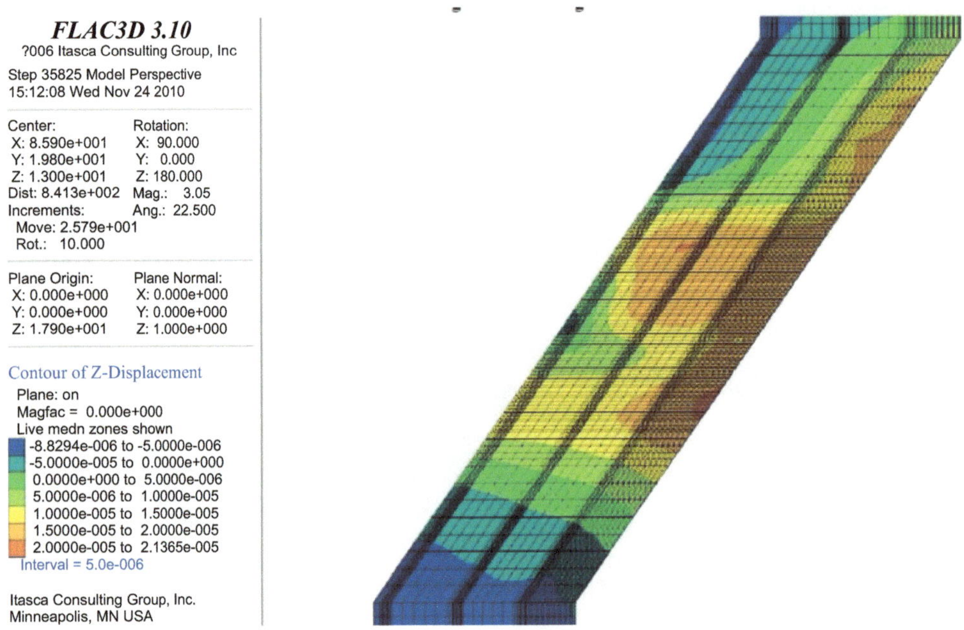

(b) Ground surface settlement at stage 10.

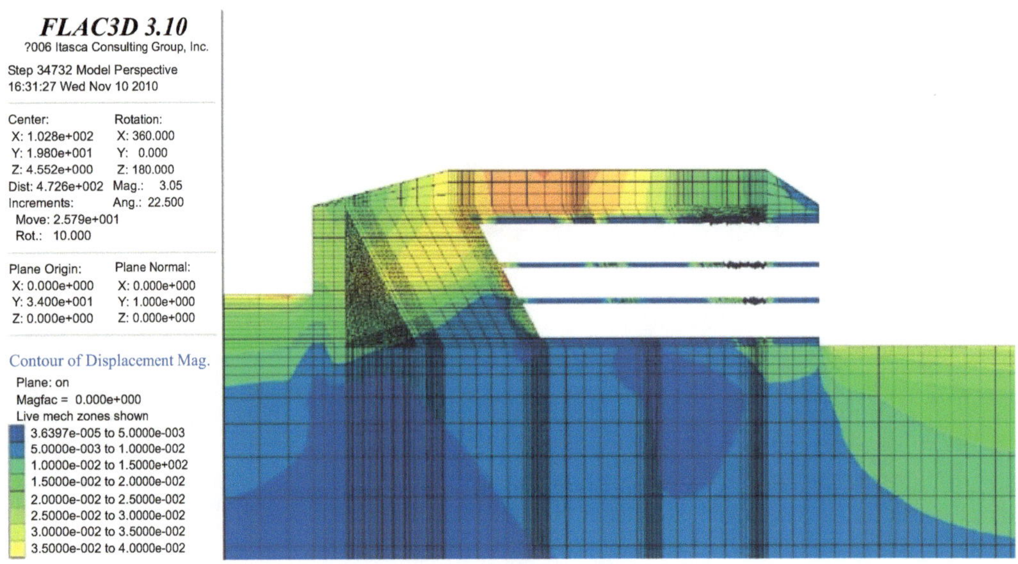

(c) Displacement contour at stage 8.

Fig. (1.3). Typical results of analysis.

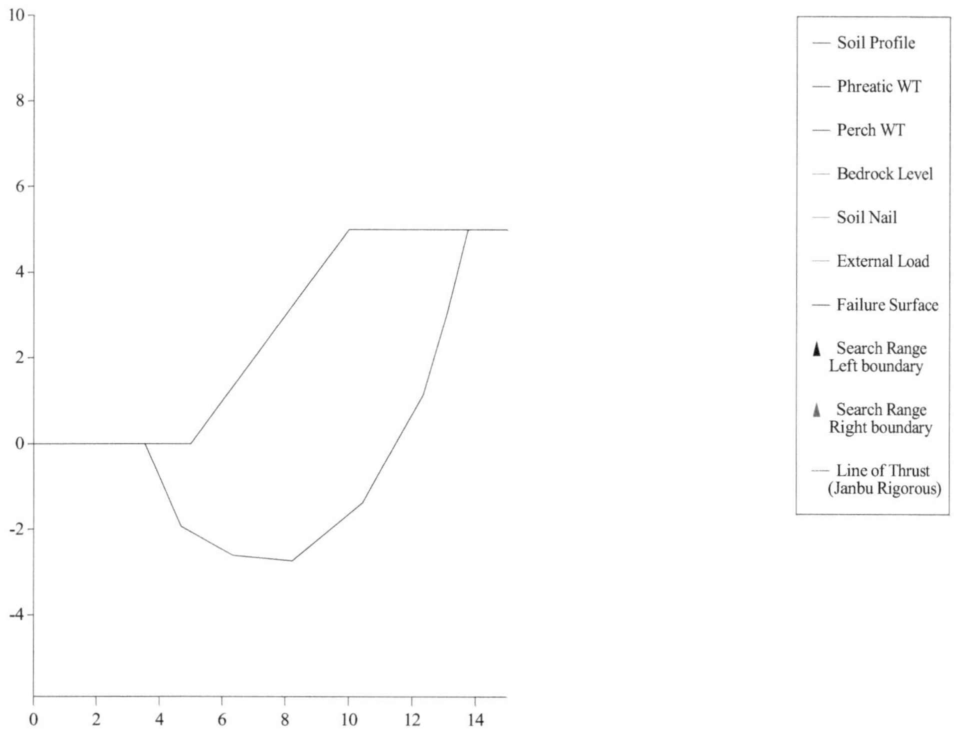

Fig. (1.4). A trial slip surface with two different factors of safety arising from the same commercial program using Janbu's simplified method.

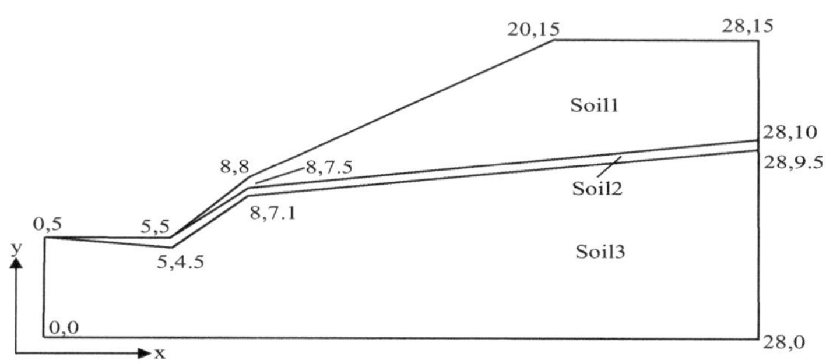

Fig. (1.5). Strange results from SRM analysis for a slope with a soft band.

From Table (**1.2**), Fig. (**1.2**) and Fig. (**1.3**), it is clear that the analysis of the present problem must relies on the use of numerical methods. No analytical solution can give the stress, displacement under all construction stages. In fact, most of the practical problems that the authors have encountered require the use of numerical methods and computer programs. Most of the engineers have adopted various engineering programs without questioning in the daily works. In particular, as the assessor for the professional engineers' examination for the Hong Kong Institution of Engineers, the authors have found that many engineers lack the basic understanding of the theory and limitations behind many engineering programs. It appears that some engineers are aware that unreasonable results may come out from the numerical analyses, and some tricks have been developed to overcome the surprising results (without any theoretical basis). However, most of the problems from computer programs are simply neglected, as most of the engineers lack the ability to assess the acceptability of numerical results from different computer programs.

The authors have carried out various internal studies on the acceptability of some major commercial geotechnical programs, and many surprising results have been found. In addition, the authors have also received many problem cases from engineers, and the assessment of these computer analysis results requires adequate fundamental knowledge of engineering as well as some knowledge about the procedures in the numerical implementation. In general, it is not easy to assess these problem cases. Some basic principles in engineering can however help to assess the acceptability of the numerical analysis. For example, the presence of a jump in the shear force for a raft foundation as given in Fig. (**1.7**) and Fig. (**1.8**) is considered to be not acceptable (even though the results come a world-famous program), as there is no point load or support at that jump location. The examples in this chapter illustrate how the authors assess the problems of different commercial programs, and the underlying principles include the use of standard problems with known solution, use of basic engineering principles, adoption of different programs or even methods. It is also true that many numerical problems cannot be assessed easily, and there is also a strong need to investigate the reasons behind these numerical problems.

An interesting example is an engineering report submitted by an engineer for the professional examination in Hong Kong. Part of the design involves a beam simply supported at two ends with several point loads and distributed load. The engineer has adopted a computer program instead of normal hand calculation in the analysis and design of such a simple problem. A very small but non-zero moment is obtained at the two ends of the beam, which contradicts the basic concept that the moments must be zero at simply supported ends. The authors feel regret that the engineer who is well experienced with the use of computer software for engineering analysis fails to explain the reason for such phenomenon, even though this problem is very minor in nature. Another interesting case is the slope stability problem as shown in Fig. (**1.4**). An engineer has adopted a commercial program for the analysis of the slope using Janbu simplified method. A factor of safety of 0.7 is given by the commercial program which is to the surprise of the engineer. The engineer then changes the default initial factor of safety from 1.0 to 1.4, and a converged factor of safety of 1.4 is obtained. The problem is later studied by the authors, and it is found that the first answer will give unreasonable internal forces which is not checked by the slope stability program (actually, most of the commercial programs do not check for this important issue). The authors have a chance to talk with that slope stability program developer about 10 years ago and have mentioned such phenomenon to them. To the author's disappointment, the software developer replies that the engineer should judge and accept the results instead of relying on the software developer to assess the acceptability of the results, while the acceptability of the results can actually be checked easily by the consistency and acceptability of the internal forces (a simple task only). The authors are also very disappointed that many engineers and some researchers (mostly research students) simply believe in the results from computer programs without adequate assessment. Although it is not easy to assess the acceptability of the computer results, the results from any computer program should be accepted with care. The authors have developed various engineering programs (include some commercial programs) for education and research purposes, and the authors find that such experience will enhance the understanding of the various problems in numerical modelling, implementation and computations. Interestingly, the authors also like to ask students to do some small projects with different commercial programs on different types of problems. Sometimes, noticeable or even completely different results are obtained from different programs using the same input parameters. The authors also like to ask the students a question "which set of results should be accepted, and why?". In general, the authors challenge the students (undergraduate and graduate students) every year for such question, and virtually this question is never answered properly. In the following, some problems as revealed by the small projects using some commercial programs are discussed.

Another interesting problem using finite element strength reduction method (SRM) is shown in Fig. (**1.5**), while the soil properties are shown in Table (**1.3**). This problem is considered by 4 different programs using 3 different domain sizes by SRM (Cheng *et al.* 2007). The critical solution using Spencer method is shown in Fig. (**1.6**) which looks reasonable, and the result is not affected by the domain size. On the other hand, the results by SRM as shown in Table (**1.4**) are disappointing, as it is not easy to conclude which result is reasonable. For some of the programs in Table (**1.4**), it is also interesting to find that different versions of the same program can give different results of analysis. For the engineers or researchers, what can they do with such results? During teaching, the authors always emphasize to the students not to believe blindly in any computer program without the knowledge and judgment. Besides these cases, the authors have also encountered different problems in programs for excavation and lateral analysis, pile driving signal analysis, plate and shell analysis.

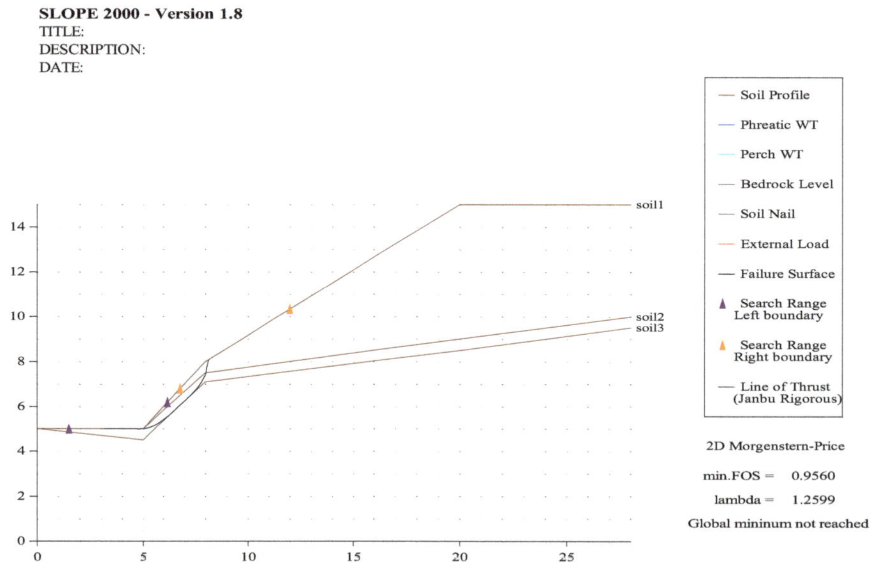

Fig. (1.6). Critical slip surface by Spencer method.

Table 1.3. Soil properties for Fig.1.5.

Soil Name	Cohesion (kPa)	Friction Angle (Degree)	Density (kN/m^3)	Elastic Modulus (MPa)	Poisson Ratio
Soil1	20	35	19	14	0.3
Soil2	0	25	19	14	0.3
Soil3	10	35	19	14	0.3

Table 1.4. FOS by SRM from different programs when c' for soft band is 0. The values in each cell are based on SRM1 (zero dilation) and SRM2 (dilation angle equal to friction angle) respectively. (min. FOS=0.927 from Spencer analysis).

Program/FOS	12m Domain	20m Domain	28m Domain
Flac3D	1.03/1.03	1.30/1.28	1.64/1.61
Phase	0.77/0.85	0.84/1.06	0.87/1.37
Plaxis	0.82/0.94	0.85/0.97	0.86/0.97

(Table 1.4) cont.....

Program/FOS	12m Domain	20m Domain	28m Domain
Flac2D	No solution	No solution	No solution

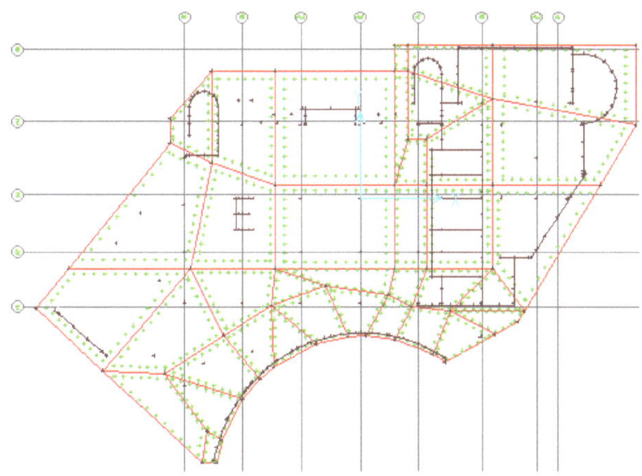

Fig. (1.7). A raft foundation modelled by irregular mesh and combination of 3 and 4 nodes thick plate element.

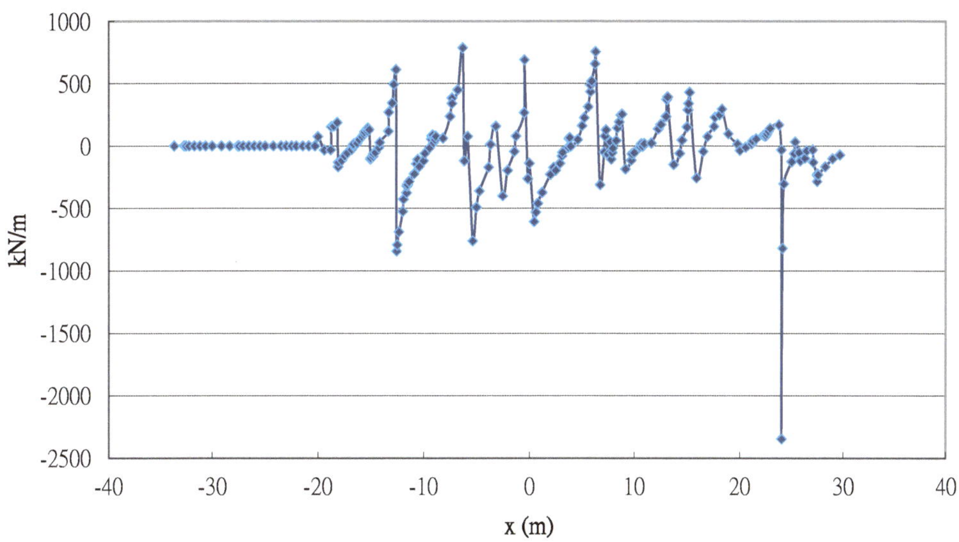

Fig. (1.8). Shear jump from the use of thick plate finite element analysis.

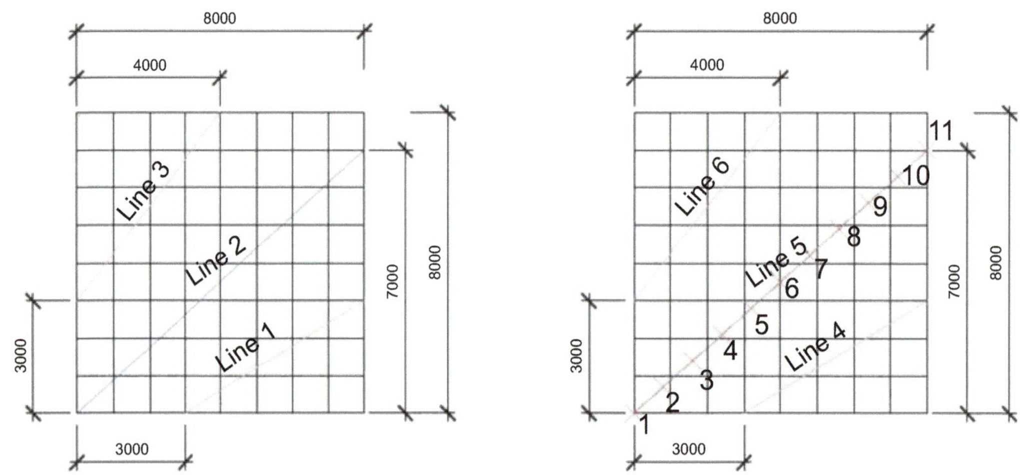

Fig. (1.9). A simple problem where the mesh is a combination of 3 and 4 nodes thick plate element.

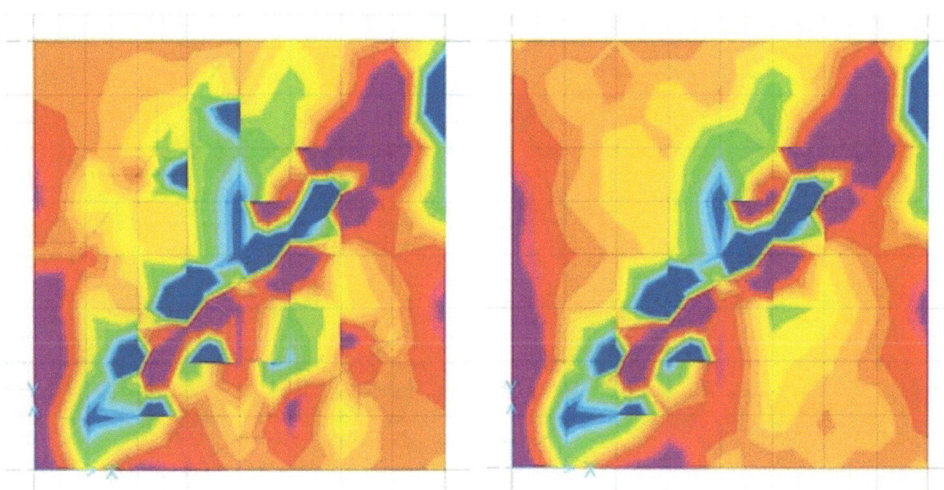

(a) (b)

Fig. (1.10). Surprising results for Mx at line 3 and line 6, where only a line load is applied at the same problem.

Another interesting problem case is given by the engineers from the Housing Department of Hong Kong (Cheng and Law 2008). With reference to the raft foundation as shown in Fig. (**1.7**), due to the transfer of loading from the use of the super-structure structural analysis program, the mesh design for the raft foundation becomes highly irregular with a combination of 3 and 4 node thick plate elements. From the use of a world-famous finite element program, there are totally three places where surprising results are obtained, and an example is shown in Fig. (**1.8**). For the given design strip, a very sharp jump in the shear force is obtained at a location where there is no point force or support. Conceptually, this is impossible but numerically possible. This problem is detected when the engineers find that the shear stress is so large that the shear capacity of the section is exceeded. Actually, the authors later find that besides the jump of shear forces, jump of moment can also occur at location where there is nothing. As shown in Fig. (**1.9**), a simple problem where the mesh is a combination of 3 and 4 nodes thick plate element is created. A simple uniform distributed load on the plate plus a line load along line 2 is applied to the problem, while no load is applied on lines 1 and 3. Some surprising jumps at the moment are obtained as shown in Fig. (**1.10**) to Fig. (**1.17**). For Fig. (**1.10a**) and Fig. (**1.10b**), they are actually the same problem, but two additional lines are added to the problem to make the mesh more distorted in Fig. (**1.10a**).

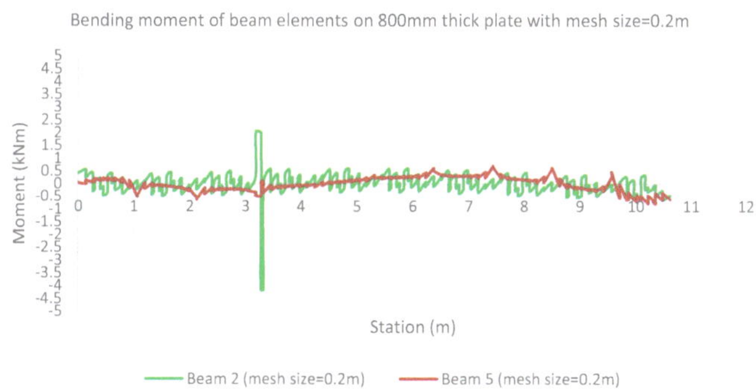

(a) Mesh size=0.2,

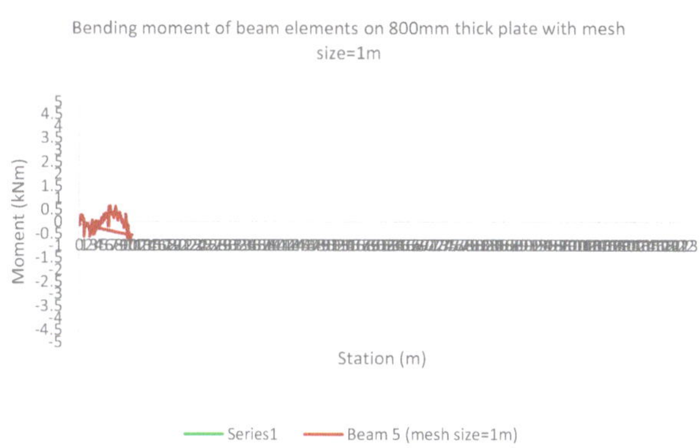

(b) Mesh size = 1.0m

Fig. (1.11). Surprising result on Mx along lines 2 and 5.

Actually, the authors have created many similar surprising results using various commercial programs, and many of these programs are world-famous and have been used by many engineers and researchers. The sense and judgment in the interpretation of the results are extremely important. For the problems as shown in Fig. (**1.4**) and Fig. (**1.15**), the authors manage to find the reasons and overcome them in the self-developed plate programs. Cheng (2018) provides series of problems for the DLO commercial program as well as some limit equilibrium slope stability analysis programs. Without the experience in programming these problems, the readers may not be aware of the technical issues and the reasons for the problems. Also, the authors like to tell that the students that if we want a detailed understanding of an engineering problem, the best way is to program the problem into a software. During the programming, the readers will find many questions for which they may never think of, and many of these questions are not mentioned in any textbook or research papers.

(a) Mesh size=1.0m

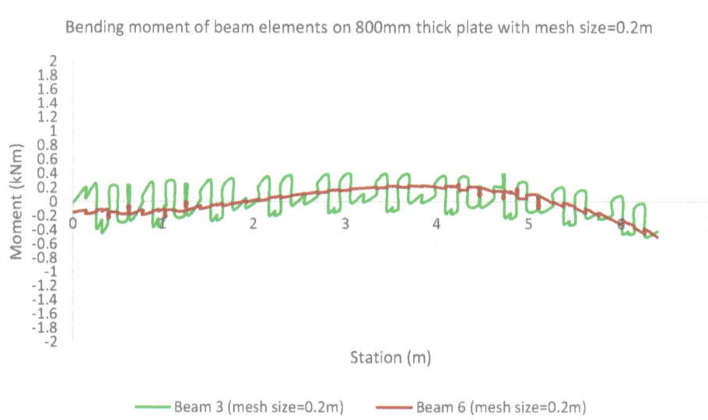

(b) Mesh size = 0.2m

Fig. (1.12). Surprising result on Mx along lines 3 and 6.

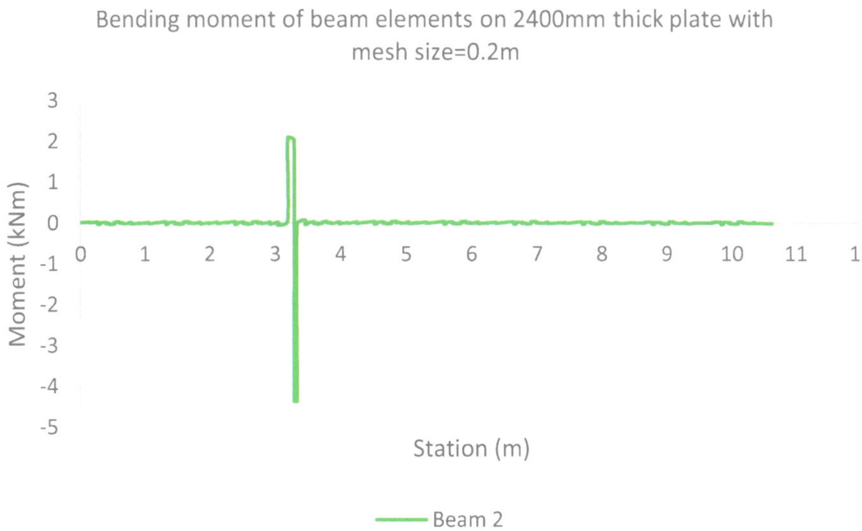

Fig. (1.13). Surprising result on Mx along line 2.

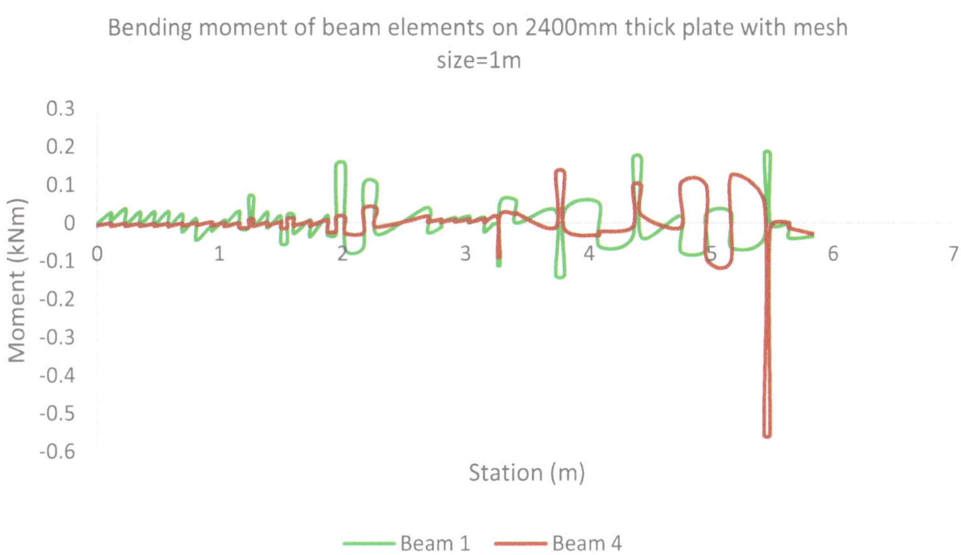

Fig. 1.14 cont.....

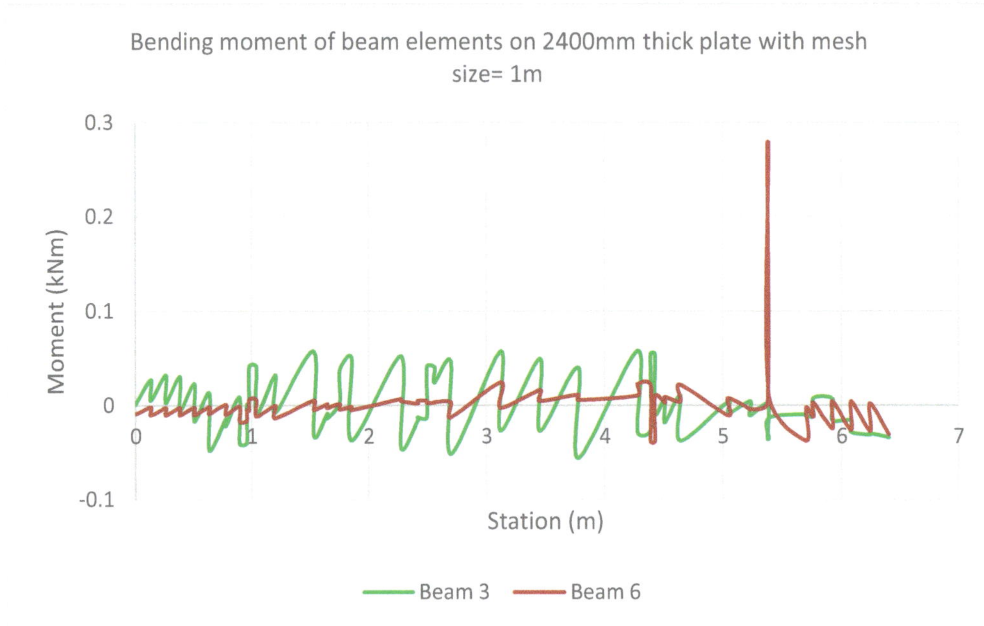

Fig. (1.14). Effect of thicker plate for the problem in Fig. (**1.9**).

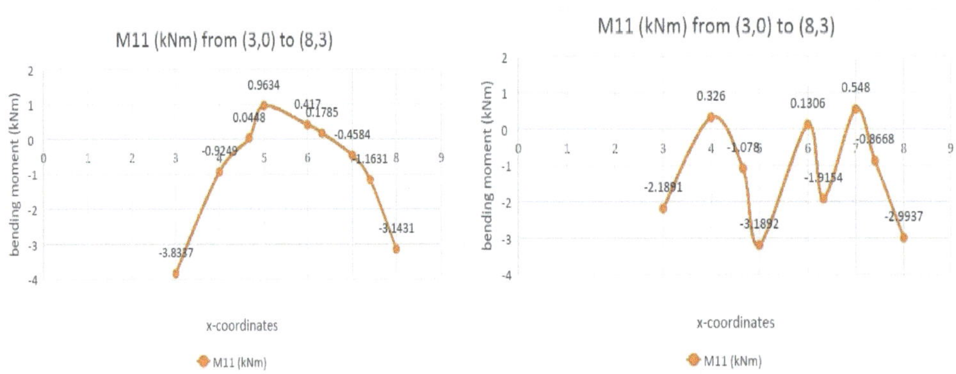

(a) Regular mesh (b) Irregular mesh

Fig. (1.15). M11 from (3,0) to (8,3) for the regular and irregular mesh, where line moment is applied along 2 or line 5, and no loading on other locations.

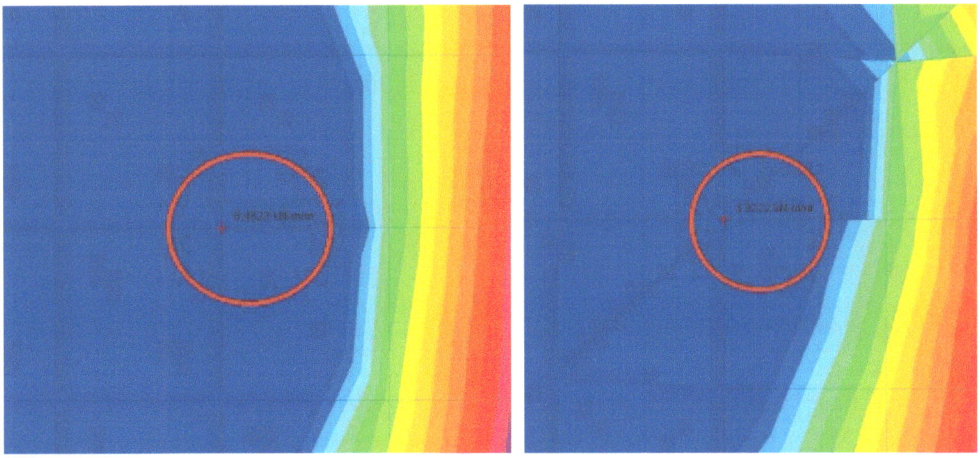

Regular mesh pattern (7.2,6.3): Irregular mesh pattern (7.2,6.3):

Moment = 6.4822kNm/m 3.922kNm/m

Fig. (1.16). Result of moment at the same point to illustrate importance of mesh design.

If we magnify to see the local effects, some interesting results are observed. When the same problem using a regular mesh and irregular mesh are considered, there are great differences at the moment at the point (7.2, 6.3) as shown in Fig. (**1.16**). The differences are actually noticeable, even when a fine irregular mesh is used. On the other hand, for the problem as shown in Fig. (**1.7**), more than 10000 elements are already used by the engineers, which are usually more than adequate for normal problems, surprising results are still obtained. For a problem without the known solution, how should an engineer assess the acceptability of the computer results can be a difficult issue. After series of analysis into the local effects, the authors find that the use of triangular and adjoining quadrilateral element is the cause for the incompatibility of the internal forces. This is due to the order of interpolation functions used in the two elements, and the derivatives of the displacement along the common side are not the same for the adjoining elements. On the other hand, the moment is a function of the derivative of the displacement, and the results on the moment can be very poor under such condition. A mesh with a mixture of different elements can sometimes make mesh generation easier, but it will create noticeable numerical inaccuracy for some problems, particularly for those problems governed by higher order differential equation.

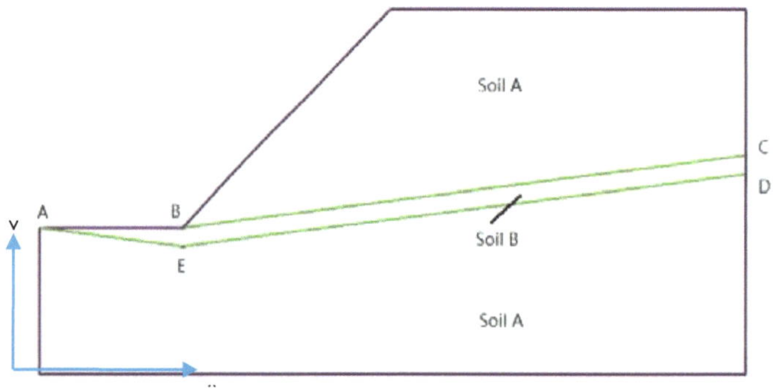

	500mm		5mm		2mm	
(m)	X	Y	X	Y	X	Y
A	0	4	0	4	0	4
B	4	4	4	4	4	4
C	20	6	20	6	20	6
D	20	5.5	20	5.995	20	5.998
E	4	3.5	4	3.995	4	3.998

Fig. (1.17). A slope with a soft band (model 1).

1.2. MORE PROBLEM CASES FROM SLOPE STABILITY ANALYSIS

In the previous section, some problem cases from the use of finite element and limit equilibrium in slope stability analysis are demonstrated. In fact, Cheng *et al.* (2007, 2008) have demonstrated more numerical problems on the limit equilibrium, finite element methods and the DLO (Cheng, 2018). In this section, the authors are going to demonstrate more cases to warn the engineers about the importance of interpretation and assessment of results instead of simply believing in any computer program.

1.2.1. A Slope with a Soft Band

A slope with a thin and weak layer is a special case and major numerical problems can exist in the analysis. For this case, the Cheng's model is used in this section. To analyze this problem, Zolfaghari *et al.* (2005) applied a genetic algorithm by using the Morgenstern-Price Method. Cheng *et al.* (2007) considered this problem with the particle swarm optimization and the Spencer Method. The slope geometry and the soil properties are shown in Fig. (**1.17**) to Fig. (**1.19**) and Table (**1.5**), respectively. It should be noted that c' is zero and ϕ' is smaller for the soft band layer. In the parametric study, different shear strength is used and DLO, LEM, SRM 1 and SRM 2 are carried out. SRM 1 is a non-associated flow rule analysis with a dilation angle = 0, while SRM 2 is an associated flow rule analysis with a dilation angle = friction angle. The thickness of soft band is set to be 500mm, 5mm and 2mm with three different slopes, while the soil properties are kept to be the same for three cases.

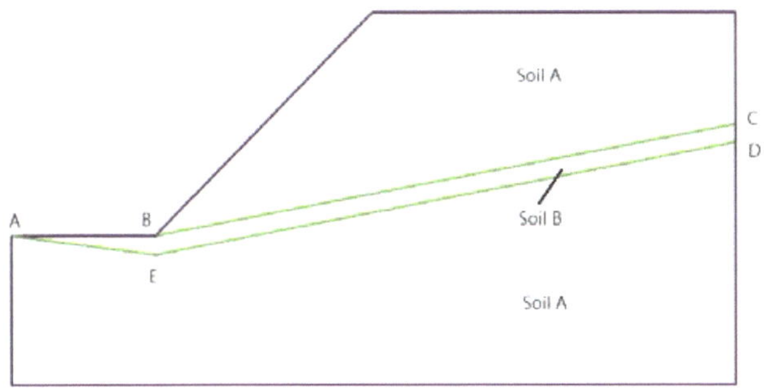

	500mm		5mm		2mm	
(m)	X	Y	X	Y	X	Y
A	0	4	0	4	0	4
B	4	4	4	4	4	4
C	20	7	20	7	20	7
D	20	6.5	20	6.995	20	6.998
E	4	3.5	4	3.995	4	3.998

Fig. (1.18). A slope with a soft band (model 2).

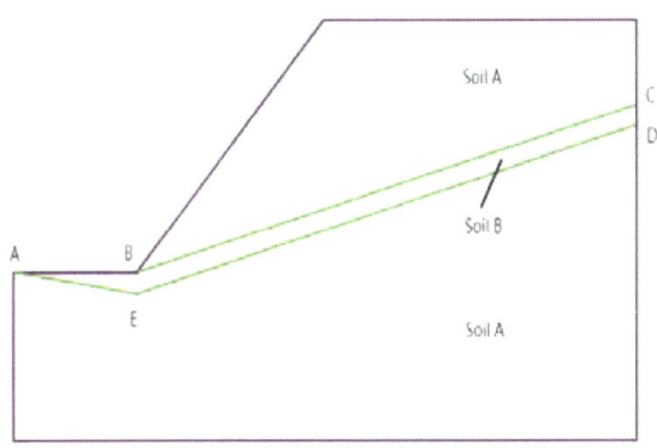

	500mm		5mm		2mm	
(m)	X	Y	X	Y	X	Y
A	0	4	0	4	0	4
B	4	4	4	4	4	4
C	20	8	20	8	20	8
D	20	7.5	20	7.995	20	7.998
E	4	3.5	4	3.995	4	3.998

Fig. (1. 19). A slope with a soft band (model 3).

Table 1.5. Soil properties for the models.

Case	Soil Name	c' (kPa)	ϕ'(°)	γ' (kN/m^3)	E' (MPa)	v'
X.1	Soil A	5	35	20	14	0.3
X.1	Soil A	0	25	20	14	0.3
X.2	Soil B	10	35	20	14	0.3
X.2	Soil B	0	25	20	14	0.3

In this analysis, the number of nodes are set to be 500, 1000, 2000 in DLO and the number of elements are set to be 2000, 5000, 10000 in SRM. For the LEM, the number of slices is set to be 50. Additionally, the number of elements in SRM is increased (maximum 10000) to resolve those cases for which the differences between SRM 1 and SRM 2 are significantly large (> 8%).

From the analysis, the FOS by SRM have great differences from those by DLO, particularly for the soft band problem with a thickness of 2mm. In model 1, the FOS are determined to be 1.163 and 1.396 from DLO (*c'* = 5 and 10 kPa for Soil A, respectively) with 1000 nodes, while the FOS are 1.23 (c' = 5 kPa for Soil A) and 1.02 (c' = 10 kPa for Soil A) in SRM 1 with 5000 and 10000 meshes respectively. For SRM 2, the FOS are found to be 1.28 and 0.74 in the same cases, and FOS are 1.129 and 1.3362 from LEM.

For the slope analysis by DLO, one thing as found is that the number of nodes can be a critical factor affecting the result. There can be no solutions for some cases in the three models. The number of nodes is required to be adjusted to fix the problem. For example, in model 2 with soft band of 2mm thickness, the result cannot be determined if the number of nodes is set to be 1000. If it is 900, the solution can be determined. This situation is highly unfavourable, and it demonstrates that the numerical implementation of DLO is not mature enough to be able to work fine under all cases.

In the SRM, the FOS obtained from SRM 2 should be larger than those from SRM 1 theoretically. Interestingly, the result from a SRM program shows that the FOS obtained from SRM 1 is larger than that from SRM 2 when the soft band becomes thinner or steeper. Apart from that, the difference between FOS from SRM 1 and SRM 2 should be small (around 2%), however, the differences are more than 2% for all the cases. In model 2 (*c'* = 10kPa for Soil A and the soft band has a thickness of 2mm), the FOS is found to be 1.05 from SRM 1 and 0.42 from SRM 2 with 10000 elements. The difference of FOS between them is 60.0% which is not believable. Again, engineers or researchers should not simply believe in the results from the use of engineering programs without assessment.

If we look careful into the results in Tables in **1.6** to **1.8**, we may be disappointed to find that some commercial programs cannot perform well under some cases, and the errors can be significant or even ridiculous. For a new problem without any known solution, how should an engineer assess and accept the results from any computer program is a difficulty issue. The reasons for these numerical problems should also be rectified in the updated versions of the commercial programs. In fact, the authors have tested several versions of the commercial programs and find that each version may give rise to different problems. So far, the authors cannot find a commercial SRM program which can be correct/reasonable under all cases. Actually, the authors have constructed more problem cases for many commercial programs, but the examples as shown here are sufficient to illustrate the difficulty to develop a numerical algorithm/computer program which can pass through all types of material parameters and geometry.

Table 1.6. Results for model 1 by DLO, LEM and SRM for case 1.

Number of Nodes	Number of Elements	Thickness of Soft Band (mm)	Soil Type	c' (kPa)	ϕ' (°)	FOS (DLO)	FOS (LEM)	FOS (SRM1)	FOS (SRM2)
500	2000	500	A	5	35	1.096	1.103	1.02	1.09
			B	0	25				
			A	10	35	1.275	1.184	1.17	1.24
			B	0	25				
		5	A	5	35	1.165	1.128	1.20	1.28
			B	0	25				
			A	10	35	1.396	1.335	0.98	0.9
			B	0	25				
		2	A	5	35	1.166	1.129	1.24	1.28
			B	0	25				
			A	10	35	1.397	1.362	0.96	0.7
			B	0	25				
1000	5000	5	A	5	35	1.162	1.128	1.22	1.28
			B	0	25				
			A	10	35	1.395	1.335	1.22	0.95
			B	0	25				
		2	A	5	35	1.163	1129	1.29	1.28
			B	0	25				
			A	10	35	1.396	1.336	0.93	0.85
			B	0	25				
2000	10000	5	A	10	35	No solution	1.335	1.21	0.96
			B	0	25				
		2	A	10	35	No solution	1.336	1.02	0.74
			B	0	25				

Table 1.7. Results for model 2 by DLO, LEM and SRM for case 2.

Number of Nodes	Number of Elements	Thickness of Soft Band (mm)	Soil Type	c' (kPa)	ϕ' (°)	FOS (DLO)	FOS (LEM)	FOS (SRM1)	FOS (SRM2)
500	2000	500	A	5	35	1.06	1.052	0.98	1.04
			B	0	25				
			A	10	35	1.223	1.194	1.11	1.17
			B	0	25				
		5	A	5	35	1.128	1.092	1.20	1.27
			B	0	25				
			A	10	35	1.388	1.283	1.16	0.94
			B	0	25				
		2	A	5	35	No solution	1.094	1.21	1.27
			B	0	25				
			A	10	35	1.34	1.283	1.05	0.4
			B	0	25				
1000	5000	5	A	5	35	1.123	1.092	1.22	1.28
			B	0	25				
			A	10	35	1.387	1.283	1.37	0.79
			B	0	25				
		2	A	5	35	No solution	1.094	1.20	1.27
			B	0	25				
			A	10	35	No solution	1.283	0.35	0.35
			B	0	25				
2000	10000	5	A	10	35	No solution	1.283	0.74	0.74
			B	0	25				
		2	A	10	35	No solution	1.283	0.42	0.42
			B	0	25				

More importantly, how should a student or engineer interpret the results from an engineering computer program is not an easy task. During teaching, the authors always emphasize the sense of engineering judgment arising from the basic theory, use of approximate methods, consideration of local effects, alternative methods or programs for assessing the results from any computer program. Again, the authors view that if the students or engineers have the experience in programming some engineering problems, they will have better background to assess and interpret the results of analysis.

1.3. STRANGE RESULTS FROM MESH REFINEMENT

It is usually believed that if a mesh is refined, the results from finite element analysis will converge to the rigorous solution. For normal cases, this is true. The authors have however also constructed some cases where the results are not true. Consider the plate as shown in Figs. (**1.20-1.24**) which are supported by columns around the perimeter. The plate is an 8mx8m plate with a thickness of 0.8m and is loaded with a uniform distributed load.

Table 1.8. Results for model 3 by DLO, LEM and SRM for case 3.

Number of Nodes	Number of Elements	Thickness of Soft Band (mm)	Soil Type	c' (kPa)	ϕ' (°)	FOS (DLO)	FOS (LEM)	FOS (SRM1)	FOS (SRM2)
500	2000	500	A	5	35	1.021	1.029	0.93	1.00
			B	0	25				
			A	10	35	1.169	1.138	1.06	1.13
			B	0	25				
		5	A	5	35	No solution	1.053	1.10	0.99
			B	0	25				
			A	10	35	No solution	1.217	1.06	0.74
			B	0	25				
		2	A	5	35	1.084	1.055	1.16	1.27
			B	0	25				
			A	10	35	No solution	1.211	1.42	0.29
			B	0	25				
1000	5000	5	A	5	35	No solution	1.053	1.10	0.98
			B	0	25				
			A	10	35	No solution	1.217	1.10	0.93
			B	0	25				
		2	A	5	35	1.082	1.055	1.00	1.26
			B	0	25				
			A	10	35	1.277	1.211	1.44	0.8
			B	0	25				

(Table 1.8) cont.....

Number of Nodes	Number of Elements	Thickness of Soft Band (mm)	Soil Type	c' (kPa)	ϕ' (°)	FOS (DLO)	FOS (LEM)	FOS (SRM1)	FOS (SRM2)
2000	10000	5	A	10	35	No solution	1.217	1.00	0.74
			B	0	25				
		2	A	10	35	No solution	1.211	1.50	1.10
			B	0	25				

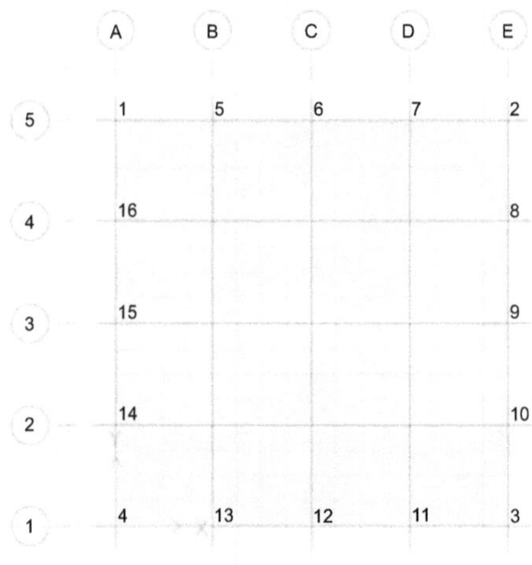

Fig. (1. 20). Case 1 for the plate analysis.

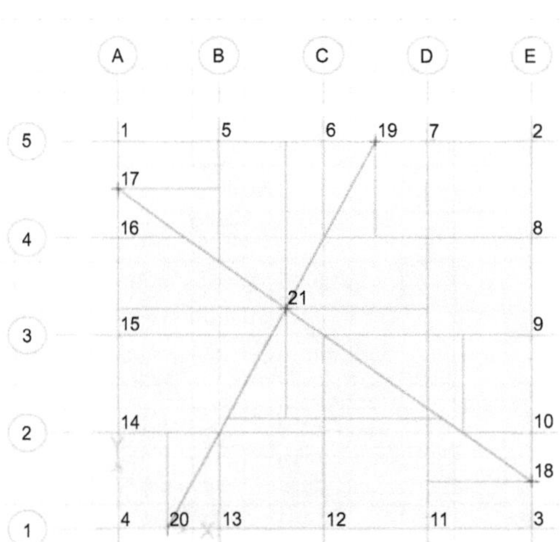

Fig. (1. 21). Case 2 for the plate analysis.

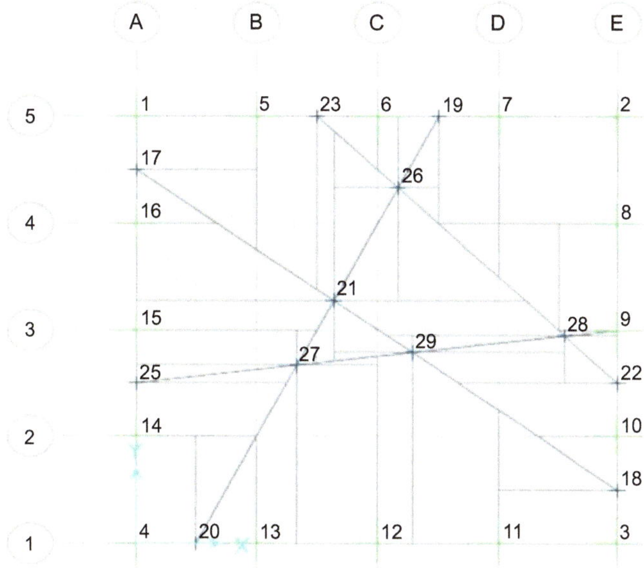

Fig. (1. 22). Case 3 for the plate analysis.

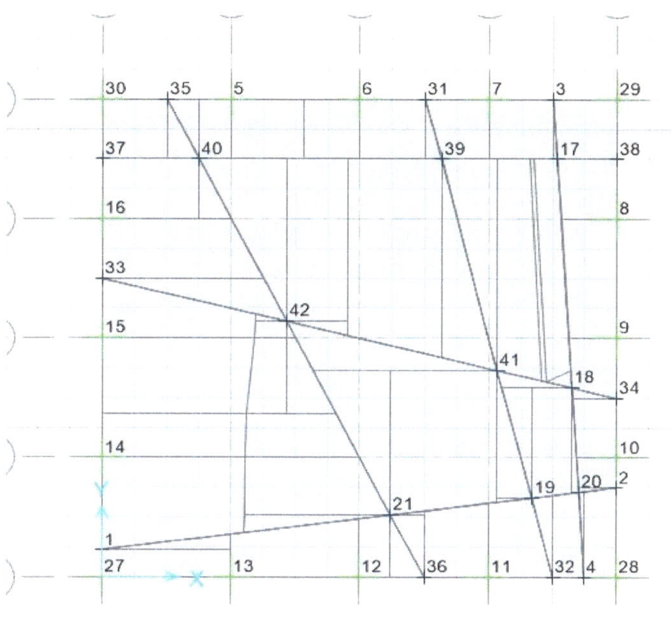

Fig. (1. 23). Case 4 for the plate analysis.

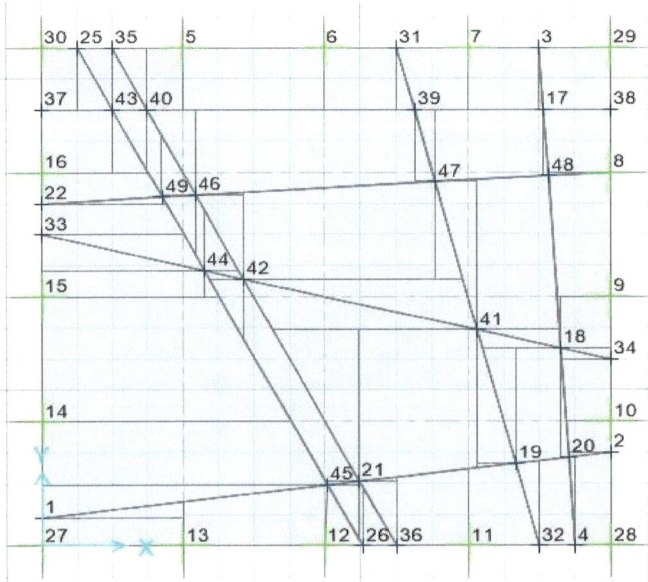

Fig. (1.24). Case 5 for the plate analysis.

The present problem is considered by the use of a world-famous plate program. By nature of symmetry, the maximum deflection and moment should be at point (4,4). For the deflection, generally the results are acceptable even if the mesh is irregular, though the result may not be too good. For the moment, very surprising results are obtained and shown in Fig. (**1.25**) and Fig. (**1.26**). The ratio between the maximum Mxx and Myy from any program to the rigorous answer should be 1.0 for case 1 which is as expected. For other cases, the ratio deviates from 1.0 by a significant amount which is highly undesirable. A very surprising result is that when the mesh is very fine, the ratio drops to 0.2-0.4, which is a very poor performance. Also, the locations for the maximum moment deviate from the point (4,4) by a noticeable amount, for which the error appears to be not greatly affected by the fineness of the mesh. The local moment from different cases at some specific points also fluctuate significantly with the mesh density, and mesh refinement appears to be helpless for this situation. After assessment, the authors find that the generation of triangular elements and adjoining quadrilateral elements is one of the major factors for such a poor result. If the moment contour is examined which are shown in Fig. (**1.27**) and Fig. (**1.28**), it is observed that there are jumps in the moments in Fig. (**1.28**) but not in Fig. (**1.27**). These jumps at the moment are not reasonable, as there are no point moment at those jump locations.

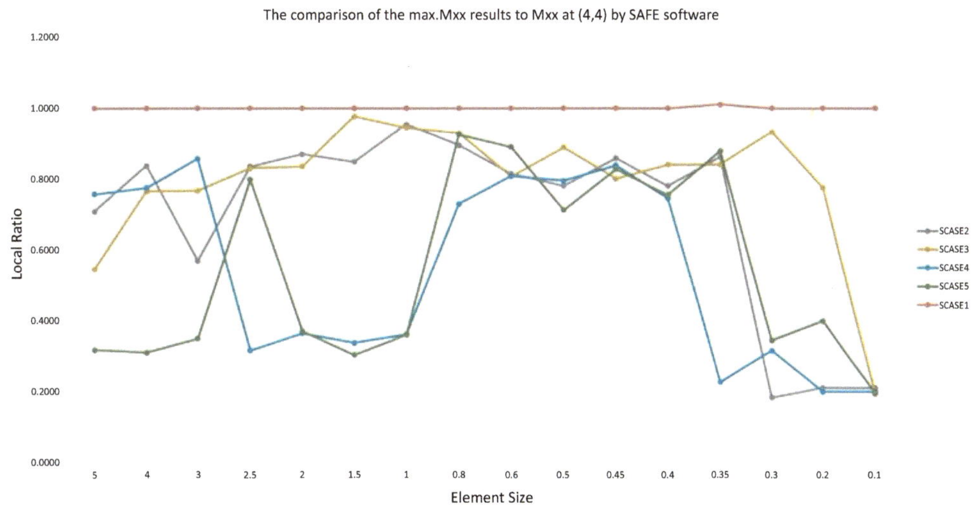

Fig. (1.25). Ratio between the maximum computed Mxx to rigorous answer.

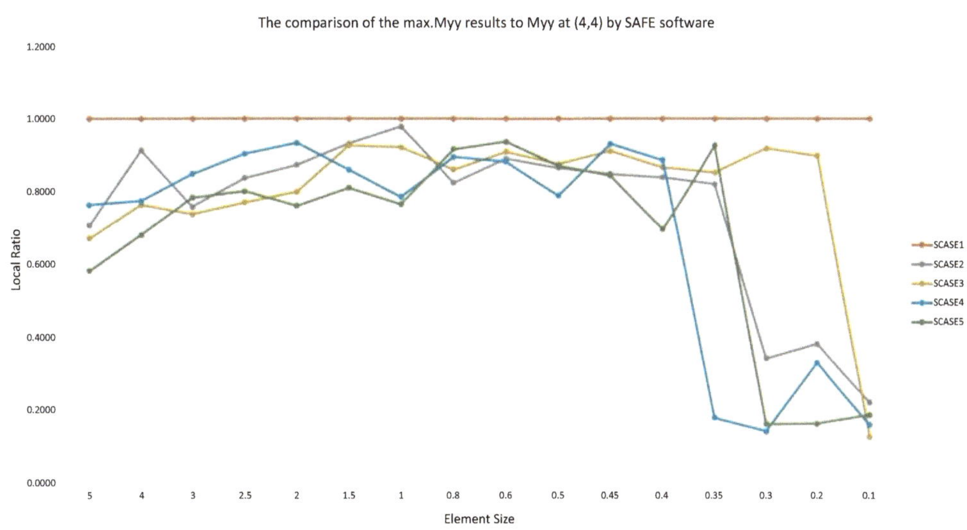

Fig. (1.26). Ratio between the maximum computed Myy to rigorous answer.

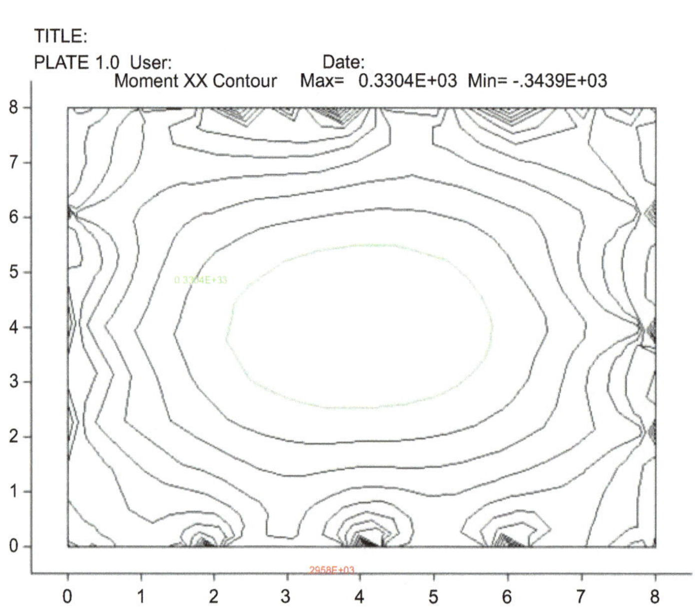

Fig. (1.27). Results of moments from PLATE by Cheng using only quadrilateral elements.

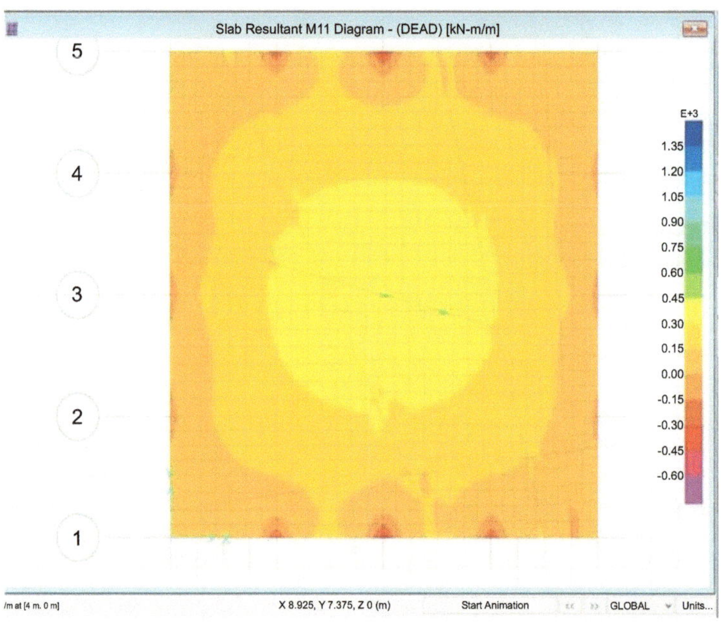

Fig. (1.28). Results of moments from a commercial program.

1.4. GOVERNING EQUATIONS FOR SOME ENGINEERING PROBLEMS

Many of the engineering problems are governed by different types of differential equations. For example, the beam on elastic ground problem is controlled by eq. **(1.1)**

$$EI\frac{d^4y}{dx^4} = q = -k_s^{'}y \tag{1.1}$$

where EI is the bending stiffness of the beam, q is the distributed load, ks' is the soil Winkler's spring constant multiply with the width of the beam and y is the vertical displacement. If EI and k_s' are constant, this ordinary differential equation can be solved easily, and analytical solutions are also available for this case. Otherwise, this equation must be solved by means of finite element method (finite difference method is less convenient for this case). For a thin plate raft foundation problem, the problem can be modelled as a plate on elastic ground by eq. (1.2) as

$$D\nabla^4\omega = q(x,y) - k_s\omega \tag{1.2}$$

Where D is $Eh^3/12(1\text{-}v^2)$, and h, v, q and ω are the thickness of the plate, Poisson ratio of the plate material, distributed load and vertical displacement of the plate. Solution to eq. (1.2) in form of Navier or Levy solutions are available only for very limited cases. If the plate is thick enough which is the common case for most of the foundation problem, the shear deformation will not be negligible. The governing equation will then be much more complicated which is given by eq. (1.3).

$$\begin{gathered}\frac{-K_sEh}{2(1+v)}\left(\nabla^2\omega + \frac{M}{D}\right) = q(x,y)\\ -D(1-v)\nabla^2\varphi_x - (1+v)\frac{\partial M}{\partial x} + \frac{K_sEh}{(1+v)}\left(\frac{\partial\omega}{\partial x} + \varphi_x\right) = 0\\ -D(1-v)\nabla^2\varphi_y - (1+v)\frac{\partial M}{\partial y} + \frac{K_sEh}{(1+v)}\left(\frac{\partial\omega}{\partial y} + \varphi_y\right) = 0\end{gathered} \tag{1.3}$$

where K_s is the shear correction factor, φ_x and are the rotation along x and y axes, respectively and M=(Mxx+Myy)/(1+v). Analytical solution to eq. (1.3) is not possible, and the use of numerical method is necessary, which form the basis of many commercial software used by the engineers. As indicated in the previous section, due to the high order of the differential equation and the interaction between Mxx, Myy and Mxy, accurate solution to eq. (1.3) is not easy, particularly when there is fixed boundary condition or the use of irregular mesh. These are some of the reasons for the poor results as given by some commercial programs.

For the stress equilibrium of a soil medium, the governing equation is given by eq. (1.4) as:

$$\begin{aligned}\frac{\partial \sigma_{xx}}{\partial x}+\frac{\partial \tau_{xy}}{\partial y}+\frac{\partial \tau_{xz}}{\partial z}&=fx\\ \frac{\partial \tau_{xy}}{\partial x}+\frac{\partial \sigma_{yy}}{\partial y}+\frac{\partial \tau_{yz}}{\partial z}&=fy\\ \frac{\partial \tau_{xz}}{\partial x}+\frac{\partial \tau_{yz}}{\partial y}+\frac{\partial \sigma_{zz}}{\partial z}&=fz\end{aligned} \tag{1.4}$$

where f is the body forces in different directions, while σ and τ are the normal and shear stresses acting on the element. Analytical solutions to eq. (1.4) are available for many simple cases, and Poulos and Davis (1980) and many others have given detailed collections of these elastic solutions. Many design charts, figures and tables are also available for the design of simple foundation problems. Currently, many engineers prefer the use of commercial programs instead of these design tables/figures, as most of the foundations are not simple in geometry or loadings.

For two-dimensional seepage problem, it is governed by the Laplace equation as given by eq. (1.5).

$$\begin{aligned}&\frac{\partial^2 \phi}{\partial x^2}+\frac{\partial^2 \phi}{\partial y^2}=0 \quad \text{for isotropic soil}\\ &k_x\frac{\partial^2 \phi}{\partial x^2}+k_y\frac{\partial^2 \phi}{\partial y^2}=0 \quad \text{for anisotropic soil}\end{aligned} \tag{1.5}$$

where ϕ is the total head, while k_x and k_y are the permeabilities of the soil along x and y directions. For consolidation in clay, it is governed by the dispersion equation (1.6) as

$$\frac{\partial u}{\partial t} = c_v \frac{\partial^2 u}{\partial z^2} \quad \text{one-dimensional consolidation}$$
$$\frac{\partial u}{\partial t} = c_v \frac{\partial^2 u}{\partial z^2} + c_r (\frac{\partial^2 u}{\partial r^2} + \frac{1}{r}\frac{\partial u}{\partial r}) \quad \text{axisymmetric consolidation} \tag{1.6}$$

where u and c_v are the pore water pressure and coefficient of consolidation. For refined analysis, the three-dimensional Biot consolidation as given by eq. (1.7) will be more appropriate.

$$-G\nabla^2 u_x + \frac{G}{1-2\nu}\frac{\partial}{\partial x}(\frac{\partial u_x}{\partial x} + \frac{\partial u_y}{\partial y} + \frac{\partial u_z}{\partial z}) + \frac{\partial u}{\partial x} = 0$$
$$-G\nabla^2 u_y + \frac{G}{1-2\nu}\frac{\partial}{\partial y}(\frac{\partial u_x}{\partial x} + \frac{\partial u_y}{\partial y} + \frac{\partial u_z}{\partial z}) + \frac{\partial u}{\partial y} = 0 \tag{1.7}$$
$$-G\nabla^2 u_z + \frac{G}{1-2\nu}\frac{\partial}{\partial z}(\frac{\partial u_x}{\partial x} + \frac{\partial u_y}{\partial y} + \frac{\partial u_z}{\partial z}) + \frac{\partial u}{\partial z} = 0$$

where G and ν are the shear modulus Poisson ratio of the soil. For many geotechnical problems governed by the limit state, the hyperbolic differential slip-line equations as given by equation (1.8) apply. Unless the weight of the soil is negligible, analytical solutions to eq. (1.8) cannot be derived, by nature of the equation.

$$\alpha\ characteristic: \quad -\frac{\partial p}{\partial S_\alpha}\sin 2\mu + 2R\frac{\partial \theta}{\partial S_\alpha} + \gamma[\sin(\alpha + 2\mu)\frac{\partial y}{\partial S_\alpha} + \cos(\alpha + 2\mu)\frac{\partial x}{\partial S_\alpha}] = 0$$
$$\beta\ characteristic: \quad \frac{\partial p}{\partial S_\beta}\sin 2\mu + 2R\frac{\partial \theta}{\partial S_\beta} + \gamma[\sin(\alpha - 2\mu)\frac{\partial y}{\partial S_\beta} + \cos(\alpha - 2\mu)\frac{\partial x}{\partial S_\beta}] = 0 \tag{1.8}$$

In eq. (1.8), γ is the unit weight of soil; μ= (π/4-*φ*/2), c and *φ* are the cohesive strength and friction angle of soil, θ is the angle between the direction of major principal stress and the y-axis, S_α and S_β are the arcs along the α and β lines, α is the angle between the direction of the body force and the vertical direction (y-axis), *p* and *R* are the mean stress and radius of the Mohr-circle at the state of failure defined by:

$$p = \frac{\sigma_1 + \sigma_3}{2} \quad ; \quad R = \frac{\sigma_1 - \sigma_3}{2} = p\sin\varphi + c\cos\varphi \tag{1.9}$$

Another equation which is commonly encountered is the pile driving equation as given by eq. (1.10), where u, c, f, l, ρ and A represent the axial displacement of pile during driving, velocity of force wave, skin friction, perimeter of pile, density of pile material and sectional area of pile. The forward and backward analyses of eq. (1.10) govern the set calculation as well as the PDA analysis of pile driving.

$$\frac{\partial^2 u}{\partial t^2} - c^2 \frac{\partial^2 u}{\partial z^2} + \frac{fl}{\rho A} = 0 \tag{1.10}$$

There are actually much more governing differential equations for different types of problems, and the authors only present some of the equations commonly faced by the engineers. Furthermore, the equations as given in this section also form the basis for the development of many commercial programs, using mostly finite element analysis. In general, it can be viewed that the purposes of the commercial engineering programs are to solve these governing differential equations, under general geometry, soil conditions, loadings, constitutive model, boundary conditions and loading history.

1.5. CLOSED-FORM SOLUTIONS

As discussed in previous section, close-form solutions are not available for most of the practical engineering problems. For some simple and special cases, closed-form but non-trivial solutions do exist. These are very important results because apart from the academically value, they can provide insight into the nature of the problem as well as forming the backbone to assess the acceptability of engineering programs. Actually, in the user guides of many engineering programs, benchmark problems based on the analytical solutions of some simple cases are usually used for assessment purposes. It should also be noted that in the in-house assessment of some commercial programs, many commercial programs actually cannot reproduce the analytical solutions fully, though the errors are usually not major.

A bearing capacity problem can be seen as a slope with a very gentle slope angle but with substantial surcharge loading. Yet most of the commercial programs cannot reproduce the exact plasticity solutions or even the ultimate load precisely under all cases (Cheng *et al.* 2010). The more critical issue is that a well-defined failure mechanism is usually not generated from many commercial programs. From the authors' internal studies using different finite element programs, the classical failure mechanism is obtained only for limited cases, and most the results are actually not clear or precise enough to compare with the classical plasticity solutions. Some finite element programs even cannot give a failure mechanism. Similarly, for the earth pressure problems, both active and passive, would also be suitable check for the computer code. Here, the slope has a slope angle of 90 degrees. By applying an active or passive pressure at the vertical face, the computer program should yield FOS=1.0 for both cases. The authors view that all computer codes should be benchmarked and validated by subjecting themselves to solving the classical problems where "closed-form" solutions exist for comparison. Hopefully, the comparison would reveal both their respective strengths and limitations so that users can put things into perspective when using the code for design in real life.

1.6. LAYOUT OF THIS BOOK

The authors have used Fortran 77, 90, 95 in developing series of computer programs for geotechnical problems (and c or c++ for some special problems). Based on the authors' experience, a short discussion about the programming approach and style will be given in next chapter, which is followed by simple engineering programs with detailed discussion and explanations so that the readers will appreciate the approaches in developing an engineering programs. More detailed discussion about advanced numerical geotechnical methods will then be introduced in the later section. Interested readers can email to the authors for the programs which are mentioned but not given in this book for their own development.

CHAPTER 2

Numerical Methods in Geotechnical Engineering

Abstract: Finite element method is the most popular and important numerical method in geotechnical engineering analysis and design. In this chapter, the basic formulation for finite element method will be introduced, followed by the various procedures in the actual implementation of the method which are illustrated by several sample Fortran programs. The style of Fortran programming, mesh generation, bandwidth/profile minimizer as well as the development of modern thick plate element are used for the illustration of the various techniques required in developing a finite element program.

Keyword: Fortran, Finite element, Isoparametric element, Programming style, Plate element, Shape functions.

2.1. INTRODUCTION TO PROGRAMMING

Many books on finite element method or numerical methods have provided sample programs for the readers to learn the actual implementation of the methods. Very good references are provided by Smith *et al.* (2014), Owen and Hinton (1980), Hinton and Owen (1977), Zienkiewicz *et al.* (2011). During learning the finite element method and the techniques in programming, the authors have learnt a lot from these references and many other technical papers. Many of these references on finite element method target at simple input data format and standard problems, but not much on other types of problems. Some open source programs or commercial programs are too fruitful with many libraries and functions which are not easy to modify or to learn. For example, the program Abaqus comes with series of program manuals which require a steep learning curve, and to develop a special add-in subroutine to Abaqus can be a tedious task. If a reader prefers to carry a new analysis and adopt his preferred pre-processor or post-processor, it is extremely time-consuming to spend the effort to learn some large scale commercial engineering programs. The present book actually targets at such a group of readers, which are most probably research students who need to develop computer codes for some new algorithms.

There are many numerical methods that are being used for various purposes, which include finite element method, boundary element method, distinct element

Y.M. Cheng, J. H. Wang, L. Liang, W. H. Fung Ivan (Eds.)

method, meshless method and others. Out of these methods, the boundary element method will not be discussed in this book, as there are many practical limitations to the boundary element method in general geotechnical engineering problems. Fortran language is adopted in this book, as many numerical programs are actually developed in Fortran because this language is targeted towards mathematical and matrix operations, and the relatively simple language format has avoided many bugs that may be found in C/C++. In views of that, the authors will only present the results in Fortran in this book, even though some engineering programs are developed in C by the authors.

2.1.1. Management of Input Data

As mentioned in chapter 1, most of the practical problems require the use of numerical methods and computer software for the analysis and design. Throughout the years, the authors have developed series of computer programs for teaching and research, and some of the experience and results will be shared in this book. To begin with, some skill commonly adopted by the authors are first discussed.

The authors have used Fortran 90/95 for the development of many structural and geotechnical programs. The authors have not adopted the more advanced Fortran 2003, 2008 or later Fortran versions, as Fortran 90/95 (Chapman, 2007) is adequate for most applications. There are some important utilities for which the authors adopt in various programs development, and some of these utilities will be discussed below. Throughout this book, the subroutines/programs are developed by mainly Lahey Fortran while GFortran and Absoft Fortran are also used for some cases, but the codes should also work under other Fortran compilers. The readers should be able to modify the codes to comply with different compilers, or to remove some old style programming formats. The authors assume the readers to have sufficient knowledge of basic Fortran language, and the works by Chapman (2007) and others can be referred if necessary.

Classically, Fortran, C other computer languages have only relatively simple methods in reading input files which are usually text file for ease of transportation between different systems. Even with the refined input standard in the latest Fortran or C language, the way a program reads the input file is still basically sequential, and a well-defined data structure is required. To allow for more flexible input approach, the authors have adopted the open source library FLIB with some modifications, and the revised version of FLIB can be obtained freely from the authors. FLIB adopts the operator overloading function available in Fortran 90/95, and only limited functions will be discussed in this section. With reference to the subroutine as shown below, the input module uses the functions in FLIB through the modules STRPAK and FIOPAK, while the data required by the main program is communicated through module DATA.

```
SUBROUTINE INPUT

USE STRPAK ; USE FIOPAK ; USE DATA

Integer :: ndat, i, j, k, ndat, bore_n, bore_s, errar

Real :: Toler

Open (25, file='input.txt')

call vi_getvar(25,'1', 'TOLER', ndat, errvar)

if (ndat > 0) then ! toler is defined with a value in the input file

call vi_data(1, TOLER, errvar) ; CALL VI_ERASE

! read the value of toler as ndat > 0, else, do not read

endif

CALL VI_GETVAR(25,'1','bore_n', ndat,errar)

if (ndat > 0) then

CALL VI_DATA(1, bore_n, errvar) ; CALL VI_ERASE

endif

CALL VI_GETVAR(25,'1','bore_s',ndat,errar)

if (ndat > 0) then

CALL VI_DATA(1, bore_s, errvar) ; CALL VI_ERASE

endif

call vi_getvar(25,'1','bore_data',ndat,errvar)

if (ndat == bore_n*bore_s) then

allocate (bore(bore_n,bore_s))

k=0
```

```
do i = 1,bore_n

do j = 1,bore_s

k=k+1 ; call vi_data(k,bore(j,i),errvar)

end do

end do

call vi_erase

endif
```

Format of input data in the Input.txt

```
bore_n 4

bore_s 3

Toler 0.001

bore_data

1.1 2.1 12.6 4.3 8.2

11.3 7.1 5.4

9.8 4.5 6.9 13.2

End
```

In this subroutine, an input file defined by the filename input.txt is opened which is given above (the authors always use notepad or equivalent to generate the input ASCII files). The keywords bore_n, bore_s, toler, bore_data in the input file control the input process. The numbers which follow the keyword will be used to define how the data are to be read. For example, there is only one number which follows the keyword ‘Toler’, and the keyword ‘Toler’ define a data which is read by the statement call vi_data(1, TOLER, errvar). The variable toler will be assigned the value of 0.001. If such keyword is not found, toler will not be read in. It should be noted that even though there are sequences in reading the variables in the input subroutine, there is not such a requirement in the actual input data format. As long as the variable ‘Toler’ appears in the input file, the function vi_data will search from top to bottom of the input file for the keyword, which is controlled by the parameter ‘1’ in the function call vi_getvar(25,'1', 'TOLER', ndat, errvar). If the keyword is specified as '0', the subroutine will start searching from the current position of the input file during the input process. It should also be noted that the keyword and the actual name of the variable need not bear the same name, but the authors usually adopt the same names for them for ease of reading the program. The integer 'ndat' in vi_getvar(25,'1', 'TOLER', ndat, errvar) returns the number of data following the keyword 'Toler'. For the keyword 'bore_n', 'bore_s', these two integers are read, and a matrix is allocated accordingly (defined by the module DATA to be allocatable matrix). There are now 12 data to be defined for the matrix bore, and the 12 data appear after the keyword bore_data. As long as there are 12 data after the keyword, there is no requirement on how they appear in the input file. A regular input file is hence not necessary. If the variable errvar returns a negative value, it means that the keyword is not found in the input. Since 12 data are defined, the computer memory required to store and read the data will be assigned by the use of dynamic memory function ALLOCATE in Fortran 90/95. Static memory is hence seldom adopted by the authors in writing a program. To define the end of the input file, the authors use the word 'end' at the end of the input file as shown above. Such input file format is convenient to be defined, when the users define the input data gradually but not sequentially. There is no need for the user to search for the right locations for the input data in the input file, as compared to the programs as provided by Smith *et al.* (2014), Owen and Hinton (1980), Hinton and Owen (1977), Zienkiewicz *et al.* (2011). In general, the authors also define a default subroutine so that any variable which is not given in the input file will be automatically assigned.

In some cases, the keyword may appear more than once. In that case, the control parameter in the vi_getvar may take '1' and '0', and an example is shown below, which is taken out from a finite element program developed by the authors. By using the function backspace, the cursor position will be backed by 1 line, hence the search for the data bearing the same keyword can start from the current position in the input file. Various modifications and variations can be combined by using the FLIB library so that a flexible input format can be developed. It should however be noted that for the classical approach where the data is read line by line, the input process will be much faster. In the actual implementation of some large scale programs, both approaches are used by the authors to achieve effectiveness and efficiency simultaneously. Some data (coordinates, element connectivity or similar) can be defined by the classical approach using conventional approach, which is much faster in operation. For loadings, boundary conditions, specification of material zones or similar, it is more convenient to use the current approach. A program can read the input in both format, using two or more input files.

```
IF (N6 > 0) THEN

CALL VI_GETVAR(25,'1','LOCAL_PRESSURE',II,ERRVAR)

CALL VI_DATA(1,J,ERRVAR) ; CALL VI_DATA (2,POINT(1),ERRVAR)

DO I = 1,(II-2)/2

CALL VI_DATA (I*2-1+2,XO,ERRVAR) ; CALL VI_DATA (I*2+2,YO,ERRVAR)

CALL CONDENSE(XO,YO,X1,Y1)

END DO

CALL VI_ERASE

91 BACKSPACE(25) ! back one line for the input file

CALL VI_GETVAR(25,'0','LOCAL_PRESSURE',II,ERRVAR)

IF (II == 0) GO TO 68

CALL VI_DATA(1,J,ERRVAR) ; CALL VI_DATA (2,POINT(1),ERRVAR)

DO I = 1,(II-2)/2

CALL VI_DATA (I*2-1+2,XO,ERRVAR) ; CALL VI_DATA (I*2+2,YO,ERRVAR)

CALL CONDENSE(XO,YO,X1,Y1)

END DO

CALL VI_ERASE

GO TO 91

68 CALL VI_ERASE

ENDIF
```

A sample grid program is given in Appendix 2-1 for illustration of the operator overloading input approach as given in this section. This program has been used by the authors for teaching on the analysis of raft foundation with the Winkler soil support, and is flexible enough to be expanded to accept more complicated load cases. There is no problem limit to the program, as the computer memory is allocated upon reading the basic control parameters of the problem. This kind of grid analysis program has been popular in Hong Kong and many countries in the past for the analysis of bridge decks, pile caps, raft foundation and transfer plates. Recently, most of the engineers adopt the plate element program for such purposes, but there are still some engineers who prefer the use of grid method in the analysis and design.

2.1.2. Some Geometry File

The authors have developed many finite element programs, and it is frequently required to determine whether a given point lie on, inside or outside of an element. This requirement come from the specification of point loads/displacements where the coordinates instead of node number is specified by the user. Some commercial programs (SAFE by CSI and others) require the user to define grids for a problem, and the loadings are specified along the grid lines or grid points, but many engineers have found this requirement annoying, and sometimes strange mesh will be defined under such requirement. The authors allow point loads/line loads to be defined inside an element or even cross element, and during the assemblage of the global load vector, it is required to find the element number for distributing the point/line load. Towards this, the authors have adopted an open source subroutine locpt with modifications (see Appendix 2-4). This subroutine will check whether a point lies on, inside or outside an element, which is required for the assemblage of the global load vector and assigning different material properties. In the seepage analysis program developed by the authors, the material properties of each element is determined by using the centroid coordinates of each element and check against the material domains as defined. For nonhomogeneous problems, this subroutine can be very useful for many applications.

```
SUBROUTINE locpt (x0, y0, xy, n, l, m)

!----------------------------------------------------------------------

! GIVEN A POLYGONAL LINE CONNECTING THE VERTICES (X(I),Y(I)) (I
= 1,...,N)

! TAKEN IN THIS ORDER. IT IS ASSUMED THAT THE POLYGONAL
PATH IS A LOOP,

! WHERE (X(N),Y(N)) = (X(1),Y(1)) OR THERE IS AN ARC FROM
(X(N),Y(N)) TO
```

```
! (X(1),Y(1)). N.B. The polygon may cross itself any number of times.

! (X0,Y0) IS AN ARBITRARY POINT AND L AND M ARE VARIABLES.

! On output, L AND M ARE ASSIGNED THE FOLLOWING VALUES ...

! L = -1 IF (X0,Y0) IS OUTSIDE THE POLYGONAL PATH

! L = 0 IF (X0,Y0) LIES ON THE POLYGONAL PATH

! L = 1 IF (X0,Y0) IS INSIDE THE POLYGONAL PATH

! M = 0 IF (X0,Y0) IS ON OR OUTSIDE THE PATH. IF (X0,Y0) IS INSIDE THE

! PATH THEN M IS THE WINDING NUMBER OF THE PATH AROUND THE POINT (X0,Y0).

IMPLICIT NONE

INTEGER, INTENT(IN) :: n

REAL(KIND=8), INTENT(IN) :: x0, y0

REAL(KIND=8), DIMENSION (N,2) :: xy

INTEGER, INTENT(OUT) :: l, m

INTEGER :: i, n0

REAL(KIND=8) :: angle, eps, pi, pi2, sum, theta, theta1, thetai, tol, u, v
```

Another subroutine for which the authors commonly adopt is the find_node, which is used to find the node number which is closest to the specified coordinate in the input file. Actually, in the mesh generation program developed by the authors, the actual coordinates of the nodes are slightly shifted to suit for the applied loadings, subject to the condition that the element is not highly distorted to preserve accuracy in the calculation. Using such geometry subroutines, the authors have developed many finite element programs where the specification of loads, displacements, boundary conditions and other requirements are given in terms of purely coordinates (instead of grid lines). In most cases, the specified loads and displacements can be assigned correctly to the specific nodes easily. In special case where mesh distortion is too serious so that the nodal coordinates cannot be shifted to suit the applied loads, the element which contain the specified point load will be determined, and the load will then be distributed to the element nodes according to the interpolation functions. This subroutine is used in many programs developed by the authors, and a good illustration is given in Appendix 2-4.

```
SUBROUTINE FIND_NODE(X0,Y0,NODE)

USE DATA

INTEGER :: NODE,I

REAL(KIND=8) :: X0,Y0,DIST1,DIST

DIST1=(X0-COORD(1,1))**2+(Y0-COORD(1,2))**2

NODE=1

DO I=2,NPOIN

DIST=(X0-COORD(I,1))**2+(Y0-COORD(I,2))**2

IF (DIST < DIST1) THEN

NODE=I ; DIST1=DIST ! update node number for NODE

ENDIF

END DO

RETURN

END SUBROUTINE FIND_NODE
```

2.1.3. Mesh Generation

Mesh generation or block generation is very important in finite element or distinct element analysis. This topic is a very big issue with many methods developed and used over the years. Various methods have been proposed for mesh generation (Frey and George, 2000, Lo, 2015). For two-dimensional problems, the use of triangular elements which is easily generated can well fulfilled the needs of most practical problems, while the generation of quadrilateral or higher elements for general domain remains not universal. This problem is even more difficult for three-dimensional problems. It is not surprising that many commercial programs still adopt the triangular or tetrahedral elements for mesh generation. Some commercial mesh generation programs can generate good mesh for most problems, but the learning curves for using those programs can be steep. In practice, most of the engineers still prefer to use the low order simple elements for practical works. Except for triangular mesh, generation of good quadrilateral or other element shapes for an arbitrary domain can be tedious and difficult. One of the advantages of the meshless method is actually the avoidance in generating mesh, which is particularly difficult for some problems with difficult geometry. Many mesh generation programs for triangular elements are available on web (Easymesh, Triangle are also used by the authors), and the authors will not repeat these works in this book. The authors seldom adopt triangular element in application due to the low accuracy of the interpolation function. The use of 4 and 8 nodes elements are commonly used by the authors for developing research type finite element programs. So far, the authors have not tried to develop a code which is versatile under all cases (Cheng and Tsui 1993), but have developed a program MESH for generating 3 and 4 nodes element, which form part of the sub-module in a Windows based program PLATE (thin and thick plate analysis for raft foundation, pile cap and transfer plate). Standalone version of the mesh generation program has been prepared by the authors for coupling with other self-developed geotechnical programs. MESH in program PLATE is sufficiently good for analysis of many raft foundation, pile cap and transfer plate, and it is designed mainly for applications in geotechnical/foundation problems instead of being a versatile program suitable for all types of problems. Nevertheless, MESH can provide a good mesh generation in nearly all the cases that the authors have tried. The use of adaptive mesh refinement is an alternative in which more elements will be generated at those regions where there are stress concentrations (Plewa *et al.* 2003). The authors have adopted such approach for finite element limit analysis with error estimator to refine the mesh automatically. If mesh refinement is adopted properly, the accuracy of analysis can be highly improved without the use of intensive computations.

In many practical engineering problems, the solution domain can be easily discretized into several major regions, and the authors adopt quadrilateral sub-domains as shown in Fig. (**2.1**) for a Hong Kong Housing Department raft foundation problem. There is no requirement to the sequence of the numbering of these sub-domains, and there is also no node to node connection requirement for these sub-domains. The flexibility in this mesh generation scheme is well favoured by some engineers for practical engineering works. Furthermore, there is no requirement for defining grid in MESH, but the users can define grid if necessary. The output can be viewed at any place by interpolation of the results at nodes closest to the specified point, without the requirement of viewing the results at grid as required by some commercial programs.

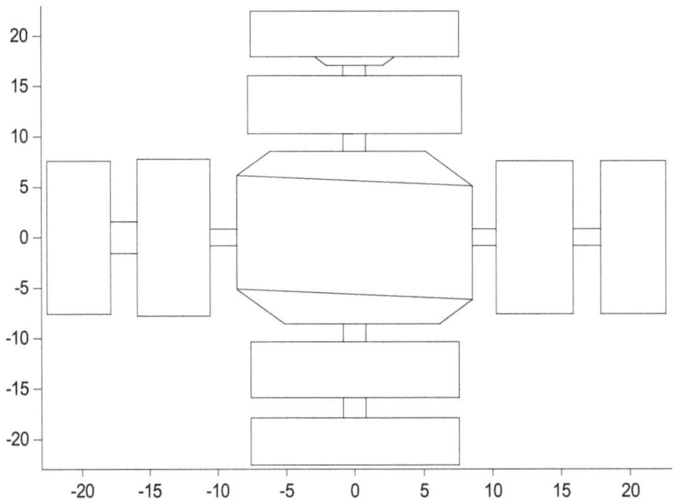

Fig. (2.1). A true raft foundation problem in Hong Kong, defined by a series of sub-domains.

Based on the maximum element size as defined by the engineer, a finite element mesh is generated as shown in Fig. (**2.2**) within a short time. The quadrilateral sub-domains can be non-rectangular, which is illustrated in Fig. (**2.3**). Actually, the sub-domains need not be connected in the mesh generation process, as shown in Fig. (**2.4**), if the engineers really have to define a mesh for such case.

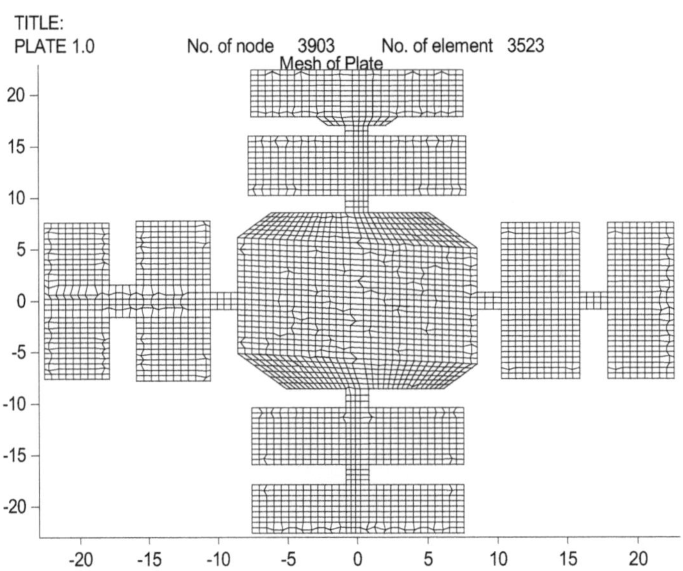

Fig. (2.2). Mesh generated for Fig. (**2.1**) based on a maximum element size of 0.5m.

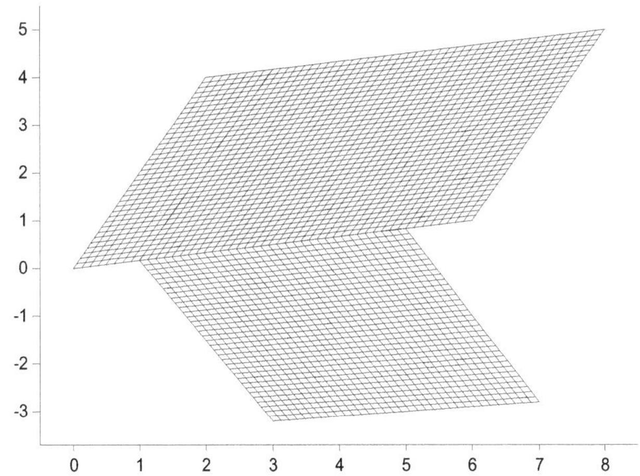

Fig. (2.3). Non-rectangular sub-domains in mesh generation.

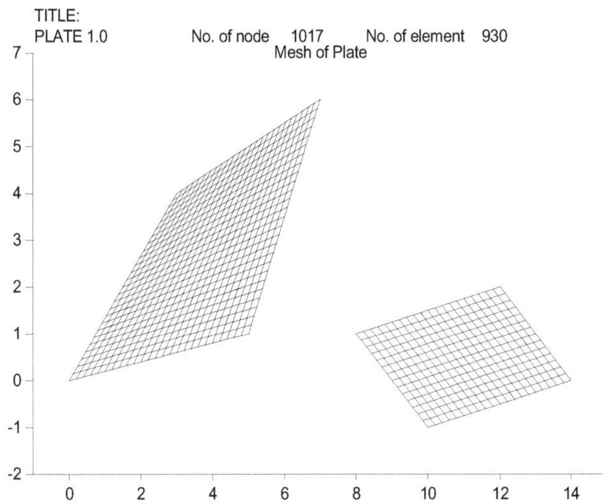

Fig. (2.4). Sub-domains need not be connected in mesh generation.

Although quadrilateral elements are generated from the sub-domains, triangular elements can actually be generated by subdividing each quadrilateral elements to two triangular elements, which is an option available in MESH. It should be noted that in Fig. (**2.2**), some of the node coordinates are shifted in order to match with the applied loadings, and the distortion of the elements are checked for acceptability before acceptance. This requirement is necessary because the mesh is generated without the consideration of the applied loads. In most cases, the final nodal coordinates from the mesh will comply with the coordinates of the applied loads exactly. There are however some cases where the distortion is so large that such shifting of nodal coordinates is not preformed. In such case, the nodal loads within an element will be distributed to the nodes of that element in the analysis.

After the generation of the mesh, a bandwidth or profile optimizer will be used to optimize the node number of the mesh so as to reduce the required computer memory storage. The authors have adopted some open source subroutines by George and Liu (1981) (see Appendix 2-3) and the GPS re-ordering scheme Algorithm 582 from ACM (1982) (http://www.netlib.org/toms-2014-06-10/582) with some modifications in program PLATE (Cheng 1993). There are many published algorithms for such applications (Livesley and Sabin 1991, Sloan and Randolph 1983, Doss and Arathi 2016, Wang and Shi 2009, Sloan 1989 and many others). More updated reordering schemes have not been adopted by the authors for implementation in computer codes. In general, the authors find these two algorithms sufficiently good for normal applications.

For the program MESH, a brief explanation to the mesh generation scheme is briefly discussed here. Refer to the problems as shown in Fig. (**2.5a**), it is internally transformed to that in Fig. (**2.5b**) so that each sub-domains are connected at the sides and nodes. This operation is transparent to the users and is carried out internally in the program. Based on Fig. (**2.5b**), the number of divisions along any direction is determined by the number of divisions on the two opposite sides of the quadrilateral, and the larger number will be adopted automatically. Elements and nodes within each sub-domain is numbered sequentially then, and the values accumulate from sub-domain 1 to the last sub-domain. The duplicate nodes along the common sides between the sub-domains are then eliminated, and the total number of nodes will decrease accordingly. It should be noted that the number of elements will not change in the elimination process. Through this simple approach, the finite element mesh for many practical engineering problems can be formed easily. If triangular elements are preferred by the users, each quadrilateral as formed will then be further sub-divided to two triangular elements.

For the mesh generation scheme and node re-ordering computer code as discussed, a simple version is given in the Appendix, with the purposes that they will be modified for personal use by the readers. The more advanced mesh generation program can be obtained from the authors free of charge upon request, and interested users can request for the computer program at natureymc@yahoo.com.hk. A simpler mesh generation program for 3/4 node element is given in Appendix 2-2 for learning purpose, which is the preliminary version of MESH. The users need to define the macro elements only with connected nodes similar to that in Fig. (**2.5b**), and the node numbering will be controlled by the parameter IDY. If IDY is given as 0, the node of the mesh will correspond to node 1 of the macro element. If IDY is given as 1, the node will be numbered from the highest y –ordinate and leftmost position. This simple code has been used by the authors in the past, and is abandoned because it is not efficient if the total number of generated elements exceed 3000. The authors develop the code MESH as a submodule of the powerful finite element program PLATE, which is effective and efficient even if the total number of generated elements exceed 100000, and is adequate for the analysis of many foundation problems and education purposes.

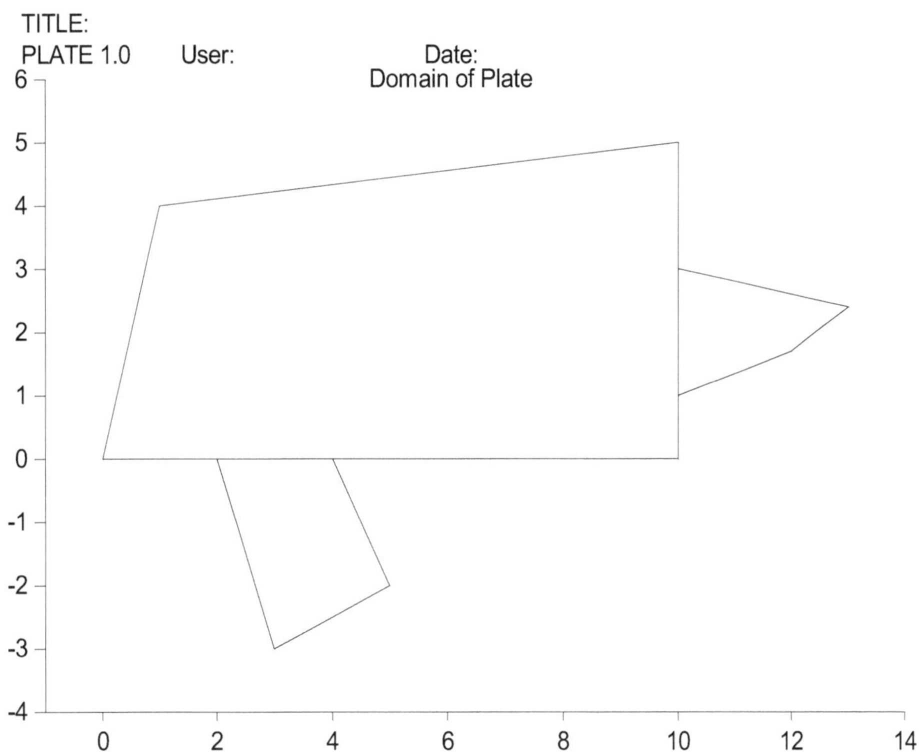

(a) Original Problem.

Fig. 2.5 cont.....

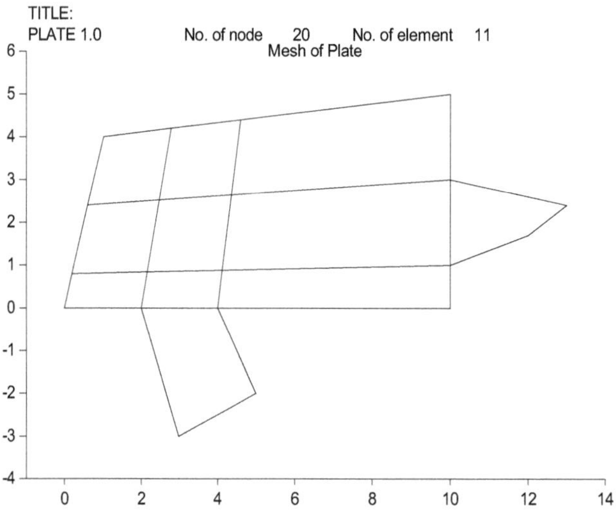

(b) Transformed domains for (a).

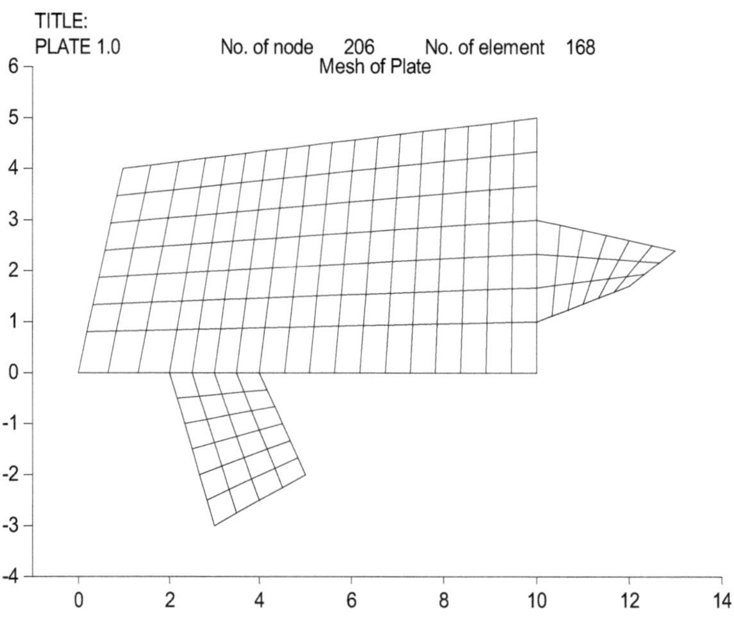

(c) Mesh generated with a maximum element size of 0.5m.

Fig. (2.5). Generation of sub-domain inside program MESH.

2.2. INTRODUCTION TO FINITE ELEMENT ANALYSIS

Most of the practical geotechnical problems are complicated in nature. In particular, the irregular geometric domain, nohomogeneous ground conditions, the presence of water, external and internal loadings and structural elements and the complex mechanical response of soil and soil-structure interaction have created extreme difficulty in the analysis of real engineering problems. Towards this, the use of numerical methods will be more appropriate. Currently, the use of finite element, finite difference, boundary element and distinct element methods are the most popular methods which are adopted by many engineering geotechnical programs. Boundary element method is most suitable for linear problem with elastic constitutive behaviour, and it is not commonly adopted in geotechnical problems. The capabilities of finite difference and finite element methods are comparable, but there are only few general finite difference programs in the world due to various technical difficulties (Itasca programs are generally based on the finite difference method). Finite difference method will only be briefly discussed in this book, but the discussion on the distinct element method in chapter 4 will provide a basis for the application of the finite difference method. The authors have also developed many programs in finite difference as well as finite element method. In general, finite difference will solve a problem in shorter time, but the application of complicated boundary conditions and loadings will be generally more tedious to consider than the finite element method. If highly irregular mesh is used, the use of finite difference method may produce a lower accuracy as compared with the use of finite element method.

Finite element method is the most popular numerical tool to solve many types of differential equations in various disciplines. Currently, there are hundreds of finite element programs available to the engineers for various applications, but there are also many limitations of FEM to geotechnical problems which include: (1) difficult to be applied for very large displacement problems, (2) generation and loss of contacts are difficult to be modelled, (3) fracture is difficult to be handled (but not impossible), (4) flow of the materials. For simplicity, two-dimensional problem will be discussed here. The extension of two-dimensional analysis to three-dimensional analysis or axi-symmetric analysis is simple, and this is covered in many books which will not be repeated here.

A problem is two-dimensional if the field quantities such as stress and displacement depend on only the two coordinates (x, y). In this sense, there are strictly no two-dimensional problems because every structure or loading system is three-dimensional in the real world. Many engineering problems can however be simplified approximately to a plane problem. Such approximation will greatly reduce the computational effort and can yield results satisfying practical requirements at the same time. Generally speaking, there are two kinds of elastic plane problems (Zienkiewicz 2011): plane strain and plane stress, which will be introduced in the following subsections.

2.2.1. Plane Strain

To illustrate the problem, Fig. (**2.6**) shows a section of a structure with its cross-section independent of its length, except for the body force. On the surface of the structure, there may be surface force or constraint. It should be noted that both the internal and external forces are parallel to the cross-section, and the values of them are invariable with the length normal to the section.

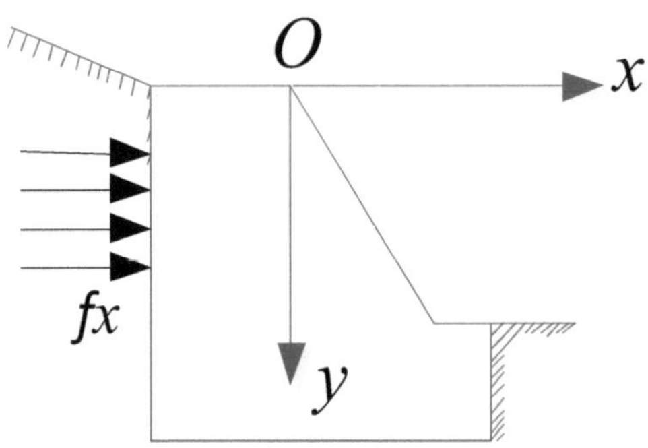

Fig. (2.6). Cross-section of an infinite long structure with external force.

Assume the length of the structure to be infinite, and consider any one of the cross-sections as the *xy*-plane, to which *z*-axis is perpendicular. All the field quantities (*e.g.*, stress, strain, and displacement, *etc.*) depend only on the two coordinates (*x, y*). Additionally, any cross-section is a symmetrical surface so that any point in this surface will only have *x*-displacement and *y*-displacement (*i.e.*, *z*-displacement is 0). For this condition, it follows that the shear strain γ and ε will be zero.

$$\gamma_{zx} = \gamma_{zy} = \varepsilon_z = 0 \tag{2.1}$$

In view of the stress-strain relations, we can deduce that

$$\tau_{zx} = \tau_{xz} = 0, \tau_{zy} = \tau_{yz} = 0, \gamma_{xz} = \gamma_{yz} = 0 \tag{2.2}$$

Therefore, all the shear strain components equal to 0 except those related to *xy*-plane (*i.e.*, ε_x, ε_y and γ_{xy}). This two-dimensional state is known as the plane strain condition.

It should be noted that while the normal strain is 0, the normal stress σ in the *z*-direction is not 0, which can be deduced according to the stress-strain relations as follows

$$0 = \varepsilon_z = \frac{1}{E}\left[\sigma_z - \nu(\sigma_x + \sigma_y)\right] \qquad \textbf{(2.3)}$$
$$\sigma_z = \nu\left(\sigma_x + \sigma_y\right)$$

Similarly, by substituting σ_{zz} to the stress-strain relations we can deduce

$$\varepsilon_x = \frac{1-\nu^2}{E}\left(\sigma_x - \frac{\nu}{1-\nu}\sigma_y\right) \qquad \textbf{(2.4)}$$

$$\varepsilon_y = \frac{1-\nu^2}{E}\left(\sigma_y - \frac{\nu}{1-\nu}\sigma_x\right) \qquad \textbf{(2.5)}$$

$$\gamma_{xy} = \frac{2(1+\nu)}{E}\tau_{xy} \qquad \textbf{(2.6)}$$

For simplicity, the stress-strain relations for 2-D plane strain problem can be expressed in a matrix form as:

$$\{\boldsymbol{\sigma}\} = [\boldsymbol{D}]\{\boldsymbol{\varepsilon}\} \qquad \textbf{(2.7)}$$

$$[\boldsymbol{D}] = \frac{E_0}{1-\nu_0^2}\begin{bmatrix} 1 & \nu_0 & 0 \\ \nu_0 & 1 & 0 \\ 0 & 0 & \frac{1-\nu_0}{2} \end{bmatrix} \quad \text{(2.8)}$$

where [D] is called the elastic matrix with $E_0 = \frac{E}{1-\nu^2}$ and $\nu_0 = \frac{\nu}{1-\nu}$.

2.2.2. Plane Stress

Plane stress is an approximate solution, in contrast to plane strain condition, which is practically exact for many problems. In other words, plane strain condition is a special solution of the complete three-dimensional equations of elasticity, whereas plane stress is only approached in the limit as the thickness of the loaded body tends to zero.

Plane stress problem is usually related with an identical thickness plate which has external force acting on the edge only, and the external force is assumed to be parallel to the plate surface and independent of the thickness of the plate as well as body force. As shown in Fig. (**2.7**), which is a deep beam plane stress problem. Suppose the thickness of the plate is *h*, and take the intermediate plane of the plate as *xy*-plane with any line perpendicular to it as *z*-axis.

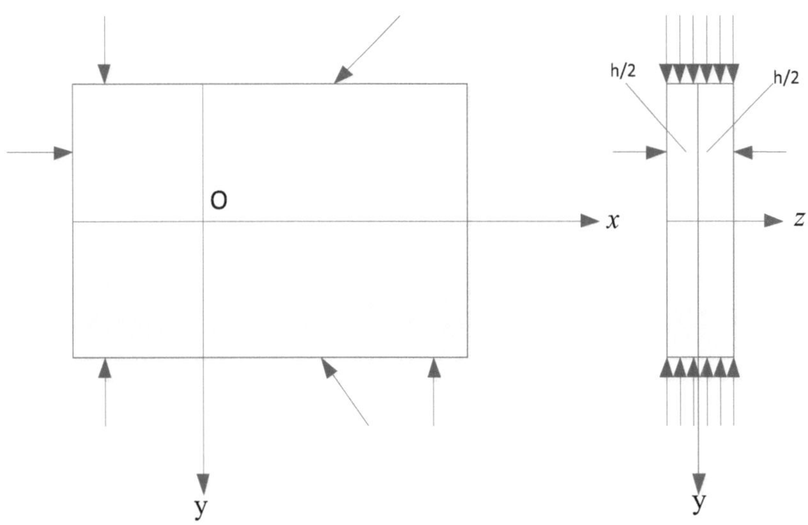

Fig. (2.7). Deep beam with identical thickness.

It is argued that if the two bounding *z*-planes of a thin plate are sufficiently close in comparison with the other dimensions, and if they are also free of tractions, the stresses on all parallel *z*-planes will be sufficiently small to be negligible, and we can write

$$\tau_{zx} = \tau_{zy} = \sigma_z = 0 \quad \mathbf{(2.9)}$$

Obviously, we can deduce that

$$\tau_{xz} = \tau_{yz} = 0 \quad \mathbf{(2.10)}$$

Thus, there are only three stress components (σ_x, σ_y and τ_{xy}) remain in plane stress condition, and all of them depend only on the two-dimensional coordinates (x, y). The stress-strain relations can be expressed as (Barber 2010)

$$\varepsilon_x = \frac{1}{E}\left(\sigma_x - \nu\sigma_y\right) \quad \mathbf{(2.11)}$$

$$\varepsilon_y = \frac{1}{E}\left(\sigma_y - \nu\sigma_x\right) \quad \mathbf{(2.12)}$$

$$\gamma_{xy} = \frac{2(1+\nu)}{E}\tau_{xy} \quad \mathbf{(2.13)}$$

For simplicity, the stress-strain relations for 2-D plane stress problem can be expressed in matrix form as follows:

$$\{\boldsymbol{\sigma}\} = [\boldsymbol{D}]\{\boldsymbol{\varepsilon}\} \quad \mathbf{(2.14)}$$

$$[\boldsymbol{D}] = \frac{E_0}{1-\nu_0^2}\begin{bmatrix} 1 & \nu_0 & 0 \\ \nu_0 & 1 & 0 \\ 0 & 0 & \frac{1-\nu_0}{2} \end{bmatrix} \quad \textbf{(2.15)}$$

where [D] is called elastic matrix with $E_0 = E$ and $\nu_0 = \nu$.

2.2.2.1. Relationship between Plane Strain and Plane Stress

The solution of a problem under either the plane strain or plane stress assumptions involves finding a two-dimensional stress field, defined in terms of $\sigma_x, \sigma_y, \tau_{xy}$, which satisfies the equilibrium equations:

$$\frac{\partial \sigma_x}{\partial x} + \frac{\partial \tau_{xy}}{\partial y} + f_x = 0 \quad \textbf{(2.16)}$$

$$\frac{\partial \tau_{yx}}{\partial x} + \frac{\partial \sigma_y}{\partial y} + f_y = 0 \quad \textbf{(2.17)}$$

or in index notation

$$\frac{\partial \sigma_{ij}}{\partial x_i} + f_i = 0 \quad \textbf{(2.18)}$$

and for which the corresponding strains $\varepsilon_x, \varepsilon_y, \gamma_{xy}$, satisfy the only non-trivial compatibility equation

$$\nabla^2(\sigma_x + \sigma_y) = 0 \quad \textbf{(2.19)}$$

The equilibrium and compatibility equations are the same in both formulations, and the only difference between these two cases is the relation between the stress and strain components, which for normal stresses are given by eqs.(2.4) and (2.5) for plane strain condition and (2.11) and (2.12) for plane stress condition. The relation between the shear stress τ_{xy} and the shear strain γ_{xy} is the same for both formulations and is given by eqs.(2.6) or (2.13). Thus, from a mathematical perspective, the plane strain solution looks like the plane stress solution for a material with different elastic constants and vice versa. Actually, it can be easily verified that equation (2.8) can be obtained from equation (2.15) by making substitutions

$$E_0 = \frac{E}{1-\nu^2}, \nu_0 = \frac{\nu}{1-\nu} \qquad \textbf{(2.20)}$$

This substitution also leaves the shear stress-shear strain relation unchanged as required.

2.2.3. Fundamentals of FEM

Except for some simple problems with regular geometry and loading, it is very difficult to solve most of the boundary value problems in engineering directly with the yield of analytical solutions. Towards this, the use of numerical method seems indispensable, and the finite element method is one of the most popular methods used by the engineers (Rao 2018, Zienkiewicz *et al.* 2011 and many other references). There are two fundamentals approaches to FEM which are the weighted residual method (WRM) and variational principle, but there are also other less popular principles which may be more effective under certain special cases. In the finite element analysis of an elastic problem, solution is obtained from the weak form of the equivalent integration for the differential equations by WRM as an approximation. Alternatively, different approximate approaches (*e.g.*, collocation method, the least square method and Galerkin method) for solving differential equations can be obtained by choosing different weights based on the WRM, and Galerkin method appears to be the most popular approach in general. Recently, the meshless methods appear to be under active research, which can be considered as variants of the least square method or the Galerkin method.

Specifically, in elasticity for instance, the principle of virtual work (including both principle of virtual displacement and virtual stress) is considered to be the weak form of the equivalent integration for the governing equilibrium equations. Furthermore, the aforementioned weak form of equivalent integration on the basis of Galerkin method can also be evolved to a variation of a functional if the differential equations have some specific properties such as linearity and self-adjointness. Principle of minimum potential energy and complementary energy are two variational approaches equivalent to the fundamental equations of elasticity.

Since displacement is usually the basic unknown quantity in FEM, only the principle of virtual displacement and minimum potential energy will be introduced in the following parts. In this case, the FEM introduced herein is also called the displacement finite element method (DFEM). There are other ways to form the basis of FEM with advantages in some cases, but these approaches are less general and will not be discussed here.

2.2.4. Principle of Virtual Displacement

Principle of virtual displacement is the weak form of the equivalent integration for equilibrium equations and force boundary conditions. Given the equilibrium equations and force boundary conditions *f* in index notation

$$\sigma_{ij,j} + f_i = 0, \quad (\text{in domain } V) \tag{2.21}$$

$$\sigma_{ij} n_j - T_i = 0, \quad (on\ domain\ boundary\ S_\sigma) \tag{2.22}$$

where n and T are the normal vector and traction along the boundary. In WRM, without loss of generality, the variation of true displacement δu_i and its boundary value (*i.e.*,-δu_i) can be selected as the weight functions in the equivalent integration

$$\int_V \delta u_i(\sigma_{ij,j} + f_i)\,\mathrm{d}V - \int_{S_\sigma} \delta u_i(\sigma_{ij} n_j - T_i)\,\mathrm{d}S = 0 \tag{2.23}$$

The weak form of equation (4.23) is given as:

$$\int_V (-\delta\varepsilon_{ij}\sigma_{ij} + \delta u_i f_i)\,\mathrm{d}V + \int_{S_\sigma} \delta u_i T_i\,\mathrm{d}S = 0 \tag{2.24}$$

It can be seen clearly from eq.(2.24) that the first item in the volume integral represents the work done by the stresses under the virtual strain (*i.e.*, internal virtual work), while the remaining items indicate the work done by the body force and surface force under the virtual displacement (*i.e.,* external virtual work). In other words, the summation of the internal and external virtual work is equal to 0, which is called the principle of virtual displacement. Under this case, we can conclude that a force system will satisfy equilibrium equations if the summation of the work done by it under any virtual displacement and strain is equal to 0.

2.2.5. Principle of Minimum Potential Energy (PMPE)

Based on equation (2.24), by substituting equations (2.14) for σ_{ij}, we can deduce that

$$\int_V (\delta\varepsilon_{ij} D_{ijkl}\varepsilon_{kl} - \delta u_i f_i)\, \mathrm{d}V + \int_{S_\sigma} \delta u_i T_i \, \mathrm{d}S = 0 \tag{2.25}$$

Due to the symmetry of the constitutive matrix D_{ijkl}, we can further obtain

$$(\delta\varepsilon_{ij}) D_{ijkl}\varepsilon_{kl} = \delta\left(\tfrac{1}{2} D_{ijkl}\varepsilon_{ij}\varepsilon_{kl}\right) = \delta U(\varepsilon_{mn}) \tag{2.26}$$

where $U(\varepsilon_{mn})$ is the unit volume strain energy. Given the assumptions in linear elasticity

$$-\delta\phi(u_i) = f_i \delta u_i, -\delta\psi(u_i) = T_i \delta u_i \tag{2.27}$$

equation (2.25) is further simplified to

$$\delta\Pi_P = 0 \tag{2.28}$$

Π_p is the total potential energy of the system, which is equal to the summation of the potential energy of deformation and external force and can be expressed as

$$\Pi_P = \Pi_P(u_i) = \int_V \left(\tfrac{1}{2} D_{ijkl}\varepsilon_{ij}\varepsilon_{kl} - f_i u_i\right) \mathrm{dV} - \int_{S_\sigma} T_i u_i \, \mathrm{d}S \tag{2.29}$$

Eq.(2.28) shows that, among all the potential displacements, the total potential energy of system will take stationary value at the real displacement, and it can be further verified that this stationary value is exactly the minimum value which is the principle of minimum potential energy.

2.3. GENERAL EXPRESSIONS AND IMPLEMENTATION PROCEDURE OF FEM

The solution of a general continuum problem by FEM always follows an orderly step-by-step process which is easy to be programmed and used by the engineers. For illustration, a three-node triangular element for plane problems is taken as an example to illustrate the general expressions and implementation procedures of FEM.

2.3.1. Discretization of Domain

The first step in the finite element method is to divide the structure or solution region into subdivisions or elements. Hence, the structure is to be modeled with suitable finite elements. In general, the number, type, size, and arrangement of the elements are critical towards good performance of the numerical analysis. A typical discretization with three-node triangular element is shown schematically in Fig. (**2.8**).

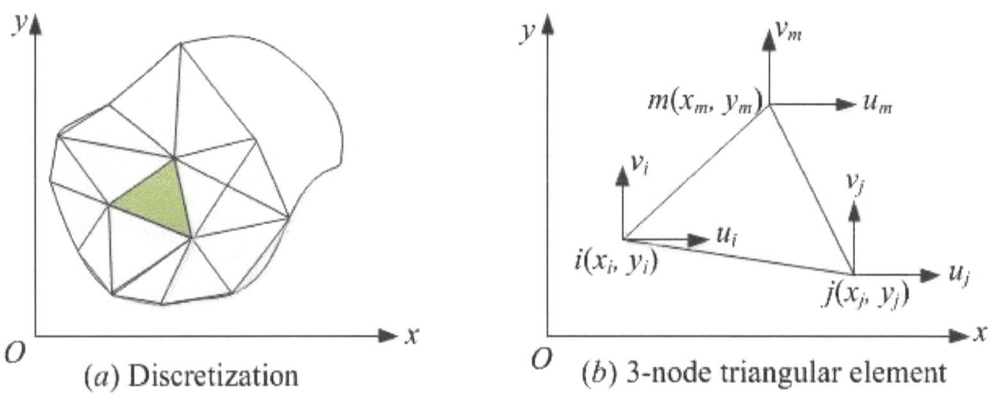

Fig. (2.8). Discretization of a 2-D domain with 3-node triangular element.

Mesh generation can be a difficult process for a general irregular domain. If only triangular element is to be generated, this is a relatively simple work, and many commercial programs can perform well in this respect. There are also some public domain codes (Easymesh or Triangle, Gmsh, Cubit, Delaundo, Geompack++ and many others) which are sufficiently good for normal academic purposes. For quadrilateral or higher elements, mesh generation is not that simple, and it is preferable to rely on the use of commercial programs for complicated problems. For normal problems, an easier solution MESH is offered by the authors.

2.3.2. Interpolation or Displacement Model

As can be seen from Fig. (**2.8b**), the nodal number of a typical 3-node triangular element is coded in anticlockwise order (*i.e.,* in the order of *i*, *j* and *m*), and each node has two degrees of freedom (DOFs) or two displacement components which is stored in a column vector in index notation as follows:

$$\boldsymbol{a}_i = \begin{bmatrix} u_i \\ v_i \end{bmatrix} \qquad (i, j, m) \tag{2.30}$$

Totally, each element has 6 nodal displacements, *i.e.,* 6 DOFs. Putting all the displacements **a** in a column vector, we can obtain the element nodal displacement column matrix as

$$\boldsymbol{a}^e = \begin{bmatrix} \boldsymbol{a}_i \\ \boldsymbol{a}_j \\ \boldsymbol{a}_m \end{bmatrix} = \begin{bmatrix} u_i & v_i & u_j & v_j & u_m & v_m \end{bmatrix}^{\mathrm{T}} \tag{2.31}$$

In FEM, nodal displacement is chosen as the basic unknowns, so interpolation at any arbitrary point is based on the three nodal displacements of each element, which is called a displacement mode. For a 3-node triangular element, linear polynomial is utilized, and the element displacement in both *x*-direction and *y*-direction are

$$u = \beta_1 + \beta_2 x + \beta_3 y \tag{2.32}$$

$$v = \beta_4 + \beta_5 x + \beta_6 y \tag{2.33}$$

where β is coefficient to be determined. Obviously, the displacements of all the three nodes should satisfy equations (2.32) and (2.33). By substituting the six nodal displacement components into these equations, it is easy to obtain another form of displacement mode as

$$u = N_i u_i + N_j u_j + N_m u_m \tag{2.34}$$

$$v = N_i v_i + N_j v_j + N_m v_m \tag{2.35}$$

where

$$N_i = \frac{1}{2A}(a_i + b_i x + c_i y) \qquad (i, j, m). \tag{2.36}$$

In equation (2.34), N_i, N_j and N_m denote the interpolation function or shape function for the three nodes, respectively. A is the area of the element, and a_i, b_i, c_i … , c_m are constants related to the coordinates of the three nodes. Similarly, equations (2.34) and (2.35) can also be expressed in the form of matrix as

$$\boldsymbol{u} = \begin{bmatrix} u \\ v \end{bmatrix} = \begin{bmatrix} N_i & 0 & N_j & 0 & N_m & 0 \\ 0 & N_i & 0 & N_j & 0 & N_m \end{bmatrix} \begin{bmatrix} u_i \\ v_i \\ u_j \\ v_j \\ u_m \\ v_m \end{bmatrix} = \boldsymbol{N a^e} \tag{2.37}$$

where N is shape function matrix and a^e is the element nodal displacement vector. For the geometric equations, element strains $\boldsymbol{\varepsilon}$ are

$$\begin{aligned} \boldsymbol{\varepsilon} = \begin{bmatrix} \varepsilon_x \\ \varepsilon_y \\ \gamma_{xy} \end{bmatrix} &= \mathbf{L}\boldsymbol{u} = \mathbf{L}\boldsymbol{N a^e} = \mathbf{L}[\boldsymbol{N}_i \quad \boldsymbol{N}_j \quad \boldsymbol{N}_m]\boldsymbol{a^e} \\ &= [\boldsymbol{B}_i \quad \boldsymbol{B}_j \quad \boldsymbol{B}_m]\boldsymbol{a^e} = \boldsymbol{B a^e} \end{aligned} \tag{2.38}$$

where L is the differential operator and B is the element strain displacement matrix which can be given as

$$\boldsymbol{B}_i = \mathbf{L}\boldsymbol{N}_i = \begin{bmatrix} \frac{\partial}{\partial x} & 0 \\ 0 & \frac{\partial}{\partial y} \\ \frac{\partial}{\partial y} & \frac{\partial}{\partial x} \end{bmatrix} \begin{bmatrix} N_i & 0 \\ 0 & N_i \end{bmatrix} = \begin{bmatrix} \frac{\partial N_i}{\partial x} & 0 \\ 0 & \frac{\partial N_i}{\partial y} \\ \frac{\partial N_i}{\partial y} & \frac{\partial N_i}{\partial x} \end{bmatrix} \quad (i,j,m) \tag{2.39}$$

Substitute equation (2.38) in equation (2.7),

$$\boldsymbol{\sigma} = \begin{bmatrix} \sigma_x \\ \sigma_y \\ \tau_{xy} \end{bmatrix} = \boldsymbol{D\varepsilon} = \boldsymbol{DBa}^e = \boldsymbol{Sa}^e \tag{2.40}$$

where

$$\boldsymbol{S} = \boldsymbol{DB} = \boldsymbol{D}[\boldsymbol{B}_i \quad \boldsymbol{B}_j \quad \boldsymbol{B}_m] = [\boldsymbol{S}_i \quad \boldsymbol{S}_j \quad \boldsymbol{S}_m] \tag{2.41}$$

S is called the element stress matrix. It should be noted that both the strain and stress matrices are constant for each element, because the displacement mode is a first order function in 3-nodes triangular element, and differentiate this function will give a constant function.

2.3.3. Stiffness Equilibrium Equation (SEE) of FEM Derived from PMPE

For elastic plane problems, the total potential energy Π_p in equation (2.29) can be expressed in matrix formulation as follows:

$$\Pi_P = \int_\Omega \frac{1}{2}\boldsymbol{\varepsilon}^{\mathrm{T}}\boldsymbol{D\varepsilon}t\,\mathrm{d}x\mathrm{d}y - \int_\Omega \boldsymbol{u}^{\mathrm{T}}\boldsymbol{f}t\,\mathrm{d}x\mathrm{d}y - \int_{S_\sigma} \boldsymbol{u}^{\mathrm{T}}\boldsymbol{T}t\,\mathrm{d}S \tag{2.42}$$

where ***t***, ***f***, and ***T*** denote the thickness, body force and surface force, respectively. For a FEM problem, the total potential energy is the summation of that from all the elements. Therefore, substituting equations (2.37) and (2.38) to equation (2.42) gives

$$\Pi_P = \sum_e \Pi_P^e = \sum_e \left(\boldsymbol{a}^{e\mathrm{T}} \int_{\Omega_e} \frac{1}{2} \boldsymbol{B}^{\mathrm{T}} \boldsymbol{D} \boldsymbol{B} t \, \mathrm{d}x\mathrm{d}y \, \boldsymbol{a}^e \right) - \sum_e \left(\boldsymbol{a}^{e\mathrm{T}} \int_{\Omega_e} \boldsymbol{N}^{\mathrm{T}} \boldsymbol{f} t \, \mathrm{d}x\mathrm{d}y \right) - \sum_e \left(\boldsymbol{a}^{e\mathrm{T}} \int_{S_\sigma^e} \boldsymbol{N}^{\mathrm{T}} \boldsymbol{T} t \, \mathrm{d}x\mathrm{d}y \right) \quad (2.43)$$

Equation (2.43) can be viewed as

$$\boldsymbol{K}^e = \int_{\Omega_e} \boldsymbol{B}^{\mathrm{T}} \boldsymbol{D} \boldsymbol{B} t \, \mathrm{d}x\mathrm{d}y \, , \boldsymbol{P}_f^e = \int_{\Omega_e} \boldsymbol{N}^{\mathrm{T}} \boldsymbol{f} t \, \mathrm{d}x\mathrm{d}y$$

$$\boldsymbol{P}_S^e = \int_{S_\sigma^e} \boldsymbol{N}^{\mathrm{T}} \boldsymbol{T} t \, \mathrm{d}x\mathrm{d}y \, , \boldsymbol{P}^e = \boldsymbol{P}_f^e + \boldsymbol{P}_S^e \quad (2.44)$$

where K^e and P^e are named as the element stiffness matrix and equivalent element nodal load matrix, respectively. Substitute equations (2.44)-(2.47) to equation (2.43), the total potential energy of the structure can be simplified as

$$\Pi_P = \boldsymbol{a}^T \frac{1}{2} \sum_e (\boldsymbol{K}^e) \, \boldsymbol{a} - \boldsymbol{a}^T \sum_e (\boldsymbol{P}^e) \quad (2.45)$$

Given

$$\boldsymbol{K} = \sum_e \boldsymbol{K}^e \, , \boldsymbol{P} = \sum_e \boldsymbol{P}^e \quad (2.46)$$

equation (2.45) is further simplified as

$$\Pi_P = \frac{1}{2} \boldsymbol{a}^T \boldsymbol{K} \, \boldsymbol{a} - \boldsymbol{a}^T \boldsymbol{P} \, \boldsymbol{a} \quad (2.47)$$

where K and P are global stiffness matrix and global nodal load matrix, respectively.

For PMPE, the variation of Π_P is equal to 0 and the unknown variable is a, thus equation (2.28) gives

$$\frac{\partial \Pi_P}{\partial \boldsymbol{a}} = 0 \tag{2.48}$$

which finally comes to the SEE of FEM as

$$\boldsymbol{K}\,\boldsymbol{a} = \boldsymbol{P} \tag{2.49}$$

From equation (2.46), we know that the global stiffness matrix and the global load matrix are the assemblage of the element stiffness matrices and equivalent element nodal load matrices respectively. Specifically, in order to solve equation (2.49), element stiffness matrix, element equivalent nodal load vector, global stiffness matrices and global nodal load vector are all determined together with some given displacement boundary conditions. Without the provision of adequate boundary condition, the system is singular as rigid body motion will produce no stress in the system.

2.3.4. Derivation of Element Stiffness Matrices (ESM)

For a 3-nodes triangular element, the element strain matrix $\boldsymbol{B}$ is constant, thus equation (2.44) gives

$$\boldsymbol{K}^e = \boldsymbol{B}^{\mathrm{T}}\boldsymbol{D}\boldsymbol{B}tA = \begin{bmatrix} \boldsymbol{K}_{ii} & \boldsymbol{K}_{ij} & \boldsymbol{K}_{im} \\ \boldsymbol{K}_{ji} & \boldsymbol{K}_{jj} & \boldsymbol{K}_{jm} \\ \boldsymbol{K}_{mi} & \boldsymbol{K}_{mj} & \boldsymbol{K}_{mm} \end{bmatrix} \tag{2.50}$$

of which the sub-matrix

$$\boldsymbol{K}_{ij} = \begin{bmatrix} k_{ij}{}^{xx} & k_{ij}{}^{xy} \\ k_{ij}{}^{yx} & k_{ij}{}^{yy} \end{bmatrix} \tag{2.51}$$

K_{ij} indicates the ith nodal force along x and y-directions in the Cartesian coordinate system when the displacement of the jth node is unit along x and y-directions, which can be easily obtained. Moreover, the element stiffness matrix is symmetric, and the computational memory required in a FEM program can be reduced by using this property.

It should be noted that for a higher order triangular element (*e.g.*, 6 nodes triangular element) or quadrilateral element for which higher order terms are involved, the strain matrix $\boldsymbol{B}$ is not constant any more so that the element stiffness matrix need to be evaluated by numerical integration (direct integration is seldom adopted). Towards this, numerical integration methods such as the Gaussian integration or the Newton-Cotes integration can be utilized.

2.3.5. Assembling of ESMs and ENLMs

For a FEM process, we need to solve equation (2.49) which is the global equilibrium equation. Most of the elements in the matrix K are 0 simply because each node is only shared by a few surrounding elements. In view of that, a rectangular matrix can represent the global stiffness matrix (which is a square matrix), and the half band width *bandwidth* can be defined as

$$bandwidth = (1 + NDIF) \times NDOF \qquad \textbf{(2.52)}$$

where *NDIF* denotes the largest absolute difference between the element node numbers among all the element in the finite element mesh. Besides the bandwidth storage, the use of profile or skyline storage is also popular in program development. If the problem size is so large that the memory of the system will be exceeded, the use of out of core solver is popular in the past (Cheng 1989). Besides the use of direct solver, the use of iterative solver can be useful and effective (Cheng 1996) in some large problems. The DDA or Manifold will be discussed in chapter 1 of vol.4 of this series. The use of iterative solver appears to be important, as the equivalent matrix is typically three times larger than the equivalent finite element formulation.

In conclusion, the properties of the global stiffness matrix can be summarized as: symmetric, banded distribution, singularity and sparsity. Among all the properties, singularity will vanish by introducing appropriate boundary conditions to eq.(2.49) to eliminate the rigid body motion. Also, other properties like banded distribution should be fully taken into consideration to reduce the computational memory and enhance the computation efficiency.

2.3.6. Isoparametric Element and Numerical Integration

Most of the engineering structure is not regular in shape, and some of them even have very complicated boundary shapes. Although the use of triangular element can always fit a complicated boundary, the accuracy of this element is low in general. To cope with the irregular boundary shape with a higher accuracy in analysis, one of the most common approaches is the use higher order element, and the isoparametric formulation is the most commonly used at present. Consider an arbitrary 4-node quadrilateral element as an example which is schematically shown in Fig. (**2.9**). If we can find the transformation from Fig. (**2.9a**) to Fig. (**2.9b**), then it will become easier to carry numerical integration with complicated shapes for an arbitrary element. Fig. (**2.9a**) is defined in the Cartesian coordinate system, while Fig. (**2.9b**) is defined in the local coordinate system (or natural coordinate system) within a specific domain (*i.e.*, $\xi,\eta \in (-1,1)$). The relation between these two kinds of coordinate system can be described as

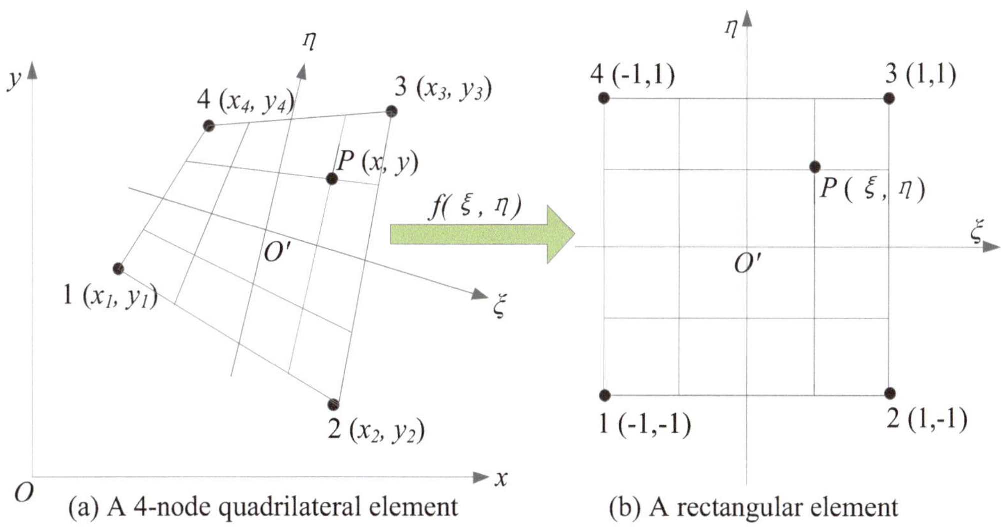

Fig. (2.9). Isoparametric transition.

$$\begin{Bmatrix} x \\ y \end{Bmatrix} = f \begin{Bmatrix} \xi \\ \eta \end{Bmatrix} \qquad \textbf{(2.53)}$$

which can be further modified by the interpolation function at nodes in the local coordinate system as follows:

$$x = \sum_{i=1}^{m} N_i' x_i \qquad y = \sum_{i=1}^{m} N_i' y_i \qquad \textbf{(2.54)}$$

where (x_i, y_i) are coordinates in Cartesian coordinate system corresponding to the *i*th node in local coordinate system, N_i is interpolation function of the *i*th node in local coordinate system, and *m* is the number of nodes chosen to transform the coordinates. Therefore, the regular element in the natural coordinate system can be transformed to the irregular element in Cartesian coordinate system. The former element is called the parent element while the latter is called the subelement. Specifically, equation (2.55) can be further expanded as

$$\begin{Bmatrix} x \\ y \end{Bmatrix} = \begin{bmatrix} N_1 & 0 & N_2 & 0 & N_3 & 0 & N_4 & 0 \\ 0 & N_1 & 0 & N_2 & 0 & N_3 & 0 & N_4 \end{bmatrix} \begin{Bmatrix} x_1 \\ y_1 \\ x_2 \\ y_2 \\ x_3 \\ y_3 \\ x_4 \\ y_4 \end{Bmatrix} \tag{2.55}$$

Using the same interpolation functions, the element displacement model can be written as

$$\begin{Bmatrix} u \\ v \end{Bmatrix} = \begin{bmatrix} N_1 & 0 & N_2 & 0 & N_3 & 0 & N_4 & 0 \\ 0 & N_1 & 0 & N_2 & 0 & N_3 & 0 & N_4 \end{bmatrix} \begin{Bmatrix} u_1 \\ v_1 \\ u_2 \\ v_2 \\ u_3 \\ v_3 \\ u_4 \\ v_4 \end{Bmatrix} \tag{2.56}$$

where J denotes the Jacobi matrix while the interpolation functions are given by

$$\begin{aligned} N_1 &= \tfrac{1}{4}(1-\xi)(1-\eta), \quad N_2 = \tfrac{1}{4}(1+\xi)(1-\eta) \\ N_3 &= \tfrac{1}{4}(1+\xi)(1+\eta), \quad N_2 = \tfrac{1}{4}(1-\xi)(1+\eta) \end{aligned} \tag{2.57}$$

As mentioned before, during the derivation of the element stiffness matrix and the equivalent load vector, the derivative of the shape function and the integration in the element surface or volume in Cartesian coordinate system are required. Since the shape functions adopted herein are expressed in the natural coordinates, therefore, the derivative and integration transformation relationships are essential when isoparametric element is used.

2.3.6.1. Derivative and Integral Transformation

According to the law of partial differential,

$$\begin{aligned} \frac{\partial N_i}{\partial \xi} &= \frac{\partial N_i}{\partial x}\frac{\partial x}{\partial \xi} + \frac{\partial N_i}{\partial y}\frac{\partial y}{\partial \xi}, \\ \frac{\partial N_i}{\partial \eta} &= \frac{\partial N_i}{\partial x}\frac{\partial x}{\partial \eta} + \frac{\partial N_i}{\partial y}\frac{\partial y}{\partial \eta}, \end{aligned} \tag{2.58}$$

or in matrix form

$$\begin{Bmatrix} \dfrac{\partial N_i}{\partial \xi} \\ \dfrac{\partial N_i}{\partial \eta} \end{Bmatrix} = \begin{bmatrix} \dfrac{\partial x}{\partial \xi} & \dfrac{\partial y}{\partial \xi} \\ \dfrac{\partial x}{\partial \eta} & \dfrac{\partial y}{\partial \eta} \end{bmatrix} \begin{Bmatrix} \dfrac{\partial N_i}{\partial x} \\ \dfrac{\partial N_i}{\partial y} \end{Bmatrix} = J \begin{Bmatrix} \dfrac{\partial N_i}{\partial x} \\ \dfrac{\partial N_i}{\partial y} \end{Bmatrix} \tag{2.59}$$

Inverse of equation (2.59) gives

$$\begin{Bmatrix} \dfrac{\partial N_i}{\partial x} \\ \dfrac{\partial N_i}{\partial y} \end{Bmatrix} = J^{-1} \begin{Bmatrix} \dfrac{\partial N_i}{\partial \xi} \\ \dfrac{\partial N_i}{\partial \eta} \end{Bmatrix} \tag{2.60}$$

where

$$J = \begin{bmatrix} \dfrac{\partial x}{\partial \xi} & \dfrac{\partial y}{\partial \xi} \\ \dfrac{\partial x}{\partial \eta} & \dfrac{\partial y}{\partial \eta} \end{bmatrix} = \begin{bmatrix} \sum_{i=1}^{4} \dfrac{\partial N_i}{\partial \xi} x_i & \sum_{i=1}^{4} \dfrac{\partial N_i}{\partial \xi} y_i \\ \sum_{i=1}^{4} \dfrac{\partial N_i}{\partial \eta} x_i & \sum_{i=1}^{4} \dfrac{\partial N_i}{\partial \eta} y_i \end{bmatrix}$$

$$= \begin{bmatrix} \dfrac{\partial N_1}{\partial \xi} & \dfrac{\partial N_2}{\partial \xi} & \dfrac{\partial N_3}{\partial \xi} & \dfrac{\partial N_4}{\partial \xi} \\ \dfrac{\partial N_1}{\partial \eta} & \dfrac{\partial N_2}{\partial \eta} & \dfrac{\partial N_3}{\partial \eta} & \dfrac{\partial N_4}{\partial \eta} \end{bmatrix} \begin{bmatrix} x_1 & y_1 \\ x_2 & y_2 \\ x_3 & y_3 \\ x_4 & y_4 \end{bmatrix} \tag{2.61}$$

For an infinitely small element, the area under the Cartesian coordinate system and the natural coordinate system are related by

$$\mathrm{d}s = \mathrm{d}x\mathrm{d}y = |\boldsymbol{J}|\mathrm{d}\xi\mathrm{d}\eta, \tag{2.62}$$

where $|\boldsymbol{J}|$ is the determinant of the Jacobian matrix $\boldsymbol{J}$. Therefore, element stiffness matrix and equivalent nodal load matrix in equation (2.44) can be transformed to

$$\boldsymbol{K}^e = \int_{\Omega_e} \boldsymbol{B}^{\mathrm{T}}\boldsymbol{D}\boldsymbol{B}\boldsymbol{t}\,|\boldsymbol{J}|\mathrm{d}\xi\mathrm{d}\eta\,, \boldsymbol{P}_f^e = \int_{\Omega_e} \boldsymbol{N}^{\mathrm{T}}\boldsymbol{f}|\boldsymbol{J}|\mathrm{d}\xi\mathrm{d}\eta$$

$$\boldsymbol{P}_S^e = \int_{S_\sigma{}^e} \boldsymbol{N}^{\mathrm{T}}\boldsymbol{T}|\boldsymbol{J}|\mathrm{d}\xi\mathrm{d}\eta \tag{2.63}$$

For the evaluation of this integral equation, usually Gaussian integration method is employed. In practice, both two and three integration points along each direction of integration are commonly used. Since the discretized system is usually over-stiff, it is commonly observed that the use of two integration points along each direction of integration will slightly reduce the stiffness of the matrix and give better results as compared with the use of three integration points. The use of exact integration is possible for some elements, but such approaches are usually tedious and are seldom adopted. The advantage in using the exact integration is that the integration is not affected by the shape of the element while the transformation as shown in eq.(2.63) may be affected if the poor shape of the element is poor.

Finite element method has evolved over the years, and many new techniques have been developed to further enhance the use and improve the accuracy of this technique. Many engineering problems can be well considered by the finite element method; hence many commercial programs have been developed for various engineering purposes. Many books and research papers have come out over the last 30 years, and the amount is too much to be mentioned here. Some of the more recent references which may be of interest to the readers include the works by Rao (2018), Zienkiewicz *et al.* (2011), Koutromanos (2018), Reddy (2015), Zhu (2018), Bathe (2014), Ganesan and Tobiska (2017), Okereke and Keates (2018), Dow (2015), Nithiarasu *et al.* (2015), Khoei (2015), Pepper (2006), Belytschko *et al.* (2014) and many others.

The authors have developed many finite element programs for teaching and research purposes which can be obtained at natureymc@yahoo.com.hk. The programs available include plane stress/strain problem, 2D/3D truss and frames, thin/thick plate bending problem, consolidation in 1D and 2D (Biot), seepage problem (confined or free surface flow), slope stability problem, pile driving problems and others.

2.4. DEVELOPMENT OF A PSEUDO 8 NODE MINDLIN QUADRILATERAL PLATE ELEMENT

There are many structures in civil engineering which are important and expensive thick plate. For example, pile cap, transfer plate and raft foundations are the most common structures which are best modelled by plate element. In fact, nearly all engineering companies in Hong Kong possess a copy of engineering program for plate analysis and design. The Kirchhoff Plate theory which has neglected the out-of-plane shear deformation is reasonably accurate for thin plate where the shear deformation is insignificant as compared with the flexural deformation. However, errors will be noticeable when the plate structure is thick and shear deformation cannot be neglected. Many new thick plate elements have been developed and used by engineers (Urugal 1999, Reddy 2000, Szilard 2004), and the use of higher order thick plate element appears to be unnecessary for practical purposes. The thick plate element by Ibrahimbegovic (1993) (termed as IB element in this book) is currently adopted by many engineers for transfer plate and pile cap analysis in many countries. The authors have recently discovered some problems for this element in several real jobs which were overlooked in the past, though this element has been used for the analysis of many thick plate structures in the past.

In view of the limitations of the plate element in some commercial plate analysis program, a new shear deformable beam free from shear locking is developed by Cheng and Law (2008). It is demonstrated by Cheng and Law (2008) that the new thick plate element can provide greater accuracy and better convergence as compared with some common plate elements, in particular for coarse mesh problem. The new thick plate is formulated initially as a rectangular element and then later extended to the general quadrilateral element. The program code of the present work is given in Appendix 2-4, and this code will be used to illustrate some programming style preferred by the authors, and the readers can understand the full structure of a finite element program, based on which more advanced features can be added. Before discussion about the development of finite element code and the thin/thick plate element, the basic formulation for the new shear deformable plate element by Cheng and Law (2008) is introduced.

2.4.1. Formulation of a New Shear Deformable Beam free from Shear Locking

Most of the thick plate elements are based on extension of shear deformable beam where shear strain is included in computation of displacement, as in contrast with the conventional beam which is assumed to be free from shear strain distortion. The relationship among the parameters (i) out-of-plane displacement ω_i; (ii) out-of-plane rotation ψ_i; (iii) out-of-plane shear strain γ for the Shear Deformable Beam is illustrated in Figs. (**2.10**) and (**2.11**).

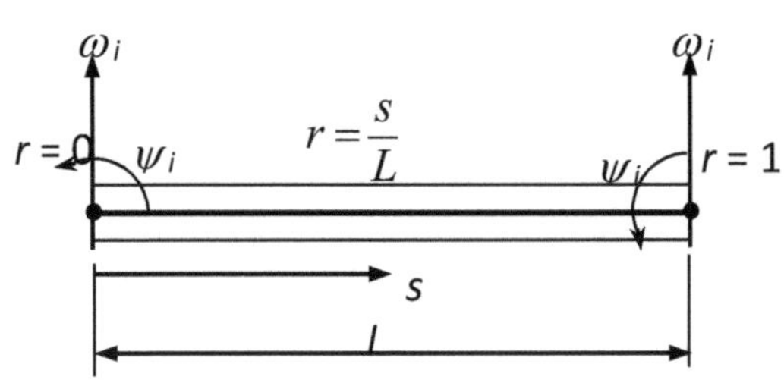

Fig. (2.10). Symbols for thick beam formulation.

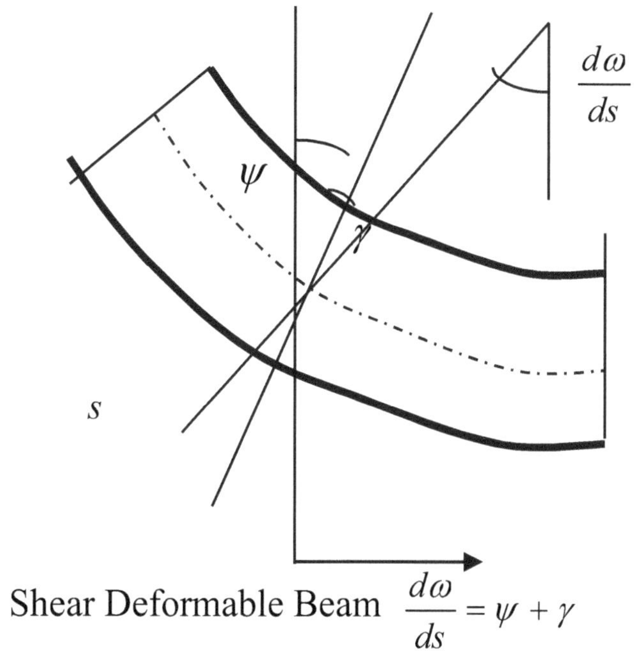

Fig. (2.11). Relationship among ω; ψ; γ; for Shear Deformable Beam.

In finite element formulation for thick plate (including shear deformation), the differences in various approaches are the order of polynomial (Bathe *et al.* 1985, Hughes *et al.* 1986, Jirousek *et al.* 1995, Zienkiewicz *et al.* 2011, Sze 2002) and the methods to determine the polynomial coefficients. In the present formulation for the beam as shown in Figs. (**2.10** and **2.11**), the deflection term ω is represented by a 3rd order polynomial. The rotation term ψ would then be a 2nd order polynomial (one order lower than deflection). The coefficients α and β terms in eq.(2.64) and (2.65) are constants to be determined from the boundary conditions and energy minimization which will be given later. r is a normalized dimensionless distance term given by $r = \frac{s}{L}$ which is in the range of 0 to 1.

$$\omega = \alpha_0 + \alpha_1 r + \alpha_2 r^2 + \alpha_3 r^3 \quad \textbf{(2.64)}$$

$$\psi = \beta_0 + \beta_1 r + \beta_2 r^2 \quad \textbf{(2.65)}$$

As illustrated in Fig. (**2.11**), the angular rotation at any section is the combined action of the bending and shear deformation given by

$$\frac{d\omega}{ds} = \frac{1}{L}\frac{d\omega}{dr} = \psi + \gamma \tag{2.66}$$

The shear strain deformation term within this short segment of beam (shear force is constant) is a constant term γ_0 given by

$$\gamma = \gamma_0 \tag{2.67}$$

Making use of the boundary conditions at $r = 0$, $\omega = \omega_i$, $\psi = \psi_i$ and at $r = 1$, $\omega = \omega_j$, $\psi = \psi_j$; α_0, α_1, β_0, β_1 in eq.(2.64) and (2.65) can be expressed in terms of the deflections at the two ends of the beam ω_i and ω_j so that eq.(2.64) and eq.(2.65) can be rewritten as:

$$\omega = (1-r)\omega_i + r\omega_j + r(-1+r)\alpha_2 + r(-1+r^2)\alpha_3 \tag{2.68}$$

$$\text{and} \quad \psi = \psi_i + (\psi_j - \psi_i - \beta_2)r + \beta_2 r^2 \tag{2.69}$$

Eq.(2.66) can now be rewritten as

$$\frac{d\omega}{dr} = L(\psi + \gamma) = -\omega_i + \omega_j + (-1+2r)\alpha_2 + (-1+3r^2)\alpha_3 \tag{2.70}$$

Substituting $r = 0$ and $r = 1$ with $\omega = \omega_i$ and $\omega = \omega_j$ in eq.(2.68), respectively and solving:

$$\alpha_2 = -3\omega_i + 3\omega_j - 2L\psi_i - L\psi_j - 3L\gamma_0; \quad \alpha_3 = 2\omega_i - 2\omega_j + L\psi_i + L\psi_j + 2L\gamma_0 \tag{2.71}$$

Back-substituting eq.(2.71) into eq.(2.68) and re-arranging

$$\omega=(1-r)^2(1+2r)\omega_i+r^2(3-2r)\omega_j+Lr(1-r)^2\psi_i+Lr^2(-1+r)\psi_j+Lr(1-r)(1-2r)\gamma_0 \quad (2.72)$$

Differentiation of eq.(2.72) with respect to s where $s = Lr$

$$\frac{1}{L}\frac{d\omega}{dr}=\frac{-6r+6r^2}{L}\omega_i+\frac{6r-6r^2}{L}\omega_j+\left(1-4r+3r^2\right)\psi_i+\left(-2r+3r^2\right)\psi_j+\left(1-6r+6r^2\right)\gamma_0 \quad (2.73)$$

Put eqs. (2.69) & (2.73) into (2.66)

$$\frac{-6r+6r^2}{L}\omega_i+\frac{6r-6r^2}{L}\omega_j+\left(1-4r+3r^2\right)\psi_i+\left(-2r+3r^2\right)\psi_j+\left(1-6r+6r^2\right)\gamma_0$$
$$=(1-r)\psi_i+r\psi_j+r(-1+r)\beta_2+\gamma_0,\quad \text{or} \quad (2.74)$$
$$r(-1+r)\beta_2=\frac{-6r+6r^2}{L}\omega_i+\frac{6r-6r^2}{L}\omega_j+\left(-3r+3r^2\right)\psi_i+\left(-3r+3r^2\right)\psi_j+\left(-6r+6r^2\right)\gamma_0$$

β_2 is now determined from eq.(2.74) so that eq.(2.69) can be re-written as

$$\psi=(1-r)\psi_i+r\psi_j+\frac{-6r+6r^2}{L}\omega_i+\frac{6r-6r^2}{L}\omega_j+\left(-3r+3r^2\right)\psi_i+\left(-3r+3r^2\right)\psi_j+\left(-6r+6r^2\right)\gamma_0$$
$$\Rightarrow\psi=\frac{6r}{L}(-1+r)\omega_i+\frac{6r}{L}(1-r)\omega_j+\left(1-4r+3r^2\right)\psi_i+r(-2+3r)\psi_j+6r(-1+r)\gamma_0 \quad (2.75)$$

Bending energy of the shear deformable beam per unit width U_B is

$$U_B=\int_0^L\frac{M^2}{2EI}dx=\frac{EIL}{2}\int_0^1\left(\frac{M}{EI}\right)^2dr=\frac{EIL}{2}\int_0^1\kappa^2dr=\frac{Et^3}{12}\frac{L}{2}\int_0^1\left(-\frac{d\psi}{Ldr}\right)^2dr \quad (2.76)$$

where $\boldsymbol{K}$ is the curvature of deflection given by $\kappa=-\frac{d\psi}{ds}=-\frac{1}{L}\frac{d\psi}{ds}$. From eq.(2.75),

$$\frac{d\psi}{Ldr}=\frac{-6+12r}{L^2}\omega_i+\frac{6-12r}{L^2}\omega_j+\frac{-4+6r}{L}\psi_i+\frac{-2+6r}{L}\psi_j+\frac{-6+12r}{L}\gamma_0$$
$$=\left(\frac{-6}{L^2}\omega_i+\frac{6}{L^2}\omega_j+\frac{-4}{L}\psi_i+\frac{-2}{L}\psi_j+\frac{-6}{L}\gamma_0\right)+\left(\frac{12}{L^2}\omega_i+\frac{-12}{L^2}\omega_j+\frac{6}{L}\psi_i+\frac{6}{L}\psi_j+\frac{12}{L}\gamma_0\right)r \quad (2.77)$$

Put $A=\left(\frac{-6}{L^2}\omega_i+\frac{6}{L^2}\omega_j+\frac{-4}{L}\psi_i+\frac{-2}{L}\psi_j+\frac{-6}{L}\gamma_0\right)$ and

$$B=\left(\frac{12}{L^2}\omega_i+\frac{-12}{L^2}\omega_j+\frac{6}{L}\psi_i+\frac{6}{L}\psi_j+\frac{12}{L}\gamma_0\right)$$

Bending energy $U_B=\frac{Et^3}{12}\frac{L}{2}\int_0^1\left(-\frac{d\psi}{Ldr}\right)^2dr=\frac{Et^3}{12}\frac{L}{2}\int_0^1(A+Br)^2dr=\frac{Et^3}{12}\frac{L}{2}\left[A^2+AB+\frac{1}{3}B^2\right]$

$$=\frac{Et^3}{12}\frac{L}{2}\left[A^2+AB+\frac{1}{4}B^2+\frac{1}{12}B^2\right]=\frac{Et^3}{12}\frac{L}{2}\left[\left(A+\frac{1}{2}B\right)^2+\frac{1}{12}B^2\right] \quad \textbf{(2.78)}$$

$$=\frac{Et^3}{12}\frac{L}{2}\left[\left(\frac{-\psi_i}{L}+\frac{\psi_j}{L}\right)^2+\frac{1}{12}\left(\frac{12}{L^2}\omega_i+\frac{-12}{L^2}\omega_j+\frac{6}{L}\psi_i+\frac{6}{L}\psi_j+\frac{12}{L}\gamma_0\right)^2\right]$$

where t is the sectional depth of the beam.

Shear energy of the beam per unit width U_s is

$$U_S=\int_0^L\frac{S^2}{2GA_r}dx=\frac{GA_rL}{2}\int_0^1\left(\frac{S}{GA_r}\right)^2dr=\frac{E5tL}{2\times2(1+\mu)\times6}\int_0^1\gamma_0{}^2dr=\frac{5EtL}{24(1+\mu)}\gamma_0{}^2 \quad \textbf{(2.79)}$$

where μ is the Poisson's ratio. In eq.(2.79), the effective area A_r used for shear energy determination is taken as 5/6 of the actual sectional area because of the parabolic shear stress distribution across the section.

$$\text{Total Energy } U=U_B+U_S \quad \textbf{(2.80)}$$

The total energy is dependent on the deflection, rotation and the shear deformation and the equilibrium point is obtained by minimizing the total energy with respect to all these dependent parameters. In the present formulation, the minimum energy principle with respect to the shear deformation is applied in addition to those for vertical deflection and rotation, *i.e.* $\frac{\partial U}{\partial\gamma_0}=0$ is also applied in the minimization which is different from many other thick plate formulations. The importance of this additional energy minimization in comparison with other classical formulations will be illustrated by the thick plate analysis in latter section. By performing partial differentiation on eq.(2.80) $\frac{\partial U}{\partial \gamma_0} = \frac{\partial U_B}{\partial \gamma_0} + \frac{\partial U_S}{\partial \gamma_0} = 0$ with substitution of U_B and U_S from eq.(2.68) and (2.69):

$$\frac{\partial}{\partial \gamma_0}\left\{\frac{Et^3}{12}\frac{L}{2}\left[\left(\frac{-\psi_i}{L}+\frac{\psi_j}{L}\right)^2+\frac{1}{12}\left(\frac{12}{L^2}\omega_i+\frac{-12}{L^2}\omega_j+\frac{6}{L}\psi_i+\frac{6}{L}\psi_j+\frac{12}{L}\gamma_0\right)^2\right]\right\}$$
$$+\frac{\partial}{\partial \gamma_0}\left\{\frac{5EtL}{24(1+\mu)}\gamma_0^{\,2}\right\}=0 \quad \textbf{(2.81)}$$

Simplifying and re-arranging

$$\gamma_0\left[\frac{5L^2+12(1+\mu)t^2}{6t^2(1+\mu)}\right]=\left[\frac{-2}{L}(\omega_i-\omega_j)-(\psi_i+\psi_j)\right] \quad \textbf{(2.82)}$$

$$\delta=\frac{6t^2(1+\mu)}{5L^2+12(1+\mu)t^2} \quad \textbf{(2.83)}$$

$$\gamma_0=\frac{-2\delta}{L}\omega_i+\frac{2\delta}{L}\omega_j-\delta\psi_i-\delta\psi_j \quad \textbf{(2.84)}$$

Let the displacement vector of the shear deformable beam be $\lfloor \omega_i \quad \psi_i \quad \omega_j \quad \psi_j \rfloor^T$. From eq.(2.84), the strain displacement B_S matrix for shear is obviously given by

$$B_S = \left[\frac{-2\delta}{L} \quad -\delta \quad \frac{2\delta}{L} \quad -\delta \right] \tag{2.85}$$

Substituting eq.(2.85) into eq.(2.77)

$$\frac{d\psi}{Ldr} = \frac{-6+12r}{L^2}\omega_i + \frac{6-12r}{L^2}\omega_j + \frac{-4+6r}{L}\psi_i + \frac{-2+6r}{L}\psi_j + \frac{-6+12r}{L}\gamma_0$$

$$\Rightarrow -\frac{d\psi}{Ldr} = \frac{6-12r}{L^2}\omega_i + \frac{-6+12r}{L^2}\omega_j + \frac{4-6r}{L}\psi_i + \frac{2-6r}{L}\psi_j$$

$$+\frac{6-12r}{L}\left[\frac{-2\delta}{L}\omega_i + \frac{2\delta}{L}\omega_j - \delta\psi_i - \delta\psi_j\right] \tag{2.86}$$

$$\Rightarrow -\frac{d\psi}{Ldr} = \frac{6(1-2r)(1-2\delta)}{L^2}\omega_i + \frac{-6(1-2r)(1-2\delta)}{L^2}\omega_j + \frac{1+3(1-2\delta)(1-2r)}{L}\psi_i$$

$$+\frac{-1+3(1-2r)(1-2\delta)}{L}\psi_j$$

From eq.(2.86), the strain displacement B_B matrix for bending is given by

$$B_B = \left[\frac{6(1-2r)(1-2\delta)}{L^2} \quad \frac{1+3(1-2\delta)(1-2r)}{L} \quad \frac{-6(1-2r)(1-2\delta)}{L^2} \quad \frac{-1+3(1-2\delta)(1-2r)}{L} \right] \tag{2.87}$$

The stiffness matrix for shear is $K_s = \int_0^1 B_S{}^T GA_r B_S \cdot L \cdot dr$ **(2.88)**

where $GA_r = \frac{E}{2(1+\mu)}\frac{5}{6}bt = \frac{5Ebt}{12(1+\mu)}$ and B is from eq.(2.85). Expanding and integrating eq.(2.88) gives the shear stiffness matrix as

$$K_s = \frac{5Ebt}{12(1+\mu)} \begin{bmatrix} \frac{4\delta^2}{L} & 2\delta^2 & -\frac{4\delta^2}{L} & 2\delta^2 \\ 2\delta^2 & L\delta^2 & -2\delta^2 & L\delta^2 \\ -\frac{4\delta^2}{L} & -2\delta^2 & \frac{4\delta^2}{L} & -2\delta^2 \\ 2\delta^2 & L\delta^2 & -2\delta^2 & L\delta^2 \end{bmatrix} \quad \textbf{(2.89)}$$

Similarly, the stiffness matrix for bending is given as follows, with B_b from (2.87)

$$K_b = \int_0^1 B_b{}^T EI \cdot B_b \cdot L \cdot dr \quad \textbf{(2.90)}$$

Expanding and integrating

$$K_b = EI \begin{bmatrix} \frac{12}{L^3}(1-2\delta)^2 & \frac{6}{L^2}(1-2\delta)^2 & -\frac{12}{L^3}(1-2\delta)^2 & \frac{6}{L^2}(1-2\delta)^2 \\ \frac{6}{L^2}(1-2\delta)^2 & \frac{4}{L}\left(1-3\delta+3\delta^2\right) & -\frac{6}{L^2}(1-2\delta)^2 & \frac{2}{L}\left(1-6\delta+6\delta^2\right) \\ -\frac{12}{L^3}(1-2\delta)^2 & -\frac{6}{L^2}(1-2\delta)^2 & \frac{12}{L^3}(1-2\delta)^2 & -\frac{6}{L^2}(1-2\delta)^2 \\ \frac{6}{L^2}(1-2\delta)^2 & \frac{2}{L}\left(1-6\delta+6\delta^2\right) & -\frac{6}{L^2}(1-2\delta)^2 & \frac{4}{L}\left(1-3\delta+3\delta^2\right) \end{bmatrix} \quad \textbf{(2.91)}$$

where $I = \frac{bt^3}{12}$

If thickness of beam tends to 0 so that δ tends to 0, K_s will tend to 0 while K_b will tend to the classical beam deflection stiffness matrix. The results as given by eqs.(2.89) and (2.91) are hence free from shear locking phenomenon.

The total stiffness of the beam will be given by $K = K_s + K_b$ **(2.92)**

At $r=\frac{1}{2}$, substituting (2.84) into (2.75) and simplifying

$$\psi_{r=\frac{1}{2}}=\frac{-3(1-2\delta)}{2L}\omega_i+\frac{3(1-2\delta)}{2L}\omega_j+\left(-\frac{1}{4}+\frac{3}{2}\delta\right)\psi_i+\left(-\frac{1}{4}+\frac{3}{2}\delta\right)\psi_j \quad \textbf{(2.93)}$$

2.4.2. Formulation of Rectangular Shear Deformable Plate based on Shear Deformable Beam

Plate is an important structure and very detailed treatments of thin and thick plates by series methods are given by Timoshenko and Krieger (1959), Reddy (2000), Wang *et al.* (2000) and Szilard (2004). For practical thick plates commonly encountered in civil engineering, the use of series methods is usually not applicable due to the geometry of plates, loading conditions and boundary conditions and the finite element method is commonly adopted instead. Over the last two decades, many thick plate elements have been developed and used, as per detailed treatments given by Zienkiewicz *et al.* (2011), Urugal (1999), Reddy (2000) and Szilard (2004). Although numerous improvements have been made to various thick plate elements over the past two decades, there are still many practical cases under which the modern thick plate elements cannot perform well and this has been well addressed by Hartmann and Katz (2004). The development of accurate and efficient thick plate element is hence still currently under active research.

The foregoing derivation for shear deformable beam is extended to the formulation of rectangular plate in this section. Consider a rectangular plate of corner nodes 1, 2, 3, 4 and side lengths $2a$ and $2b$ as indicated in Fig. (**2.12**). Fictitious mid-side nodes 5, 6, 7 and 8 are added for use in the formulation, but they will be statically eliminated so that only the corner nodes will be retained in the stiffness matrix formulation. This approach is adopted as a 4 nodes element possesses good numerical accuracy as compared with the triangular element but is more convenient to be used as compared with a 8 nodes element. Actually, most of the commonly engineering thick plate elements in various commercial softwares are also four nodes element (excluding those for research purpose).

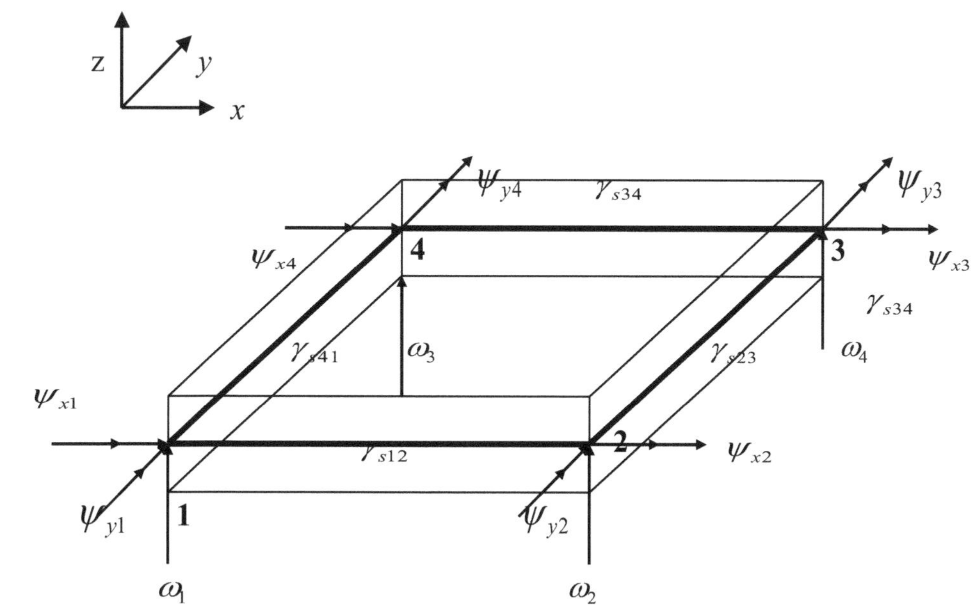

Fig. (2.12a). Degree of freedom for thick plate.

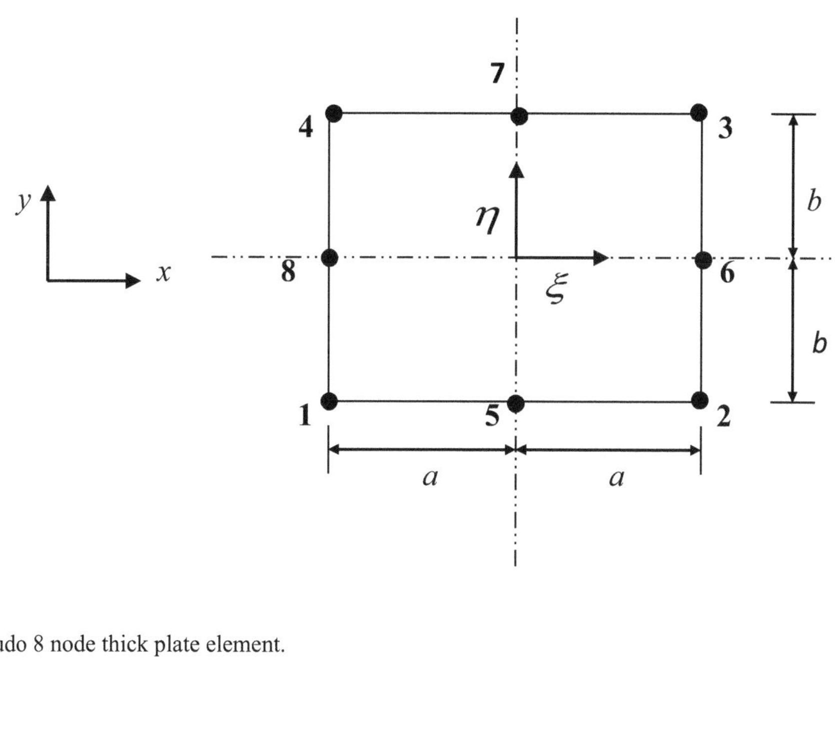

Fig. (2.12b). Pseudo 8 node thick plate element.

The special feature about this element is that it will also degenerate to the classical thin plate element if the thickness of the plate is small. For the plate as shown, the nodal displacement a^e of the corner nodes is given as:

$$a^e = \lfloor \omega_1 \quad \psi_{x1} \quad \psi_{y1} \quad \omega_2 \quad \psi_{x2} \quad \psi_{y2} \quad \omega_3 \quad \psi_{x3} \quad \psi_{y3} \quad \omega_4 \quad \psi_{x4} \quad \psi_{y4} \rfloor^T \quad \textbf{(2.94)}$$

2.4.2.1. Formulation of the Shear Stiffness Matrix of the Thick Plate Element

2.4.2.1.1. Transverse Shear Strain at Sides of Element

From eq.(2.84), the shear strains at the edges $\overline{12}, \overline{23}, \overline{34}, \overline{41}$ of the element are :

$$\begin{cases} \gamma_{s12} = -\dfrac{\delta_a}{a}\omega_1 + \dfrac{\delta_a}{a}\omega_2 + \delta_a\psi_{y1} + \delta_a\psi_{y2} \\ \gamma_{s23} = -\dfrac{\delta_b}{b}\omega_2 + \dfrac{\delta_b}{b}\omega_3 - \delta_b\psi_{x2} - \delta_b\psi_{x3} \\ \gamma_{s34} = -\dfrac{\delta_{12}}{a}\omega_4 + \dfrac{\delta_{12}}{a}\omega_3 + \delta_a\psi_{y4} + \delta_a\psi_{y3} \\ \gamma_{s41} = -\dfrac{\delta_b}{b}\omega_1 + \dfrac{\delta_b}{b}\omega_4 - \delta_b\psi_{y1} - \delta_a\psi_{y4} \end{cases} \quad \textbf{(2.95)}$$

(taking note that the rotations ψ_{y1} and ψ_{y2} are clockwise positive when viewed at edge $\overline{12}, \overline{34}$ instead of anticlockwise as derived in eq.(2.84)

$$\text{where} \quad \delta_a = \frac{\left(\dfrac{t}{2a}\right)^2 (1+\mu)}{\dfrac{5}{6} + 2\left(\dfrac{t}{2a}\right)^2 (1+\mu)} = \frac{3t^2(1+\mu)}{10a^2 + 6t^2(1+\mu)} \quad ; \delta_b = \frac{3t^2(1+\mu)}{10b^2 + 6t^2(1+\mu)} \quad \textbf{(2.96)}$$

and t = thickness of the plate, μ = Poisson's ratio. It can easily be seen that when $t \to 0$, δ_a, $\delta_b \to 0$, and $\gamma_{si} \to 0$

2.4.2.1.2. Shear Strain within Element

The shear strain $\gamma = [\gamma_x\ \gamma_y]^T$ within element can be expressed as:

$$\gamma_x = \Sigma \gamma_{xi} N_i^0 \qquad \gamma_y = \Sigma \gamma_{yi} N_i^0 \tag{2.97}$$

and N_i^0 $(i = 1, 2, 3, 4)$ are the classical bilinear interpolation functions for Q4 quadrilateral element. Substituting eq.(2.97) into (2.95) and simplifying

$$\begin{aligned}\gamma_x = &\frac{-(1-\eta)\delta_a}{2a}\omega_1 + \frac{(1-\eta)\delta_a}{2}\psi_{y1} + \frac{(1-\eta)\delta_a}{2a}\omega_2 + \frac{(1-\eta)\delta_a}{2}\psi_{y2} \\ &+ \frac{(1+\eta)\delta_a}{2a}\omega_3 + \frac{(1+\eta)\delta_a}{2}\psi_{y3} + \frac{-(1+\eta)\delta_a}{2a}\omega_4 + \frac{(1+\eta)\delta_a}{2}\psi_{y4}\end{aligned} \tag{2.98}$$

$$\begin{aligned}\gamma_y = &\frac{-(1-\xi)\delta_b}{2b}\omega_1 + \frac{-(1-\xi)\delta_b}{2}\psi_{x1} + \frac{-(1+\xi)\delta_b}{2}\omega_2 + \frac{-(1+\xi)\delta_b}{2}\psi_{x2} \\ &+ \frac{(1+\xi)\delta_b}{2b}\omega_3 + \frac{-(1+\xi)\delta_b}{2}\psi_{x3} + \frac{(1-\xi)\delta_b}{2b}\omega_4 + \frac{-(1-\xi)\delta_b}{2}\psi_{x4}\end{aligned} \tag{2.99}$$

So the B matrix for shear is given by $\gamma = B_S a^e$ and B_s=[B_{s1} B_{s2}], where

$$B_{S1} = \begin{bmatrix} \frac{-(1-\eta)\delta_a}{2a} & 0 & \frac{(1-\eta)\delta_a}{2} & \frac{(1-\eta)\delta_a}{2a} & 0 & \frac{(1-\eta)\delta_a}{2} \\ \frac{-(1-\xi)\delta_b}{2b} & \frac{-(1-\xi)\delta_b}{2} & 0 & \frac{-(1+\xi)\delta_b}{2a} & \frac{-(1+\xi)\delta_b}{2} & 0 \end{bmatrix}$$

$$B_{S2} = \begin{bmatrix} \frac{(1+\eta)\delta_a}{2a} & 0 & \frac{(1+\eta)\delta_a}{2} & \frac{-(1+\eta)\delta_a}{2a} & 0 & \frac{(1+\eta)\delta_a}{2} \\ \frac{(1+\xi)\delta_b}{2b} & \frac{-(1+\xi)\delta_b}{2} & 0 & \frac{(1-\xi)\delta_b}{2b} & \frac{-(1-\xi)\delta_b}{2} & 0 \end{bmatrix} \tag{2.100}$$

The D matrix for shear is

$$D_S = G \cdot \frac{5}{6} t \cdot \begin{bmatrix} 1 & 0 \\ 0 & 1 \end{bmatrix} = \frac{E}{2(1+\mu)} \left(\frac{5t}{6} \right) \begin{bmatrix} 1 & 0 \\ 0 & 1 \end{bmatrix} = \frac{5Et}{12(1+\mu)} \begin{bmatrix} 1 & 0 \\ 0 & 1 \end{bmatrix} \tag{2.101}$$

The stiffness matrix for shear is formed by expanding and integrating of the following expression:

$$K_s^e = \int_{-b}^{b} \int_{-a}^{a} B_s^T D_S B \cdot dxdy = ab \int_{-1}^{1} \int_{-1}^{1} B_s^T D_s B \cdot d\xi d\eta \tag{2.102}$$

2.4.2.1.3. Formulation of the Stiffness Matrix of the Thick Plate Element

In the formulation of the bending stiffness matrix, normal (in the vertical plane perpendicular to the edge) and tangential rotation (in the vertical plane containing the edge) of the mid-sides nodes 5, 6, 7, 8 at the 4 edges as shown in Fig. (**2.12b**) are first determined. For the rotations about axes parallel to the sides where these nodes are situated, the rotations are taken as average of the end nodes 1, 2, 3 and 4 for the present bilinear element as

$$\begin{cases} \psi_{x5} = \frac{1}{2}\left[\psi_{x1} + \psi_{x2}\right] \\ \psi_{y6} = \frac{1}{2}\left[\psi_{y2} + \psi_{y3}\right] \\ \psi_{x7} = \frac{1}{2}\left[\psi_{x3} + \psi_{x4}\right] \\ \psi_{y8} = \frac{1}{2}\left[\psi_{y1} + \psi_{y4}\right] \end{cases} \tag{2.103}$$

For the rotations about axes perpendicular to the sides, they are determined by eq.(2.83) as:

$$\begin{cases} \psi_{y5} = \dfrac{3(1-2\delta_a)}{4a}\omega_1 - \dfrac{3(1-2\delta_a)}{4a}\omega_2 + \left(-\dfrac{1}{4}+\dfrac{3}{2}\delta_a\right)\psi_{y1} + \left(-\dfrac{1}{4}+\dfrac{3}{2}\delta_a\right)\psi_{y2} \\ \psi_{x6} = -\dfrac{3(1-2\delta_b)}{4b}\omega_2 + \dfrac{3(1-2\delta_b)}{4b}\omega_3 + \left(-\dfrac{1}{4}+\dfrac{3}{2}\delta_b\right)\psi_{x2} + \left(-\dfrac{1}{4}+\dfrac{3}{2}\delta_b\right)\psi_{x3} \\ \psi_{y7} = \dfrac{3(1-2\delta_a)}{4a}\omega_4 - \dfrac{3(1-2\delta_a)}{4a}\omega_3 + \left(-\dfrac{1}{4}+\dfrac{3}{2}\delta_a\right)\psi_{y3} + \left(-\dfrac{1}{4}+\dfrac{3}{2}\delta_a\right)\psi_{y4} \\ \psi_{x8} = -\dfrac{3(1-2\delta_b)}{4b}\omega_1 + \dfrac{3(1-2\delta_b)}{4b}\omega_4 + \left(-\dfrac{1}{4}+\dfrac{3}{2}\delta_b\right)\psi_{x1} + \left(-\dfrac{1}{4}+\dfrac{3}{2}\delta_b\right)\psi_{x4} \end{cases} \quad \textbf{(2.104)}$$

The bending energy for the pseudo 8 node element is determined by the classical Q8 element shape functions as :

$$\begin{cases} N_1 = -\dfrac{1}{4}(1-\xi)(1-\eta)(1+\xi+\eta) & N_5 = \dfrac{1}{2}(1+\xi^2)(1-\eta) \\ N_2 = -\dfrac{1}{4}(1+\xi)(1-\eta)(1-\xi+\eta) & N_6 = \dfrac{1}{2}(1+\xi)(1-\eta^2) \\ N_3 = -\dfrac{1}{4}(1+\xi)(1+\eta)(1-\xi-\eta) & N_7 = \dfrac{1}{2}(1-\xi^2)(1+\eta) \\ N_4 = -\dfrac{1}{4}(1-\xi)(1+\eta)(1+\xi-\eta) & N_8 = \dfrac{1}{2}(1-\xi)(1-\eta^2) \end{cases} \quad \textbf{(2.105)}$$

Put $$\psi_x = \sum_{i=1}^{8} N_i \psi_{xi} \quad \textbf{(2.106)}$$

$$\psi_y = \sum_{i=1}^{8} N_i \psi_{yi} \quad \textbf{(2.107)}$$

Eliminating ψ_{x5} to ψ_{x8} and ψ_{y5} to ψ_{y8} by (2.104), eq.(2.106) and (2.107) become

$$\begin{aligned}\psi_x &= -\frac{3(1-2\delta_b)}{4b}N_8\omega_1 + \left[N_1 + \frac{1}{2}N_5 + \left(-\frac{1}{4}+\frac{3}{2}\delta_b\right)N_8\right]\psi_{x1} - \frac{3(1-2\delta_b)}{4b}N_6\omega_2 \\ &+ \left[N_2 + \frac{1}{2}N_5 + \left(-\frac{1}{4}+\frac{3}{2}\delta_b\right)N_6\right]\psi_{x2} + \frac{3(1-2\delta_b)}{4b}N_6\omega_3 + \left[N_3 + \frac{1}{2}N_7 + \left(-\frac{1}{4}+\frac{3}{2}\delta_b\right)N_6\right]\psi_{x3} \\ &+ \frac{3(1-2\delta_b)}{4b}N_8\omega_4 + \left[N_4 + \frac{1}{2}N_7 + \left(-\frac{1}{4}+\frac{3}{2}\delta_b\right)N_8\right]\psi_{x4}\end{aligned} \quad \textbf{(2.108)}$$

$$\begin{aligned}\psi_y &= \frac{3(1-2\delta_a)}{4a}N_5\omega_1 + \left[N_1 + \frac{1}{2}N_8 + \left(-\frac{1}{4}+\frac{3}{2}\delta_a\right)N_5\right]\psi_{y1} - \frac{3(1-2\delta_a)}{4a}N_5\omega_2 \\ &+ \left[N_2 + \frac{1}{2}N_6 + \left(-\frac{1}{4}+\frac{3}{2}\delta_a\right)N_5\right]\psi_{y2} - \frac{3(1-2\delta_a)}{4a}N_7\omega_3 + \left[N_3 + \frac{1}{2}N_6 + \left(-\frac{1}{4}+\frac{3}{2}\delta_a\right)N_7\right]\psi_{y3} \\ &+ \frac{3(1-2\delta_a)}{4a}N_7\omega_4 + \left[N_4 + \frac{1}{2}N_8 + \left(-\frac{1}{4}+\frac{3}{2}\delta_a\right)N_7\right]\psi_{y4}\end{aligned} \quad \textbf{(2.109)}$$

The 4 node element formulation as proposed is hence a pseudo 8 node formulation which will increase the accuracy of the solution. It should be pointed out that different orders are used for bending and shear energy terms which is better than the classical Q8 Mindlin plate element formulation (Zienkiewicz *et al.* 2011).

The curvature matrix of the element is given by

$$\kappa = \begin{Bmatrix} \kappa_x \\ \kappa_y \\ \kappa_{xy} \end{Bmatrix} = \begin{Bmatrix} \dfrac{\partial \psi_x}{\partial y} \\ \dfrac{\partial \psi_y}{\partial x} \\ \left(\dfrac{\partial \psi_y}{\partial y} - \dfrac{\partial \psi_x}{\partial x}\right) \end{Bmatrix} \quad \textbf{(2.110)}$$

Bearing in mind that
$$\begin{cases} \dfrac{\partial \psi_x}{\partial y} = \dfrac{\partial \psi_x}{\partial \eta} \cdot \dfrac{\partial \eta}{\partial y} = \dfrac{\partial \psi_x}{\partial \eta} \cdot \dfrac{1}{b} \\ \dfrac{\partial \psi_y}{\partial x} = \dfrac{\partial \psi_y}{\partial \xi} \cdot \dfrac{\partial \xi}{\partial x} = \dfrac{\partial \psi_y}{\partial \xi} \cdot \dfrac{1}{a} \\ \dfrac{\partial \psi_y}{\partial y} = \dfrac{\partial \psi_y}{\partial \eta} \cdot \dfrac{\partial \eta}{\partial y} = \dfrac{\partial \psi_y}{\partial \eta} \cdot \dfrac{1}{b} \\ \dfrac{\partial \psi_x}{\partial x} = \dfrac{\partial \psi_x}{\partial \xi} \cdot \dfrac{\partial \xi}{\partial x} = \dfrac{\partial \psi_x}{\partial \xi} \cdot \dfrac{1}{a} \end{cases} \tag{2.111}$$

and by appropriately differentiating (2.108) and (2.109), the B matrix for bending as equated by $\kappa = B_B\, a^e$ would be

$B_B = [\, B_{B1}\; B_{B2}\; B_{B3}\; B_{B4}]$, where

$$\begin{gathered}
B_{B1} = \begin{bmatrix} \dfrac{3(1-2\delta_b)(1-\xi)\eta}{4b^2} & \dfrac{(1-\xi)(3\eta-1-6\eta\delta_b)}{4b} & 0 \\ -\dfrac{3(1-2\delta_a)(1-\eta)\xi}{4a^2} & 0 & \dfrac{(1-\eta)(3\xi-1-6\delta_a\xi)}{4a} \\ \dfrac{3[(1-2\delta_a)(1-\xi^2)+(1-2\delta_b)(1-\eta^2)]}{-8ab} & \dfrac{(1-\eta)[-3\eta-1+6\delta_b(1+\eta)]}{8a} & \dfrac{(1-\xi)[3\xi+1-6\delta_a(1+\xi)]}{8b} \end{bmatrix} \\
B_{B2} = \begin{bmatrix} \dfrac{3(1-2\delta_b)(1+\xi)\eta}{4b^2} & \dfrac{(1+\xi)(3\eta-1-6\eta\delta_b)}{4b} & 0 \\ \dfrac{3(1-2\delta_a)(1-\eta)\xi}{4a^2} & 0 & \dfrac{(1-\eta)(3\xi+1-6\delta_a\xi)}{4a} \\ \dfrac{3[(1-2\delta_a)(1-\xi^2)+(1-2\delta_b)(1-\eta^2)]}{8ab} & \dfrac{(1-\eta)[3\eta+1-6\delta_b(1+\eta)]}{8a} & \dfrac{(1+\xi)[-3\xi+1-6\delta_a(1-\xi)]}{8b} \end{bmatrix} \\
B_{B3} = \begin{bmatrix} -\dfrac{3(1-2\delta_b)(1+\xi)\eta}{4b^2} & \dfrac{(1+\xi)(3\eta+1-6\eta\delta_b)}{4b} & 0 \\ \dfrac{3(1-2\delta_a)(1+\eta)\xi}{4a^2} & 0 & \dfrac{(1+\eta)(3\xi+1-6\delta_a\xi)}{4a} \\ \dfrac{3[(1-2\delta_a)(1-\xi^2)+(1-2\delta_b)(1-\eta^2)]}{-8ab} & \dfrac{(1+\eta)[-3\eta+1-6\delta_b(1-\eta)]}{8a} & \dfrac{(1+\xi)[3\xi-1+6\delta_a(1-\xi)]}{8b} \end{bmatrix} \\
B_{B4} = \begin{bmatrix} -\dfrac{3(1-2\delta_b)(1-\xi)\eta}{4b^2} & \dfrac{(1-\xi)(3\eta+1-6\eta\delta_b)}{4b} & 0 \\ -\dfrac{3(1-2\delta_a)(1+\eta)\xi}{4a^2} & 0 & \dfrac{(1+\eta)(3\xi-1-6\delta_a\xi)}{4a} \\ \dfrac{3[(1-2\delta_a)(1-\xi^2)+(1-2\delta_b)(1-\eta^2)]}{8ab} & \dfrac{(1+\eta)[3\eta-1+6\delta_b(1-\eta)]}{8a} & \dfrac{(1-\xi)[-3\xi-1+6\delta_a(1+\xi)]}{8b} \end{bmatrix}
\end{gathered} \tag{2.112}$$

$$\begin{bmatrix} M_x \\ M_y \\ M_{xy} \end{bmatrix} = D_b \kappa = D_b \begin{Bmatrix} \dfrac{\partial \psi_x}{\partial y} \\ \dfrac{\partial \psi_y}{\partial x} \\ \left(\dfrac{\partial \psi_x}{\partial x} - \dfrac{\partial \psi_y}{\partial y} \right) \end{Bmatrix}$$

The D_b matrix for bending is given by
and

$$D_b = \frac{Et^3}{12(1-\mu^2)} \begin{bmatrix} 1 & -\mu & 0 \\ -\mu & 1 & 0 \\ 0 & 0 & \dfrac{1-\mu}{2} \end{bmatrix} \tag{2.113}$$

So the stiffness matrix for bending can be formed by expanding and integrating the following expression

$$K_b^e = \int_{-b}^{b}\int_{-a}^{a} B_B{}^T D_b B_B dxdy = ab\int_{-1}^{1}\int_{-1}^{1} B_B{}^T D_b B_B d\xi d\eta \tag{2.114}$$

The total stiffness of the plate will be $K = K_s^e + K_b^e$ **(2.115)**

where K_s^e is given in eq.(2.102)

2.4.3. *Extension to General Quadrilateral Plate Element*

With minor changes to the formulation (but more complicated equations), the rectangular thick plate element as given in previous section can be extended to a general quadrilateral element as follows. From (2.95), the strains at edge ij of a general quadrilateral element is:

$$
\begin{aligned}
\gamma_{sij} &= -\frac{\delta_{ij}}{l_{ij}}\left[2(\omega_i-\omega_j)+\left(\psi_i+\psi_j\right)l_{ij}\right] \\
&= -\frac{\delta_{ij}}{l_{ij}}\left[2(\omega_i-\omega_j)+\left(\psi_{xi}\frac{y_j-y_i}{l_{ij}}-\psi_{yi}\frac{x_j-x_i}{l_{ij}}+\psi_{xj}\frac{y_j-y_i}{l_{ij}}-\psi_{yj}\frac{x_j-x_i}{l_{ij}}\right)l_{ij}\right] \\
&= -\frac{\delta_{ij}}{l_{ij}}\left[2(\omega_i-\omega_j)+\psi_{xi}\Delta y_{ji}-\psi_{yi}\Delta x_{ji}+\psi_{xj}\Delta y_{ji}-\psi_{yj}\Delta x_{ji}\right]
\end{aligned} \quad \textbf{(2.116)}
$$

Where,

$$\Delta x_{ji}=x_j-x_i \quad \Delta y_{ji}=y_j-y_i \quad \textbf{(2.117)}$$

$$\delta_{ij}=\frac{\left(\frac{t}{l_{ij}}\right)^2(1+v)}{\frac{5}{6}+2\left(\frac{t}{l_{ij}}\right)^2(1+v)} \quad (i,j=1,2,3,4) \quad \textbf{(2.118)}$$

For convenience, a term γ_{sij}^* is defined and is given $\gamma_{sij}^*=l_{ij}\gamma_{sij}$

$$\gamma_s^*=\left[\gamma_{s12}^* \quad \gamma_{s23}^* \quad \gamma_{s34}^* \quad \gamma_{s41}^*\right]^T \quad \textbf{(2.119)}$$

$$\text{then} \quad \gamma_s^*=T_s a^e \quad \textbf{(2.120)}$$

$$T_s=\begin{bmatrix}-2\delta_{12} & -\delta_{12}\Delta y_{21} & \delta_{12}\Delta x_{21} & 2\delta_{12} & -\delta_{12}\Delta y_{21} & \delta_{12}\Delta x_{21} & 0 & 0 & 0 & 0 & 0 & 0\\ 0 & 0 & 0 & -2\delta_{23} & -\delta_{23}\Delta y_{32} & \delta_{23}\Delta x_{32} & 2\delta_{23} & -\delta_{23}\Delta y_{32} & \delta_{23}\Delta x_{32} & 0 & 0 & 0\\ 0 & 0 & 0 & 0 & 0 & 0 & -2\delta_{34} & -\delta_{34}\Delta y_{43} & \delta_{34}\Delta x_{43} & 2\delta_{34} & -\delta_{34}\Delta y_{43} & \delta_{34}\Delta x_{43}\\ 2\delta_{41} & -\delta_{41}\Delta y_{14} & \delta_{41}\Delta x_{14} & 0 & 0 & 0 & 0 & 0 & 0 & -2\delta_{41} & -\delta_{41}\Delta y_{14} & \delta_{41}\Delta x_{14}\end{bmatrix} \quad \textbf{(2.121)}$$

As the axes are not along the main directions, transformation of strain from local to global axes will be required, and the nodal shear strain will become

$$\begin{Bmatrix}\gamma_{xi}\\ \gamma_{yi}\end{Bmatrix}=\frac{1}{\Delta y_{ik}\Delta x_{ji}-\Delta y_{ji}\Delta x_{ik}}\begin{bmatrix}\Delta y_{ik} & -\Delta y_{ji}\\ -\Delta x_{ik} & \Delta x_{ji}\end{bmatrix}\begin{Bmatrix}\gamma^*_{sij}\\ \gamma^*_{ski}\end{Bmatrix}=\begin{bmatrix}T_{sx}\\ T_{sy}\end{bmatrix}\begin{Bmatrix}\gamma^*_{sij}\\ \gamma^*_{ski}\end{Bmatrix} \quad \textbf{(2.122)}$$

where j=i+1 if j < 5 and = 1 if j=5, k=i-1 if k > 1, else k=4. The nodal shear strain can be expressed in terms of nodal displacement as

$$\left.\begin{aligned}\gamma_{xi}&=T_{sx}\gamma^*_s=T_{sx}T_s a^e\\ \gamma_{yi}&=T_{sy}\gamma^*_s=T_{sy}T_s a^e\end{aligned}\right\} \quad \textbf{(2.123)}$$

or $\gamma_{xi}=[\gamma_{x1}\quad \gamma_{x2}\quad \gamma_{x3}\quad \gamma_{x4}]^T$, $\gamma_{yi}=[\gamma_{y1}\quad \gamma_{y2}\quad \gamma_{y3}\quad \gamma_{y4}]^T$

Eqs. (2.98) and (2.99) will become:

$$\gamma=\begin{Bmatrix}\gamma_x\\ \gamma_y\end{Bmatrix}=\begin{bmatrix}N_sT_{sx}T_s\\ N_sT_{sy}T_s\end{bmatrix}a^e=B_sa^e \quad \textbf{(2.124)}$$

$$\text{where } B_s=\begin{bmatrix}NT_{sx}T_s\\ NT_{sy}T_s\end{bmatrix} \quad \textbf{(2.125)}$$

and the shape function matrix N is given by $N = \begin{bmatrix} N_1 & N_2 & N_3 & N_4 \end{bmatrix}$ **(2.126)**

Shear stiffness for this general quadrilateral element will still be given by eq.(2.102) but B_s as given by eq.(2.125) will be used.

The normal rotation ψ_n and tangential rotation ψ_s for mid-side nodes can be expressed as :

$$\psi_{nij} = \psi_{xij}\cos\theta_{ij} + \psi_{yij}\sin\theta_{ij} \qquad \psi_{sij} = \psi_{xij}\sin\theta_{ij} - \psi_{yij}\cos\theta_{ij} \tag{2.127}$$

where θ_{ij} is the angle of the line $\overline{ij}$ with the X-axis.

$$\Rightarrow \psi_{nij} = \psi_{xij}\frac{x_j - x_i}{l_{ij}} + \psi_{yij}\frac{y_j - y_i}{l_{ij}} = \psi_{xij}\frac{\Delta x_{ji}}{l_{ij}} + \psi_{yij}\frac{\Delta y_{ji}}{l_{ij}} \tag{2.128}$$

$$\Rightarrow \psi_{sij} = \psi_{xij}\frac{y_j - y_i}{l_{ij}} - \psi_{yij}\frac{x_j - x_i}{l_{ij}} = \psi_{xij}\frac{\Delta y_{ji}}{l_{ij}} - \psi_{yij}\frac{\Delta x_{ji}}{l_{ij}} \tag{2.129}$$

$$\text{or} \quad \begin{Bmatrix} \psi_{nij} \\ \psi_{sij} \end{Bmatrix} = \frac{1}{l_{ij}}\begin{bmatrix} \Delta x_{ji} & \Delta y_{ji} \\ \Delta y_{ji} & -\Delta x_{ji} \end{bmatrix}\begin{Bmatrix} \psi_{xij} \\ \psi_{yij} \end{Bmatrix} \tag{2.130}$$

For node 5, the normal shear strain is:

$$\psi_{n5} = \frac{1}{2}\left[(\psi_{n1})_{\overline{12}} + (\psi_{n2})_{\overline{12}}\right] = \frac{1}{2l_{12}}\left[\Delta x_{21}(\psi_{x1} + \psi_{x2}) + \Delta y_{21}(\psi_{y1} + \psi_{y2})\right] \tag{2.131}$$

Similar equations can be written for nodes 6, 7, 8 and will not be repeated here. According to eq. (2.75), the tangential rotation at mid-side nodes 5, 6, 7, 8 are:

$$\psi_{s5} = \frac{1}{2}\psi_{s1} + \frac{1}{2}\psi_{s2} + 3(1-2\delta_{12})\left[\frac{2}{l_{12}}(-\omega_1 + \omega_2) - \psi_{s1} - \psi_{s2}\right]\frac{1}{4}$$

$$= \frac{1}{2}\psi_{s1} + \frac{1}{2}\psi_{s2} + \frac{3}{4}(1-2\delta_{12})[-\psi_{s1} - \psi_{s2}] + \frac{3}{4}(1-2\delta_{12})\left[\frac{2}{l_{12}}(-\omega_1 + \omega_2)\right]$$

$$= \frac{3}{2l_{12}}(1-2\delta_{12})(\omega_2 - \omega_1) - \frac{1}{4}(\psi_{s1} + \psi_{s2}) + \frac{6}{4}\delta_{12}(\psi_{s1} + \psi_{s2}) \tag{2.132}$$

$$= \frac{3}{2l_{12}}(1-2\delta_{12})(\omega_2 - \omega_1) - \frac{1}{4}(1-6\delta_{12})(\psi_{s1} + \psi_{s2})$$

$$= \frac{3}{2l_{12}}(1-2\delta_{12})(\omega_2 - \omega_1) - \frac{1}{4l_{12}}(1-6\delta_{12})\left[\Delta y_{21}(\psi_{x1} + \psi_{x2}) - \Delta x_{21}(\psi_{y1} + \psi_{y2})\right]$$

Similar equations can also be established for nodes 6,7 and 8 and will not be repeated here. For rotation at mid-side nodes, inverting eq.(2.130) for node 5

$$\begin{Bmatrix}\psi_{x5}\\ \psi_{y5}\end{Bmatrix} = \frac{1}{l_{12}}\begin{bmatrix}\Delta x_{21} & \Delta y_{21}\\ \Delta y_{21} & -\Delta x_{21}\end{bmatrix}\begin{Bmatrix}\psi_{n5}\\ \psi_{s5}\end{Bmatrix} \tag{2.133}$$

$$\psi_{x5} = \frac{1}{l_{12}}(\Delta x_{21}\psi_{n5} + \Delta y_{21}\psi_{s5}) \tag{2.134}$$

$$\begin{aligned}
\Rightarrow \psi_{x5} &= \frac{\Delta x_{21}}{l_{12}} \times \frac{1}{2l_{12}}\left[\Delta x_{21}(\psi_{x1} + \psi_{x2}) + \Delta y_{21}(\psi_{y1} + \psi_{y2})\right] \\
&+ \frac{\Delta y_{21}}{l_{12}}\left\{\frac{3}{2l_{12}}(1-2\delta_{12})(\omega_2 - \omega_1) - \frac{1}{4l_{12}}(1-6\delta_{12})\left[\Delta y_{21}(\psi_{x1} + \psi_{x2}) - \Delta x_{21}(\psi_{y1} + \psi_{y2})\right]\right\} \\
&= -\frac{3\Delta y_{21}}{2l_{12}^2}(1-2\delta_{12})\omega_1 + \frac{1}{2l_{12}^2}\left[\Delta x_{21}^2 - \frac{\Delta y_{21}^2}{2}(1-6\delta_{12})\right]\psi_{x1} + \frac{\Delta x_{21}\Delta y_{21}}{2l_{12}^2}\left[1 + \frac{1}{2}(1-6\delta_{12})\right]\psi_{y1} \\
&+ \frac{3\Delta y_{21}}{2l_{12}^2}(1-2\delta_{12})\omega_2 + \frac{1}{2l_{12}^2}\left[\Delta x_{21}^2 - \frac{\Delta y_{21}^2}{2}(1-6\delta_{12})\right]\psi_{x2} + \frac{\Delta x_{21}\Delta y_{21}}{2l_{12}^2}\left[1 + \frac{1}{2}(1-6\delta_{12})\right]\psi_{y2}
\end{aligned} \tag{2.135}$$

$$\psi_{y5} = \frac{1}{l_{12}}\left(\Delta y_{21}\psi_{n5} - \Delta x_{21}\psi_{s5}\right) \tag{2.136}$$

$$\Rightarrow \psi_{y5} = \frac{\Delta y_{21}}{l_{12}} \times \frac{1}{2l_{12}}\left[\Delta x_{21}(\psi_{x1}+\psi_{x2}) + \Delta y_{21}(\psi_{y1}+\psi_{y2})\right]$$

$$\begin{aligned}
&-\frac{\Delta x_{21}}{l_{12}}\left\{\frac{3}{2l_{12}}(1-2\delta_{12})(\omega_2-\omega_1) - \frac{1}{4l_{12}}(1-6\delta_{12})\left[\Delta y_{21}(\psi_{x1}+\psi_{x2}) - \Delta x_{21}(\psi_{y1}+\psi_{y2})\right]\right\} \\
&= \frac{3\Delta x_{21}}{2l_{12}^{2}}(1-2\delta_{12})\omega_1 + \frac{\Delta x_{21}\Delta y_{21}}{2l_{12}^{2}}\left[1+\frac{1}{2}(1-6\delta_{12})\right]\psi_{x1} + \frac{1}{2l_{12}^{2}}\left[\Delta y_{21}^{2} - \frac{\Delta x_{21}^{2}}{2}(1-6\delta_{12})\right]\psi_{y1} \\
&-\frac{3\Delta x_{21}}{2l_{12}^{2}}(1-2\delta_{12})\omega_2 + \frac{\Delta x_{21}\Delta y_{21}}{2l_{12}^{2}}\left[1+\frac{1}{2}(1-6\delta_{12})\right]\psi_{x2} + \frac{1}{2l_{12}^{2}}\left[\Delta y_{21}^{2} - \frac{\Delta x_{21}^{2}}{2}(1-6\delta_{12})\right]\psi_{y2}
\end{aligned} \tag{2.137}$$

Similar results can be written for other mid-side nodes and will not be repeated here. It is now possible to write shear strain within the general quadrilateral element by extending eqs.(2.106) and (2.107) with equations eqs.(2.135) and (2.137). The authors have adopted the symbolic algebra capability from Mathcad for generating the resulting strain-displacement matrix B_B which is extremely lengthy and will not be reproduced here. The tedious and lengthy algebra for the various matrices are coded into a Mathcad program and then later a Fortran 95 program for research and teaching, and the coding for the formulation of the plate program is given in Appendix 2-4 of this chapter.

9 nodes or 8 nodes plate elements can provide good accuracy. They are however not convenient to be used in practice, and the accuracy of "PLATE" which is practically a node element is actually as good as those 9 nodes elements (see Cheng and Law 2008). For practical applications, most of the engineers and computer softwares adopt 4 nodes elements to suit the superstructure layout and loadings. Cheng and Law (2008) found that the accuracy of "PLATE" is far better than many 4 nodes elements, and the present thick plate element can be useful to large transfer plate or pile cap where a fine mesh can be very time consuming in the analysis.

2.4.4. Structure of Program PLATE as Given in Appendix 2-4

The authors have developed a finite element program PLATE based on the theory discussed in previous sections. The structure of the program will be discussed in this section, and the readers can learn how to develop a finite element computer code for an engineering problem by reading this relatively simple program. In fact, the authors have noticed that most engineers do not have adequate knowledge in the structure of an engineering software as well as the actual implementation of the numerical method. The program in Appendix 2-4 can serve as an introduction towards this major problem.

In this program, the input format is similar to that in Appendix 2-1 and will not be repeated here. It should be noted that there are 3 degree of freedom for each node, and 5 internal stress which correspond to $M_{xx,}$ $M_{yy,,}$ M_{xy} , $F_{xz,}$and F_{xy} . In the input subroutine, if a rectangular problem is defined, the user is required to define the grid lines along x and y directions. If the problem is non-rectangular in shape, the mesh should be generated by an external program, and the mesh and connectivity of the problem is read in directly. In the advanced Windows version PLATE, mesh generation is automatically carried out if macro-elements are defined. Voids can be defined, and the elements and nodes within the void zone will be removed then.

The definition of the applied loads, prescribed boundary conditions, nodal vertical and rotational springs are all controlled by keywords. These data are put into the global stiffness matrix (stored as a band form) or the global load vector.

2.4.5. Numerical Implementation of Mindlin Plate Bending Program

A relatively complicated thin and thick program is given in Appendix 2-4 for the illustration of some of techniques used by the authors for developing various educational finite element program. A rather complicated Windows based version of PLATE (with about 100000 lines for Fortran codes with interface library Winteracter) with many options and advanced features can be obtained from the authors for research at natureymc@yahoo.com.hk. Originally, the authors would like to put forward a seepage program with free surface flow analysis for illustrating how to develop a finite element program, but such program can be found in many book on finite element method. On the other hand, the authors are not aware of any book which has given sufficient account on thick plate analysis (simple thick plate computer code is available in some book). Furthermore, the structure of the program is more complicated than that for seepage, and it can better reflect that various skills are required to formulate a commercial finite element program.

To begin with, the data structure is examined. As shown in Appendix 2-4, all the memory required for the analysis is made dynamic. That means, as long as there is computer memory, there is no limit to the actual size of the program. The readers can add the code to control the case where the memory required for dynamic allocation exceeds that available from the operation system. This is a simple exercise by adding an error control code in the allocation statement. The subroutine using the Dos interrupt as well as the graphics library Dislin in this program can be removed by the readers easily, or the readers can adopt Dos environment (or emulation) to further develop the present program. The input process should be simple enough for the readers to read, and the authors will only concentrate on those topics which are more related to engineering and numerical implementation of finite element method. The readers should also have adequate knowledge in basic finite element analysis before reading Appendix 2-4.

To determine the global stiffness matrix, the elemental stiffness matrix must be evaluated. The stiffness matrix equations as given by eq.(2.90) and (2.102) are evaluated by numerical integration using the Gaussian method. The Gaussian points and weighting factors are declared in subroutine GAUSSQ while the constitutive matrix D is defined by subroutine MODPB. Depends on the number of Gaussian points to be used for the numerical integration, the memory required to store the Gaussian point stresses after the analysis will be determined. In general, 2x2 integration rule will be sufficiently accurate for normal application. The shape functions and the derivatives of the shape functions as given by eq.(2.61) to carry out the Jacobian transformation in the numerical integration are given in subroutine SFR2. The Jacobian transformation and the evaluate of the determinant as required in eq.(2.61) and eq.(2.62) is given by the subroutine JACOB and JACOB2. The strain displacement matrix as given by eq.(2.125) is given by subroutine BMATPB. The determination of the global stiffness matrix starts with the determination of the elemental stiffness matrix as given by eq.(2.90) and (2.102), and numerical integration loops over the four Guassian points which is given by subroutine STIFF. The element stiffness is then assembled into the banded global stiffness through the subroutine ASSEMK. After that, the boundary condition is defined by the subroutine BOUND. For prescribed boundary condition, the wipe row and column method is adopted, and the corresponding diagonal term will then be 1.0. The global load vector will also be modified with the given boundary conditions. Another simple method to implement the prescribed boundary condition is the big M method, where only the corresponding diagonal terms and load vectors are modified. This method is simple in implementation, but will induce a very small error in theory (but negligible in practice).

In this program, the grillage member as given in Appendix 2-1 is actually coupled inside the present code. That means, a plate element can be connected to a grid member by using the subroutine GRILLAGE. Such coupling is possible and easy to implement, as a grid member has the same degree of freedom as a plate element. Point loads and distributed loads can be defined in subroutine LOADPB. In the present relatively simple version, the point loads must lie exactly on the nodal points in the mesh, while such requirement is removed in the more advanced version of program PLATE which is a Windows based program. Furthermore, the distributed load in this version must fill the whole area of a plate element, while such requirement is relaxed in the more advanced version of PLATE.

The boundary conditions that can be defined by PLATE include point spring supports, Winkler's spring and pile/pile group. For nodal spring, it is controlled by the keyword nodal_spring, and vertical, rotation along x and along y spring constants can be defined. The rotational spring can define the effect of pile connected to the pile cap, and a separate lateral load analysis of the pile is required to define this spring constant. In general, the pile along its length can be modelled as beam element connected with soil Winkler nodal springs. A unit moment is applied to the top of the pile, and the rotation of the pile is determined. The ratio between the unit moment and the rotation at top of the pile will then be the rotational spring constant, which can be defined along the x and y axes respectively. For soil Winkler's spring, it is controlled by the keyword Winkler_spring, and the equivalent nodal spring will then be determined by $\int[N]^T Ks\ dA$. In the updated PLATE program, Winkler spring constant can be defined by individual region, which is useful to model a raft foundation on nonhomogeneous medium. Actually, some engineers model the vertical response of large diameter bore piles by defining an equivalent Winkler's constant instead of a point stiffness of the pile. This refined model can better reflect the size effect of a pile, and the bending moment and deformation of the pile cap will usually be smaller. In the present version of PLATE, piles can be defined by the keyword Pile_Properties. The pile response will then be modelled by the ring shear theory by Randolph (Fleming *et al.* 2009). The flexibility coefficient f_{ii} of a pile under vertical load can be given by

$$f_{ii} = \left[1 + \frac{K_2}{A_p E_p} \cdot \frac{\tanh(ul)}{u}\right]\left[K_2 + K_1 \frac{\tanh(ul)}{u}\right]^{-1} \tag{2.138}$$

Where A_p and E_p are the sectional area and Young's modulus of the pile. K_1 and K_2 are given by:

$$K_1 = 2\pi G/\ln(r_m/r_0) \tag{2.139}$$

$$K_2 = \frac{4r_b G_b}{(1-\mu)} \tag{2.140}$$

G is the average shear modulus of soil over the pile shaft area, r_0 and r_m are the radius of pile shaft and maximum radius of influence of pile. r_b and G_b are the radius of the pile base and the shear modulus of the pile base material, μ is the Poisson ratio while u is given by

$$u^2 = \frac{K_1}{A_p E_p} \tag{2.141}$$

The inverse of the flexibility coefficient will give the stiffness of the pile (or equivalent pile spring) which can be assembled into the global stiffness matrix. If the pile group interaction effect is to be considered, the flexibility coefficient f_{ij} can be obtained from eq.(2-138) by replacing K_1 and K_2 with K'_1 and K'_2, which are given by

$$K_1' = K_1 \ln \frac{r_m}{r_o} \bigg/ \ln \frac{r_m}{R} \tag{2.142}$$

$$K_2' = K_2 \frac{\pi R}{2r_b} \tag{2.143}$$

Where R is the distance between two piles. The flexibility matrix of the pile group is formed by subroutine PILEG, and the inverse of the flexibility matrix (subroutine GJNV) will then be the stiffness matrix corresponding to the pile group. This stiffness matrix will be assembled into the global stiffness matrix, and a complete pile group interaction analysis can be defined using a two-dimensional model. The authors have used program SAP2000 to define a true three-dimensional model with soil, pile caps and pile groups, and compared the results from that by using the Randolph model coupled with PLATE. The results from both analyses are nearly the same, which means that a true three-dimensional model is not required, unless the problem under consideration is highly complicated. For highly complicated case, the Windows version PLATE should be used, as boreholes can be defined to model the three-dimensional ground condition. The time required to form the computer model using PLATE is however much less than that required for a true three-dimensional analysis. The user interface of the updated PLATE using the pile option is given in Fig. (**2.13**). The effect of soil Winkler's spring can further be superimposed into the complete analysis if required. In the updated program PLATE, elastic half space option can also be defined for the soil response, and the flexibility coefficients f_{ii} is defined by the Boussinesq or Mindlin's solution, plus the option of a cut-off distance. Beyond a cut-off distance as given by the users, the flexibility coefficient of the pile group interaction or elastic half space effect will be taken as 0! This option is useful in controlling the over-interaction which is observed from analysis. From actual field measurement, it is also observed that the effect of a pile can extend only to 5 to 10 pile radius. The assumption of infinite distance of response from theory of elasticity is actually not really observed, but it can create the problem of over-interaction in the coupled analysis.

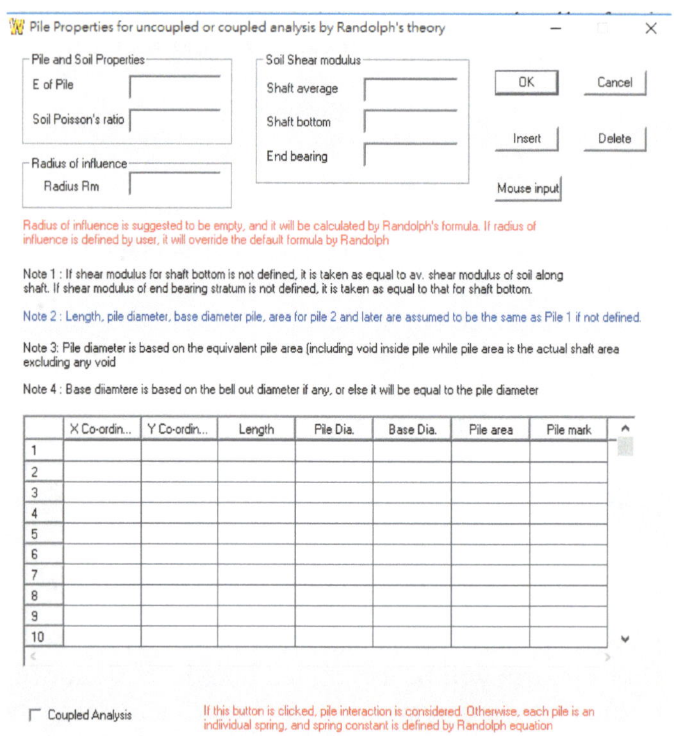

Fig. (2.13). Defining pile and pile group in the updated PLATE (the group interaction effect can be activated by clicking the check bottom at bottom of this menu).

In the updated program PLATE, a special option has been added. The ground condition beneath a raft foundation can be defined through a borehole option as shown below. The yield stress for individual load case can also be if there is any yielding of soil beneath the foundation. The options are shown in Fig. (**2.14**). In this option, the three-dimensional soil condition is formed by interpolation method to define the three-dimensional geometry of the underground condition. A separate three-dimensional analysis program (transparent to the users) is used to generate the flexibility coefficients of the soil support, based on the mesh layout of the raft foundation. The inverse of this flexibility matrix will be assembled into the global stiffness matrix for the analysis of the raft foundation. Using such approach, the engineer can save lots of time in defining a true three-dimensional problem, but the results from such coupled analysis will be virtually the same as that from a true three-dimensional analysis.

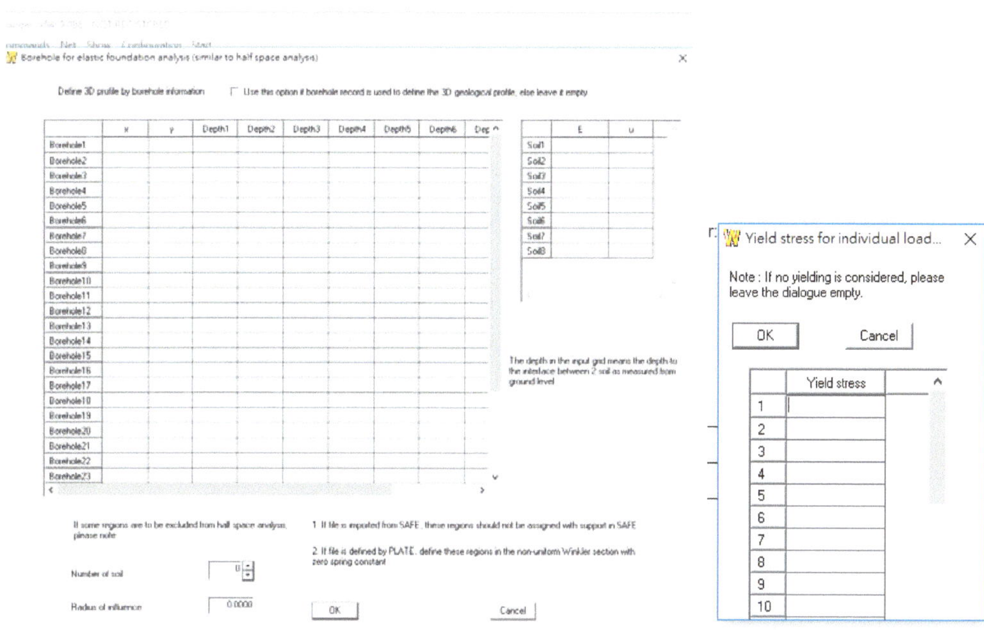

Fig. (2.14). Defining the soil condition beneath a raft foundation through the Borehole option in the updated PLATE.

A special issue about the definition of a moment along a shear wall should be discussed in this section. Classically, a moment has no size, which is sufficiently good for columns. For shear wall and core wall which take a large wind moment, the moment is carried by the long shear wall along its length. The actual length of a shear wall is long and cannot be modelled as a point moment. On the other hand, no program in the world allow a size to be assigned to a moment. To overcome such dilemma, the moment can be transformed to stress along the wall by the formula σ=P/A+Mx/Ix (for shear wall) and σ=P/A+Mx/Ix +My/Iy for core wall. A trapezoidal stress distribution will be obtained for the stress on the shear wall, which can then be distributed as point loads along the nodes. Along the side of a plate element where the stress is evaluated, the total area of the loading become point loads at the two ends of an edge can be distributed according to Fig. (**2.15**). Through such approach, the total vertical load and moment will be preserved. In the limit where there are many point loads in a computer model, the overall effect will then be equivalent to a line moment, and the moment can be given a size. In Fig. (**2.16**), a size can be defined to a line moment, and the transformation of a line moment to point loads will be automatic.

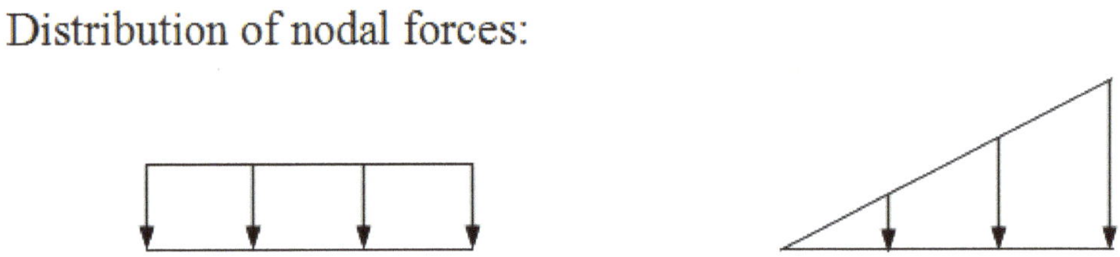

Fig. (2.15). Distribution of stresses along a shear wall arising from moment as point loads.

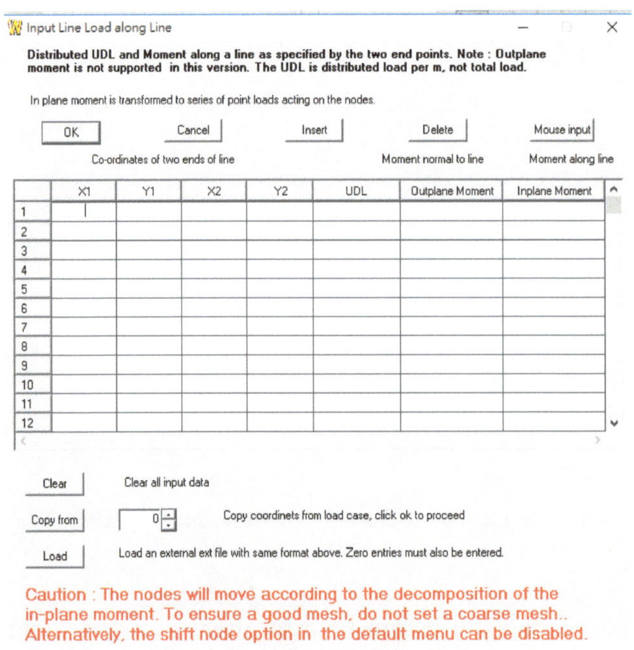

Fig. (2.16). A size can be assigned to a moment in the updated PLATE.

For pile cap or raft foundation, most engineers simply apply the loadings from the superstructure and neglect the effect of the superstructure in the foundation analysis. This rationale is based on the assumption of hinge connection between the columns and pile cap which is actually not correct, but this will greatly simplify the work load required to carry out the foundation analysis and design. In the updated PLATE, this limitation can be overcome by two approaches. The addition of grillage elements and nodal springs can partly represent the effect of walls and columns. For a more realistic analysis, the flexibility or stiffness matrix $[K_{ij}]$ of the superstructure can be determined from a structural analysis program. This stiffness matrix can then be coupled into the PLATE analysis program for a more refined analysis.

2.4.6. Development of a Pseudo 9 Node Mindlin Quadrilateral Plate Element

The pseudo 8 node plate element as given in section 2.3 and Appendix 2-4 has been shown to give very good accuracy by Cheng and Law (2008), and is free from shear locking effect. The accuracy of this element is actually better than most of the plate element available in many commercial finite element programs in the internal studies by the authors.

As described in the foregoing for the pseudo high order element, the 4 mid-side nodes have been "absorbed" into the 4 corner nodes in the formulation of the stiffness matrix. To increase the accuracy further without creating more computation, a "centre node" is added to form a new pseudo-higher order element in this section. It is well known that 9 node element can perform much better than 8 node element (or lesser) when the element is highly distorted (see Zienkiewicz *et al.* Taylor 2011). It will be shown in the latter sections that this type of element performs even better than the pseudo 8 nodes "PLATE-Q8" element developed by Cheng and Law in the previous studies (2008a, 2008b). In this section, the formulation of the rectangular element is first discussed. For the rectangular thick plate as shown in Fig. (**2.17**), the corner nodes are denoted by 1, 2, 3, 4; mid-side nodes are denoted by 5, 6, 7, 8 and a centre node 9 is added in the element. The side lengths of this rectangle are given by $2a$ and $2b$.

In Fig. (**2.17**), the global coordinate system is x and y as shown whilst ξ and η are the normalized values given by $\frac{x}{a}=\xi$ and $\frac{y}{b}=\eta$. The moments (M_x and M_y) are taken as positive when they act in the positive directions along the x and y axes which is the popular convention adopted in finite element analysis. In the stiffness matrix formulation, the mid-side nodes and center node will be statically eliminated in the stiffness matrix. This approach will give a 4 nodes element which possesses good numerical accuracy similar to a 9 nodes, but is a 4 node element to the engineers so that it can avoid many practical precautions for engineering use required for higher element as discussed by Cheng and Law (2008). The special feature about this element is that it will degenerate to the classical thin plate element if the thickness of the plate approaches zero as demonstrated in latter parts of this paper. For the plate as shown in Fig. (**3**), the nodal displacement vector a^e of the corner nodes are given as:

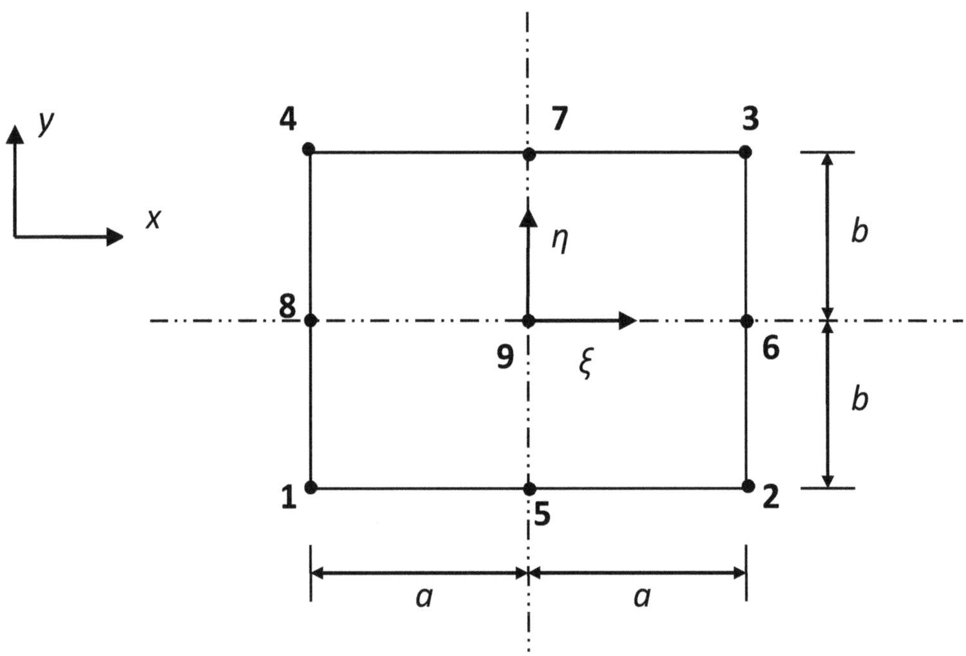

Fig. (2.17). The Psuedo 9-Node Thin/Thick Plate Element.

$$a^e = \begin{bmatrix} \omega_1 & \psi_{x1} & \psi_{y1} & \omega_2 & \psi_{x2} & \psi_{y2} & \omega_3 & \psi_{x3} & \psi_{y3} & \omega_4 & \psi_{x4} & \psi_{y4} \end{bmatrix}^T \quad \textbf{(2.144)}$$

The stiffness matrix of the element will be the sum of the shear stiffness matrix and the bending stiffness matrix as

$$K^e = K_s^e + K_b^e \quad \textbf{(2.145)}$$

2.4.6.1. Shear Stiffness Matrix

The transverse (out-of-plane) shear strains along the edges $\overline{12}, \overline{23}, \overline{34}, \overline{41}$ are constants. From eq.(2.84), we obtain (see Cheng and Law(2008b)) :

$$\begin{cases} \gamma_{s12} = -\dfrac{\delta_a}{a}\omega_1 + \dfrac{\delta_a}{a}\omega_2 + \delta_a\psi_{y1} + \delta_a\psi_{y2} \\ \gamma_{s23} = -\dfrac{\delta_b}{b}\omega_2 + \dfrac{\delta_b}{b}\omega_3 - \delta_b\psi_{x2} - \delta_b\psi_{x3} \\ \gamma_{s34} = -\dfrac{\delta_a}{a}\omega_4 + \dfrac{\delta_a}{a}\omega_3 + \delta_a\psi_{y4} + \delta_a\psi_{y3} \\ \gamma_{s41} = -\dfrac{\delta_b}{b}\omega_1 + \dfrac{\delta_b}{b}\omega_4 - \delta_b\psi_{y1} - \delta_a\psi_{y4} \end{cases} \quad \textbf{(2.146)}$$

Where:

$$\delta_a = \frac{\left(\dfrac{t}{2a}\right)^2 (1+\mu)}{\dfrac{5}{6} + 2\left(\dfrac{t}{2a}\right)^2 (1+\mu)} = \frac{3t^2(1+\mu)}{10a^2 + 6t^2(1+\mu)} \quad ; \delta_b = \frac{3t^2(1+\mu)}{10b^2 + 6t^2(1+\mu)}$$

t = thickness of the plate, μ = Poisson's ratio, when $t \rightarrow 0$, δ_a, $\delta_b \rightarrow 0$, and $\gamma \rightarrow 0$ and no-shear locking will happen.

The nodal shear strain are therefore

$$\begin{matrix} \gamma_{x1} = \gamma_{s12} & \gamma_{y1} = \gamma_{s41} \\ \gamma_{x2} = \gamma_{s12} & \gamma_{y2} = \gamma_{s23} \\ \gamma_{x3} = \gamma_{s34} & \gamma_{y3} = \gamma_{s23} \\ \gamma_{x4} = \gamma_{s34} & \gamma_{y4} = \gamma_{s41} \end{matrix} \quad \textbf{(2.147)}$$

The shear strain within an element can be expressed as:

$$\gamma = \begin{bmatrix} \gamma_x \\ \gamma_y \end{bmatrix} \tag{2.148}$$

where
$$\begin{cases} \gamma_x = \gamma_{x1}N_1^0 + \gamma_{x2}N_2^0 + \gamma_{x3}N_3^0 + \gamma_{x4}N_4^0 \\ \gamma_y = \gamma_{y1}N_1^0 + \gamma_{y2}N_2^0 + \gamma_{y3}N_3^0 + \gamma_{y4}N_4^0 \end{cases} \tag{2.149}$$

and $N_i^0 (i = 1, 2, 3, 4)$ are bilinear interpolation functions for Q4 element given as

$$N_1^0 = \frac{1}{4}(1-\xi)(1-\eta); \quad N_2^0 = \frac{1}{4}(1+\xi)(1-\eta); \quad N_3^0 = \frac{1}{4}(1+\xi)(1+\eta)$$
$$N_4^0 = \frac{1}{4}(1-\xi)(1+\eta)$$

By substituting eqs.(2-146) and (2-147) into (2-149) and after simplification, γ can be expressed in terms of the nodal displacement a^e such that $\gamma = B_{Sa^e}$, where

$$B_S = \begin{bmatrix} B_{S1} & B_{S2} \end{bmatrix} \tag{2.150}$$

and

$$B_{S1} = \begin{bmatrix} \frac{-(1-\eta)\delta_a}{2a} & \frac{(1-\eta)\delta_a}{2} & 0 & \frac{(1-\eta)\delta_a}{2a} & \frac{(1-\eta)\delta_a}{2} & 0 \\ \frac{-(1-\xi)\delta_b}{2b} & 0 & \frac{(1-\xi)\delta_b}{2} & \frac{-(1+\xi)\delta_b}{2b} & 0 & \frac{(1+\xi)\delta_b}{2} \end{bmatrix}$$

$$B_{S2} = \begin{bmatrix} \frac{(1+\eta)\delta_a}{2a} & \frac{(1+\eta)\delta_a}{2} & 0 & \frac{-(1+\eta)\delta_a}{2a} & \frac{(1+\eta)\delta_a}{2} & 0 \\ \frac{(1+\xi)\delta_b}{2b} & 0 & \frac{(1+\xi)\delta_b}{2} & \frac{(1-\xi)\delta_b}{2b} & 0 & \frac{(1-\xi)\delta_b}{2} \end{bmatrix}$$

The D matrix for shear is given by

$$D_S = G \cdot \frac{5}{6} t \cdot \begin{bmatrix} 1 & 0 \\ 0 & 1 \end{bmatrix} = \frac{E}{2(1+\mu)} \left(\frac{5t}{6} \right) \begin{bmatrix} 1 & 0 \\ 0 & 1 \end{bmatrix} = \frac{5Et}{12(1+\mu)} \begin{bmatrix} 1 & 0 \\ 0 & 1 \end{bmatrix} \tag{2.151}$$

We can obtain the stiffness matrix for shear as:

$$K_s^e = \int_{-1}^{1}\int_{-1}^{1} B_s^T D_s B_S |J| \cdot d\xi d\eta = ab \int_{-1}^{1}\int_{-1}^{1} B_s^T D_s B_S \cdot d\xi d\eta \tag{2.152}$$

2.4.6.2. Bending Stiffness Matrix

For the derivation of pseudo bending stiffness matrix K_b^e as indicated in eq.(2.145), the normal (in the vertical plane perpendicular to the edge) and tangential rotations (in the vertical plane containing the edge) at mid-sides nodes and the centre node 5, 6, 7, 8 and 9 in Fig. (**3**) have to be expressed in terms of the displacements of the corner nodes. Whilst the tangential rotation determined easily as previous, the normal rotation has to be taken as average of that of the adjacent corner nodes. So the followings can be listed as:

$$\psi_{y5} = \frac{1}{2}(\psi_{y1} + \psi_{y2});\ \psi_{x6} = \frac{1}{2}(\psi_{x2} + \psi_{x3});\ \psi_{y7} = \frac{1}{2}(\psi_{y3} + \psi_{y4});\ \psi_{x8} = \frac{1}{2}(\psi_{x1} + \psi_{x4}) \tag{2.153}$$

$$\begin{aligned}
\psi_{x5} &= \frac{3(1-2\delta_a)}{4a}\omega_1 - \frac{3(1-2\delta_a)}{4a}\omega_2 + \left(-\frac{1}{4} + \frac{3}{2}\delta_a\right)\psi_{x1} + \left(-\frac{1}{4} + \frac{3}{2}\delta_a\right)\psi_{x2} \\
\psi_{y6} &= \frac{3(1-2\delta_b)}{4b}\omega_2 - \frac{3(1-2\delta_b)}{4b}\omega_3 + \left(-\frac{1}{4} + \frac{3}{2}\delta_b\right)\psi_{y2} + \left(-\frac{1}{4} + \frac{3}{2}\delta_b\right)\psi_{y3} \\
\psi_{x7} &= \frac{3(1-2\delta_a)}{4a}\omega_4 - \frac{3(1-2\delta_a)}{4a}\omega_3 + \left(-\frac{1}{4} + \frac{3}{2}\delta_a\right)\psi_{x3} + \left(-\frac{1}{4} + \frac{3}{2}\delta_a\right)\psi_{x4} \\
\psi_{y8} &= \frac{3(1-2\delta_b)}{4b}\omega_1 - \frac{3(1-2\delta_b)}{4b}\omega_4 + \left(-\frac{1}{4} + \frac{3}{2}\delta_b\right)\psi_{y1} + \left(-\frac{1}{4} + \frac{3}{2}\delta_b\right)\psi_{y4}
\end{aligned} \tag{2.154}$$

It should be noted that the displacements of node 9 are first expressed in terms of that of nodes 5, 6, 7 and 8 by eq.(2.93) and by substituting (2-154) and using (2-

72), they are finally expressed in terms of the displacements of the center nodes only.

$$\psi_{x9} = \frac{3(1-2\delta_b)}{8b}\omega_1 + \left(-\frac{1}{8}+\frac{3}{4}\delta_b\right)\psi_{x1} - \frac{3a(1-2\delta_b)}{16b}\psi_{y1} - \frac{3(1-2\delta_b)}{8b}\omega_2 + \left(-\frac{1}{8}+\frac{3}{4}\delta_b\right)\psi_{x2}$$

$$+\frac{3a(1-2\delta_b)}{16b}\psi_{y2} - \frac{3(1-2\delta_b)}{8b}\omega_3 + \left(-\frac{1}{8}+\frac{3}{4}\delta_b\right)\psi_{x3} - \frac{3a(1-2\delta_b)}{16b}\psi_{y3} + \frac{3(1-2\delta_b)}{8b}\omega_4 \quad \textbf{(2.155)}$$

$$\left(-\frac{1}{8}+\frac{3}{4}\delta_b\right)\psi_{x4} + \frac{3a(1-2\delta_b)}{16b}\psi_{y4}$$

$$\psi_{y9} = \frac{3(1-2\delta_a)}{8a}\omega_1 - \frac{3b(1-2\delta_a)}{16a}\psi_{x1} + \left(-\frac{1}{8}+\frac{3}{4}\delta_a\right)\psi_{y1} + \frac{3(1-2\delta_a)}{8a}\omega_2 + \frac{3b(1-2\delta_a)}{16a}\psi_{x2}$$

$$+\left(-\frac{1}{8}+\frac{3}{4}\delta_a\right)\psi_{y2} - \frac{3(1-2\delta_a)}{8a}\omega_3 - \frac{3b(1-2\delta_a)}{16a}\psi_{x3} + \left(-\frac{1}{8}+\frac{3}{4}\delta_a\right)\psi_{y3} - \frac{3(1-2\delta_a)}{8a}\omega_4 \quad \textbf{(2.156)}$$

$$+\frac{3b(1-2\delta_a)}{16a}\psi_{x4} + \left(-\frac{1}{8}+\frac{3}{4}\delta_a\right)\psi_{y4}$$

The shape functions of the 9 node element can be listed as:

$$\begin{cases} N_1 = \frac{1}{4}\xi\eta(1-\xi)(1-\eta) & N_5 = -\frac{1}{2}(1-\xi^2)\eta(1-\eta) & N_9 = (1-\xi^2)(1-\eta^2) \\ N_2 = -\frac{1}{4}\xi\eta(1+\xi)(1-\eta) & N_6 = \frac{1}{2}\xi(1+\xi)(1-\eta^2) & \\ N_3 = \frac{1}{4}\xi\eta(1+\xi)(1+\eta) & N_7 = \frac{1}{2}(1-\xi^2)\eta(1+\eta) & \\ N_4 = -\frac{1}{4}\xi\eta(1-\xi)(1+\eta) & N_8 = -\frac{1}{2}\xi(1-\xi)(1-\eta^2) & \end{cases} \quad \textbf{(2.157)}$$

The bending energy for the pseudo 9-node element is determined by the classical element shape function as:

$$\psi_x = \sum_{i=1}^{9} N_i \psi_{xi}; \qquad \psi_y = \sum_{i=1}^{9} N_i \psi_{yi} \tag{2.158}$$

Substituting from eq.(2-153) to (2-156) into (2-158), we may list

$$\begin{aligned}\psi_x =& \frac{3(1-2\delta_a)}{4a}\left(N_5+\frac{N_9}{2}\right)\omega_1 + \left[N_1+\frac{N_8}{2}+\left(N_5+\frac{N_9}{2}\right)\left(-\frac{1}{4}+\frac{3}{2}\delta_a\right)\right]\psi_{x1} - \frac{3b(1-2\delta_a)}{16a}N_9\psi_{y1}\\ &-\frac{3(1-2\delta_a)}{4a}\left(N_5+\frac{N_9}{2}\right)\omega_2 + \left[N_2+\frac{N_6}{2}+\left(N_5+\frac{N_9}{2}\right)\left(-\frac{1}{4}+\frac{3}{2}\delta_a\right)\right]\psi_{x2} + \frac{3b(1-2\delta_a)}{16a}N_9\psi_{y2}\\ &-\frac{3(1-2\delta_a)}{4a}\left(N_7+\frac{N_9}{2}\right)\omega_3 + \left[N_3+\frac{N_6}{2}+\left(N_7+\frac{N_9}{2}\right)\left(-\frac{1}{4}+\frac{3}{2}\delta_a\right)\right]\psi_{x3} - \frac{3b(1-2\delta_a)}{16a}N_9\psi_{y3}\\ &+\frac{3(1-2\delta_a)}{4a}\left(N_7+\frac{N_9}{2}\right)\omega_4 + \left[N_4+\frac{N_8}{2}+\left(N_7+\frac{N_9}{2}\right)\left(-\frac{1}{4}+\frac{3}{2}\delta_a\right)\right]\psi_{x4} + \frac{3b(1-2\delta_a)}{16a}N_9\psi_{y4}\end{aligned} \tag{2.159}$$

$$\begin{aligned}\psi_y =& \left(N_8+\frac{N_9}{2}\right)\frac{3(1-2\delta_b)}{4b}\omega_1 - \frac{3a(1-2\delta_b)}{16b}N_9\psi_{x1} + \left[N_1+\frac{N_5}{2}+\left(N_8+\frac{N_9}{2}\right)\left(-\frac{1}{4}+\frac{3}{2}\delta_b\right)\right]\psi_{y1}\\ &+\left(N_6+\frac{N_9}{2}\right)\frac{3(1-2\delta_b)}{4b}\omega_2 + \frac{3a(1-2\delta_b)}{16b}N_9\psi_{x2} + \left[N_2+\frac{N_5}{2}+\left(N_6+\frac{N_9}{2}\right)\left(-\frac{1}{4}+\frac{3}{2}\delta_b\right)\right]\psi_{y2}\\ &-\left(N_6+\frac{N_9}{2}\right)\frac{3(1-2\delta_b)}{4b}\omega_3 - \frac{3a(1-2\delta_b)}{16b}N_9\psi_{x3} + \left[N_3+\frac{N_7}{2}+\left(N_6+\frac{N_9}{2}\right)\left(-\frac{1}{4}+\frac{3}{2}\delta_b\right)\right]\psi_{y3}\\ &-\left(N_8+\frac{N_9}{2}\right)\frac{3(1-2\delta_b)}{4b}\omega_4 + \frac{3a(1-2\delta_b)}{16b}N_9\psi_{x4} + \left[N_4+\frac{N_7}{2}+\left(N_8+\frac{N_9}{2}\right)\left(-\frac{1}{4}+\frac{3}{2}\delta_b\right)\right]\psi_{y4}\end{aligned} \tag{2.160}$$

The curvature of the element is given by:

$$\kappa = \begin{Bmatrix}\kappa_x\\ \kappa_y\\ \kappa_{xy}\end{Bmatrix} = \begin{Bmatrix}-\dfrac{\partial \psi_x}{\partial x}\\ -\dfrac{\partial \psi_y}{\partial y}\\ -\dfrac{\partial \psi_y}{\partial x}-\dfrac{\partial \psi_x}{\partial y}\end{Bmatrix} \tag{2.161}$$

Using $\dfrac{\partial \psi_x}{\partial y} = \dfrac{\partial \psi_x}{\partial \eta} \cdot \dfrac{\partial \eta}{\partial y} = \dfrac{\partial \psi_x}{\partial \eta} \cdot \dfrac{1}{b}$, $\dfrac{\partial \psi_y}{\partial x} = \dfrac{\partial \psi_y}{\partial \xi} \cdot \dfrac{\partial \xi}{\partial x} = \dfrac{\partial \psi_y}{\partial \xi} \cdot \dfrac{1}{a}$ *etc.*

and by appropriately differentiating eq.(2-159) and (2-160), the B matrix for bending as equated by $\kappa = B_B a^e$ can be presented as $B_B = \begin{bmatrix} B_{B1} & B_{B2} & B_{B3} & B_{B4} \end{bmatrix}$ where

$$
B_{B1} = \begin{bmatrix}
\dfrac{3(1-2\delta_b)(1-\xi)\eta}{4b^2} & \dfrac{(1-\xi)(-1+3\eta-6\delta_b\eta)}{4b} & \dfrac{3a(1-2\delta_b)(1-\xi^2)\eta}{8b^2} \\
-\dfrac{3(1-2\delta_a)\xi(1-\eta)}{4a^2} & \dfrac{3b(1-2\delta_a)\xi(1-\eta^2)}{8a^2} & \dfrac{(1-\eta)(-1+3\xi-6\delta_a\xi)}{4a} \\
\dfrac{-3\left[(1-2\delta_a)(1-\xi^2)+(1-2\delta_b)(1-\eta^2)\right]}{8ab} & \dfrac{3(1-2\delta_a)(1-\xi^2)\eta+(1-\eta)\left[-1-3\eta+6\delta_b(1+\eta)\right]}{8a} & \dfrac{(1-\xi)\left[1+3\xi-6\delta_a(1+\xi)\right]-3(1-2\delta_b)\xi(1-\eta^2)}{8b}
\end{bmatrix}
$$

$$
B_{B2} = \begin{bmatrix}
\dfrac{3(1-2\delta_b)(1+\xi)\eta}{4b^2} & \dfrac{(1+\xi)(-1+3\eta-6\delta_b\eta)}{4b} & -\dfrac{3a(1-2\delta_b)(1-\xi^2)\eta}{8b^2} \\
\dfrac{3(1-2\delta_a)\xi(1-\eta)}{4a^2} & -\dfrac{3b(1-2\delta_a)\xi(1-\eta^2)}{8a^2} & \dfrac{(1-\eta)(1+3\xi-6\delta_a\xi)}{4a} \\
\dfrac{3\left[(1-2\delta_a)(1-\xi^2)+(1-2\delta_b)(1-\eta^2)\right]}{8ab} & \dfrac{-3(1-2\delta_a)(1-\xi^2)\eta+(1-\eta)\left[1+3\eta-6\delta_b(1+\eta)\right]}{8a} & \dfrac{(1+\xi)\left[1-3\xi-6\delta_a(1-\xi)\right]+3(1-2\delta_b)\xi(1-\eta^2)}{8b}
\end{bmatrix}
$$

$$
B_{B3} = \begin{bmatrix}
-\dfrac{3(1-2\delta_b)(1+\xi)\eta}{4b^2} & \dfrac{(1+\xi)(1+3\eta-6\eta\delta_b)}{4b} & \dfrac{3a(1-2\delta_b)(1-\xi^2)\eta}{8b^2} \\
\dfrac{3(1-2\delta_a)\xi(1+\eta)}{4a^2} & \dfrac{3b(1-2\delta_a)\xi(1-\eta^2)}{8a^2} & \dfrac{(1+\eta)(1+3\xi-6\delta_a\xi)}{4a} \\
\dfrac{-3\left[(1-2\delta_a)(1-\xi^2)+(1-2\delta_b)(1-\eta^2)\right]}{8ab} & \dfrac{3(1-2\delta_a)(1-\xi^2)\eta+(1+\eta)\left[1-3\eta-6\delta_b(1-\eta)\right]}{8a} & \dfrac{(1+\xi)\left[-1+3\xi+6\delta_a(1-\xi)\right]-3(1-2\delta_b)\xi(1-\eta^2)}{8b}
\end{bmatrix} \tag{2.162}
$$

$$
B_{B4} = \begin{bmatrix}
-\dfrac{3(1-2\delta_b)(1-\xi)\eta}{4b^2} & \dfrac{(1-\xi)(1+3\eta-6\eta\delta_b)}{4b} & -\dfrac{3a(1-2\delta_b)(1-\xi^2)\eta}{8b^2} \\
-\dfrac{3(1-2\delta_a)\xi(1+\eta)}{4a^2} & -\dfrac{3b(1-2\delta_a)\xi(1-\eta^2)}{8a^2} & \dfrac{(1+\eta)(-1+3\xi-6\delta_a\xi)}{4a} \\
\dfrac{3\left[(1-2\delta_a)(1-\xi^2)+(1-2\delta_b)(1-\eta^2)\right]}{8ab} & \dfrac{-3(1-2\delta_a)(1-\xi^2)\eta+(1+\eta)\left[-1+3\eta+6\delta_b(1-\eta)\right]}{8a} & \dfrac{(1-\xi)\left[-1-3\xi+6\delta_a(1+\xi)\right]+3(1-2\delta_b)\xi(1-\eta^2)}{8b}
\end{bmatrix}
$$

$$
\begin{bmatrix} M_x \\ M_y \\ M_{xy} \end{bmatrix} = D_b \kappa \quad \text{and} \quad D_b = \frac{Et^3}{12(1-\mu^2)} \begin{bmatrix} 1 & -\mu & 0 \\ -\mu & 1 & 0 \\ 0 & 0 & \dfrac{1-\mu}{2} \end{bmatrix} \tag{2.163}
$$

So the stiffness matrix for bending is

$$K_b^e = \int_{-b}^{b}\int_{-a}^{a} B_B{}^T D_b B_B dxdy = \int_{-1}^{1}\int_{-1}^{1} B_B{}^T D_b B_B |J| d\xi d\eta = ab\int_{-1}^{1}\int_{-1}^{1} B_B{}^T D_b B_B d\xi d\eta \quad \textbf{(2.164)}$$

To calculate the equivalent load vector, we should first determine the out-of-plane displacement ω. So a simple form of $[F]$ may be as follows:

$$[F] = \int [N_i] \cdot q dA \quad \textbf{(2.165)}$$

$$[N_i] = [N_1^0 \ \ 0 \ \ 0 \ \ N_2^0 \ \ 0 \ \ 0 \ \ N_3^0 \ \ 0 \ \ 0 \ \ N_4^0 \ \ 0 \ \ 0] \quad \textbf{(2.166)}$$

where N_i^0 $(i = 1, 2, 3, 4)$ are the bilinear interpolation functions for Q4 element.

The rectangular element formulation as shown above is hence a pseudo 9 node formulation which will give a very high accuracy but still retain the simplicity of a Q4 element which is good enough for practical use. The total stiffness of the plate element will be obtained by Eq.(2-152) and (2-154). The present pseudo 9 node element behaves as a 4 node element to the engineers externally and the problems of classical higher order elements are avoided, but internally it retains nearly all the accuracy of a 9 node element. It is also well known that 9 node element can perform much better than 8 node element when the element is highly distorted (see Zienkiewicz *et al.* 2011), so the present element possesses addition advantage over the 8 node element for problems with highly distorted mesh which is unavoidable for many practical projects, and a project in Hong Kong will be used for illustration of this important feature of a good thick plate element.

2.4.7. Extension of "PLATE-Q9" to General Quadrilateral Element

The formulation rectangular thick plate element "PLATE-Q9" can be extended to a general element as shown in Figs. (**2.18** and **2.19**).

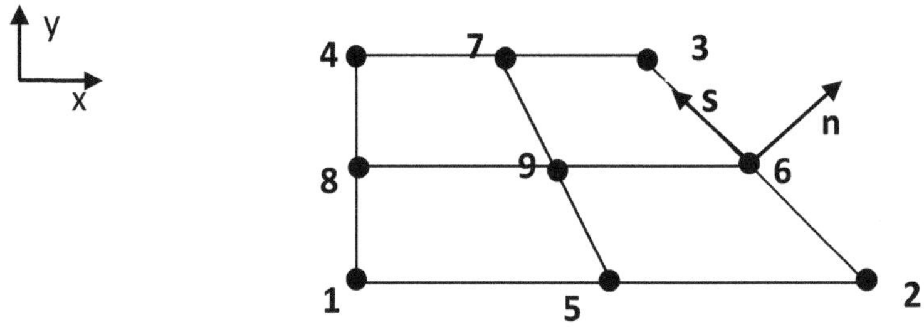

Fig. (2.18). The trapezoidal element.

From eq.(2-146), the strains at the edges of a general trapezoidal element are:

$$\begin{cases} \gamma_{s12} = -\dfrac{2\cdot\delta_{12}}{l_{12}}\omega_1 + \dfrac{2\cdot\delta_{12}}{l_{12}}\omega_2 + \delta_{12}\psi_{y1} + \delta_{12}\psi_{y2} \\ \gamma_{s23} = -\dfrac{2\cdot\delta_{23}}{l_{23}}\omega_2 + \dfrac{2\cdot\delta_{23}}{l_{23}}\omega_3 - \dfrac{\delta_{23}\cdot\Delta y_{32}}{l_{23}}\psi_{x2} + \dfrac{\delta_{23}\cdot\Delta x_{32}}{l_{23}}\psi_{y2} - \dfrac{\delta_{23}\cdot\Delta y_{32}}{l_{23}}\psi_{x3} + \dfrac{\delta_{23}\cdot\Delta x_{32}}{l_{23}}\psi_{y3} \\ \gamma_{s34} = -\dfrac{2\cdot\delta_{34}}{l_{34}}\omega_3 + \dfrac{2\cdot\delta_{34}}{l_{34}}\omega_4 + \delta_{34}\psi_{y4} + \delta_{34}\psi_{y3} \\ \gamma_{s41} = -\dfrac{2\cdot\delta_{41}}{l_{41}}\omega_4 + \dfrac{2\cdot\delta_{41}}{l_{41}}\omega_1 - \delta_{41}\psi_{y1} - \delta_{41}\psi_{y4} \end{cases} \quad (2.167)$$

Where : $\Delta x_{32} = x_3 - x_2 \quad \Delta y_{32} = y_3 - y_2$

$$\delta_{ij} = \frac{\left(\dfrac{t}{l_{ij}}\right)^2 (1+v)}{\dfrac{5}{6} + 2\left(\dfrac{t}{l_{ij}}\right)^2 (1+v)} \quad (i, j = 1,2,3,4) \quad (2.168)$$

where t = thickness of the plate, μ = Poisson's ratio, When t → 0, δ_a , δ_b → 0, and γ_{si} → 0, the element can hence avoid shear locking. For convenience, a term γ_{sij}^* is defined and given $\gamma_{sij}^* = l_{ij}\gamma_{sij} \quad (i = 1, 2, 3, 4)$.

$$\gamma_s^* = \begin{bmatrix} \gamma_{s12}^* & \gamma_{s23}^* & \gamma_{s34}^* & \gamma_{s41}^* \end{bmatrix}^T \tag{2.169}$$

then
$$\gamma_s^* = T_s a^e \tag{2.170}$$

$$T_s = \begin{bmatrix} -2\delta_{12} & 0 & \delta_{12}l_{12} & 2\delta_{12} & 0 & \delta_{12}l_{12} & 0 & 0 & 0 & 0 & 0 & 0 \\ 0 & 0 & 0 & -2\delta_{23} & -\delta_{23}\Delta y_{32} & \delta_{23}\Delta x_{32} & 2\delta_{23} & -\delta_{23}\Delta y_{32} & \delta_{23}\Delta x_{32} & 0 & 0 & 0 \\ 0 & 0 & 0 & 0 & 0 & 0 & -2\delta_{34} & 0 & \delta_{34}l_{34} & 2\delta_{34} & 0 & \delta_{34}l_{34} \\ 2\delta_{41} & -\delta_{41}l_{41} & 0 & 0 & 0 & 0 & 0 & 0 & 0 & -2\delta_{41} & -\delta_{41}l_{41} & 0 \end{bmatrix} \tag{2.171}$$

As the axes are not along the main directions, transformation of strain from local to global axes will be required, and the nodal shear strain will become:

$$\begin{aligned} \begin{Bmatrix} \gamma_{xi} \\ \gamma_{yi} \end{Bmatrix} &= \frac{1}{\Delta y_{ik}\Delta x_{ji} - \Delta y_{ji}\Delta x_{ik}} \begin{bmatrix} \Delta y_{ik} & -\Delta y_{ji} \\ -\Delta x_{ik} & \Delta x_{ji} \end{bmatrix} \begin{Bmatrix} \gamma^*_{sij} \\ \gamma^*_{ski} \end{Bmatrix} \\ &= \begin{bmatrix} T_{sx} \\ T_{sy} \end{bmatrix} \begin{Bmatrix} \gamma^*_{sij} \\ \gamma^*_{ski} \end{Bmatrix} \end{aligned} \tag{2.172}$$

where $j = i + 1$ if $j < 5$ and $=1$ if $j = 5$; $k = i - 1$ if $k > 1$ else $k = 4$.

The nodal shear strain can be expressed in terms of the nodal displacement as

$$\left. \begin{aligned} \gamma_{xi} &= T_{sx}\gamma_s^* = T_{sx}T_s a^e \\ \gamma_{yi} &= T_{sy}\gamma_s^* = T_{sy}T_s a^e \end{aligned} \right\} \tag{2.173}$$

Shear strain within element is given by:

$$\gamma_x = \Sigma\gamma_{xi}N_i^0 \qquad \gamma_y = \Sigma\gamma_{yi}N_i^0 \tag{2.174}$$

And N_i^0 (i = 1, 2, 3, 4) are the classical bilinear interpolation functions for Q4 quadrilateral element. Eqs. (2-146) and (2-147) will become:

$$\gamma = \begin{Bmatrix} \gamma_x \\ \gamma_y \end{Bmatrix} = \begin{bmatrix} N_s T_{sx} T_s \\ N_s T_{sy} T_s \end{bmatrix} a^e = B_s a^e \qquad \textbf{(2.175)}$$

The normal rotation ψ_n and tangential rotation ψ_s for edge $\overline{23}$ can be expressed as:

$$\psi_{n6} = \frac{1}{2}\left[(\psi_{n2})_{\overline{23}} + (\psi_{n3})_{\overline{23}}\right] = \frac{1}{2l_{23}}\left[\Delta x_{32}(\psi_{x2} + \psi_{x3}) + \Delta y_{32}(\psi_{y2} + \psi_{y3})\right] \qquad \textbf{(2.176)}$$

$$\psi_{s6} = \frac{3}{2l_{23}}(1-2\delta_{12})(\omega_3 - \omega_2) - \frac{1}{4l_{23}}(1-6\delta_{12})\left[\Delta y_{32}(\psi_{x2} + \psi_{x3}) - \Delta x_{32}(\psi_{y2} + \psi_{y3})\right] \qquad \textbf{(2.177)}$$

The rotations at mid-side nodes are given by:

$$\begin{Bmatrix} \psi_{x6} \\ \psi_{y6} \end{Bmatrix} = \frac{1}{l_{12}} \begin{bmatrix} \Delta x_{32} & \Delta y_{32} \\ \Delta y_{32} & -\Delta x_{32} \end{bmatrix} \begin{Bmatrix} \psi_{n6} \\ \psi_{s6} \end{Bmatrix} \qquad \textbf{(2.178)}$$

$$\begin{aligned} \psi_{x6} &= \frac{1}{l_{12}}(\Delta x_{32}\psi_{n6} + \Delta y_{32}\psi_{s6}) \\ &= -\frac{3\Delta y_{32}}{2l_{23}^2}(1-2\delta_{23})\omega_2 + \frac{1}{2l_{23}^2}\left[\Delta x_{32}^2 - \frac{\Delta y_{32}^2}{2}(1-6\delta_{23})\right]\psi_{x2} + \frac{\Delta x_{32}\Delta y_{32}}{2l_{23}^2}\left[1+\frac{1}{2}(1-6\delta_{23})\right]\psi_{y2} \\ &+ \frac{3\Delta y_{32}}{2l_{23}^2}(1-2\delta_{23})\omega_3 + \frac{1}{2l_{23}^2}\left[\Delta x_{32}^2 - \frac{\Delta y_{32}^2}{2}(1-6\delta_{23})\right]\psi_{x3} + \frac{\Delta x_{32}\Delta y_{32}}{2l_{23}^2}\left[1+\frac{1}{2}(1-6\delta_{23})\right]\psi_{y3} \end{aligned} \qquad \textbf{(2.179)}$$

$$
\begin{aligned}
\psi_{y6} &= \frac{1}{l_{23}}\left(\Delta y_{32}\psi_{n6} - \Delta x_{32}\psi_{s6}\right) \\
&= \frac{3\Delta x_{32}}{2l_{23}^{2}}\left(1-2\delta_{23}\right)\omega_2 + \frac{\Delta x_{32}\Delta y_{32}}{2l_{23}^{2}}\left[1+\frac{1}{2}\left(1-6\delta_{23}\right)\right]\psi_{x2} + \frac{1}{2l_{23}^{2}}\left[\Delta y_{32}^{2} - \frac{\Delta x_{32}^{2}}{2}\left(1-6\delta_{23}\right)\right]\psi_{y2} \\
&\quad - \frac{3\Delta x_{32}}{2l_{23}^{2}}\left(1-2\delta_{23}\right)\omega_3 + \frac{\Delta x_{32}\Delta y_{32}}{2l_{23}^{2}}\left[1+\frac{1}{2}\left(1-6\delta_{23}\right)\right]\psi_{x3} + \frac{1}{2l_{23}^{2}}\left[\Delta y_{32}^{2} - \frac{\Delta x_{32}^{2}}{2}\left(1-6\delta_{23}\right)\right]\psi_{y3}
\end{aligned} \tag{2.180}
$$

The relations at other nodes at the right angle edges can be written as follows:

$$
\begin{cases}
\psi_{x5} = \dfrac{1}{2}\left[\psi_{x1} + \psi_{x2}\right] \\
\psi_{x7} = \dfrac{1}{2}\left[\psi_{x3} + \psi_{x4}\right] \\
\psi_{y8} = \dfrac{1}{2}\left[\psi_{y1} + \psi_{y4}\right]
\end{cases} \tag{2.181}
$$

$$
\begin{cases}
\psi_{y5} = \dfrac{3\left(1-2\delta_{12}\right)}{2l_{12}}\omega_1 - \dfrac{3\left(1-2\delta_{12}\right)}{2l_{12}}\omega_2 + \left(-\dfrac{1}{4}+\dfrac{3}{2}\delta_{12}\right)\psi_{y1} + \left(-\dfrac{1}{4}+\dfrac{3}{2}\delta_{12}\right)\psi_{y2} \\
\psi_{y7} = \dfrac{3\left(1-2\delta_{34}\right)}{2l_{34}}\omega_4 - \dfrac{3\left(1-2\delta_{34}\right)}{2l_{34}}\omega_3 + \left(-\dfrac{1}{4}+\dfrac{3}{2}\delta_{34}\right)\psi_{y3} + \left(-\dfrac{1}{4}+\dfrac{3}{2}\delta_{34}\right)\psi_{y4} \\
\psi_{x8} = -\dfrac{3\left(1-2\delta_{41}\right)}{2l_{41}}\omega_1 + \dfrac{3\left(1-2\delta_{41}\right)}{2l_{41}}\omega_4 + \left(-\dfrac{1}{4}+\dfrac{3}{2}\delta_{41}\right)\psi_{x1} + \left(-\dfrac{1}{4}+\dfrac{3}{2}\delta_{41}\right)\psi_{x4}
\end{cases} \tag{2.182}
$$

$$
\begin{cases}
\psi_{x9} = \dfrac{3\left(1-2\delta_{57}\right)}{2l_{57}}\omega_5 - \dfrac{3\left(1-2\delta_{57}\right)}{2l_{57}}\omega_7 + \left(-\dfrac{1}{4}+\dfrac{3}{2}\delta_{57}\right)\psi_{x5} + \left(-\dfrac{1}{4}+\dfrac{3}{2}\delta_{57}\right)\psi_{x7} \\
\psi_{y9} = \dfrac{3\left(1-2\delta_{68}\right)}{2l_{68}}\omega_8 - \dfrac{3\left(1-2\delta_{68}\right)}{2l_{68}}\omega_6 + \left(-\dfrac{1}{4}+\dfrac{3}{2}\delta_{68}\right)\psi_{y8} + \left(-\dfrac{1}{4}+\dfrac{3}{2}\delta_{68}\right)\psi_{y6}
\end{cases} \tag{2.183}
$$

Where: $$l_{57}=\sqrt{\frac{(x_3+x_4-x_1-x_2)^2}{2}+\frac{(y_3+y_4-y_1-y_2)^2}{2}},\; l_{68}=\frac{l_{12}+l_{34}}{2} \quad \textbf{(2.184)}$$

$$\delta_{57}=\frac{\left(\frac{t}{l_{57}}\right)^2(1+v)}{\frac{5}{6}+2\left(\frac{t}{l_{57}}\right)^2(1+v)},\; \delta_{68}=\frac{\left(\frac{t}{l_{68}}\right)^2(1+v)}{\frac{5}{6}+2\left(\frac{t}{l_{68}}\right)^2(1+v)} \quad \textbf{(2.185)}$$

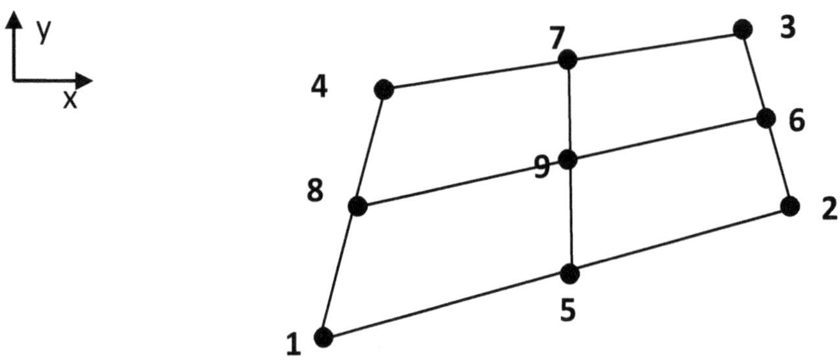

Fig. (2.19). The general qquadrilateral element.

For plate analysis, the use of rectangular element will give the best result in general. Due to the geometric requirement, the use of rectangular element is not always possible for practical problem. A general domain can however be always decomposed to the combination of rectangular element, trapezoidal element and general quadrilateral element. Actually, some commercial programs generate a mesh composed of rectangular element and triangular element, but due to the difference in the order of the interpolation function, shear jump may be generated across the boundary between rectangular element and triangular element which has been discussed by Cheng and Law (2008b). The combined use of rectangular/trapezoidal/quadrilateral elements will be adequate for problems with arbitrary geometry and can provide high accuracy for the engineers.

For a general quadrilateral element as shown in Fig. (**2.19**), points 5,6,7,8 are the mid-points of the four sides. Line 78 is parallel to line 13, similarly, line 56 is parallel to line 13, hence line 78 is parallel to line 56. Similarly, line 76 is parallel to line 58, hence 5678 is a parallelogram. Point 9 is hence the mid-point between line 75 and 68. The strains at the edges of a general quadrilateral element are:

$$=-\frac{3\Delta y_{21}}{2l_{12}^{2}}(1-2\delta_{12})\omega_1+\frac{1}{2l_{12}^{2}}\left[\Delta x_{21}^{2}-\frac{\Delta y_{21}^{2}}{2}(1-6\delta_{12})\right]\psi_{x1}+\frac{\Delta x_{21}\Delta y_{21}}{2l_{12}^{2}}\left[1+\frac{1}{2}(1-6\delta_{12})\right]\psi_{y1} \qquad \textbf{(2.186)}$$

$$\text{Where:}\Delta x_{ji}=x_j-x_i \qquad \Delta y_{ji}=y_j-y_i \qquad \textbf{(2.187)}$$

in eq.(2-186) is the same as that in eq. (2-168). Similar to the derivation of the right trapezoid element, we can arrive at:

$$T_s=\begin{bmatrix} -2\delta_{12} & -\delta_{12}\Delta y_{21} & \delta_{12}\Delta x_{21} & 2\delta_{12} & -\delta_{12}\Delta y_{21} & \delta_{12}\Delta x_{21} & 0 & 0 & 0 & 0 & 0 & 0 \\ 0 & 0 & 0 & -2\delta_{23} & -\delta_{23}\Delta y_{32} & \delta_{23}\Delta x_{32} & 2\delta_{23} & -\delta_{23}\Delta y_{32} & \delta_{23}\Delta x_{32} & 0 & 0 & 0 \\ 0 & 0 & 0 & 0 & 0 & 0 & -2\delta_{34} & -\delta_{34}\Delta y_{43} & \delta_{34}\Delta x_{43} & 2\delta_{34} & -\delta_{34}\Delta y_{43} & \delta_{34}\Delta x_{43} \\ 2\delta_{41} & -\delta_{41}\Delta y_{14} & \delta_{41}\Delta x_{14} & 0 & 0 & 0 & 0 & 0 & 0 & -2\delta_{41} & -\delta_{41}\Delta y_{14} & \delta_{41}\Delta x_{14} \end{bmatrix} \qquad \textbf{(2.188)}$$

According to eqs. (2-186) and (2-187),the normal rotation ψ_n and tangential rotation ψ_s for edge of $\overline{12}$ can be expressed as

$$\psi_{n5}=\frac{1}{2}\left[(\psi_{n1})_{\overline{12}}+(\psi_{n2})_{\overline{12}}\right]=\frac{1}{2l_{12}}\left[\Delta x_{21}(\psi_{x1}+\psi_{x2})+\Delta y_{21}(\psi_{y1}+\psi_{y2})\right] \qquad \textbf{(2.189)}$$

$$\psi_{s5}=\frac{3}{2l_{12}}(1-2\delta_{12})(\omega_2-\omega_1)-\frac{1}{4l_{12}}(1-6\delta_{12})\left[\Delta y_{21}(\psi_{x1}+\psi_{x2})-\Delta x_{21}(\psi_{y1}+\psi_{y2})\right] \qquad \textbf{(2.190)}$$

For rotation at mid-side nodes, inverting Eq. (2-189) and (2-190) for node 5 gives:

$$\begin{Bmatrix}\psi_{x5}\\ \psi_{y5}\end{Bmatrix}=\frac{1}{l_{12}}\begin{bmatrix}\Delta x_{21} & \Delta y_{21}\\ \Delta y_{21} & -\Delta x_{21}\end{bmatrix}\begin{Bmatrix}\psi_{n5}\\ \psi_{s5}\end{Bmatrix} \qquad \textbf{(2.191)}$$

$$
\begin{aligned}
\psi_{x5} &= \frac{1}{l_{12}}\left(\Delta x_{21}\psi_{n5} + \Delta y_{21}\psi_{s5}\right) \\
&= -\frac{3\Delta y_{21}}{2l_{12}^2}\left(1-2\delta_{12}\right)\omega_1 + \frac{1}{2l_{12}^2}\left[\Delta x_{21}^2 - \frac{\Delta y_{21}^2}{2}\left(1-6\delta_{12}\right)\right]\psi_{x1} + \frac{\Delta x_{21}\Delta y_{21}}{2l_{12}^2}\left[1+\frac{1}{2}\left(1-6\delta_{12}\right)\right]\psi_{y1} \\
&\quad + \frac{3\Delta y_{21}}{2l_{12}^2}\left(1-2\delta_{12}\right)\omega_2 + \frac{1}{2l_{12}^2}\left[\Delta x_{21}^2 - \frac{\Delta y_{21}^2}{2}\left(1-6\delta_{12}\right)\right]\psi_{x2} + \frac{\Delta x_{21}\Delta y_{21}}{2l_{12}^2}\left[1+\frac{1}{2}\left(1-6\delta_{12}\right)\right]\psi_{y2}
\end{aligned}
\tag{2.192}
$$

$$
\begin{aligned}
\psi_{y5} &= \frac{1}{l_{12}}\left(\Delta y_{21}\psi_{n5} - \Delta x_{21}\psi_{s5}\right) \\
&= \frac{3\Delta x_{21}}{2l_{12}^2}\left(1-2\delta_{12}\right)\omega_1 + \frac{\Delta x_{21}\Delta y_{21}}{2l_{12}^2}\left[1+\frac{1}{2}\left(1-6\delta_{12}\right)\right]\psi_{x1} + \frac{1}{2l_{12}^2}\left[\Delta y_{21}^2 - \frac{\Delta x_{21}^2}{2}\left(1-6\delta_{12}\right)\right]\psi_{y1} \\
&\quad - \frac{3\Delta x_{21}}{2l_{12}^2}\left(1-2\delta_{12}\right)\omega_2 + \frac{\Delta x_{21}\Delta y_{21}}{2l_{12}^2}\left[1+\frac{1}{2}\left(1-6\delta_{12}\right)\right]\psi_{x2} + \frac{1}{2l_{12}^2}\left[\Delta y_{21}^2 - \frac{\Delta x_{21}^2}{2}\left(1-6\delta_{12}\right)\right]\psi_{y2}
\end{aligned}
\tag{2.193}
$$

Similar results can be formulated for the other mid-side nodes, which will not be repeated here. Finally, we obtain the rotation of the center node 9 as follows:

$$
\begin{aligned}
\psi_{x9} &= -\frac{3\Delta y_{75}}{2l_{57}^2}\left(1-2\delta_{57}\right)\omega_5 + \frac{1}{2l_{57}^2}\left[\Delta x_{75}^2 - \frac{\Delta y_{75}^2}{2}\left(1-6\delta_{57}\right)\right]\psi_{x5} + \frac{\Delta x_{75}\Delta y_{75}}{2l_{57}^2}\left[1+\frac{1}{2}\left(1-6\delta_{57}\right)\right]\psi_{y5} \\
&\quad + \frac{3\Delta y_{75}}{2l_{57}^2}\left(1-2\delta_{57}\right)\omega_7 + \frac{1}{2l_{57}^2}\left[\Delta x_{75}^2 - \frac{\Delta y_{75}^2}{2}\left(1-6\delta_{57}\right)\right]\psi_{x7} + \frac{\Delta x_{75}\Delta y_{75}}{2l_{57}^2}\left[1+\frac{1}{2}\left(1-6\delta_{57}\right)\right]\psi_{y7} \\
\psi_{y9} &= \frac{3\Delta x_{86}}{2l_{68}^2}\left(1-2\delta_{68}\right)\omega_6 + \frac{\Delta x_{86}\Delta y_{86}}{2l_{68}^2}\left[1+\frac{1}{2}\left(1-6\delta_{68}\right)\right]\psi_{x6} + \frac{1}{2l_{68}^2}\left[\Delta y_{86}^2 - \frac{\Delta x_{86}^2}{2}\left(1-6\delta_{68}\right)\right]\psi_{y6} \\
&\quad - \frac{3\Delta x_{86}}{2l_{68}^2}\left(1-2\delta_{68}\right)\omega_8 + \frac{\Delta x_{86}\Delta y_{86}}{2l_{68}^2}\left[1+\frac{1}{2}\left(1-6\delta_{68}\right)\right]\psi_{x8} + \frac{1}{2l_{68}^2}\left[\Delta y_{86}^2 - \frac{\Delta x_{86}^2}{2}\left(1-6\delta_{68}\right)\right]\psi_{y8}
\end{aligned}
\tag{2.194}
$$

Where:

$$
l_{57} = \sqrt{\frac{\left(x_3+x_4-x_1-x_2\right)^2}{2} + \frac{\left(y_3+y_4-y_1-y_2\right)^2}{2}}, \quad l_{68} = \sqrt{\frac{\left(x_2+x_3-x_4-x_1\right)^2}{2} + \frac{\left(y_2+y_3-y_4-y_1\right)^2}{2}}
\tag{2.195}
$$

Where δ_{ij} is the same as that in eq. (2.185).

The authors have developed a FORTRAN program using this general quadrilateral element, and the readers can obtain a demonstration copy of the program from the authors for testing and research if required. Based on various studies, it is found that this element possesses various advantages over the classical thick plate element, particularly when the number of element is small. Unlike the classical Q4 elements where the internal forces follow a bilinear distribution within an element, the present Q9 plate element possess higher order distribution of internal forces within an element but possess the advantage of being q Q4 element to the users. Interested readers can also follow the stiffness matrix subroutine in Appendix 2-4 to develop the stiffness matrix of this advanced Q9 element.

2.4.8. Performance of the "PLATE-Q9" Demonstrated by Numerical Examples

Cheng and Law (2008) have carried out a detailed numerical study about the performance of the PLATE-Q8 elements for a series of standard problems, and have also carried out comparisons with many other thick plate elements under different conditions. It is found that PLATE-Q8 possesses very high accuracy and is ranked the best or near to the best in most of the comparisons under different plate thicknesses and mesh densities. The authors have also carried out similar studies for the PLATE-Q9 element, and the accuracy of the PLATE-Q9 is even better than the PLATE-Q8 element in all cases. The improvements over these normal and simple cases with the use of rectangular elements are not important and will not be shown here. The most important advantage of the PLATE-Q9 element is that it can maintain a good accuracy for a highly irregular mesh or for a coarse mesh with rapid changes in the internal forces which is useful for highly irregular problems. In this section, the special advantages of the PLATE-Q9 element will be demonstrated.

For illustration, a cantilever plate element of width 1.8m, length 1.5m, thickness 0.2m formulated in accordance with (28) to (30) is fixed at two adjacent nodes (Nodes 1 and 2) as shown in Fig. (**2.20**). Two point loads of 150kN are acting at the two free corner nodes 3 and 4 at case (1) and additional moments are added in case (2).

This plate is analyzed by the PLATE-Q8 element and PLATE-Q9 Element as well as the DKT element in Ansys, and only one element is used for this problem. As shown in Figs. (**2.21** and **2.22**), the moment contours for the moment in the direction of the cantilever span of the PLATE-Q9 Element are curved near to the corners. Due to the limitations of the basic formulation of the PLATE-Q8 Element, only linear variation of stress can be demonstrated whilst that of the PLATE-Q9 can demonstrate non-linear variations even when one element is used. The differences in the results are more pronounced in case (2) where there are end moments being applied at the free corners. For the DKT elements, unless 64 elements are used, the variation of the internal forces inside this simple plate cannot be reflected.

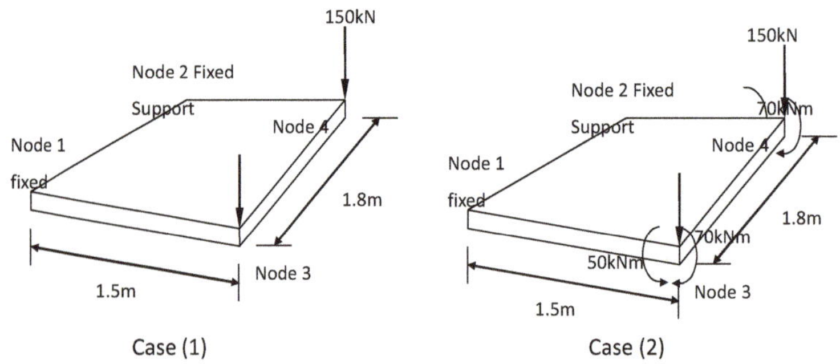

Fig. (2.20). Example for demonstration of the good performance of the PLATE Q9 element as compared with PLATE Q8 node element.

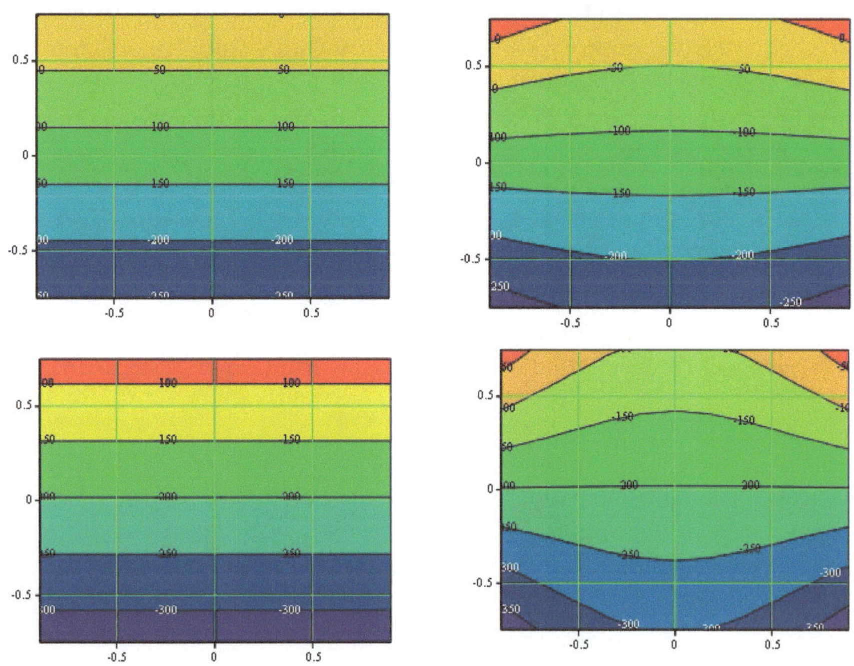

Fig. (2.21a). (a) – Moment along Span of Y–axis of Cantilever for Case (1) of Fig. (**2.20**) – PLATE Q8 on the Left and PLATE Q9 on the Right.**Fig. (2.21b).** (b) – Moment along Span of Y-axis of Cantilever for Case

(2) of Fig. (**2.20**) – PLATE Q8 on the Left and PLATE Q9 on the Right.

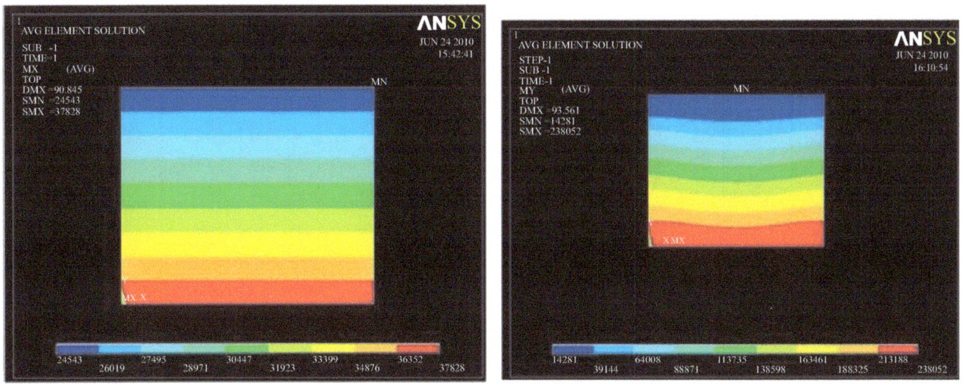

(a) Results using 4 DKT elements (b) Results using 64 DKT elements

Fig. (2.22). Results of analysis using Ansys with discrete Kirchhoff triangle elements.

For this problem, the authors have derived the analytical solution (approximate). For comparison purpose, an approximate analytical thin plate solution has been derived by the authors. This cantilever slab is treated as a thin plate and analyzed by the Ritz Method with its displacement function as

$$w(x,y)=\sum_{i=1}^{n} c_i f_i(x,y) \tag{2.196}$$

As the cantilever slab is symmetrical about the centroidal axis, the deflection curve can be expressed as:

$$w(x,y)=X(x)\cdot Y(y) \tag{2.197}$$

where x and y are coordinates defined in Fig. (**8**) and the functions $X(x)$ and $Y(y)$ are

$$X(x)=c_1+c_2\left(\frac{x}{A}-\frac{2x^3}{A^3}+\frac{x^4}{A^4}\right) \tag{2.198}$$

$$Y(y)=\frac{3y^2}{B^2}-\frac{y^3}{B^3} \tag{2.199}$$

The $X(x)$ function is so chosen so that it is curved and symmetrical about the axis $x = A/2$. The $Y(y)$ function is in fact the deflection shape of a cantilever beam. c_1 and c_2 are the coefficients to be determined. The authors have actually tried a higher order function for $Y(y)$ but have found that the coefficients from 4th order upwards will be zero from the Ritz energy minimization analysis. Ge and Liu (1999) and Yuan (1993) have demonstrated that for cantilever slab under uniformly distributed load, the use of eq.(2-197) will be adequate, and such approximation is accurate when compared with infinite series solution (error around 2%).

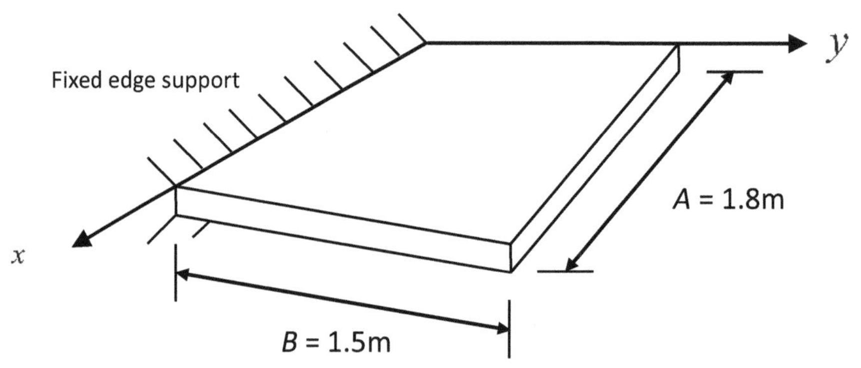

Fig. (2.23). Plate Model for Analysis by Ritz Method.

The Ritz Method is carried out with minimization of the sum of the internal energy and external potential energy. For the plate structure, the former is the sum of the bending and twisting energies given as:

$$U=\frac{1}{2D}\int_{y_1}^{y_2}\int_{x_1}^{x_2}M_x^2dxdy+\frac{1}{2D}\int_{y_1}^{y_2}\int_{x_1}^{x_2}M_y^2dxdy+\frac{1}{2D}\int_{y_1}^{y_2}\int_{x_1}^{x_2}M_{xy}^2dxdy \tag{2.200}$$

Where: $D = \frac{Et^3}{12(1-\mu)}$ with t as the plate thickness, E and μ as the Young's Modulus and Poisson ratio of the material, respectively and

$$M_x = -D\left(\frac{\partial^2 w}{\partial x^2} + \mu\frac{\partial^2 w}{\partial y^2}\right); \quad M_y = -D\left(\frac{\partial^2 w}{\partial y^2} + \mu\frac{\partial^2 w}{\partial x^2}\right) \text{ and}$$
$$M_{xy} = -D(1-\mu)\frac{\partial^2 w}{\partial x \partial y^2} \qquad \textbf{(2.221)}$$

The external potential energy for case (a) of Fig. (**2.20**) will be

$$V = Pw(0,b) + Pw(a,b) \text{ and for case (b)} \qquad \textbf{(2.222)}$$

$$V = Pw(0,b) + M_X \times \left.\frac{\partial w}{\partial x}\right|_{(0,b)} + M_X \times \left.\frac{\partial w}{\partial x}\right|_{(a,b)} + Pw(a,b) + M_Y \times \left.\frac{\partial w}{\partial y}\right|_{(0,b)} + M_Y \times \left.\frac{\partial w}{\partial y}\right|_{(a,b)} \qquad \textbf{(2.223)}$$

Where: P, M_x and M_y are the applied loads and moments at the free corners. To minimize the total energy $\Pi = U + V$, differentiation is performed with respect to the coefficients C_1 and C_2 in (2-198) such that

$$\frac{\partial \Pi}{\partial c_1} = 0 \quad \text{and} \quad \frac{\partial \Pi}{\partial c_2} = 0 \qquad \textbf{(2.224)}$$

On solving eq.(2-224), C_1 and C_2 can be determined for cases (a) and (b) in Fig. (**2.23**) and the complete deflection, moment fields can be obtained. The moment contours along the Y axis are plotted as in Fig. (**2.24**).

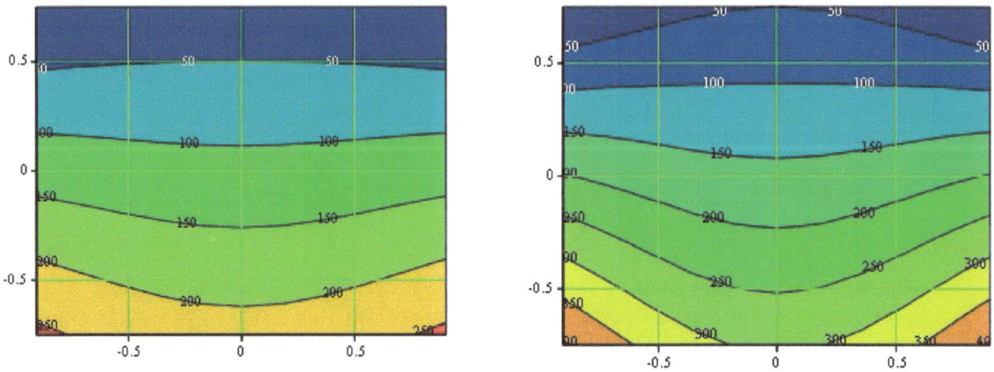

Fig. (2.24). Moment along Span of Y-axis for Cases (1) and (2) of Fig. (**2.20**) – by the Ritz Method with Case (1) on the left and Case (2) on the right.

Since only 1 element is used for the present study, the performance of the present element near to locations with high variation in internal forces is clearly illustrated. It can be seen readily that the behaviour of PLATE-Q9 element (with curving in the moment contours similar to that in Fig. (**2.21**) is extremely good, keeping in mind that only 1 element has been used in the analysis.

The authors have also tried a commercial program SAFE by CSI, and the results of the analysis are shown in Figs. (**2.25** and **2.26**). Even with a mesh size of 20x20, the results from this famous program are still not entirely satisfactory.

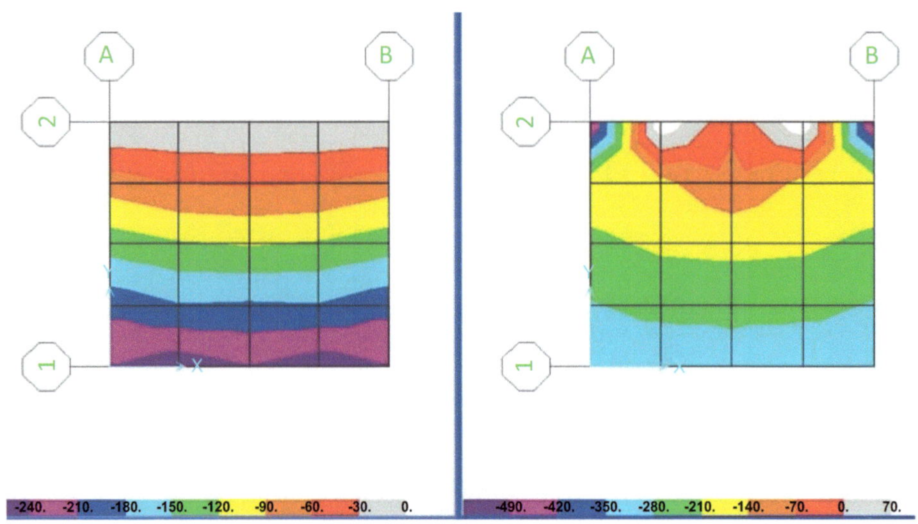

Fig. (2.25). Moment along Span of Y-axis for Cases (1) and (2) of Fig. (**2.20**) – by the IB element of mesh size 4′4 with Case (1) on the left and Case (2) on the right.

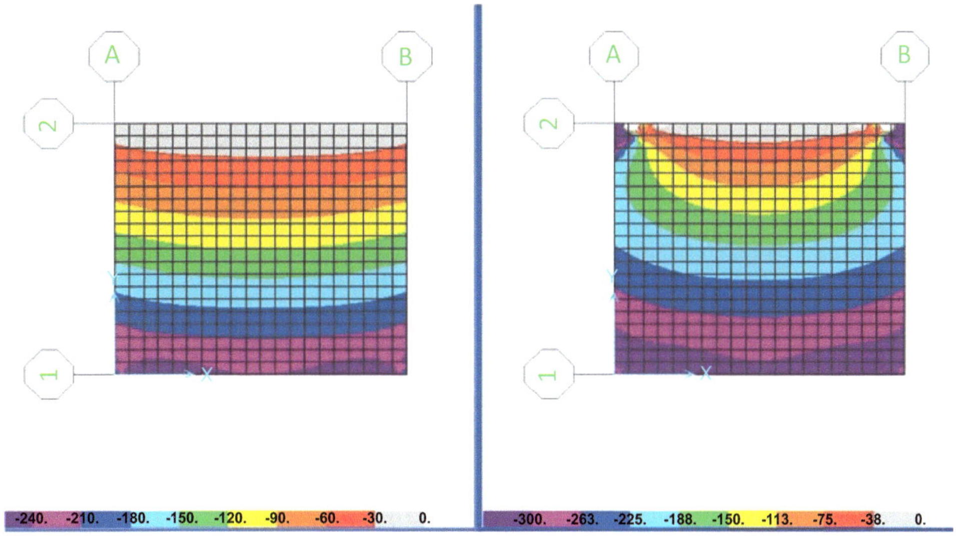

Fig. (2.26). Moment along Span of Y-axis for Cases (1) and (2) of Fig. (**2.20**) – by the IB element of mesh size 20´20 with Case (1) on the left and Case (2) on the right.

2.4.9. Patch Test for PLATE-Q9

PLATE-Q9 can be viewed as a non-conforming element in that there are only 4 nodes with higher order interpolation function. To confirm that the element is stable, patch tests are performed on this element for the cases of pure bending, shear, or twist under both thick and thin cases based on the studies by Naganarayana and Prathap, 1989.

The patch-test problems are presented as follows:

2.4.9.1. Pure Bending Patch Test

The plate shown in Fig. (**2.27**) is considered for a pure bending patch test. A distributed edge couple of constant intensity is applied at the free edge of the patch model and the plate which is restrained at the left-hand edge. The illustration of bending patch test is shown in Fig. (**2.28**).

The model as shown in Fig. (**2.27**) is considered for the case of thicknesses t=1.0m and 0.001m. The elastic properties of the plate are and μ=0.0. The Poisson ratio is set to zero so that the results can be compared directly with elementary beam theory solutions. The non-dimensional deflection at the free edge according to beam theory should be 6.00. The results of patch test are shown in Table **2.1**.

Table 2.1. Tip deflections wEt^3/ML^2 for pure bending patch test.

t/L	Node	PLATE-Q9	Theory
0.1	2	6.00	**6.00**
	3	6.00	
0.01	2	6.00	
	3	6.00	
0.001	2	6.00	
	3	6.00	

PLATE-Q9 can exactly reproduce the bending stress resultants $M_x = M$over all the test cases, and the element can pass through this patch test.

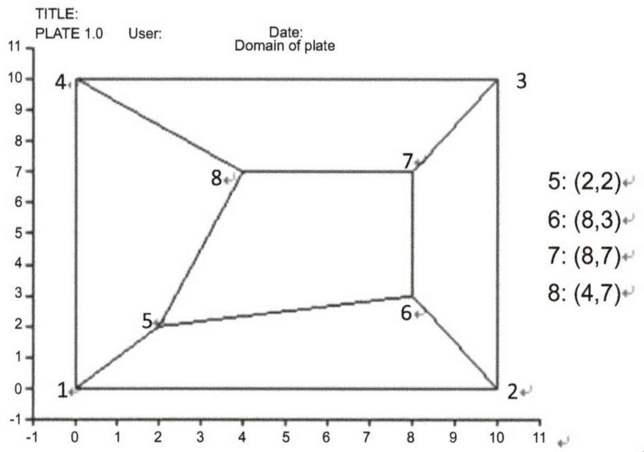

Fig. (2.27). The first patch test of PLATE-Q9.

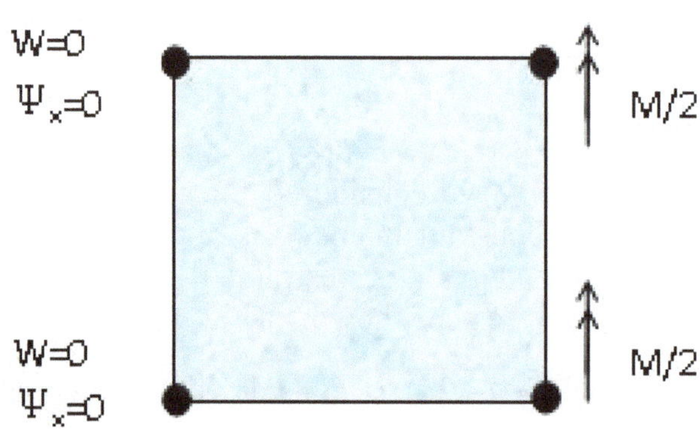

Fig. (2.28). Pure bending model to the patch test in Fig. (**2.27**).

2.4.9.2. Patch Test for Shear

The plate patch for pure shear is shown in Fig. (**2.29**), where a constant intensity shear is applied at right-hand edge of the model, and all the nodal rotations are set to zero to prevent any bending effect. The non-dimensional tip deflection is wEt / P = 2.0. The results in Table **2.2** indicate that this element passes through the shear patch test, and can give correctly in the shear test.

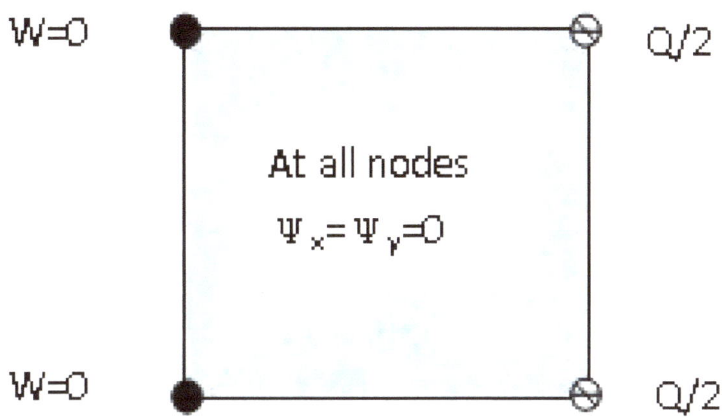

Fig. (2.29). Patch test for pure shear model.

Table 2.2. Tip deflections *wEt/PL* for pure shear patch test.

t/L	Node	PLATE	Theory
0.1	2	2.00	**2.00**
	3	2.00	
0.01	2	2.00	
	3	2.00	
0.001	2	2.00	
	3	2.00	

2.4.9.3. Patch Test for Shear and Bending

In the patch for shear and bending as shown in Fig. (**2.30**), a constant intensity shear is applied at right-hand edge of the model and rotations is allowed. The results in Table **2.3** demonstrate that the element can pass through this patch test.

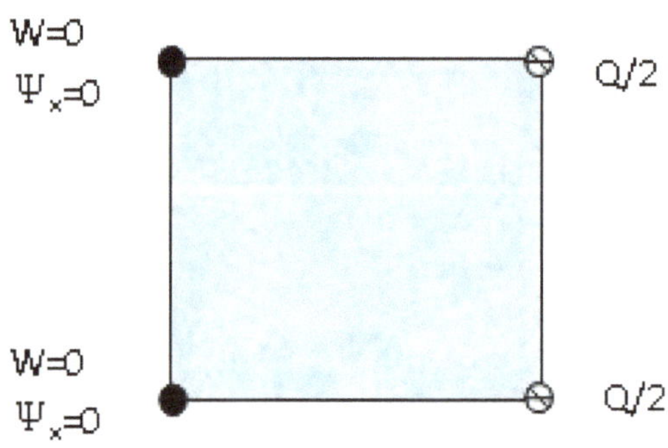

Fig. (2.30). Constant shear and linear bending model.

Table 2.3. Tip deflections wEt^3/PL^2 for combined bending and shear patch test.

t/L	Node	PLATE	Theory
0.1	2	4.0	**4.00**
	3	4.0	
0.01	2	4.0	
	3	4.0	
0.001	2	4.0	
	3	4.0	

2.4.9.4. Patch Test for Twist

For this patch test, the test problem is supported on three corners and loaded with a concentrated force at the fourth corner as shown in Fig. (**2.31**) with a constant Kirchhoff twist. The results for a thin plate also agree well with the present element in this test (Table **2.4**).

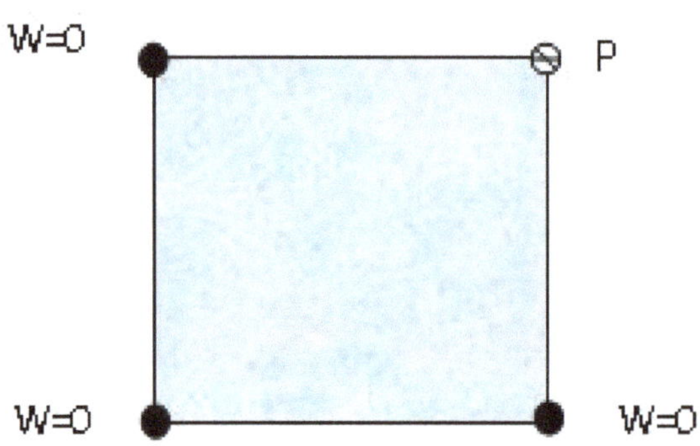

Fig. (2.31). Constant twist patch test.

Fig. (2.31). Constant twist patch test.

Table 2.4. Tip deflections wEt^3/ML for pure bending patch test.

t/L	Node	PLATE	Theory
0.1	2	0.00	0.00
	3	6.0	6.00
0.01	2	0.00	0.00
	3	6.00	6.00
0.001	2	0.00	0.00

The previous pseudo PLATE-Q8 element by Cheng and Law (2008b) has been demonstrated to give very high accuracy and is convenient to be used for practical irregular problems. This element has also received applications in Hong Kong and some countries. In this section, the authors have developed a new higher order pseudo 9 node thick plate element by an extension of the PLATE-Q8 formulation. The PLATE-Q9 element differs from the conventional 8 or 9 node elements by having the mid-side nodes and centre nodes "statically condensed" into the corner nodes. Such "static condensation" has the merit of retaining the higher order of polynomial functions in the formulation of the shape functions (thus higher accuracy) but with considerable reduction in the size of the final stiffness matrix of an element, which is different from the classical four node element by Bathe and Dvorkin (1985). The "PLATE-Q9" element is formulated by energy minimization which is the similar with "PLATE-Q8", which is more direct to the engineers as compared with the use of bubble functions, substitute shear strain or stabilization matrices. With the increased order of polynomials in the formulation of the shape functions, a much better performance as compared with the "PLATE-8" in terms of internal forces change, edge effort, and rate of convergence has been proven in the previous worked examples and the comparison with the Ritz Method. PLATE-Q9 appears to be 4 node element externally which is more suitable for engineering application, but is a higher order function in the internal calculation. Although the formulation of the stiffness matrix is more complicated, there is no significant increase in computation time as compared with the other plate elements or PLATE-Q8 elements with the same meshing. In this respect, this element is an element which is more suitable for engineering applications for complicated problems.

Numerical examples have demonstrated that "PLATE-Q9" element shows much better performance than "PLATE-Q8" in terms of internal forces change, edge effort, and rate of convergence. In particular, this new element possesses very high accuracy for a coarse mesh or at locations where there are rapid changes of internal forces. This element is a special form of non-conforming element which is also demonstrated to pass through 4 different patch tests. The suitability of this new element is clearly demonstrated in this paper, and the "PLATE-Q9" element can be recommended for the analysis of complicated thick plate problems.

2.5. FINAL DISCUSSION

Besides the use of minimum energy and variational principle which is usually adopted for structural engineering problems, many engineering problems governed by differential equations can be solved by means of Galerkin formulation, where an equivalent stiffness matrix can be obtained. A typical example is the seepage problem governed by a Laplace equation, and the equivalent stiffness matrix is a simple matrix defined by the permeability of the soil. There are many problems which can be discretized and solved by similar Galerkin formulation, which is well covered in many books on the use of finite element method for solution of partial differential equations. This topic is too wide and deep to be discussed here, and interested readers can refer to other publications for such needs. The authors have developed programs for pile driving wave equation analysis, Biot consolidation using elastic and Cam-clay model, seepage and others applications, and interested readers can try to read these codes for learning, if necessary. A sample program for refined one-dimensional consolidation analysis is given in the Appendix 2-5 of this chapter, and some interesting results are also found.

There are also many other numerical methods that have evolved over the time, including boundary element method, finite volume method, meshfree methods and many other methods. There are also tremendous development in every numerical method, including coupling and extensions to various conditions. Readers are suggested to consult other references for these advanced topics.

APPENDIX 2-1. ILLUSTRATION OF REFINE INPUT FORMAT BY A GRID ANALYSIS PROGRAM.

Grid analysis has been popular in the past for the analysis of plate type two-dimensional structures. Many bridge analysis and design are also carried out by the use of grillage method in the past. Grillage system is one for which all members lie on a plane, and it can be viewed as a special case of a space frame member where two deflections and one rotation can be neglected. This type of member can be used for floor, bridge deck, raft foundation, pile caps or similar structures. It has the advantage of simple in operation, low computer memory and is fast in computation. In fact, some engineers code the grid analysis programs in some calculators with simple basic or c interpreters in the past. The degrees of freedom of the grid element are the vertical deflection, rotation along and about the element at the two ends which are given by eq.(A-1). The stiffness matrix of a grid element is hence a 6x6 matrix, and assemblage of the matrix has to take care of the fact that θ_x for a member can be equal to θ_y for the orthogonal connected member. For practical purpose of designing for steel reinforcement, most of the engineers will design the grids to be 90° to each other as shown in Fig. (**A-1a**). There are however some cases where the grid members may preferably be not orthogonal to each other as shown in Fig. (**A-1b**). In that case, the stiffness matrix as given by eq.(A-1) along the member own coordinate system needs to be transformed to the global x/y coordinate system. The transformed global stiffness will then be given by $[K_{global}]=[T]^T[K][T]$. The subroutine SATMAK in this appendix gives the stiffness matrix of a grid member along the global axis system, and the subroutine assemble the global stiffness matrix. The subroutine BANDWIDTH returns the bandwidth of the global stiffness matrix, based on which the element stiffness matrix is assembled into the global stiffness matrix. The subroutine BOUND put the boundary condition into the global stiffness matrix and the load vector, after which the subroutine GAUSSB solve the system of equations for the nodal displacements. The subroutine OUTPUT compute the member internal forces and then output the nodal displacement and reactions to the users. Although this program is simple in structure, it has illustrated the main concept behind a large scale structural and finite element program. The understanding of this simple program can help the readers to understand the more complicated finite element program in Appendix 2-4.

This program analyze a system of grid members which can connect with each other at any angle, not limited to 90°. By using the keywords to control how a program will read the input file, the readability and maintainability of the input file will be greatly enhanced. This program is written many years ago by the authors under Dos environment, and the use of Dos interrupts have been adopted in part of the computer code. The readers can however bypass the Dos interrupts and supplied a data file to the program for analysis. This program can also be modified easily for accepting more complicated loadings and support conditions. A sample input file is given after the grid program. Interested readers can remove the subroutine readkey and let the program read in the input files directly with very minor modifications of the code. This program actually start with opening the input file which is preferably prepared by Notepad or similar text editor. It should be notes that some editors may insert hidden control codes which are transparent to the users, and this should be avoided.

$$[K]=\begin{bmatrix} \frac{12EI}{L^3} & 0 & \frac{6EI}{L^2} & -\frac{12EI}{L^3} & 0 & \frac{6EI}{L^2} \\ 0 & \frac{GJ}{L} & 0 & 0 & -\frac{GJ}{L} & 0 \\ \frac{6EI}{L^2} & 0 & \frac{4EI}{L} & -\frac{6EI}{L^2} & 0 & \frac{2EI}{L} \\ -\frac{12EI}{L^3} & 0 & -\frac{6EI}{L^2} & \frac{12EI}{L^3} & 0 & -\frac{6EI}{L^2} \\ 0 & -\frac{GJ}{L} & 0 & 0 & \frac{GJ}{L} & 0 \\ \frac{6EI}{L^2} & 0 & \frac{2EI}{L} & -\frac{6EI}{L^2} & 0 & \frac{4EI}{L} \end{bmatrix} \quad \textbf{(A-1)}$$

Degree of freedom vector=(δ_1, θ_{x1}, θ_{y1}, δ_2, θ_{x2}, θ_{y2})

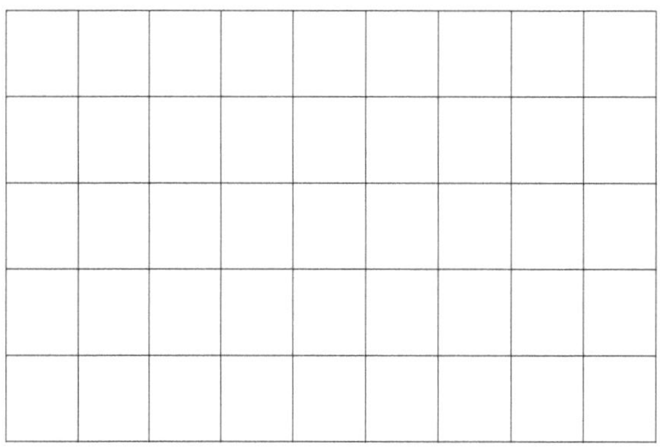

(a) Rectangular grid pattern.

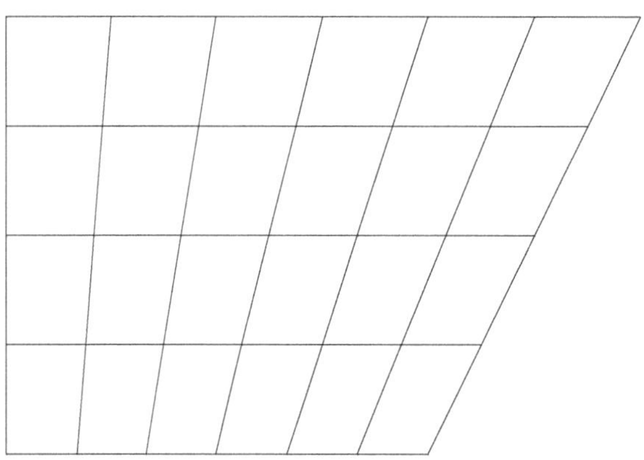

(b) Non-rectangular pattern.

Fig. (A-1). A typical grillage layout for the analysis of raft foundation.

In this program, the computer memory for the coordinates of the nodes, connectivity of the element, global stiffness matrix and load vector are not defined until the main control parameters for the number of nodes and elements. The program will read the main control parameters which are the total number of nodes and elements for the given problem. Based on these information, the memory required for analysis is allocated for most of the variables, except for a few variables which may not be present for a given problem. For example, nodal spring is not necessarily present in all problems, and the computer memory is allocated to the variables only if the keywords vertical_spring, spring_x, spring_y which correspond to the vertical nodal spring, rotational springs for rotation along x and y directions are defined.

Some improvements can be applied to this program to further enhance the usability. For example, the users can add additional keywords to control the input of loads and boundary conditions by using the subroutine Find_node, and there will be not a need to define the node number in the input process. Furthermore, for most of the practical grillage problem which is defined by a quadrilateral domain, a simple mesh generation program can be developed by subdividing the domain into series of quadrilaterals. One more suggested enhancement is the definition of the nodal spring support. If a raft foundation is supported on soil, the nodal spring can be automatically determined by evaluating the effective area for each node and multiply with the Winkler's spring constant for the soil support. A single input of the soil Winkler's spring will then sufficient to define all the support conditions instead of defining the individual nodal spring by the users. Furthermore, the coordinates of the nodes can be shifted so as to match with the superstructure column/line loads. These improvements can all be accomplished easily by the readers, but the authors have not carry out such improvement work because a much more versatile and advanced thin and thick plate analysis program coupled with grillage member has been developed later. All the improvements as suggested above have actually been implemented into the PLATE program.

```
module data

real, allocatable, dimension(:) :: TORCON(:),UDL(:),SMA(:),SPRING(:),OLEN(:)

REAL,ALLOCATABLE, DIMENSION(:,:) :: CORD

REAL(KIND=8), ALLOCATABLE, DIMENSION(:,:) :: BK

REAL(KIND=8), ALLOCATABLE, DIMENSION(:) :: RW,RW1,BB
```

```
integer, allocatable, dimension(:,:) :: MCON(:,:)

integer, allocatable, dimension(:) :: NSS(:),IFIX(:)

REAL :: WV(8),A(6,3),S(3,3),SAT(3,6),ASAT(6,6),SR(3),R4N

INTEGER ::
IFLG,NDAT,I4N,NSPRING,pos,NSPRINGV,nspringx,nspringy,SIZE

CHARACTER(len=24) :: FILDAT,FILOUT,str

integer :: JCT, kb, NM

end module data

program main

USE DATA

call startup

call open

call input ! read in input data

CALL STIFF ! generate element stiffness matrix, then global stiffness matrix, boundary condition

CALL SOLVE ! solve stiffness matrix equation

DEALLOCATE(BK)

SIZE=3*JCT+3*NM

ALLOCATE(BB(SIZE))

call output(BB,SIZE) ! output member forces

close(5) ; close(6)

call view

end

SUBROUTINE OPEN

use data
```

```
integer :: CODE
logical :: ALIVE
character(len=32) :: command
write (*,*)
write (*,*) ' To open an existing data file, press 1. '
write (*,*) ' To open a new datafile, press 2 '
write (*,*) ' To quit, press 3'
write (*,*)
15 CALL READKEY(CODE)
if (CODE == 6) STOP
IF (CODE < 4 .OR. CODE > 5) GO TO 15
IF (CODE == 5) THEN
INQUIRE(FILE='grid_new.DAT',EXIST=ALIVE)
IF (.NOT. ALIVE) STOP
ENDIF
PRINT*, 'What is your DATA file name ? '
READ(*,'(A)') FILDAT
PRINT*, 'What is your OUTPUT file name ? '
READ(*,'(A)') FILOUT
IF (CODE == 4) THEN
OPEN(5, FILE=FILDAT, STATUS='OLD')
open(6,file=filout)
ELSE
COMMAND(1:18)='COPY grid_new.DAT '
```

```
COMMAND(19:)=FILDAT
CALL SYSTEM(COMMAND)
COMMAND(1:32)=''
COMMAND(1:5)='EDIT '
COMMAND(6:)=FILDAT
CALL SYSTEM(COMMAND)
OPEN(5, FILE=FILDAT, STATUS='OLD')
open(6,file=filout)
ENDIF
WRITE (6,2) FILDAT
2 FORMAT (' THE DATA IS READ FROM FILE - ',A//)
write (*,*) 'Proceed with computation y/n ? '
call readkey(code)
if (code /=2) stop
RETURN
END SUBROUTINE OPEN
subroutine ReadKey(code)
implicit none
integer*2 KeyType, CharCode, ShiftCode
integer*4 Regs( 9), L, code
integer*2 ScanCode, ASCII_Char
integer*2 KTT( 163), CCT( 163), SCT( 163)
data KTT / 4, 57*1, 10*2, 15*3, 30*2, 6*3, 12*1, 3, 8*2, &
& 7*3, 1, 15*3 /
```

```
data CCT / 27, 12*0, 8, 9, 81, 87, 69, 82, 84, 89, 85, 73, 79, &
& 80, 91, 93, 13, 0, 65, 83, 68, 70, 71, 72, 74, 75, 76, &
& 59, 39, 96, 0, 92, 90, 88, 67, 86, 66, 78, 77, 44, 46, 47, &
& 5*0, 1, 2, 3, 4, 5, 6, 7, 8, 9, 10, 2*0, &
& 7, 2, 9, 0, 3, 0, 1, 0, 8, 4, 10, 5, 6, 1, 2, 3, 4, 5, 6, 7, 8, &
& 9, 10, 1, 2, 3, 4, 5, 6, 7, 8, 9, 10, 1, 2, 3, 4, 5, 6, 7, 8, &
& 9, 10, 0, 3, 1, 8, 10, 7, &
& 49, 50, 51, 52, 53, 54, 55, 56, 57, 48, 45, 61, 9, 11, 12, &
& 11, 12, 11, 12, 11, 12, 2, 3*0, 4, 5, 6, 9, 2*0, &
& 7, 2, 9, 0, 3, 0, 1, 0, 8, 4, 10, 5, 6 /
data SCT / 14*3, 1, 43*3, 25*0, 10*1, 10*2, 10*3, 6*2, 12*3, 2, &
& 0, 0, 1, 1, 2, 2, 3, 3, 10*2, 13*3 /
code = 1000
1000 Regs( 1) = 16*256
call INTRUP( Regs, 22)
L = Regs( 1)
ASCII_Char = iand( L, 255)
ScanCode = iand( ishft( L, -8), 255)
if( ASCII_Char .eq. 0 .or. ASCII_Char .eq. 224 ) goto 2000
CharCode = ASCII_Char
if( CharCode .eq. 3 ) then
KeyType = -5
goto 9999
end if
```

```
KeyType = -1
if( 65 .le. ASCII_Char .and. ASCII_Char .le. 90 ) then
ShiftCode = 1
elseif( ASCII_Char .le. 26 .and. ASCII_Char .ne. 9 ) then
ShiftCode = 2
elseif( ASCII_Char .eq. 27 ) then
KeyType = -4
ShiftCode = 0
else
ShiftCode = 0
end if
goto 9999
2000 continue
if( ScanCode .eq. 0 ) then
KeyType = -6
goto 9999
end if
if( ScanCode .gt. 163 ) goto 1000
CharCode = CCT( ScanCode)
if( CharCode .eq. 0 ) goto 1000
KeyType = -KTT( ScanCode)
ShiftCode = SCT( ScanCode)
9999 continue
if (keytype == -4 .and. charcode == 27 .and. shiftcode == 0) code=0 !esc
```

```
if (keytype == -1 .and. charcode == 69 .and. shiftcode == 1) code=1 !E
if (keytype == -1 .and. charcode == 101 .and. shiftcode == 0) code=1
if (keytype == -1 .and. charcode == 121 .and. shiftcode == 0) code=2 !y
if (keytype == -1 .and. charcode == 89 .and. shiftcode == 1) code=2
if (keytype == -1 .and. charcode == 110 .and. shiftcode == 0) code=3 !N
if (keytype == -1 .and. charcode == 78 .and. shiftcode == 1) code=3
if (keytype == -1 .and. charcode == 49 .and. shiftcode == 0) code=4
if (keytype == -1 .and. charcode == 50 .and. shiftcode == 0) code=5
if (keytype == -1 .and. charcode == 51 .and. shiftcode == 0) code=6
return
end
subroutine input
USE DATA
USE STRPAK
USE FIOPAK
integer :: i,j1,j2
real :: xo,yo
CALL VI_GETVAR(5,'1','No_of_node',NDAT,IFLG)
CALL VI_DATA(1,I4N,IFLG) ; JCT = I4N ; CALL VI_ERASE
CALL VI_GETVAR(5,'1','No_of_element',NDAT,IFLG)
CALL VI_DATA(1,I4N,IFLG) ; NM = I4N ; CALL VI_ERASE
ALLOCATE
(CORD(JCT,2),TORCON(NM),UDL(NM),SMA(NM),OLEN(NM),MCON(NM,
2),IFIX(3*JCT))
IFIX=0 ; UDL=0. ; RW=0 ; RW1=0.
```

```
ALLOCATE(RW(JCT*3),RW1(3*JCT))
CALL VI_GETVAR(5,'1','NODAL_DATA',NDAT,IFLG)
DO I = 1,NDAT
IF (I-I/3*3 ==1) THEN
CALL VI_DATA(I,I4N,IFLG) ; K = I4N
ENDIF
IF (I-I/3*3 ==2) THEN
CALL VI_DATA(I,R4N,IFLG) ; CORD(K,1) = R4N
ENDIF
IF (I-I/3*3 ==0) THEN
CALL VI_DATA(I,R4N,IFLG) ; CORD(K,2) = R4N
ENDIF
END DO
CALL VI_ERASE
CALL VI_GETVAR(5,'1','MEMBER_DATA',NDAT,IFLG)
DO I = 1,NDAT
IF (I-I/5*5 == 1) THEN
CALL VI_DATA(I,I4N,IFLG) ; K = I4N
ENDIF
IF (I-I/5*5 ==2) THEN
CALL VI_DATA(I,I4N,IFLG) ; MCON(K,1) = I4N
ENDIF
IF (I-I/5*5 ==3) THEN
CALL VI_DATA(I,I4N,IFLG) ; MCON(K,2) = I4N
```

```
ENDIF
IF (I-I/5*5 ==4) THEN
CALL VI_DATA(I,R4N,IFLG) ; SMA(K) = R4N
ENDIF
IF (I-I/5*5 ==0) THEN
CALL VI_DATA(I,R4N,IFLG) ; TORCON(K) = R4N
ENDIF
END DO
CALL VI_ERASE
CALL VI_GETVAR(5,'1','NODAL_LOAD',NDAT,IFLG)
if (ndat > 0) then
write (6,*)
write (6,*) " Nodal loads applied"
write (6,*) " node V Mx My"
DO I = 1,NDAT
IF (I-I/4*4 == 1) THEN
CALL VI_DATA(I,K,IFLG)
ENDIF
IF (I-I/4*4 ==2) THEN
CALL VI_DATA(I,RW(3*(K-1)+1),IFLG)
ENDIF
IF (I-I/4*4 ==3) THEN
CALL VI_DATA(I,RW(3*(K-1)+2),IFLG)
ENDIF
```

```
IF (I-I/4*4 ==0) THEN
CALL VI_DATA(I,RW(3*K),IFLG)
ENDIF
write (6,'(15x,i4,3(1x,f10.3))') k,rw(3*(k-1)+1:3*k)
END DO
CALL VI_ERASE
endif
! READ SPRING STIFFNESS
call vi_getvar(5,'1','member_udl',ndat,iflg)
if (ndat > 0) then
write (6,*)
write (6,*) " Member UDL applied"
write (6,*) " no. UDL"
do i = 1,ndat
if (i-i/2*2 /= 0) call vi_data(i,k,iflg)
if (i-i/2*2 == 0) call vi_data(i,udl(k),iflg)
write (6,'(20x,i4,2x,f10.3)') k,udl(k)
end do
call vi_erase
write (6,*)
endif
call vi_getvar(5,'1','V_FIX',ndat,iflg)
if (ndat > 0) then
do i = 1,ndat
```

```
if (i-i/2*2 /= 0) call vi_data(i,k,iflg)
if (i-i/2*2 == 0) call vi_data(i,RW(3*(k-1)+1),iflg)
IFIX(3*(K-1)+1)=1
end do
call vi_erase
endif
call vi_getvar(5,'1','RX_FIX',ndat,iflg)
if (ndat > 0) then
do i = 1,ndat
if (i-i/2*2 /= 0) call vi_data(i,k,iflg)
if (i-i/2*2 == 0) call vi_data(i,RW(3*(k-1)+2),iflg)
IFIX(3*(K-1)+2)=1
end do
call vi_erase
endif
call vi_getvar(5,'1','RY_FIX',ndat,iflg)
if (ndat > 0) then
do i = 1,ndat
if (i-i/2*2 /= 0) call vi_data(i,k,iflg)
if (i-i/2*2 == 0) call vi_data(i,RW(3*k),iflg)
IFIX(3*K)=1
end do
call vi_erase
endif
```

```
CALL VI_GETVAR(5,'1','VERTICAL_SPRING',NDAT,IFLG) ; NSPRINGV =
NDAT/2

CALL VI_GETVAR(5,'1','R_SPRING_x',NDAT,IFLG) ; NSPRINGX = NDAT/2

CALL VI_GETVAR(5,'1','R_SPRING_y',NDAT,IFLG) ; NSPRINGY = NDAT/2

NSPRING=NSPRINGV+NSPRINGX+NSPRINGY

IF (NSPRING > 0) ALLOCATE(NSS(NSPRING),SPRING(NSPRING))

CALL VI_GETVAR(5,'1','VERTICAL_SPRING',NDAT,IFLG)

if (NDAT > 0) then

DO I = 1,NDAT

IF (I-I/2*2 ==1) THEN

CALL VI_DATA(I,I4N,IFLG) ; NSS(I/2+1) = I4N

ENDIF

IF (I-I/2*2 ==0) THEN

CALL VI_DATA(I,R4N,IFLG) ; SPRING(I/2) = R4N

ENDIF

END DO

CALL VI_ERASE

endif

CALL VI_GETVAR(5,'1','R_SPRING_X',NDAT,IFLG)

if (NDAT > 0) then

DO I = 1,NDAT

IF (I-I/2*2 ==1) THEN

CALL VI_DATA(I,I4N,IFLG) ; NSS(I/2+1+NSPRINGV) = I4N

ENDIF
```

```
IF (I-I/2*2 ==0) THEN
CALL VI_DATA(I,R4N,IFLG) ; SPRING(I/2+NSPRINGV) = R4N
ENDIF
END DO
CALL VI_ERASE
endif
CALL VI_GETVAR(5,'1','R_SPRING_Y',NDAT,IFLG)
if (NDAT > 0) then
DO I = 1,NDAT
IF (I-I/2*2 ==1) THEN
CALL VI_DATA(I,I4N,IFLG) ; NSS(I/2+1+NSPRINGV+NSPRINGX) = I4N
ENDIF
IF (I-I/2*2 ==0) THEN
CALL VI_DATA(I,R4N,IFLG) ; SPRING(I/2+NSPRINGV+NSPRINGX) = R4N
ENDIF
END DO
CALL VI_ERASE
endif
DO I = 1,NM
J1 = MCON(I,1) ; J2 = MCON(I,2)
XO = CORD(J2,1) - CORD(J1,1)
YO = CORD(J2,2) - CORD(J1,2)
OLEN(I) = SQRT (XO *XO + YO * YO)
End do
```

```
CALL BANDWIDTH ! determine bandwidth of the global stiffness matrix

ALLOCATE(BK(jct*3,kb)) ! allocate memory for the banded global stiffness matrix

! SIZE=3*JCT*KB

!

! ---------- OUTPUT THE GEOMETRIC DATA ---------------------------------

!

WRITE (6,2350) JCT,NM,NSPRING

2350 FORMAT (' JOINT AND MEMBER DATA', &

& ' (',I3,' JOINTS',TR1, I3,' MEMBERS',TR1,I3,' SPRINGS)'// &

& ' NODE X-ORDINATE Y-ORDINATE'/)

DO I = 1,JCT

WRITE (6,2360) I,(CORD(I,J),J=1,2)

end do

WRITE (6,2370)

DO I = 1,NM

WRITE (6,2380) I,(MCON(I,J),J=1,2),SMA(I),TORCON(I),OLEN(I)

end do

2360 FORMAT (I4,1X,2F11.3)

2370 FORMAT (/'MEMBER CONNECTIVITY',3X,'INERTIA',2X,' TORSION CONST', &

& 4X,'LENGTH'/)

2380 FORMAT (2I5,' TO',I4,4E14.5)

IF (NSPRINGV > 0) THEN

WRITE (6,2800)
```

```
DO I = 1,NSPRINGV
WRITE (6,2810) NSS(I),SPRING(I)
END DO
ENDIF
IF (NSPRINGX > 0) THEN
WRITE (6,2801)
DO I = 1,NSPRINGX
WRITE (6,2810) NSS(I),SPRING(I)
END DO
ENDIF
IF (NSPRINGY > 0) THEN
WRITE (6,2802)
DO I = 1,NSPRINGY
WRITE (6,2810) NSS(I),SPRING(I)
END DO
ENDIF
2800 FORMAT (//' VERTICAL SPRING VALUES'//' NODE ',' SPRING')
2801 FORMAT (//' ROTATION SPRING X VALUES'//' NODE ',' SPRING')
2802 FORMAT (//' ROTATION SPRING Y VALUES'//' NODE ',' SPRING')
2810 FORMAT (2X,I4,2X,E16.5)
end
subroutine stiff
use data
integer :: i,j,j1,j2
```

```
DO I = 1,NM
CALL SATMAK (I)
asat=0.
asat=matmul(a,sat)
CALL ASSEMK(I)
j1=mcon(i,1) ; j2=mcon(i,2)
do j=1,3
rw(3*(j1-1)+j)=rw(3*(j1-1)+j)+wv(j)
rw(3*(j2-1)+j)=rw(3*(j2-1)+j)+wv(3+j)
end do
END DO
END
SUBROUTINE SOLVE
USE DATA
DO I = 1,NSPRINGV
BK(3*(NSS(I)-1)+1,1) = BK(3*(NSS(I)-1)+1,1) + SPRING(I)
END DO
DO I = 1,NSPRINGX
BK(3*(NSS(I)-1)+2,1) = BK(3*(NSS(I)-1)+2,1) + SPRING(I)
END DO
DO I = 1,NSPRINGY
BK(3*NSS(I),1) = BK(3*NSS(I),1) + SPRING(I)
END DO
RW1=RW
```

```
CALL BOUND
CALL GAUSSB(BK,RW,3*JCT,KB)
END
SUBROUTINE SATMAK (I)
use data
integer :: i,j1,j2
real :: XO,YO,SA,CA,W1,W2
J1 = MCON(I,1) ; J2 = MCON(I,2)
XO = CORD(J2,1) - CORD(J1,1)
YO = CORD(J2,2) - CORD(J1,2)
SA = YO/OLEN(I) ; CA = XO/OLEN(I)
WV = 0.0
IF (UDL(I) .LE. 1E-12) GOTO 2830
W1 = UDL(I) * OLEN(I)/2.0
W2 = W1 * OLEN(I)/6.0
WV(1) = W1 ; WV(2) = W2 * SA
WV(3) = -W2 * CA ; WV(4) = W1
WV(5) = -W2 * SA ; WV(6) = W2 * CA
WV(7) = -W2 ; WV(8) = W2
2830 CONTINUE
S(1,1) = TORCON(I)/OLEN(I) ; V4 = 4.0 * SMA(I)/OLEN(I)
V2 = 0.5 * V4 ; V3 = 0.75 * V4
S(2,2) = V4 ; S(2,3) = V2
S(3,2) = V2 ; S(3,3) = V4
```

```
S(1,2) = 0.0 ; S(1,3) = 0.0
S(2,1) = 0.0 ; S(3,1) = 0.0
A = 0.0 ; A(2,1) = CA ; A(5,1) = -CA
A(1,2) = -1.0/OLEN(I) ; A(1,3) = A(1,2)
A(4,2) = -A(1,2) ; A(4,3) = A(4,2)
A(2,2) = -SA ; A(5,3) = -SA
A(3,1) = SA ; A(3,2) = CA
A(6,1) = -SA ; A(6,3) = CA
sat=0.
sat=matmul(s,transpose(a))
RETURN
END
subroutine startup
use data
write (*,*)
WRITE (*,*) ' Plane grillage analysis for raft foundation, pile'
write (*,*) ' cap or similar structures by Dr. Y.M. Cheng for'
write (*,*) ' teaching. Commercial use is prohibited '
write (*,*) ' 10/10/2002'
write (*,*)
end
subroutine bandwidth
use data
integer :: i
```

```
KB = 0
DO I =1,NM
KB1 = IABS(MCON(I,1) - MCON(I,2))
IF (KB1 .GT. KB) KB = KB1
end do
KB = (KB +1) * 3
end
SUBROUTINE ASSEMK(IELEM)
USE DATA
INTEGER                                                          ::
I,J,IELEM,INODE,IDOFN,JNODE,JDOFN,NROWS,NCOLS,NODEJ
DO I = 1,6
DO J = 1,I
INODE=(I-1)/3+1 ; IDOFN=I-3*(INODE-1) ; INODE=MCON(IELEM,INODE)
JNODE=(J-1)/3+1 ; JDOFN=J-3*(JNODE-1) ; JNODE=MCON(IELEM,JNODE)
NROWS=3*(INODE-1)+IDOFN ; NCOLS=3*(JNODE-1)+JDOFN
IF (NROWS > NCOLS) THEN
NODEJ=NROWS ; NROWS=NCOLS ; NCOLS=NODEJ
ENDIF
BK(NROWS,NCOLS-NROWS+1)= BK(NROWS,NCOLS-NROWS+1)+ &
ASAT(J,I)
END DO
END DO
END
```

```
SUBROUTINE GAUSSB(GK,B,NT,NW)
INTEGER :: NT,NW,K,IK,I,J,JK,II,JJ
REAL(KIND=8) :: GK(NT,NW),B(NT),X,CN
DO 30 I=1,NT
X=GK(I,1)
IF(DABS(X)<= 0.00000000001) THEN
WRITE(6,2) I
2 FORMAT(' **** SINGULARITY IN ROW', I5)
STOP
ENDIF
IK=I
DO 20 J=2,NW
IK=IK+1
CN=GK(I,J)/X
JK=0
DO 10 K=J,NW
JK=JK+1
10 GK(IK,JK)=GK(IK,JK)-CN*GK(I,K)
GK(I,J)=CN
20 B(IK)=B(IK)-CN*B(I)
30 B(I)=B(I)/X
DO 50 J=2,NT
II=NT-J+1
DO 40 K=2,NW
```

```
JJ=II+K-1
40 B(II)=B(II)-GK(II,K)*B(JJ)
50 CONTINUE
RETURN
END SUBROUTINE GAUSSB
subroutine output(BKK,NN)
use data
integer :: i,J1,J2,H,NN
REAL(8) :: BKK(NN)
write (6,2590)
2590 FORMAT (//' RESULT OF GRILLAGE ANALYSIS '// &
& ' JOINT Z-MOVEMENT X-ROTATION Y-ROTATION')
do i=1, jct
write (6,2610) i,rw(3*(i-1)+1:3*i)
end do
bkk=0.d0
2610 format (1x,i4,4x,3e13.5)
DO I=1,NM
CALL SATMAK (I)
J1=MCON(I,1) ; J2=MCON(I,2)
ID=0 ; M=0
1650 JJ=J1-1
IF (JJ.GT.0) THEN
DO JK=1,JJ
```

```
ID=ID+3
END DO
END IF
DO IC=1,3
LSD=ID+1
ASAT(M+IC,1)=RW(LSD)
ID=ID+1
END DO
IF (M.LE.0) THEN
M=3 ; J3=J1 ; J1=J2 ; ID=0 ; GOTO 1650
END IF
DO II=1,3
SR(II)=0.0
DO K=1,6
SR(II)=SR(II)+SAT(II,K)*ASAT(K,1)
END DO
END DO
SR(2)=SR(2)-WV(7)
SR(3)=SR(3)-WV(8)
K1=3*JCT+3*I-2
BKK(K1)=SR(1)
BKK(K1+1)=SR(2)
BKK(K1+2)=SR(3)
SR(2)=SR(2)+WV(7)
```

```
SR(3)=SR(3)+WV(8)
DO J=1,6
ASAT(J,1)=0.0
DO LM=1,3
ASAT(J,1)=ASAT(J,1)+A(J,LM)*SR(LM)
END DO
END DO
M=J3
DO IC=1,2
DO H=1,3
NB=3*M-3+H
IJ=3*IC-3+H
BKK(NB)=BKK(NB)+ASAT(IJ,1)
BKK(NB)=BKK(NB)-WV(IJ)
END DO
M=J2
END DO
END DO
!
! ----------- OUTPUT MOMENTS AND TORQUES ------------------------------
!
WRITE (6,2140)
2140 FORMAT (' ---- ----- ----- ----- ----- ----- ----- ----- ---- ')
2620 FORMAT (/' MEMBER MOMENTS FOR TWO ENDS', &
```

```
& ' FOR NODES TORQUE')
2600 FORMAT (' ')
WRITE (6,2620)
WRITE (6,2600)
DO J =1,NM
J1 = MCON(J,1)
J2 = MCON(J,2)
K = 3 * JCT + 3 * J - 1
WRITE (6,2630) J,BKK(K),BKK(K+1),J1,J2,BKK(K-1)
END DO
!
! ----------- OUTPUT THE SHEAR FORCES ----------------------------------
!
WRITE (6,2140)
WRITE (6,2640)
2640 FORMAT (/' MEMBER ', &
& ' SHEAR FORCE FOR NODES')
WRITE (6,2600)
DO 1940 J = 1,NM
J1 = IABS (MCON(J,1))
J2 = IABS (MCON(J,2))
K = 3 * JCT + 3 * J - 1
V1 = -(BKK(K) + BKK(K+1))/OLEN(J) + UDL(J) * OLEN(J)/2.0
V2 = V1 - UDL(J) * OLEN(J)
```

```
WRITE (6,2630) J,V1,V2,J1,J2
2630 FORMAT (I5,TR3,2E13.5,I4,' AND',I4,E16.5)
1940 CONTINUE
!
! ------------- OUTPUT THE JOINT FORCES AND REACTIONS -
-----------------
!
2650 FORMAT (/' NODE VERTICAL FORCE', &
& ' MOMENT ABT. X-X MOMENT ABT. Y-Y')
WRITE (6,2140)
WRITE (6,2650)
WRITE (6,2600)
NB = 0
DO 1960 I = 1,JCT
IC1 = NB + 3 * I - 2
IC2 = IC1 + 1
IC3 = IC2 + 1
WRITE (6,2660) I,BKK(IC1),BKK(IC2),BKK(IC3)
1960 CONTINUE
2660 FORMAT (I5,TR5,3E16.6)
end
SUBROUTINE BOUND
USE DATA
INTEGER :: K,KV,J,NN
```

```
! Define the prescribed boundary conditions by using the wipe row and column method

NN=3*JCT

LOOP1: DO K=1,NN

IF (IFIX(K) == 0) CYCLE LOOP1

DO J=2,KB

KV=K+J-1

IF (NN-KV >= 0) THEN

RW(KV)=RW(KV)-BK(K,J)*RW1(K)

BK(K,J)=0.D0

ENDIF

KV=K-J+1

IF (KV > 0) THEN

RW(KV)=RW(KV)-BK(KV,J)*RW1(K)

BK(KV,J)=0.D0

ENDIF

BK(K,1)=1.D0 ; RW(K)=RW1(K)

END DO

END DO LOOP1

RETURN

END SUBROUTINE BOUND

SUBROUTINE view

use data

integer :: CODE
```

```
character(len=32) :: command
write (*,*)
write (*,*) ' To view the output file, press 1, other keys to quit '
write (*,*)
15 CALL READKEY(CODE)
if (CODE /= 4) STOP
CALL SYSTEM(COMMAND)
COMMAND(1:24)=''
COMMAND(1:5)='EDIT '
COMMAND(6:)=FILout
CALL SYSTEM(COMMAND)
RETURN
END
```

Sample Input

```
! This mark means this line is not interpreted as data
no_of_node=25 ! 25 nodes for this problem
no_of_element=40 ! 40 elements for this problem
nodal data means node, x/y coordinates
nodal_data
1 0 0 ! node number, x and y coordinates
2 0.6 0 ! nodes need not in order
3 2 0
4 3.4 0
5 4 0
```

6 0 0.6

7 0.6 0.6

8 2 0.6

9 3.4 0.6

10 4 0.6

11 0 2.05

12 0.6 2.05

13 2 2.05

14 3.4 2.05

15 4 2.05

16 0 3.4

17 0.6 3.4

18 2 3.4

19 3.4 3.4

20 4 3.4

21 0 4

22 0.6 4

23 2 4

24 3.4 4

25 4 4

member data means member no., node nos. at two ends, EI and GJ

member_data

1, 1, 2, 125000, 312500 ! member number, first node, second node

2, 2, 3, 291667, 729167 ! EI and GJ for the member

3, 3, 4, 291667, 729167

4, 4, 5, 125000, 729167

5, 6, 7, 125000, 729167

6, 7, 8, 291667, 729167

7, 8, 9, 291667, 729167

8, 9, 10, 125000, 312500

9, 11, 12, 125000, 312500

10, 12, 13, 291667, 729167

11, 13, 14, 291667, 729167

12, 14, 15, 125000, 312500

13, 16, 17, 125000, 312500

14, 17, 18, 291667, 729167

15, 18, 19, 291667, 729167

16, 19, 20, 125000, 312500

17, 21, 22, 125000, 312500

18, 22, 23, 291667, 729167

19, 23, 24, 291667, 729167

20, 24, 25, 125000, 312500

21, 5, 10, 125000, 312500

22, 10, 15, 302083, 755208

23, 15, 20, 281250, 703125

24, 20, 25, 125000, 312500

25, 4, 9, 125000, 312500

26, 9, 14, 302083, 755208

```
27, 14, 19, 281250, 703125
28, 19, 24, 125000, 312500
29, 3, 8, 125000, 312500
30, 8, 13, 302083, 755208
31, 13, 18, 281250, 703125
32, 18, 23, 125000, 312500
33, 2, 7, 125000, 312500
34, 7, 12, 302083, 755208
35, 12, 17, 281250, 703125
36, 17, 22, 125000, 312500
37, 1, 6, 125000, 312500
38, 6, 11, 302083, 755208
39, 11, 16, 281250, 703125
40, 16, 21, 125000, 312500
! nodal load means node, vertical force and moments x and y
nodal_load
7, -2650, 0, 0 ! user can modify to define coordinates for input !
12, -500930, 0, 0
17, -2088575, 0, 0
18, -2487150, 0, 0
19, -2088575, 0, 0
14, -500930, 0, 0
9, -2650, 0, 0
! member udl means node, udl
```

```
! member_udl
! V fix means node and prescribed value for vertical direction
! v_fix
! Rx fix means node and prescribed value for rotation x
! rx_fix
! Ry fix means node and prescribed value for rotation y
! ry_fix
vertical_spring
1 25000000 ! node number, spring constant
5 25000000
21 25000000
25 25000000
! r_spring_x
! r_spring_y
END
```

In this sample input, each group of input data is controlled by a keyword. After the number of elements and nodes are read in, the main program will allocate the computer memory required for the analysis. The details of the member properties, loading and support conditions will then be read sequentially in the program, but the input data can follow any order without restriction. The use of operator overloading in Fortran 90/95 has helped to make this flexible input format possible. The major disadvantage of the use of operator overloading input format is the computer resources overhead, which can be a concern to those old computer (IBM XT or AT type) with a slow hard disk, and these are the hardware that the authors having been using during the development of this code.

APPENDIX 2-2. A SIMPLE TWO-DIMENSIONAL MESH GENERATION PROGRAM

Mesh generation is not a simple task, particularly for arbitrary irregular problems.

Most of the actual problems (particularly those related to research) can be discretized into several major domains, hence a very versatile mesh generation may not be necessary for research and education purpose. The authors have developed a simple computer code many years ago, which relies on the super-elements as defined by the users (see Fig. **2.5b**). This code has also a simple node renumbering scheme so as to minimize the bandwidth required for analysis. This code is very simple with lots of limitations, but it has inspired the authors to develop a refined and faster mesh generation program which can be obtained from the authors. The present code is adequate for simple research purpose, where the number of elements is less than 3000. Beyond 3000 elements, the speed of computation is slow, and interested readers can debug and improve the present program if necessary, or use the refined program MESH which will be much faster in operation and produce good mesh in general. There is a major limitation to the present program which is the master program size. When the authors develop the code, the authors have only a Fortran 77 compiler to work with. Dynamic allocation of memory is not available until Fortran 90, and the authors deal with this limitations by defining a large vector, and the vector is partitioned into series of vectors and matrices for uses (dynamic dimensioning). This is possibly the best approach at the time of Fortran 77. Later, a more sophisticated mesh generation MESH is developed by the authors using totally dynamic memory approach. The new Mesh program under PLATE is completely different from the present simple mesh generation program, and is quite tedious to be illustrated in this book.

The input to this program is very simple. The users define the number of macro-elements which must be connected at the nodes, and the number of divisions along the two-directions (which can be made automatic easily by defining a maximum element size). After that, the coordinates for the macro-element nodes are defined by the users. During the mesh generation, the element sizes can be increased or decreased with a ratio QL or QM to generate a graded mesh.

This program start by forming the mesh for individual macro-element, as the number of division is specified by the users. The generation of elements and nodes for a quadrilateral macro-element is very simple, and the accumulation of elements from different macro-elements can be tedious. A simple method is adopted by the users many years ago. Suppose the number of individual nodes in each macro-element is given by Ni, and the last node number in macro-element 1 in N1. The first node in macro-element 2 is accumulated to 1+N1, while the last node in macro-element 2 is accumulated to node N1+N2. This process can repeated until all the elements in the last macro-element is numbered. It is clear that there are many duplicate nodes in such numbering system, and they must be condensed out. The exterior nodes of each macro-element is compared with the exterior nodes of other macro-element then. If the coordinates of different exterior nodes are the same, one node is condensed, and all the node number will be decreased by one, and the connectivity of each element will be changed. In the end, all the duplicated nodes are eliminated by subroutine CHECK, and the node numbering will be post-processed for minimum profile or bandwidth. Some old Fortran IV code similar to the present program is available in some journal, and the authors have actually adopted such codes during learning of finite element method. There are however some drawback to the layout of the macro-element which is not mentioned in the journal paper, but is found by the authors during mesh generation for a seepage problem. In view of that, the authors have developed the present code for research and teaching (Cheng and Tsui 1993).

```
PROGRAM GMESH

IMPLICIT REAL*4 (A-H,O-Z)

PARAMETER (MTOT=150000)

INTEGER IA(MTOT)

REAL A

COMMON/NUMROAD/IAUX,KGEO,N0,N01,N02,N03,IDY,NELEMAUX,NO
DEAUX,ITWO

CHARACTER*20 INPUT

COMMON A(MTOT)

EQUIVALENCE (A(1),IA(1))

a=0. ; ia=0

WRITE (*,*) 'INPUT DATA FILENAME ? '

READ (*,*) INPUT

OPEN(UNIT=IAUX,FILE=INPUT)

OPEN(1,FILE='MESH')

! ITWO = 1, SINGLE PRECISION

! ITWO = 2, DOUBLE PRECISION

!
```

```
ITWO=1

CALL DATIN

CALL CLOSEF

END

BLOCK DATA

IMPLICIT INTEGER*4 (I-N)

COMMON/NUMROAD/IAUX,KGEO,N0,N01,N02,N03,IDY,NELEMAUX,NO
DEAUX,ITWO

DATA IAUX,KGEO /11,31/

END

SUBROUTINE DATIN

IMPLICIT REAL*4(A-H,O-Z)

COMMON A(1)

REAL A

INTEGER IA(1)

EQUIVALENCE(A(1),IA(1))

COMMON/NUMROAD/IAUX,KGEO,N0,N01,N02,N03,IDY,NELEMAUX,NO
DEAUX,ITWO

READ(IAUX,*) NODEAUX,NELEMAUX,IDY,ITYPE

! NODEAUX = NODE NO. OF SUPER-ELEMENT

! NELEMAUX = NO. OF SUPER-ELEMENT

! IDY = 1 NODE NUMBER START FROM NODE 1

! IDY = 0 NODE NUMBER START FROM HIGHEST Y

! ITYPE = 1 FOR 4 NODES, ELSE 3 NODES MESH GENERATION

N0=1
```

```
N01=N0+2*NODEAUX*ITWO

N02=N01+6*NELEMAUX

N03=N02+2*NELEMAUX*ITWO

N04=N03+3*(NELEMAUX+1)

N1=N04

IF(IDY.GE.0)IDY=1

IF(IDY.LT.0)IDY=-1

IF(NODEAUX.EQ.0.OR.NELEMAUX.EQ.0) THEN

STOP

END IF

CALL INITIAL(A(N0),IA(N01),A(N02),IA(N03),N1,ITYPE)

RETURN

END

SUBROUTINE INITIAL(XYZAUX,NEAUX,QLM,NEPR,N1,ITYPE)

IMPLICIT REAL*4(A-H,O-Z)

COMMON A(1)

REAL A

INTEGER IA(1)

EQUIVALENCE(A(1),IA(1))

COMMON/NUMROAD/IAUX,KGEO,N0,N01,N02,N03,IDY,NELEMAUX,NO
DEAUX,ITWO

DIMENSION XYZAUX(2,*),NEAUX(6,*),QLM(2,*),NEPR(3,*)

DIMENSION XYZARFA(2,4)

CHARACTER YESNO
```

```
DO I=1,NODEAUX

!xyzaux = coordinates of nodes of super-element

READ(IAUX,*) II,(XYZAUX(III,II),III=1,2) ! read in coordinates of super-
element

IF(II.NE.I) THEN

STOP

END IF

END DO

YESNO="N"

DSMIN=1.0D0

DO I=1,NODEAUX

DO J=I+1,NODEAUX

DX=XYZAUX(1,I)-XYZAUX(1,J)

DY=XYZAUX(2,I)-XYZAUX(2,J)

DS=DX*DX+DY*DY

IF(DS.LT.DSMIN) DSMIN=DS

IF(DS.EQ.0.0D0) stop

END DO

END DO

DSMIN=DSMIN/100000.0

IF(YESNO.EQ."Y") THEN

CALL CLOSEF

STOP

END IF
```

```
! ----------------------------------------------------------------
! Reading element information of auxiliary file and making mesh
! ----------------------------------------------------------------
NEPR(1,1)=N1 ; NN = 0
DO I=1,NELEMAUX
! neaux = node numbers of super-element ielemaux
READ(IAUX,*) IELEMAUX,(NEAUX(II,I),II=1,4), LL,MM,QL,QM
NN = NN + (LL+1)*(MM+1)
! LL,MM DIVISION IN 1ST AND 2ND DIRECTION
! QL, QM =1 FOR EQUAL DIVISION, ELSE RATIO OF DIVISION
! USEFUL FOR GRADING OF MESH
IF(IELEMAUX.NE.I) THEN
! WRITE(KGEO,'(1X,2(A,I5))')
! & '!!!! Error: Element',IELEMAUX,' not equal to',I
! CALL CLOSEF
STOP
END IF
! LL = 4, MM =4
! QL = 1, QM =1
L=LL+2 ; M=MM+2
NEAUX(5,I)=L ; NEAUX(6,I)=M
QLM(1,I)=QL ; QLM(2,I)=QM
NEPR(2,I)=NEPR(1,I)+2*L*M*ITWO
NEPR(3,I)=NEPR(2,I)+L*M
```

```
N1=NEPR(1,I) ; N2=NEPR(2,I)

N3=NEPR(3,I)

N4=NEPR(3,I)+4*(L-1)*(M-1)

N5=N4+L*ITWO ; N6=N5+M*ITWO

write (1,*) n1,n2,n3,n4,n5,n6

! write(KGEO,'(7I5,2F10.2)')IELEMAUX,(NEAUX(II,I),II=1,4),

! & LL,MM,QL,QM

!

! A(N4) AND A(N5) SAVE CSAII AND ETARJ VALUES

!

CALL MESH(A(N4),A(N5),L,M,QL,QM)

DO J=1,4

DO K=1,2

XYZARFA(K,J)=XYZAUX(K,NEAUX(J,I))

END DO

END DO

CALL
DFINFNM(A(N4),A(N5),A(N1),IA(N2),IA(N3),L,M,XYZARFA,NODEREC)

CALL
CHECK(IELEMAUX,A(N1),IA(N2),IA(N03),NODEREC,NODEFAL,DSMIN,N
EAUX)

NEPR(1,I+1)=N4

END DO

!

! RENUMBERING NODES
```

```
!
100 KPERMU=0
DO I=1,NELEMAUX
N1=NEPR(1,I) ; N2=NEPR(2,I)
N3=NEPR(3,I)
DO J=1,N3-N2
CALL GETNOD(A(N1),IA(N2),XI,YI,J,NODEI)
DO I1=I,NELEMAUX
N4=NEPR(1,I1) ; N5=NEPR(2,I1)
N6=NEPR(3,I1)
DO J1=1,N6-N5
CALL GETNOD(A(N4),IA(N5),XJ,YJ,J1,NODEJ)
KCHANGE=0
IF(NODEI.NE.NODEJ)THEN
IF(YI.LT.IDY*YJ-DSMIN.AND.NODEI.GT.NODEJ)KCHANGE=1
IF(YI.GT.IDY*YJ+DSMIN.AND.NODEI.LT.NODEJ)KCHANGE=1
IF(ABS(YI-YJ).LE.DSMIN)THEN
IF(XI.LT.XJ.AND.NODEI.GT.NODEJ)KCHANGE=1
IF(XI.GT.XJ.AND.NODEI.LT.NODEJ)KCHANGE=1
END IF
END IF
IF(KCHANGE.NE.0)THEN
KPERMU=KPERMU+1
DO I2=1,NELEMAUX
```

```
N8=NEPR(2,I2)
N9=NEPR(3,I2)
DO J2=N8,N9-1
IF(IA(J2).EQ.NODEI)THEN
IA(J2)=NODEJ
ELSE IF(IA(J2).EQ.NODEJ)THEN
IA(J2)=NODEI
END IF
END DO
END DO
KCHANGE=NODEI
NODEI=NODEJ
NODEJ=KCHANGE
END IF
END DO
END DO
END DO
END DO
IF(KPERMU.NE.0) GOTO 100
WRITE(1,'(3(1X,I6))') NODEFAL,NN ! node,element
CALL WRGEOTL(IA(N03),NODEFAL,NELEMAUX,ITYPE,NN)
RETURN
END
SUBROUTINE MESH(CSAII,ETARJ,L,M,QL,QM)
```

```
IMPLICIT REAL*4 (A-H,O-Z)

DIMENSION CSAII(*),ETARJ(*)

COMMON/NUMROAD/IAUX,KGEO,N0,N01,N02,N03,IDY,NELEMAUX,NO
DEAUX,ITWO

CSAIL=0.0

DO I=1,L-1

CSAIL=CSAIL+QL**(I-1)

END DO

ETARM=0.0

DO J=1,M-1

ETARM=ETARM+QM**(J-1)

END DO

CSAII(1)=-1.0

DO I=2,L

CSAII(I)=CSAII(I-1)+2.0*(QL**(I-2))/CSAIL

END DO

ETARJ(1)=-1.0

DO J=2,M

ETARJ(J)=ETARJ(J-1)+2.0*(QM**(J-2))/ETARM

END DO

RETURN

END

!*********************************************

! Define node and element information
```

```
!*********************************************

SUBROUTINE
DFINFNM(CSAII,ETARJ,XY,NODNUM,NODSEL,L,M,XYZARFA,NODERE
C)

IMPLICIT REAL*4(A-H,O-Z)

DIMENSION CSAII(*),ETARJ(*),XY(2,*),NODSEL(4,*),NODNUM(*)

DIMENSION XYZ(2),XYZARFA(2,4)

COMMON/NUMROAD/IAUX,KGEO,N0,N01,N02,N03,IDY,NELEMAUX,NO
DEAUX,ITWO

DO J=1,M

DO I=1,L

IJNOD=L*(J-1)+I

NODNUM(IJNOD)=IJNOD

css=CSAII(I)

tnn=ETARJ(J)

CALL COORD(XYZ,css,tnn,XYZARFA)

XY(1,IJNOD)=XYZ(1) ; XY(2,IJNOD)=XYZ(2)

END DO

END DO

DO J=1,M-1

DO I=1,L-1

MIJK=(L-1)*(J-1)+I

NODSEL(1,MIJK)=L*(J-1)+I ; NODSEL(2,MIJK)=L*(J-1)+I+1

NODSEL(3,MIJK)=L*J+I+1 ; NODSEL(4,MIJK)=L*J+I

END DO
```

```
END DO
NODEREC=L*M
RETURN
END
!*********************************************
! Calculate coordinate
!*********************************************
SUBROUTINE COORD(XYZ,CSAI,ETAR,XYZARFA)
IMPLICIT REAL*4 (A-H,O-Z)
DIMENSION XYZ(2),XYZARFA(2,4)
DO I=1,2
XYZ(I)=0.0
DO J=1,4
XYZ(I)=XYZ(I)+FAIARFA(J,CSAI,ETAR)*XYZARFA(I,J)
END DO
END DO
RETURN
END
!*********************************************
! Shape function
!*********************************************
REAL*4 FUNCTION FAIARFA(J,CSAI,ETAR)
IMPLICIT INTEGER*4 (I-N)
IMPLICIT REAL*4 (A-H,O-Z)
```

```
DIMENSION CSAIARFA(4),ETARARFA(4)

DATA CSAIARFA /-1.0,1.0,1.0,-1.0/

DATA ETARARFA /-1.0,-1.0,1.0,1.0/

FAIARFA=0.25*(1+CSAI*CSAIARFA(J))*(1+ETAR*ETARARFA(J))

RETURN

END

!*********************************************

! Define node and element information

!*********************************************

SUBROUTINE
CHECK(IELEMAUX,XY,NODNUM,NEPR,NODEREC,NODEFAL,DSMIN,
NEAUX)

IMPLICIT REAL*4(A-H,O-Z)

COMMON A(1)

REAL A

INTEGER IA(1)

EQUIVALENCE(A(1),IA(1))

COMMON/NUMROAD/IAUX,KGEO,N0,N01,N02,N03,IDY,NELEMAUX,NO
DEAUX,ITWO

DIMENSION XY(2,*),NODNUM(*),NEPR(3,*)

INTEGER NEAUX(6,*)

LOGICAL need_merge

IF (IELEMAUX.EQ.1) THEN

NODEFAL=0 ; NEWNODE=NODEREC

GOTO 100
```

```
END IF
NEWNODE=0
DO I=1,NODEREC
XI=XY(1,I) ; YI=XY(2,I) ; KCUNT=0
DO J=1,IELEMAUX-1
need_merge = .false.
IF ((I .le. NEAUX(5, IELEMAUX)) .AND. &
& (NEAUX(1, IELEMAUX) .EQ. NEAUX(4, J)) .AND. &
& (NEAUX(2, IELEMAUX) .EQ. NEAUX(3, J))) THEN
need_merge = .true.
END IF
IF ((mod(I, NEAUX(5, IELEMAUX)) .EQ. 0) .AND. &
& (NEAUX(2, IELEMAUX) .EQ. NEAUX(1, J)) .AND. &
& (NEAUX(3, IELEMAUX) .EQ. NEAUX(4, J))) THEN
need_merge = .true.
END IF
IF ((mod(I, NEAUX(5, IELEMAUX)) .EQ. 1) .AND. &
& (NEAUX(1, IELEMAUX) .EQ. NEAUX(2, J)) .AND. &
& (NEAUX(4, IELEMAUX) .EQ. NEAUX(3, J))) THEN
need_merge = .true.
END IF
IF (((NODEREC + 1 - I) .le. NEAUX(5, IELEMAUX)) .AND. &
& (NEAUX(3, IELEMAUX) .EQ. NEAUX(2, J)) .AND. &
& (NEAUX(4, IELEMAUX) .EQ. NEAUX(1, J))) THEN
```

```
need_merge = .true.
END IF
IF (.not. need_merge) THEN
cycle
END IF
N1=NEPR(1,J) ; N2=NEPR(2,J) ; N3=NEPR(3,J)
NODEREC1=N3-N2
CALL NODCHECK(A(N1),IA(N2),XI,YI,NODEREC1,KCUNT,DSMIN)
IF(KCUNT.GT.0)GOTO 10
END DO
NEWNODE=NEWNODE+1
NODNUM(I)=NODEFAL+NEWNODE
GOTO 20
10 CONTINUE
NODNUM(I)=KCUNT
20 CONTINUE
END DO
100 CONTINUE
NODEFAL=NODEFAL+NEWNODE
RETURN
END
SUBROUTINE NODCHECK(XY,NODNUM,XI,YI,N,KCUNT,DSMIN)
IMPLICIT REAL*4 (A-H,O-Z)
DIMENSION XY(2,*),NODNUM(*)
```

```
DO J=1,N
XJ=XY(1,J) ; YJ=XY(2,J)
DSREC=(XI-XJ)*(XI-XJ)+(YI-YJ)*(YI-YJ)
IF(DSREC.LT.DSMIN) THEN
KCUNT=NODNUM(J)
EXIT
END IF
END DO
RETURN
END
!*********************************************
SUBROUTINE GETNOD(XY,NODNUM,XI,YI,I,NODEI)
IMPLICIT REAL*4 (A-H,O-Z)
DIMENSION XY(2,*),NODNUM(*)
XI=XY(1,I) ; YI=XY(2,I) ; NODEI=NODNUM(I)
RETURN
END
!*********************************************
! Writing GEOTL.DAT file
!*********************************************
SUBROUTINE WRGEOTL(NEPR,NB11,NB18,ITYPE,NELS)
IMPLICIT REAL*4(A-H,O-Z)
COMMON A(1)
REAL A,xy
```

```
INTEGER IA(1),NN

EQUIVALENCE(A(1),IA(1))

ALLOCATABLE XY(:,:)

COMMON
/NUMROAD/IAUX,KGEO,N0,N01,N02,N03,IDY,NELEMAUX,NODEAUX,IT
WO

DIMENSION NEPR(3,*)

ALLOCATE (XY(2,NB11))

IF (ITYPE ==1) NOD=4

IF (ITYPE /=1) NOD=3

! ----------------------------------------------------

! Writing node information into GEOTL.DAT file

! ----------------------------------------------------

NN=0

DO I=1,NB11

DO I1=1,NELEMAUX

N1=NEPR(1,I1) ; N2=NEPR(2,I1) ; N3=NEPR(3,I1)

DO J=1,N3-N2

CALL GETNOD(A(N1),IA(N2),XI,YI,J,NODEI)

IF (I.EQ.NODEI)GOTO 10

END DO

END DO

10 CONTINUE

XY(1,I) = XI ; XY(2,I) = YI

END DO
```

```
DO I = 1,NB11

WRITE(1,'(2X,I5,2X,F10.4,2X,F10.4)') I,XY(1:2,I)

END DO

! -------------------------------------------------------

! Writing element information into GEOTL.DAT file

! -------------------------------------------------------

! WRITE(*,'(1X,A)') '**** Writing element data . . .'

!C

DO I1=1,NELEMAUX

N1=NEPR(1,I1) ; N2=NEPR(2,I1)

N3=NEPR(3,I1) ; N4=NEPR(1,I1+1)

CALL WRITELE(IA(N2),IA(N3),N4-N3,NN,XY,NB11,ITYPE,NOD,NELS)

END DO

RETURN

END

!C*********************************************

SUBROUTINE
WRITELE(NODNUM,NODSEL,N,NN,XY,NB11,ITYPE,NOD,NELS)

COMMON
/NUMROAD/IAUX,KGEO,N0,N01,N02,N03,IDY,NELEMAUX,NODEAUX,IT
WO

DIMENSION NODNUM(*),NODSEL(4,*),XY(2,NB11)

NN=0

DO I=1,N/4

NN=NN+1
```

```
WRITE(1,'(2X,I5,4(2X,I5))') NN, (NODNUM(NODSEL(J,I)),J=1,4)

END DO

RETURN

END

SUBROUTINE CLOSEF

IMPLICIT INTEGER*4 (I-N)

COMMON/NUMROAD/IAUX,KGEO,N0,N01,N02,N03,IDY,NELEMAUX,NO
DEAUX,ITWO

CLOSE(UNIT=IAUX)

RETURN

END
```

APPENDIX 2-3. BANDWIDTH/PROFILE MINIMIZER

In the development of the mesh generation program MESH, the authors have adopted two bandwidth/profile minimizers (Cheng 1993). A simple one based on the reverse-Cuthill method by George and Liu (1981) is slightly modified and given here for illustration. The modifications to the code by George and Liu is minimal, except with the use of some modern dynamic memory management. The performance of this method is only moderate, but is sufficiently simple to be adopted for research purposes. The authors have also developed other bandwidth minimizer using other algorithms, which can be obtained freely from the authors. An algorithm JCL (equation.com) which is based on heuristic algorithm appears to function very well for the bandwidth

Before introducing the mechanism of the skyline profile minimizer, some graph theory concerning the organization of the mesh have to be introduced. The mesh generated from the mesh generation program is a set of node (nodal point) linked by lines (element boundaries). The mathematical relation that define how the mesh lines and nodes are located and joined can be expressed by several algorithms. In the mesh generation program, the graph relation is stored in an array called JOT(NMEM,NENODE), where

NMEM : number of element in the mesh

NENODE : number of node per element. In simple words, the nth row of the array JOT contains the boundary node numbers of the nth element. For the simplified mesh below:

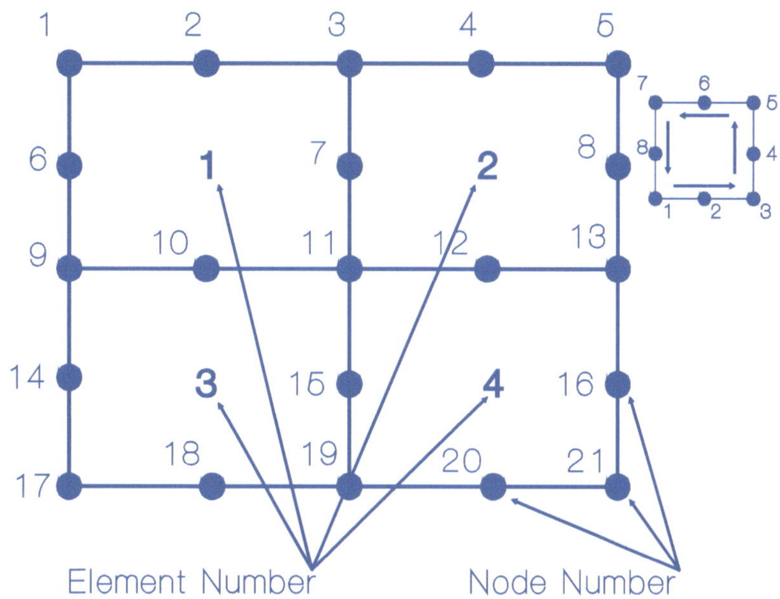

Fig. (A3-1). Elements and nodes for a simple mesh problem.

The element generated is 8-node isoparametric quadrilateral (NENODE=8). Total number of element counted is 4. Therefore, the element-based connectivity array is a 8 x 4 array. In element 1, the boundary nodes (define by the small element on the top right) are:

Element 1: 9 10 11 7 3 2 1 6, similarly,

Element 2: 11 12 13 8 5 4 3 7

Element 3: 17 18 19 15 11 10 9 14

Element 4: 19 20 21 16 13 12 11 15

In node-based adjacency array storage methodology, the adjacent node(s) with respect to a particular node is evaluated. The mesh is stored using the integer array pair (XADJ,ADJNCY). For a mesh of N nodes, the adjacency of x is the list containing all the nodes adjacent to node x. Such a structure can be implemented quite simply and economically by storing the adjacency list sequentially in a one-dimensional array ADJNCY along with an index array XADJ of length N+1 containing pointers to the beginning of each adjacency list in ADJNCY. The total storage for the node-based adjacency array is:

{ NNODE + 1 } + {(XADJ(NNODE+1)) -1}

XADJ ADJNCY

For the mesh below, the array ADJNCY & XADJ is evaluated as follow:

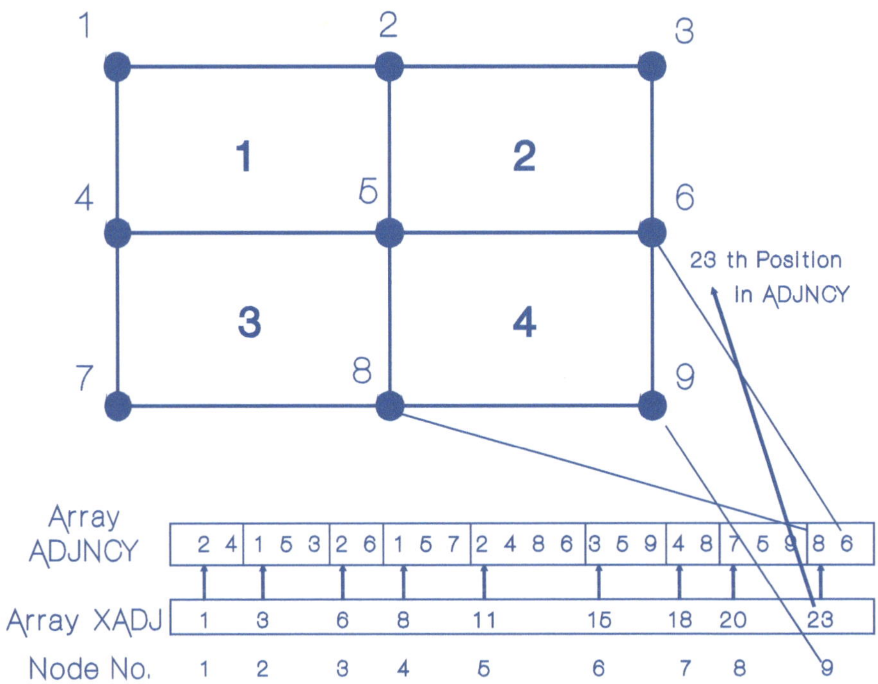

Fig. (A3-2). The graph relation for a simple mesh problem.

The corresponding graph matrix is expressed as:

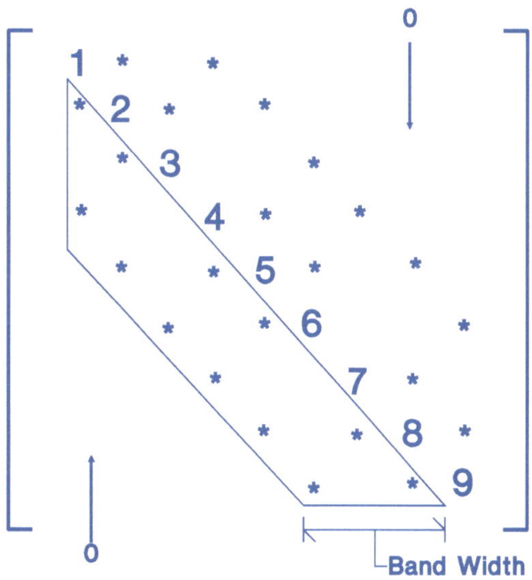

Fig. (A3-3). Graph matrix for the simple mesh problem.

In the skyline profile minimizer, an algorithm (Reverse CutHill McKee) is adopted to reduce the bandwidth of the above graph matrix. In addition to the global reduction of the bandwidth, the algorithm is also able to reduce the local bandwidth (individual row and column). This facilitates a slightly more sophisticated graphic storage scheme known as envelope or profile method. For the traditional storage scheme, the whole band in the graphic matrix is stored in the memory. However, in the skyline envelope method, it takes the advantage of the variation of band width with each individual node and the bandwidth are stored individually.

The profile reduction ordering scheme used in this program is the variant of the Cuthill_McKee ordering. In 1969, Cuthill and McKee published their algorithm which was primarily designed to reduce the bandwidth of a sparse symmetric matrix. The scheme makes use of the following observation. Let y be a labelled node, and z an unlabelled neighbour of y. To minimize the bandwidth of the row associated with z, it is apparent that the node z should be ordered as soon as possible after y. The following figures illustrate this point.

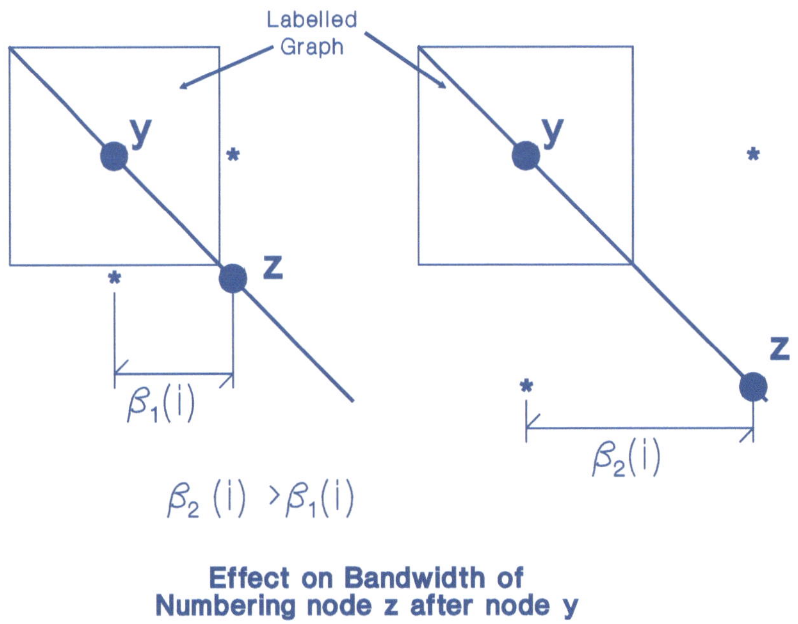

Fig. (5145). Renumbering of nodes in Reverse Cuthill McKee algorithm.

The Cuthill-McKee scheme may be regarded as a method that reduces the bandwidth of a matrix via a local minimization of the β_i's. This suggests that the scheme can be used as a method to reduce the profile $\Sigma\beta_i$ of a matrix. George (1971), in his study of the profile methods, discovered that the ordering obtained by reversing the Cuthill-McKee ordering often turns out to be much superior to the original ordering in terms of profile reduction, although the bandwidth remains unchanged. He called this the reverse Cuthill-McKee ordering (RCM). It has since then be called the reverse Cuthill-McKee ordering (RCM). It has since been proved that the reverse scheme is never inferior, as far as envelope storage and envelope operation counts are concerned with compared with the Cuthill-Mckee method.

Diagrammatically, the Reverse Cuthill McKee algorithm can be summarised as:

STEP 1:Determine a staring node r and assign $x_1 \leftarrow r$.

STEP 2:(Main loop) For i=1,....N, find all the unnumbered neighbour of the node x_i and number them in increasing order of degree.

STEP 3:(Reverse ordering) The reverse Cuthill-McKee ordering is given by y_1, y_2, ... , y_N where $y_i = x_{N-i+1}$ for i = 1 ,.... ,N

In the case when the graph is disconnected, we can apply the above algorithm to each connected component of the graph. In order to illustrate how the algorithm works, we work through Step 1 to Step 3 for the example.

Consider the graph on the right. In the first step of the algorithm, we are required to work out the starting node and assign $x_1 \leftarrow r$. The mechanism of evaluation of the starting node will be discussed later. For simplicity, node "g" is picked as the starting node, that is $x_1 = g$. After the completion of Step 1, Step 2 of the algorithm is executed to evaluate how the nodes are numbered. The method and result are listed in the following table:

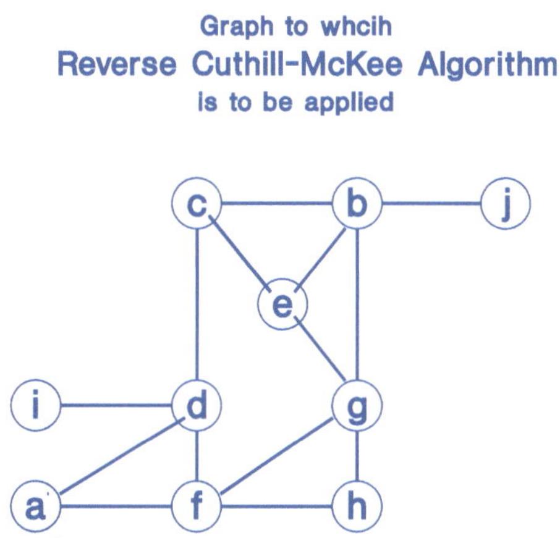

Fig. (A3-5). Example graph example.

i	Node x_i	Unnumbered neighbours in increasing order of degree
1	g	h,e,b,f
2	h	-
3	e	c
4	b	j
5	f	a,d
6	c	-
7	j	-
8	a	-

9 d i

10 i -

The resulting reverse Cuthill-McKee ordering is given in below diagram. The envelope size is 22.

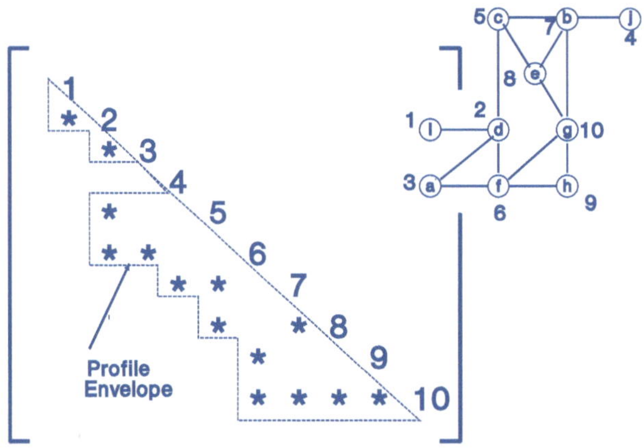

The final ordering & its matrix structure

Fig. (A3-6). A36 Reordered graph for Fig. A3-5.

Fif.A3-6 Reordered graph for Fig.A3-5.

The effectiveness of the ordering algorithm depends quite crucially on the choice of the starting node. In the example, if we pick node "a" as the starting node, we get a smaller profile of 18. In finding a starting node for the RCM algorithm, the main objective is to find a pair of nodes which are at maximum or near maximum "distance" apart (defined below). Substantial experience suggested that such nodes are good starting nodes for several ordering algorithms including RCM algorithm.

A path of length *k* from node x_o to x_k is another set of distinct vertices (x_o, x_1, ... , x_k), where

$x_i \in$ Adj (x_{i+1}) for $0 \leq i \leq k-1$ where Adj(x_k) = adjacent node of x_k

The *distance d(x,y)* between two nodes x and y in the connected graph G = (X,E) is simply the length of a shortest path joining nodes x and y. Following Berge (1962), the eccentricity of a node x is defined as:

$l(x) = \max \{ d(x,y) \mid y \in X \}$

The *diameter* of graph G is given by:

$\delta(G) = \max \{ l(x) \mid x \in X \}$, or equivalently

$\delta(G) = \max \{ d(x,y) \mid x,y \in X \}$

A node $x \in X$ is said to be a *peripheral* node if its eccentricity is equal to the diameter of the graph, that is, if $l(x) = \delta(G)$.

In the following figure, a graph of 8 nodes with a diameter of 5 is shown. The nodes x_2, x_5, x_7 are the peripheral nodes.

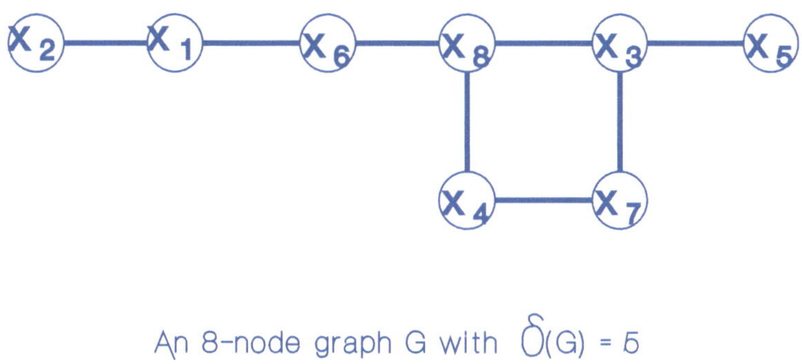

Fig. (A3-7). Determination of peripheral nodes.

With the terminology established, the objective in the following section is to describe an efficient heuristic algorithm for finding nodes of high eccentricity. The algorithm is not guaranteed to find a peripheral node, or even one that is close to being peripheral. Nevertheless, the nodes found usually do have high eccentricity, and are good starting nodes for the algorithms (*e.g.* RCM) that employ them. Furthermore, except for some fairly trivial situations, there seems to be no reason to expect that peripheral nodes are any better as starting nodes than those found by this algorithm. Finally, in many situations, it is probably too expensive to find peripheral nodes even if it were known to be desirable to use them, since the best known algorithm for finding them has a high time complexity bound (ie. tend to be time consuming in run time execution). In what follows, the nodes produced by this algorithm is referred as *pseudo-peripheral* nodes.

Before the algorithm is explained, some notation and terminology which is useful in description of the algorithm have to be introduced. A key construction in the algorithm is the *rooted level structure* (Arany 1972[15]). Given a node $x \in X$, the level structure rooted at x is the partitioning _(x) of X satisfying

_(x) = { L_o(x), L_2(x), ... , $L_{l(x)}$(x) } ,

where L_o(x) = {x}, L_1(x) = Adj(L_o(x)),

and L_i(x)=Adj(L_{i-1}(x)) - L_{i-2}(x), i = 2, 3, ... , l(x).

The eccentricity l(x) of x is called the *length* of _(x), and the *width* w(x) of _(x) is defined by

w(x) = max { $|L_i(x)|$ | $0 \leq i \leq$ l(x) }

In the following figure, a rooted level structure of the graph in preceding page is shown, rooted at x_6. Note that l(x_6) = 3 and w(x_6) = 3.

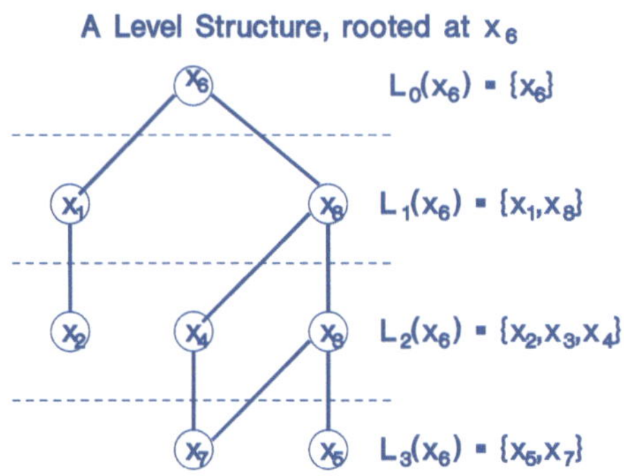

Fig. (A3-8). Root level structure.

After the necessary terminology is introduced, the pseudo-peripheral node finding algorithm which is essentially a modification of an algorithm due to *Gibbs et al.* can be described. Adapting the level structure notation just introduced, the algorithm can be presented as follows.

Algorithm of finding the *Pseudo-peripheral node*

STEP 1:(Initialization) - Choose an arbitrary node r in X.

STEP 2:(Generate a level structure) - Construct the level structure rooted at r : _(r) = { L_o(r), L_1(r), ... , $L_{l(r)}$(r) }.

STEP 3:(Shrink last level) - Choose a node x in $L_{l(r)}$(r) of minimum degree

STEP 4:(Generate a level structure) -

a)Construct the level structure rooted at x:

_(x) = { $L_o(x), L_2(x), \ldots, L_{l(x)}(x)$ }

b)If l(x) > l(r), set r ← x and go to step 3.

STEP 5:(Finished) - The node x is a pseudo-peripheral node.

The diagram below shows the schematic layout of the skyline minimizer (SKYMIN11.EXE).

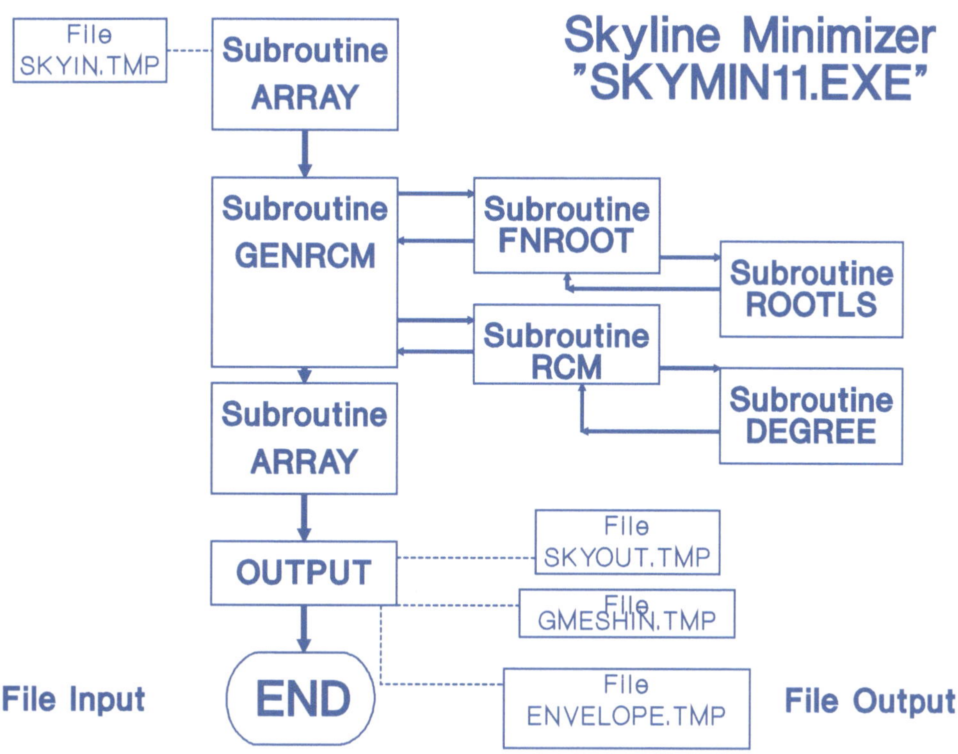

Fig. (A3-9). Structure of the skyline profile minimizer.

Appendix 2-3.1. Functions of Some Subroutines

Appendix 2-3.1.1. Subroutine ARRAY

The subroutine "ARRAY" is the skyline minimizer convert if the *element based connectivity array (JOT)* into the *node based adjacency array pair (XADJ,ADJNCY)*. The task of conversion is perform by the following procedures:

"For each node in the mesh, check whether the node number equal to that prescribe by JOT(NMEM,NENODE). For instance, if the node number i was found identical to JOT(x,y), then node JOT(x,y+1) and JOT(x,y-1) will be the adjacent node number of node i (N.B. there may be as many as 12 adjacent nodes in case of triangular element). Suppose "z" nodes that found to be adjacent to node i, then all of these nodes will be declared as the adjacency of node i. They will be stored at the end of the array "ADJNCY" and the new pointer (Old pointer + z) will be stored at the i th position of the array "XADJ". The above process will be repeated until all node in the mesh have been checked."

The adjacency array pair will be used as input in the subroutine GENRCM for node number re-ordering. The subroutine ARRAY will be called again to generate the re-ordered adjacency array pair. The re-ordered array pair will be stored in an output file "GMESHIN.TMP" and will be used when the mesh have to graphically displayed on a monitor.

Appendix 2-3.1.2. Subroutine GENRCM (GENeral RCM)

This subroutine finds the RCM ordering of a general disconnected graph. It proceeds through the graph, and calls the subroutine RCM to number each connected component. The inputs to the subroutine are the number of nodes (or equations) NEQNS, and the graph in the array pair (XADJ,ADJNCY). The arrays MASK and XLS are working arrays, used by the subroutines FNROOT and RCM, which are called by GENRCM.

The subroutine begins by setting all values of MASK to 1 (loop DO 100 i= ...). It then loops through MASK until it finds an i for which MASK(i) =1; node i along with MASK, XADJ, and ADJNCY will specify a connected subgraph of the original graph G. The subroutines FNROOT and RCM are then called to order the nodes of that subgraph. Note that NUM points to the first free position in the array PERM, and is updated after each call to RCM. The actual parameter in GENRCM corresponding to PERM in RCM is PERM(NUM); that is, PERM in RCM corresponds to the last NEQNS - NUM + 1 elements of PERM in GENRCM. Note also that these same elements of PERM are used to store the level structure in FNROOT. They correspond to the array LS in the execution of that subroutine.

After the component is ordered, the search for another *i* for which MASK(i)≠0 resumes, until either the loop is exhausted, or NEQNS nodes have been numbered.

Appendix 2-3.1.3. Subroutine FNROOT (FiNd ROOT)

This subroutine finds a pseudo-peripheral node of a connected component of a given graph. The subroutine operates on the connected component specified by the input arguments ROOT, MASK, XADJ, and ADJNCY, as described before.

The first call to ROOTLS corresponds to STEP 2 of the algorithm. If the component consists of a single node or a chain with ROOT as its endpoint, then ROOT is a peripheral node and LS contains it corresponding rooted level structure, so execution terminates. Otherwise, a node of minimum degree in the last level is found (STEP 3 of the algorithm; DO 300 ... loop of the subroutine). The new level structure rooted at this node is generated (the call to ROOTLS with label 400) and the termination test (STEP 4.b of the algorithm) is performed. If the test fails, control transfers to statement 100 and the procedure is repeated. On exit, ROOT is the node number of the pseudo-peripheral node, and the array pair (XLS, LS) contains the corresponding rooted level structure.

Appendix 2-3.1.4. Subroutine ROOTLS (ROOTed Level Structure)

The purpose of this subroutine is to generate a level structure of the connected component specified by the input parameters ROOT, MASK, XADJ, and ADJNCY. On exit from the subroutine, the level structure generated is rooted at ROOT, and is contained in the array pair (XLS,LS), with nodes at level k given by LS(j), XLS(k)≤j<XLS(k+1). The number of levels is provided by the variable NLVL. Note that since Fortran does not allow zero subscripts, we cannot have a "zero level," so k here corresponds to level L_{k-1} in the level structure _(ROOT). Thus, NLVL is one greater than the eccentricity of ROOT.

The subroutine finds the nodes level by level; a new level is obtained for each execution of the loop DO 400 .... As each new node is found (in executing the loop DO 300 ...), the node number is placed in the array LS more than once. After the level structure has been generated, the values of MASK for the nodes in the level structure are reset to 1 (by executing the loop DO 500 ...).

Appendix 2-3.1.5. Subroutine RCM (Reverse Cuthill-McKee)

This subroutine applies the RCM algorithm. It operates on a connected component specified by the input parameters ROOT, MASK, XADJ, and ADJNCY. The starting node is ROOT.

Since the algorithm requires the degrees of the nodes in the component, the first step is to compute those degrees by calling the subroutine DEGREE. The nodes are found and ordered by a level fashion; a new level is numbered each time the loop DO 600 I = ... is executed. The loop DO 200 I = ... finds the unnumbered neighbours of a node, and the remainder of the DO 600 loop implements a linear insertion sort to order those neighbours in increasing order of degree. The new ordering is recorded in the array PERM. The final loop (DO 700 I= ...) reverse the ordering, so that the reverse Cuthill-McKee ordering, rather than the standard Cuthill-McKee ordering is obtained.

Note that just as in the subroutine ROOTLS, MASK(i) is set to zero as node *i* is recorded. However, unlike ROOTLS, the subroutine RCM does not restore MASK to its original input state. The values of MASK corresponding to the nodes of the connected component that has been numbered remain set to zero on exit from the subroutine.

Appendix 2-3.1.6. Subroutine DEGREE

This subroutine computes the degrees of the nodes in a connected component of a graph. The subroutine operates on the connected component specified by the input parameters ROOT, MASK, XADJ and ADJNCY.

Beginning with the first level (containing only ROOT), the degrees of the nodes are computed one level at a time (loop DO 400 I= ...). As the neighbours of these nodes are examined (loop DO 200 J+ ...), those which are not already recorded in LS are put in that array, thus generating the next level of nodes. When a node is put in LS, its corresponding value of XADJ has it sign changed, so that the node will only be recorded once. This function was performed using MASK in the subroutine ROOTLS, but here MASK must be maintained in its input form so that the degree will be computed correctly. The variable CCSIZE contains the number of nodes currently in LS. After all node have been found, and their degrees have been computed, the nodes in LS are used to reset the signs of the corresponding elements of XADJ to their original values (loop DO 500 I = ...).

PROGRAM SKYMIN11

APPENDIX 2-4. THIN AND THICK PLATE FINITE ELEMENT PROGRAM

For the illustration and learning in developing a finite element program, a fairly complicated thin and thick Mindlin plate program is demonstrated. The plate element which is developed by Cheng and Law (2008) has demonstrated very high accuracy as compared with other published plate element. This program is developed under Dos environment with the support of graphics by Dislin library. The use of Dos interrupt is also employed. A highly improved Windows program PLATE 1.0 has been developed by Cheng with the funding support from the Housing Department of the Hong Kong Government. The Windows version has many functions not available in this primitive program, but the present Dos program is a better and simpler version for illustration of the development of a finite element program. It should be noticed that the present program allow a coupled analysis, and a full stiffness matrix will be required if the user specified a coupled analysis in the input file.

```
MODULE DATA

INTEGER NPOIN,NELEM,IBS,NSTRE,NGAUB,NGAUV,NPILE

INTEGER NNODE,NEVAB,NGAUS2,NN,NSPRG,HG,BAND,ERRORN

INTEGER JCT,NM

REAL (KIND=8),ALLOCATABLE :: COORD(:,2)

REAL (KIND=8),ALLOCATABLE :: GPCOD(:,2,:),DBMAT(:,:),CARTD(2,:),
&

SMATX(:,:,:,:),ELCOD(2,:),DERIV(3,:),BMATX(5,:),UDL(:)

REAL (KIND=8),ALLOCATABLE :: SHAPE(:),REACT(:),SPRING(:)

INTEGER, ALLOCATABLE :: LNODS(:,:),PILEN(:),LOAD_NODE(:)

INTEGER, ALLOCATABLE :: NSPRING(:),DRAW(:,:),IFIX(:)

REAL(KIND=8), ALLOCATABLE :: GSTIF(:,:),ESTIF(:,:,:),GRID_X(:)

REAL(KIND=8), ALLOCATABLE ::
GLOAD(:),ELOAD(:),GLOAD1(:),GRID_Y(:)

REAL(KIND=8) :: PROPS(4),POSGP(3),WEIGP(3),RW,RT,KZERO

REAL(KIND=8), ALLOCATABLE :: PIPROP(:,6), PSTIF(:,:), PFLEX(:,:),
PDISPT(:)

REAL(KIND=8), ALLOCATABLE ::
CORD(:,2),TORCON(:),SMA(:),OLEN(:),GSTIF1(:,:)

INTEGER, ALLOCATABLE :: GNODE(:),MCON(:,2)

CHARACTER(LEN=1) :: COUPLING, RECTANGULAR_GRID,SHIFT

REAL(KIND=8) :: EP,GS,GB,VS,PA
```

```
CHARACTER(24) FILDAT, FILOUT, command
END MODULE DATA

!------------------------------ MAIN PROGRAM -----------------------------

PROGRAM MAIN
USE DATA
IMPLICIT NONE
character(1) :: key
!---------------------------- CALL SUBROUTINES ---------------------------
! npoin = total no. of points
! nelem = total no. of element
! ngaus = no. of Gaussian points
! nnode = no. of nodes per element
! nsprg = no. of spring supports
! nevab = total no. of unknowns per element = 3xnnode
! IBS = integration scheme - B for bending; S for shear
Call start
CALL OPEN
5 CALL INPUT(NPOIN,NELEM,NSTRE,PROPS,NNODE,NEVAB,NSPRG,NPILE)
IF (ERRORN > 0) THEN
WRITE (*,*) 'There are errors in the input file. Press q to quit, else edit '
read (*,*) key
if (key == "Q" .or. key == "q") stop
```

```
close(5) ; close(6)
command(1:5)='edit '
command(6:)=fildat
call system(command)
OPEN(6, FILE=FILOUT)
OPEN(5,FILE=FILDAT,STATUS='OLD')
go to 5
endif
close(5) ; close(6)
call drawmesh(2,npoin,nnode,nelem,1)
write (*,*) 'Is the mesh o.k. ? Press n if not acceptable.'
write (*,*) 'Press e for edit, else other key to proceed. '
read (*,*) key
if (key == "N" .or. key == "n") stop
if (key == "e" .or. key == "E") then
command(1:5)='edit '
command(6:)=fildat
call system(command)
OPEN(6, FILE=FILOUT)
OPEN(5,FILE=FILDAT,STATUS='OLD')
go to 5
ELSE
OPEN(5, FILE=FILDAT, STATUS='OLD')
open(6,file=filout,position='append')
```

```
ENDIF

ALLOCATE(SHAPE(NNODE),DERIV(2,NNODE),ELCOD(2,NNODE),CART
D(2,NNODE), &

DBMAT(NSTRE,NEVAB),ESTIF(NELEM,NEVAB,NEVAB), &

ELOAD(NEVAB),REACT(NN),GLOAD1(NN))

GLOAD1=0.D0 ; DBMAT=0.0D0 ; REACT=0.D0

IF (COUPLING == "N") CALL BANDWIDTH ! bandwidth for normal problem

IF (COUPLING == "N") THEN

ALLOCATE(GSTIF(NN,BAND))

ELSE

ALLOCATE(GSTIF(NN,NN),GSTIF1(NN,NN))

! for coupled analysis, a full stiffness matrix is required

ENDIF

CALL STIFPB(NELEM,NNODE,NEVAB,IBS,NSTRE,NPOIN,NGAUS2,&

POSGP,WEIGP,PROPS)

CALL GRILLAGE ! read in grillage member

CALL INSPRG(NN,NSPRG) ! read in nodal spring

CALL
LOADPB(NELEM,PROPS,NEVAB,NGAUB,POSGP,WEIGP,NNODE,NPOIN,
NN)

CALL PILEG(PSTIF,PFLEX) ! read in piles, using Randolph theory

CALL BOUND ! read in boundary condition

IF (COUPLING == "N" .OR. COUPLING == "n") THEN

CALL GAUSSB(GSTIF,GLOAD,NN,BAND) ! banded matrix solver

ELSE
```

```
CALL GAUSSJ(NN,GSTIF,GLOAD) ! full matrix solver

ENDIF

CALL DISPT(NN,NPOIN,GLOAD) ! output displacement

CALL REACTION(NEVAB,NNODE,NELEM,NPOIN) ! output support reaction

CALL
STREPB(NELEM,NGAUV,NGAUB,NSTRE,NNODE,NEVAB,NGAUS2)

! output element stress

CALL PREACT(GLOAD,PDISPT) ! output pile reaction

PRINT*, 'The analysis has completed successfully !!'

close(5) ; close(6)

call drawmesh(2,npoin,nnode,nelem,3) ! draw mesh of the problem

END PROGRAM MAIN

SUBROUTINE START

WRITE (*,*) 'Thick plate finite element analysis (Minlin plate) 1.0'

write (*,*) 'suitable for pile cap, raft foundation, bridge, slab structure'

write (*,*) 'or similar. Include grillage element in the present version'

write (*,*) 'Demo version for teaching purpose. Commercial use is'

write (*,*) 'strictly prohibited. Program similar to SAFE but include'

write (*,*) 'pile group interactions effects and is more flexible.'

write (*,*) 'Dr. Y.M. Cheng, Department of Civil and Structural Engineering'

write (*,*) 'Hong Kong Polytechnic University, 4/1/2002'

write (*,*)

end
```

```
!------------------------------ OPEN FILES ------------------------------

SUBROUTINE OPEN

USE DATA

PRINT*, 'What is your DATA file name ? '

READ(*,'(A)') FILDAT

OPEN(5, FILE=FILDAT, STATUS='OLD')

PRINT*, 'What is your OUTPUT file name ? '

READ(*,'(A)') FILOUT

OPEN(6, FILE=FILOUT)

WRITE (6,10) FILDAT

10 FORMAT(2X,'INPUT FILENAME IS ',A24)

RETURN

END SUBROUTINE OPEN

!------------------ INPUT CONTROL AND GEOMETRICAL DATA -
------------------

SUBROUTINE
INPUT(NPOIN,NELEM,NSTRE,PROPS,NNODE,NEVAB,NSPRG, NPILE)

USE STRPAK ; USE FIOPAK

USE DATA, ONLY: COORD,LNODS,SPRING,NSPRING,SHIFT,PIPROP, &

& PSTIF,PFLEX,HG,RW,RT,COUPLING,RECTANGULAR_GRID, GRID_X,
&

& GRID_Y,NN,KZERO,IFIX,GLOAD,UDL, GCORD,TORCON,SMA,OLEN,
&

& MCON,JCT,NM,GNODE,PIPROP,
EP,GS,GB,VS,PA,PILEN,errorn,LOAD_NODE
```

```
REAL(KIND=8) :: PROPS(4),PRES,L, R4N,XO,YO ,POINT(3)

REAL(KIND=8), ALLOCATABLE :: XY(:,:)

INTEGER, ALLOCATABLE :: A(:),B(:)

INTEGER                                                          ::
NPOIN,NELEM,NNODE,IPOIN,NSPRG,NPILE,ERRVAR,NDAT,NDAT1

INTEGER :: L2,L3,COUNTER,I,J,K,I2,K1,I4N,NGASH,NODE

CHARACTER(LEN=80) :: NLINE

ERRORN=0

CALL VI_GETVAR(5,'1', 'NNODE', ndat, errvar)

CALL VI_DATA(1, NNODE, errvar) ; CALL VI_ERASE

IF (ERRVAR == -3 .OR. ERRVAR == -2) NNODE = 4 ! DEFAULT IS 4 NODE
ELEMENT

IF (NNODE.NE.4.AND.NNODE.NE.8) THEN

WRITE (*,*) 'No. of node per element must be either 4 or 8'

errorn=1 ; return

endif

CALL VI_GETFN(5,'1', 'COUPLING', COUPLING, errvar) ; CALL VI_ERASE

if (errvar == -3 .OR. ERRVAR == -2) COUPLING="N" ! DEFAULT USE
BAND MATRIX

CALL DEFAULT ! default setting for the analysis


! nstre = 5 internal unknown, two vertical shear, two moments, one torsion

CALL
VI_GETFN(5,'1','RECTANGULAR_GRID',RECTANGULAR_GRID,ERRVAR)

IF (ERRVAR == -3 .OR. ERRVAR == -2) RECTANGULAR_GRID="N"


IF (RECTANGULAR_GRID == "N" .OR. RECTANGULAR_GRID == "n")
```

```
THEN
! read in non-rectangular domain, the nodal coordinates must be defined by user using mesh
! generation program
CALL VI_GETVAR(5,'1', 'NPOIN', ndat, errvar)
CALL VI_DATA(1, NPOIN, errvar) ; CALL VI_ERASE
CALL VI_GETVAR(5,'1', 'NELEM', ndat, errvar)
CALL VI_DATA(1, NELEM, errvar) ; CALL VI_ERASE
IF (ALLOCATED(COORD)) DEALLOCATE(COORD,LNODS)
ALLOCATE(COORD(NPOIN,2),LNODS(NELEM,NNODE))
CALL FIO_LOCVAR(5,'1', 'ELEMENT_NODE', NDAT, NLINE, errvar)
DO IELEM=1,NELEM
READ(5,*) NUMEL,(LNODS(NUMEL,INODE),INODE=1,NNODE)
END DO
COORD=0.0
! read nodal coordinates, finishing with the last node of all
CALL FIO_LOCVAR(5,'1', 'NODE_COORDINATES', NDAT, NLINE, errvar)
DO I=1,NPOIN
READ(5,*) IPOIN,(COORD(IPOIN,IDIME),IDIME=1,2)
END DO
ELSE ! read in rectangular domain by grids, grid x and y need not in order
CALL VI_GETVAR(5,'1','GRID_X',NDAT,ERRVAR)
IF (ALLOCATED(GRID_X)) DEALLOCATE(GRID_X)
ALLOCATE(GRID_X((NDAT+NPILE)))
```

```
DO I = 1,NDAT
CALL VI_DATA(I,GRID_X(I),ERRVAR)
END DO
CALL VI_ERASE
CALL VI_GETVAR(5,'1','GRID_Y',NDAT1,ERRVAR)
IF (ALLOCATED(GRID_Y)) DEALLOCATE(GRID_Y)
ALLOCATE(GRID_Y((NDAT1+NPILE)))
DO I = 1,NDAT1
CALL VI_DATA(I,GRID_Y(I),ERRVAR)
END DO
CALL VI_ERASE
! SORT THE GRID_X AND GRID_Y
DO I = 1,(NDAT+K-1)
DO J = I,(NDAT+K)
IF (GRID_X(I).GE.GRID_X(J)) THEN
L = GRID_X(I) ; GRID_X(I) = GRID_X(J) ; GRID_X(J)=L
END IF
END DO
END DO
DO I = 1,(NDAT1+K1-1)
DO J = I,(NDAT1+K1)
IF (GRID_Y(I).GE.GRID_Y(J)) THEN
L = GRID_Y(I) ; GRID_Y(I) = GRID_Y(J) ; GRID_Y(J)=L
END IF
```

```
END DO

END DO

NPOIN=(NDAT+K)*(NDAT1+K1) ; NELEM=(NDAT+K-1)*(NDAT1+K1-1)

IF (ALLOCATED(COORD)) DEALLOCATE(COORD,LNODS)

ALLOCATE(COORD(NPOIN,2),LNODS(NELEM,NNODE))

CALL FORM_RECTANGULAR((NDAT+K),(NDAT1+K1)) ! form rectangular
mesh

K=0 ; K1=0 ; I2=0

ENDIF

!--------- add void in mesh --------------

CALL VI_GETVAR(5,'1','VOID',NDAT,ERRVAR) ! add void in mesh

IF (NDAT == 0) GO TO 34 ! exit if no void

if (ndat/2*2-ndat /= 0) then

write (*,*) 'No. of data to define x/y ordinates of void must be even'

errorn=1 ; return

endif

33 NDAT=NDAT/2

IF (ALLOCATED(XY)) DEALLOCATE(XY,A,B)

ALLOCATE(XY(NDAT,2),A(NPOIN),B(NELEM)) ! store A to store void points

DO I = 1,NDAT

CALL VI_DATA(2*I-1,XY(I,1),ERRVAR)

CALL VI_DATA(2*I,XY(I,2),ERRVAR)

END DO

CALL VI_ERASE
```

```
J=NPOIN ; K=NELEM ! store initial point and element numbers
A=0
DO I = 1,J
CALL LOCPT(COORD(I,1),COORD(I,2),XY,NDAT,L2,L3)
IF (L2 == 1) A(I)=1 ! count all points inside voids
END DO
! eliminate elements and nodes within void
B=0
DO I = 1,K
COUNTER=0
DO I1 = 1,NNODE
DO I2 = 1,J
IF ((LNODS(I,I1)==I2).AND.(A(I2)==1)) COUNTER=1 ! COUNTER indicates no. of node in void in each element
END DO
END DO
IF (COUNTER==1) B(I)=1 ! counter element inside void
END DO
COUNTER=0
DO I = 1,K
IF (B(I)==1) THEN
NELEM=NELEM-1
DO I2 = (I-COUNTER),(K-1-COUNTER)
LNODS(I2,:)=LNODS((I2+1),:) ! eliminate the element inside void
```

```
END DO
COUNTER = COUNTER + 1
END IF
END DO
COUNTER=0
DO I = 1,J
IF (A(I)==1) THEN
NPOIN=NPOIN-1
DO I2 = (I-COUNTER),(J-1-COUNTER)
COORD(I2,:)=COORD((I2+1),:) ! eliminate coordinates
END DO
COUNTER = COUNTER + 1
END IF
END DO
DO IELEM = 1,NELEM ! eliminate element
DO I2 = 1,NNODE
COUNTER=0
DO I = 1,J
IF ((LNODS(IELEM,I2).GE.I).AND.(A(I)==1)) COUNTER=COUNTER+1
END DO
LNODS(IELEM,I2)=LNODS(IELEM,I2)-COUNTER ! shift the node number
END DO
END DO
```

```
DEALLOCATE(XY) ; DEALLOCATE(A) ; DEALLOCATE(B)
! INPUT PILE INFORMATION
! PIPROP(I,1) = X COORD OF PILE
! 2 Y COORD OF PILE
! 3 FOR LENGTH OF PILE
! 4 FOR DIAMETER OF PILE
! 5 FOR DIAMETER OF BASE
! 6 FOR AREA OF PILE
34 CALL VI_GETVAR(5,'1', 'PILE_PROPERTIES', ndat, errvar)
IF (NDAT > 0) THEN
NPILE=NDAT/6 ; COUPLING="Y" ! use full matrix when there is pile coupling
IF (ALLOCATED(PIPROP)) DEALLOCATE(PIPROP,PILEN)
ALLOCATE (PIPROP(NPILE,6),PILEN(NPILE))
DO IPILE=1,NPILE
DO I=1,6
CALL VI_DATA((I+(IPILE-1)*6),PIPROP(IPILE,I), errvar)
END DO
IF (SHIFT == "Y" .OR. SHIFT == "y") THEN
CALL FIND_NODE(piprop(ipile,1),piprop(ipile,2),NODE)
COORD(NODE,1)=PIPROP(IPILE,1) ; COORD(NODE,2)=PIPROP(IPILE,2)
PILEN(IPILE)=NODE
ENDIF
END DO
CALL VI_ERASE
```

```
! EP = YOUNG MODULUS OF PILE GS = SHEAR MODULUS OF SHAFT SOIL GB = SHEAR MODULUS OF BASE SOIL

! VS = POISSON RATIO OF SOIL PA = AREA COVERED BY PILE GROUP

CALL VI_GETVAR(5,'1', 'E_PILE', ndat, errvar)

CALL VI_DATA(1, EP, errvar) ; CALL VI_ERASE

CALL VI_GETVAR(5,'1', 'G_SHAFT_SOIL', ndat, errvar)

CALL VI_DATA(1, GS, errvar) ; CALL VI_ERASE

CALL VI_GETVAR(5,'1', 'G_BASE_SOIL', ndat, errvar)

CALL VI_DATA(1, GB, errvar) ; CALL VI_ERASE

IF (ERRVAR == -3) GB = GS ! DEFAULT SHEAR MODULUS OF BASE SOIL = SHEAR MODULUS OF SHAFT SOIL

CALL VI_GETVAR(5,'1', 'POISSON_RATIO_SOIL', ndat, errvar)

CALL VI_DATA(1, VS, errvar) ; CALL VI_ERASE

CALL VI_GETVAR(5,'1', 'AREA_PILE_GROUP', ndat, errvar)

CALL VI_DATA(1, PA, errvar) ; CALL VI_ERASE

! READ(5,*) EP,GS,GB,VS,PA

IF (ALLOCATED(PSTIF)) DEALLOCATE(PSTIF,PFLEX)

ALLOCATE (PSTIF(NPILE,NPILE),PFLEX(NPILE,NPILE))

ELSE

NPILE=0 ! DEFAULT NO PILE

ENDIF

!----------------- input nodal spring ---------------

CALL VI_GETVAR(5,'1', 'NODAL_SPRING', NDAT, errvar)

IF (NDAT > 0) THEN
```

```
NSPRG=NDAT/3
IF (ALLOCATED(NSPRING)) DEALLOCATE(NSPRING,SPRING)
ALLOCATE(NSPRING(NSPRG),SPRING(NSPRG))
CALL FIO_LOCVAR(5,'1','NODAL_SPRING',NDAT,NLINE,ERRVAR)
DO ISPRG=1,NSPRG
READ(5,*) XO,YO,SPRING(ISPRG)
CALL FIND_NODE(XO,YO,NODE) ! search for the node number
IF (SHIFT == "Y" .OR. SHIFT == "y") THEN ! shift node coordinates of mesh
COORD(NODE,1)=XO ; COORD(NODE,2)=YO
endif
NSPRING(ISPRG)=NODE
END DO
ELSE
NSPRG=0
ENDIF
call vi_erase
!--------------------- print shape --------------------------------------
WRITE(6,905) NNODE,NPOIN,NELEM,NPILE,NSPRG
WRITE (6,910)
NN=3*NPOIN
IF (ALLOCATED(GLOAD)) DEALLOCATE(GLOAD,IFIX)
ALLOCATE (GLOAD(NN),IFIX(NN))
GLOAD=0.D0 ; IFIX = 0
DO IELEM=1,NELEM
```

```
WRITE(6,915) IELEM,(LNODS(IELEM,INODE),INODE=1,NNODE)
END DO
CALL VI_GETVAR(5,'1', 'NODAL_LOADS', NDAT, errvar)
IF (NDAT == 0) GO TO 911
WRITE(6,980)
980 FORMAT(//' NODAL LOADS',/' -----------',/' NODE POINT LOAD',3X,&
'X-Z MOMENT Y-Z MOMENT')
ALLOCCATE(LOAD_NODE(NDAT/5))
DO I=1,NDAT/5 ! read in point load coordinates and loading
CALL VI_DATA ((I-1)*5+1,XO,ERRVAR)
CALL VI_DATA ((I-1)*5+2,YO,ERRVAR)
CALL VI_DATA ((I-1)*5+3,POINT(1),ERRVAR)
CALL VI_DATA ((I-1)*5+4,POINT(2),ERRVAR)
CALL VI_DATA (I*5,POINT(3),ERRVAR)
CALL FIND_NODE(XO,YO,NODE)
if (shift == "Y" .or. shift == "y") then
COORD(NODE,1)=XO ; COORD(NODE,2)=YO
endif
LOAD_NODE(I)=NODE
DO IDOFN=1,3
NGASH=(NODE-1)*3+IDOFN
GLOAD(NGASH)=GLOAD(NGASH) + POINT(IDOFN) ! save into global force
array
END DO
```

```
WRITE(6,925) NODE,(POINT(IDOFN),IDOFN=1,3)

925 FORMAT(I5,3F12.5)

END DO

CALL VI_ERASE

911 WRITE (6,920)

DO I =1,NPOIN

WRITE(6,935) I,(COORD(I,IDIME),IDIME=1,2)

END DO

IF (NPILE > 0) THEN ! write pile information

WRITE (6,720)

DO IPILE=1,NPILE

WRITE(6,725)
PIPROP(IPILE,1),PIPROP(IPILE,2),(PIPROP(IPILE,JPILE),JPILE=3,6)

END DO

WRITE(6,710)

WRITE(6,715) EP,GS,GB,VS,PA

ENDIF

IF (NSPRG > 0) THEN

WRITE (6,961)

DO ISPRG=1,NSPRG

WRITE(6,971) NSPRING(ISPRG),SPRING(ISPRG) ! write nodal spring

END DO

ENDIF

!--------------------- GRILLAGE --------------------------
```

```
JCT=0 ; NM=0

CALL VI_GETVAR(5,'1','GRILLAGE_NODE_NO',NDAT,ERRVAR)

CALL VI_DATA(1,I4N,ERRVAR) ; JCT = I4N

if (errvar /= -3 .AND. ERRVAR /= -2) COUPLING="Y"

CALL VI_ERASE

CALL VI_GETVAR(5,'1','GRILLAGE_ELEMENT_NO',NDAT,ERRVAR)

CALL VI_DATA(1,I4N,ERRVAR) ; NM = I4N ; CALL VI_ERASE

IF (ALLOCATED(GCORD)) DEALLOCATE(GCORD,GNODE,TORCON,SMA,OLEN,MCON)

ALLOCATE(GCORD(JCT,2),GNODE(JCT),TORCON(NM),SMA(NM),OLEN(NM),MCON(NM,2))

GCORD=0. ; GNODE=0 ; TORCON=0. ; SMA=0. ; OLEN=0. ; MCON=0

CALL VI_GETVAR(5,'1','GRILLAGE_NODE_DATA',NDAT,ERRVAR)

DO I = 1,NDAT ! read grillage member information

IF (I-I/3*3 ==1) THEN

CALL VI_DATA(I,I4N,ERRVAR) ; K = I4N

ENDIF

IF (I-I/3*3 ==2) THEN

CALL VI_DATA(I,R4N,ERRVAR) ; GCORD(K,1) = R4N

ENDIF

IF (I-I/3*3 ==0) THEN

CALL VI_DATA(I,R4N,ERRVAR) ; GCORD(K,2) = R4N

ENDIF

END DO

DO I = 1,JCT
```

```
CALL FIND_NODE(GCORD(I,1),GCORD(I,2),GNODE(I)) ! GNODE store global node number of grillage node

END DO

CALL VI_ERASE

! search for nearest nodes

! GNODE(K)=NPOIN store the equivalent node number

CALL VI_GETVAR(5,'1','GRILLAGE_MEMBER_DATA',NDAT,ERRVAR)

DO I = 1,NDAT

IF (I-I/5*5 == 1) THEN

CALL VI_DATA(I,I4N,ERRVAR) ; K = I4N

ENDIF

IF (I-I/5*5 ==2) THEN

CALL VI_DATA(I,I4N,ERRVAR) ; MCON(K,1) = I4N

ENDIF

IF (I-I/5*5 ==3) THEN

CALL VI_DATA(I,I4N,ERRVAR) ; MCON(K,2) = I4N

ENDIF

IF (I-I/5*5 ==4) THEN

CALL VI_DATA(I,R4N,ERRVAR) ; SMA(K) = R4N

ENDIF

IF (I-I/5*5 ==0) THEN

CALL VI_DATA(I,R4N,ERRVAR) ; TORCON(K) = R4N

ENDIF

END DO
```

```
CALL VI_ERASE
DO I = 1,NM
J1 = IABS (MCON(I,1))
J2 = IABS (MCON(I,2))
IF (J2 .LT. J1) THEN
M = MCON(I,1)
MCON (I,1) = MCON (I,2)
MCON (I,2) = M
J1 = IABS (MCON(I,1))
J2 = IABS (MCON(I,2))
END IF
XO = GCORD(J2,1) - GCORD(J1,1)
YO = GCORD(J2,2) - GCORD(J1,2)
OLEN(I) = SQRT (XO *XO + YO * YO) ! length of each grillage member
END DO

10 WRITE(6,940)
!------------------ read the fixed values -----------
CALL FIO_LOCVAR(5,'1', 'NODE_FIXITY', NDAT, NLINE, errvar)
IF (ERRVAR == 0) THEN
ELSE
!----------------- input v fix------------------
CALL VI_GETVAR(5,'1','V_FIX',NDAT,ERRVAR)
IF (NDAT == 0) GO TO 37 ! prescribed vertical displacement input
```

```
36 CALL VI_DATA(1,PRES,ERRVAR) ; NDAT=(NDAT-1)/2
IF (ALLOCATED(XY)) DEALLOCATE(XY)
ALLOCATE(XY(NDAT,2))
WRITE(6,*)
WRITE(6,*) ' VERTICAL Z DISPLACEMENT RESTRAINED'
DO I = 1,NDAT
CALL VI_DATA(2*I,XY(I,1),ERRVAR)
CALL VI_DATA(2*I+1,XY(I,2),ERRVAR)
END DO
CALL VI_ERASE
DO I = 1,NPOIN
CALL LOCPT(COORD(I,1),COORD(I,2),XY,NDAT,L2,L3)
IF (L2 == 1) THEN
GLOAD(3*(I-1)+1)=PRES
IFIX(3*(I-1)+1)=1
WRITE (6,955) I,COORD(I,1),COORD(I,2),PRES
ENDIF
END DO
BACKSPACE(5)
CALL VI_GETVAR(5,'0','V_FIX',NDAT,ERRVAR)
IF (NDAT == 0) DEALLOCATE(XY)
IF (NDAT == 0) GO TO 37
DEALLOCATE(XY) ; GO TO 36
!------------------ V coor fix ----------------
```

```
37 CALL VI_GETVAR(5,'1','V_FIX_COOR',NDAT,ERRVAR)
IF (NDAT == 0) GO TO 39
38 CALL VI_DATA(1,PRES,ERRVAR) ; NDAT=(NDAT-1)/2
IF (ALLOCATED(XY)) DEALLOCATE(XY)
ALLOCATE(XY(NDAT,2))
WRITE(6,*)
WRITE(6,*) ' VERTICAL Z DISPLACEMENT RESTRAINED'
DO I = 1,NDAT
J=0
CALL VI_DATA(2*I,XY(I,1),ERRVAR)
CALL VI_DATA(2*I+1,XY(I,2),ERRVAR)
CALL FIND_NODE(XY(I,1),XY(I,2),J)
GLOAD(3*(J-1)+1)=PRES
IFIX(3*(J-1)+1)=1
WRITE (6,955) J,COORD(J,1),COORD(J,2),PRES
END DO
CALL VI_ERASE

BACKSPACE(5)
CALL VI_GETVAR(5,'0','V_FIX_COOR',NDAT,ERRVAR)
IF (NDAT == 0) DEALLOCATE(XY)
IF (NDAT == 0) GO TO 39
DEALLOCATE(XY) ; GO TO 38
!----------------- input Rx fix ---------------
```

```
39 CALL VI_GETVAR(5,'1','RX_FIX',NDAT,ERRVAR)
IF (NDAT == 0) GO TO 41
40 CALL VI_DATA(1,PRES,ERRVAR) ; NDAT=(NDAT-1)/2
IF (ALLOCATED(XY)) DEALLOCATE(XY)
ALLOCATE(XY(NDAT,2))
WRITE(6,*)
WRITE(6,*) ' ROTATIONAL X DISPLACEMENT RESTRAINED'
DO I = 1,NDAT ! prescribed rotation x input
CALL VI_DATA(2*I,XY(I,1),ERRVAR)
CALL VI_DATA(2*I+1,XY(I,2),ERRVAR)
END DO
CALL VI_ERASE
DO I = 1,NPOIN
CALL LOCPT(COORD(I,1),COORD(I,2),XY,NDAT,L2,L3)
IF (L2 == 1) THEN
GLOAD(3*(I-1)+2)=PRES
IFIX(3*(I-1)+2)=1
WRITE (6,955) I,COORD(I,1),COORD(I,2),PRES
ENDIF
END DO
BACKSPACE(5)
CALL VI_GETVAR(5,'0','RX_FIX',NDAT,ERRVAR)
IF (NDAT == 0) DEALLOCATE(XY)
IF (NDAT == 0) GO TO 41
```

```
DEALLOCATE(XY) ; GO TO 40
!------------------ input Rx coor fix ----------------
41 CALL VI_GETVAR(5,'1','RX_FIX_COOR',NDAT,ERRVAR)
IF (NDAT == 0) GO TO 43
42 CALL VI_DATA(1,PRES,ERRVAR) ; NDAT=(NDAT-1)/2
IF (ALLOCATED(XY)) DEALLOCATE(XY)
ALLOCATE(XY(NDAT,2))
WRITE(6,*)
WRITE(6,*) ' ROTATIONAL X DISPLACEMENT RESTRAINED'
DO I = 1,NDAT
J=0
CALL VI_DATA(2*I,XY(I,1),ERRVAR)
CALL VI_DATA(2*I+1,XY(I,2),ERRVAR)
CALL FIND_NODE(XY(I,1),XY(I,2),J)
GLOAD(3*(J-1)+2)=PRES
IFIX(3*(J-1)+2)=1
WRITE (6,955) J,COORD(J,1),COORD(J,2),PRES
END DO
CALL VI_ERASE
BACKSPACE(5)
CALL VI_GETVAR(5,'0','RX_FIX_COOR',NDAT,ERRVAR)
IF (NDAT == 0) DEALLOCATE(XY)
IF (NDAT == 0) GO TO 43
DEALLOCATE(XY) ; GO TO 42
```

```
!------------------ input Ry fix ----------------
43 CALL VI_GETVAR(5,'1','RY_FIX',NDAT,ERRVAR)
IF (NDAT == 0) GO TO 45
44 CALL VI_DATA(1,PRES,ERRVAR) ; NDAT=(NDAT-1)/2
ALLOCATE(XY(NDAT,2))
WRITE(6,*)
WRITE(6,*) ' ROTATIONAL Y DISPLACEMENT RESTRAINED'
DO I = 1,NDAT ! prescribed rotation y input
CALL VI_DATA(2*I,XY(I,1),ERRVAR)
CALL VI_DATA(2*I+1,XY(I,2),ERRVAR)
END DO
CALL VI_ERASE
DO I = 1,NPOIN
CALL LOCPT(COORD(I,1),COORD(I,2),XY,NDAT,L2,L3)
IF (L2 == 1) THEN
GLOAD(3*I)=PRES
IFIX(3*I)=1
WRITE (6,955) I,COORD(I,1),COORD(I,2),PRES
ENDIF
END DO
BACKSPACE(5)
CALL VI_GETVAR(5,'0','RY_FIX',NDAT,ERRVAR)
IF (NDAT == 0) DEALLOCATE(XY)
IF (NDAT == 0) GO TO 45
```

```
DEALLOCATE(XY) ; GO TO 44

!------------------ input Ry coor fix ----------------

45 CALL VI_GETVAR(5,'1','RY_FIX_COOR',NDAT,ERRVAR)
IF (NDAT == 0) GO TO 35
46 CALL VI_DATA(1,PRES,ERRVAR) ; NDAT=(NDAT-1)/2
ALLOCATE(XY(NDAT,2))
WRITE(6,*)
WRITE(6,*) ' ROTATIONAL Y DISPLACEMENT RESTRAINED'
DO I = 1,NDAT
J=0
CALL VI_DATA(2*I,XY(I,1),ERRVAR)
CALL VI_DATA(2*I+1,XY(I,2),ERRVAR)
CALL FIND_NODE(XY(I,1),XY(I,2),J)
GLOAD(3*J)=PRES
IFIX(3*J)=1
WRITE (6,955) J,COORD(J,1),COORD(J,2),PRES
END DO
BACKSPACE(5)
CALL VI_GETVAR(5,'0','RY_FIX_COOR',NDAT,ERRVAR)
IF (NDAT == 0) DEALLOCATE(XY)
IF (NDAT == 0) GO TO 35
DEALLOCATE(XY) ; GO TO 46

35 ENDIF ! end NODE_FIXITY
```

```
!---------------------- Plate Properties ---------------------------

CALL VI_ERASE

CALL VI_GETVAR(5,'1', 'E_PLATE', ndat, errvar) ! Young modulus of plate

CALL VI_DATA(1,PROPS(1), errvar) ; CALL VI_ERASE

CALL VI_GETVAR(5,'1', 'POISSON_RATIO_PLATE', ndat, errvar) ! Poisson
ratio of plate

CALL VI_DATA(1,PROPS(2) , errvar) ; CALL VI_ERASE

CALL VI_GETVAR(5,'1', 'THICKNESS_PLATE', ndat, errvar) ! thickness of
plate

CALL VI_DATA(1, PROPS(3), errvar) ; CALL VI_ERASE

CALL VI_GETVAR(5,'1', 'GLOBAL_PRESSURE', ndat, errvar) ! UDL on plate

IF (ERRVAR /= -3) THEN

CALL VI_DATA(1,PROPS(4), errvar) ; CALL VI_ERASE

ELSE

PROPS(4)=0. ! DEFAULT IS ZERO GLOBAL PRESSURE

ENDIF

NEVAB=NNODE*3

!---------------------- WINKLER'S SPRING --------------------------

21 CALL VI_GETVAR(5,'1', 'WINKLER_SPRING', ndat, errvar)

! input for elastic spring over entire plate

IF (NDAT /= 0) THEN

CALL VI_DATA(1, KZERO, errvar) ; CALL VI_ERASE

ELSE

KZERO=0.D0 ! DEFAULT IS ZERO
```

```
ENDIF

WRITE(6,960) (PROPS(IPROP),IPROP=1,4),KZERO

IF (ALLOCATED(UDL)) DEALLOCATE(UDL)

ALLOCATE(UDL(NELEM))

UDL=PROPS(4)

!---------------------- LOCAL PRESSURE ---------------------------

CALL VI_GETVAR(5,'1','LOCAL_PRESSURE',NDAT,ERRVAR) ! pressure
within a region

IF (NDAT == 0) GO TO 62

61 CALL VI_DATA(1,PRES,ERRVAR) ; NDAT=(NDAT-1)/2

IF (ALLOCATED(XY)) DEALLOCATE(XY)

ALLOCATE(XY(NDAT,2))

WRITE(6,956)

DO I = 1,NDAT

CALL VI_DATA(2*I,XY(I,1),ERRVAR)

CALL VI_DATA(2*I+1,XY(I,2),ERRVAR)

END DO

CALL VI_ERASE

DO I = 1,NELEM

COUNTER=0

DO J = 1,NNODE

CALL
LOCPT(COORD(LNODS(I,J),1),COORD(LNODS(I,J),2),XY,NDAT,L2,L3)

IF (L2 == 1) COUNTER=COUNTER+1

END DO
```

```
IF (COUNTER == NNODE) THEN
UDL(I)=UDL(I)+PRES
WRITE(6,957) I,PRES
END IF
END DO
BACKSPACE(5)
CALL VI_GETVAR(5,'0','LOCAL_PRESSURE',NDAT,ERRVAR)
IF (NDAT == 0) DEALLOCATE(XY)
IF (NDAT == 0) GO TO 62
DEALLOCATE(XY) ; GO TO 61
62 CONTINUE
!-----------------------------------------------------
710 FORMAT(//,' PILES',/' -----')
715 FORMAT(/36H Young's Modulus of Pile Material =,E10.3,/&
36H Shear Modulus of Shaft Soil =,E10.3,/&
36H Shear Modulus of Base Soil =,E10.3,/&
36H Poisson's Ratio of Soil =,F6.3,/&
36H Area Covered by Pile Group =,E10.3)
720 FORMAT(/7HX-COORD,2X,7HY-COORD,3X,10H Length,3X,10HShaft Dia.,&
3X,10H Base Dia.,3X,' Area')
725 FORMAT(F7.3,2X,F7.3,3X,F10.3,3X,F10.3,3X,F10.3,3X,F10.3)
905 FORMAT(//30HNO. OF NODES IN EACH ELEMENT =,I4,4X,30H TOTAL NO. OF NODES =,I4,4X,/30H TOTAL NO. OF ELEMENTS =,I4,4X,30H NO. OF PILES =,I4,4X, /30H NO. OF SPRINGS =,I4)
```

```
910 FORMAT(//' NODE SEQUENCE',/' -------------',/8H ELEMENT, &
6X,12HNODE NUMBERS)
915 FORMAT(1X,I4,6X,8I5)
920 FORMAT(//' NODAL POINT COORDINATES',/' -----------------------', &
/6H NODE,7X,1HX,9X,1HY)
935 FORMAT(1X,I5,3F10.3)
940 FORMAT(//' RESTRAINED NODES',/' ----------------',/5H NODE,2X, &
7HX-COORD,2X,7HY-COORD,2X,12HFIXED VALUES)
955 FORMAT(1X,I4,2X,F7.3,2X,F7.3,2X,3F10.6)
956 FORMAT(//' UDL AT ELEMENTS',/' ----------------',/2X,8H
ELEMENT,3X,3HUDL)
957 FORMAT(3X,I4,3X,E12.5)
960 FORMAT(//' MATERIAL PROPERTIES',/' -------------------', &
/' ELASTIC MODULUS =',E14.6,/' POISSONS RATIO =',E14.6, &
/' THICKNESS OF PLATE =',E14.6,/' UDL ON PLATE =',E14.6, &
/' WINKLER SPRING =',E14.6)
961 FORMAT(//' SPRING',/' ------',/1X,'NODE SPRING CONSTANT')
971 FORMAT(1X,I4,5X,E16.5)
985 FORMAT( /' H-G CONTROL:',I2)
RETURN
20 END SUBROUTINE INPUT

!-------------------- INPUT DEFAULT VALUES --------------------
SUBROUTINE DEFAULT
USE DATA
```

```
IF (NNODE == 4) then
NGAUV = 1 ; NGAUB = 2
ENDIF
IF (NNODE == 8) THEN
NGAUV = 2 ; NGAUB = 3
ENDIF
NSTRE=5
RECTANGULAR_GRID="N"
SHIFT = "Y"
END SUBROUTINE DEFAULT

!------------------------FORM_RECTANGULAR----------------------------------
SUBROUTINE FORM_RECTANGULAR(NDAT,NDAT1)
USE DATA
INTEGER :: I,J,K,NDAT,NDAT1
DO I = 1,NDAT1
DO J = 1,NDAT
K=(I-1)*NDAT+J
COORD(K,1)=GRID_X(J) ; COORD(K,2)=GRID_Y(I)
END DO
END DO
DO I=1,NDAT1-1
DO J = 1,NDAT-1
K=(I-1)*(NDAT-1)+J
```

```
LNODS(K,1)=(I-1)*NDAT+J ; LNODS(K,2)=(I-1)*NDAT+J+1

LNODS(K,3)=I*NDAT+J+1 ; LNODS(K,4)=I*NDAT+J

END DO

END DO

END SUBROUTINE FORM_RECTANGULAR

!--------------------------------------------------------------------------------

SUBROUTINE locpt (x0, y0, xy, n, l, m)

!---------------------------------------------------------------------

! GIVEN A POLYGONAL LINE CONNECTING THE VERTICES (X(I),Y(I)) (I = 1,...,N)

! TAKEN IN THIS ORDER. IT IS ASSUMED THAT THE POLYGONAL PATH IS A LOOP,

! WHERE (X(N),Y(N)) = (X(1),Y(1)) OR THERE IS AN ARC FROM (X(N),Y(N)) TO

! (X(1),Y(1)). N.B. The polygon may cross itself any number of times.

! (X0,Y0) IS AN ARBITRARY POINT AND L AND M ARE VARIABLES.

! On output, L AND M ARE ASSIGNED THE FOLLOWING VALUES ...

! L = -1 IF (X0,Y0) IS OUTSIDE THE POLYGONAL PATH

! L = 0 IF (X0,Y0) LIES ON THE POLYGONAL PATH

! L = 1 IF (X0,Y0) IS INSIDE THE POLYGONAL PATH

! M = 0 IF (X0,Y0) IS ON OR OUTSIDE THE PATH. IF (X0,Y0) IS INSIDE THE

! PATH THEN M IS THE WINDING NUMBER OF THE PATH AROUND THE POINT (X0,Y0).
```

```
IMPLICIT NONE

INTEGER, INTENT(IN) :: n

REAL(KIND=8), INTENT(IN) :: x0, y0

REAL(KIND=8), DIMENSION (N,2) :: xy

INTEGER, INTENT(OUT) :: l, m

! Local variables

INTEGER :: i, n0

REAL(KIND=8) :: angle, eps, pi, pi2, sum, theta, theta1, thetai, tol, u, v

! ****** EPS IS A MACHINE DEPENDENT CONSTANT. EPS IS THE

! SMALLEST NUMBER SUCH THAT 1.0 + EPS > 1.0


eps = 0.0000001 ! THE ACCURACY OF A POINT LYING ON A LINE CAN
BE ADJUSTED BY THIS PARAMETER


n0 = n

IF (xy(1,1) == xy(n,1) .AND. xy(1,2) == xy(n,2)) n0 = n - 1

pi = DATAN2(0.0D0, -1.0D0)

pi2 = 2.0*pi ; tol = 4.0*eps*pi

l = -1 ; m = 0

u = xy(1,1) - x0 ; v = xy(1,2) - y0

IF (u == 0.0 .AND. v == 0.0) GO TO 20

IF (n0 < 2) RETURN

theta1 = DATAN2(v, u)

sum = 0.0 ; theta = theta1

DO i = 2, n0
```

```
u = xy(i,1) - x0

v = xy(i,2) - y0

IF (u == 0.0 .AND. v == 0.0) GO TO 20

thetai = DATAN2(v, u)

angle = DABS(thetai - theta)

IF (DABS(angle - pi) < tol) GO TO 20

IF (angle > pi) angle = angle - pi2

IF (theta > thetai) angle = -angle

sum = sum + angle

theta = thetai

END DO

angle = DABS(theta1 - theta)

IF (DABS(angle - pi) < tol) GO TO 20

IF (angle > pi) angle = angle - pi2

IF (theta > theta1) angle = -angle

sum = sum + angle

! SUM = 2*PI*M WHERE M IS THE WINDING NUMBER

m = DABS(sum)/pi2 + 0.2

IF (m == 0) RETURN

l = 1

IF (sum < 0.0) m = -m

RETURN
```

```
! (X0, Y0) IS ON THE BOUNDARY OF THE PATH

20 l = 0

RETURN

END SUBROUTINE locpt

!------------ FIND_NODE --------------------------------------

SUBROUTINE FIND_NODE(X0,Y0,NODE)

USE DATA

INTEGER :: NODE,I

REAL(KIND=8) :: X0,Y0,DIST1,DIST

DIST1=(X0-COORD(1,1))**2+(Y0-COORD(1,2))**2

NODE=1

DO I=2,NPOIN

DIST=(X0-COORD(I,1))**2+(Y0-COORD(I,2))**2

IF (DIST < DIST1) THEN

NODE=I ; DIST1=DIST

ENDIF

END DO

RETURN

END SUBROUTINE FIND_NODE

!----------------------- BANDWIDTH ---------------------------

SUBROUTINE BANDWIDTH

USE DATA

INTEGER :: I,J,MAXN,MINN
```

```
BAND=0
DO I = 1,NELEM
MAXN=LNODS(I,1) ; MINN=LNODS(I,1)
DO J = 2,NNODE
IF (LNODS(I,J) > MAXN) MAXN=LNODS(I,J)
IF (LNODS(I,J) < MINN) MINN=LNODS(I,J)
END DO
IF (BAND < MAXN-MINN+1) BAND=MAXN-MINN+1
END DO
BAND=BAND*3 ! BAND*3 BECAUSE EACH NODE HAS 3 DOF
RETURN
END SUBROUTINE BANDWIDTH

!-------------------------- FORM STIFFNESS MATRIX ------------------------

SUBROUTINE STIFPB(NELEM,NNODE,NEVAB,IBS,NSTRE,NPOIN, &
& NGAUS2,POSGP,WEIGP,PROPS)
USE DATA, ONLY: SMATX,ELCOD,GPCOD,DERIV,SHAPE,CARTD,BMATX,DBMAT, &
& LNODS,COORD,GSTIF,ESTIF,NGAUB,NGAUV,HG,RW,RT,KZERO
INTEGER NELEM,LNODE,IBS,NNODE,NEVAB,NPOIN,KGASP,IGAUS,JGAUS
REAL(KIND=8) :: POSGP(3),DAREA,WEIGP(3),DJACB,PROPS(4)
REAL(KIND=8) :: EXISP,ETASP,DMATX(5,5),K!,KH(3)
NGAUS2=NGAUB**2+NGAUV**2
ALLOCATE
```

```
(SMATX(NELEM,NSTRE,NEVAB,NGAUS2),GPCOD(NELEM,2,NGAUS2),
&

& BMATX(5,NEVAB))

! initialize the global stiffness matrix

GSTIF=0.0D0 ; ESTIF=0.0D0

! calculates element stiffness matrix for plate bending element

! loop over each element

! initialize the gauss point coordinates

GPCOD=0.0D0

DO IELEM=1,NELEM

! evaluate the coordinates of the element nodal points

DO INODE=1,NNODE

LNODE=LNODS(IELEM,INODE)

ELCOD(1:2,INODE)=COORD(LNODE,1:2)

END DO

! --------------------------------------

! calculate D matrix of PLATE

! --------------------------------------

CALL MODPB(PROPS,DMATX,ELCOD)

KGASP=0

! ------------------------------------------------

! enter loops for numerical integration -- bending

! ------------------------------------------------

CALL GAUSSQ(NGAUB,POSGP,WEIGP)
```

```
DO IGAUS=1,NGAUB
EXISP=POSGP(IGAUS)
K=1.
IF (IGAUS.EQ.2) THEN
K=-1.
END IF
DO JGAUS=1,NGAUB
ETASP=POSGP(JGAUS)*K
KGASP=KGASP+1 ! Gaussian point no. for the element
! evaluate the shape function, element area, etc.
CALL SFR2(ETASP,EXISP,NNODE)
CALL JACOB(KGASP,IELEM,NNODE)
CALL JACOB2(DJACB,NNODE,IELEM)
DAREA=DJACB*WEIGP(IGAUS)*WEIGP(JGAUS)
! evaluate the B and DB matrices
CALL BMATPB(NNODE,1,0)
CALL DBE(DMATX,NSTRE,NEVAB,NNODE,N)
! calculate the element stiffness by eq.(2.114)
DO IEVAB=1,NEVAB
DO JEVAB=IEVAB,NEVAB
DO ISTRE=1,NSTRE
ESTIF(IELEM,IEVAB,JEVAB)=ESTIF(IELEM,IEVAB,JEVAB)&
+BMATX(ISTRE,IEVAB)&
*DBMAT(ISTRE,JEVAB)*DAREA
```

```
END DO
END DO
END DO
! store the components of the DB matrix for the element
DO ISTRE=1,NSTRE
DO IEVAB=1,NEVAB
SMATX(IELEM,ISTRE,IEVAB,KGASP)=DBMAT(ISTRE,IEVAB)
END DO
END DO
END DO
END DO
!-----------------------------------------------
! enter loops for numerical integration -- shear
!-----------------------------------------------
CALL GAUSSQ(NGAUV,POSGP,WEIGP)
DO IGAUS=1,NGAUV
EXISP=POSGP(IGAUS)
K=1.
IF (IGAUS.EQ.2) THEN
K=-1.
END IF
DO JGAUS=1,NGAUV
ETASP=POSGP(JGAUS)*K
KGASP=KGASP+1 ! Gaussian point no. for the element
```

```
! evaluate the shape function, element area, etc.
CALL SFR2(ETASP,EXISP,NNODE)
CALL JACOB(KGASP,IELEM,NNODE)
CALL JACOB2(DJACB,NNODE,IELEM)
DAREA=DJACB*WEIGP(IGAUS)*WEIGP(JGAUS)
! evaluate the B and DB matrices
CALL BMATPB(NNODE,0,1)
CALL DBE(DMATX,NSTRE,NEVAB,NNODE,N)
! calculate the element shear stiffness by eq.(2.102)
DO IEVAB=1,NEVAB
DO JEVAB=IEVAB,NEVAB
DO ISTRE=1,NSTRE
ESTIF(IELEM,IEVAB,JEVAB)=ESTIF(IELEM,IEVAB,JEVAB)&
+BMATX(ISTRE,IEVAB)&
*DBMAT(ISTRE,JEVAB)*DAREA
END DO
END DO
END DO
! store the components of the DB matrix for the element
DO ISTRE=1,NSTRE
DO IEVAB=1,NEVAB
SMATX(IELEM,ISTRE,IEVAB,KGASP)=DBMAT(ISTRE,IEVAB)
END DO
END DO
```

```
END DO

END DO

! ------------------------------------------

! addition of Winkler soil spring stiffness matrix

! ------------------------------------------

IF (KZERO > 0.0000001) THEN

CALL SFR2(0.D0,0.D0,NNODE)

CALL JACOB2(DJACB,NNODE,IELEM)

DAREA=DJACB*4.

! calculate the element stiffness

DO INODE=1,NNODE

DO JNODE=INODE,NNODE

ESTIF(IELEM,((INODE-1)*3+1),((JNODE-1)*3+1))=ESTIF(IELEM,((INO-
E-1)*3+1),((JNODE-1)*3+1)) &

& +SHAPE(INODE)*SHAPE(JNODE)*DAREA*KZERO

END DO

END DO

ENDIF

!---------------------------------------------------

! construct the lower triangle of the stiffness matrix

!---------------------------------------------------

50 DO IEVAB=1,NEVAB

DO JEVAB=1,NEVAB

ESTIF(IELEM,JEVAB,IEVAB)=ESTIF(IELEM,IEVAB,JEVAB)
```

```
END DO

END DO

!------------------------------------

! assemble to the global stiffness matrix

!------------------------------------

CALL assemk(nnode,ielem)

!------------------------------------

END DO

RETURN

END SUBROUTINE STIFPB

!---------------------- DETERMINE ELASTICITY MARTIX ----------------------

SUBROUTINE MODPB(PROPS,DMATX,ELCOD)

USE DATA, ONLY: NNODE

REAL(KIND=8) :: DMATX(5,5),PROPS(4), &

ELCOD,THICK

! calculates D matrix for plate bending element, thickness can be different for different element

DMATX = 0.0D0

YOUNG=PROPS(1)

POISS=PROPS(2)

THICK=PROPS(3)

DMATX(1,1)=YOUNG*THICK*THICK*THICK/(12.0*(1.0-POISS*POISS))

DMATX(1,2)=POISS*DMATX(1,1)
```

```
DMATX(2,2)=DMATX(1,1)

DMATX(2,1)=DMATX(1,2)

DMATX(3,3)=(1.0-POISS)*DMATX(1,1)/2.0

DMATX(4,4)=YOUNG*THICK/(2.4*(1.0+POISS)) ! shear correction incorporated

DMATX(5,5)=DMATX(4,4) ! hence 2.4 instead of 2

RETURN

END SUBROUTINE MODPB


!------------------------- DETERMINE GAUSSIAN POINTS ------------------------

SUBROUTINE GAUSSQ(NGAUS,POSGP,WEIGP)

REAL(KIND=8) :: POSGP(3),WEIGP(3)

INTEGER NGAUS

POSGP=0.0D0; WEIGP=0.0D0

IF (NGAUS == 3) THEN ! 3x3 integration scheme

POSGP(1)=-0.774596669241483

POSGP(2)=0.0D0

POSGP(3)=-1.*POSGP(1)

WEIGP(1)=0.555555555555556

WEIGP(2)=0.888888888888889

WEIGP(3)=WEIGP(1)

ELSEIF (NGAUS == 2) THEN ! 2x2 integration scheme

POSGP(1)=-0.577350269189626

POSGP(2)=-1.*POSGP(1)
```

```
WEIGP(1)=1.0

WEIGP(2)=WEIGP(1)

ELSEIF (NGAUS == 1) THEN

POSGP(1)=0.0D0 ; WEIGP(1) = 2.0

END IF

RETURN

END SUBROUTINE GAUSSQ

!-------------------- Shape Function -------------------------------------

SUBROUTINE SFR2(S,T,NNODE)

USE DATA, ONLY: DERIV,SHAPE

REAL(KIND=8) :: S,T,S2,T2,SS,ST,TT,SST,STT,ST2

DERIV=0.0D0

! calculates shape functions and their derivatives for 2D elements for 4 and 8
nodes element

S2=S*2.0

T2=T*2.0

SS=S*S

ST=S*T

TT=T*T

SST=S*S*T

STT=S*T*T

ST2=S*T*2.0

IF (NNODE.EQ.4) THEN
```

```
! shape functions
SHAPE(1) = 0.25*(1.-T)*(1.-S)
SHAPE(2) = 0.25*(1.-T)*(1.+S)
SHAPE(3) = 0.25*(1.+T)*(1.+S)
SHAPE(4) = 0.25*(1.+T)*(1.-S)
! shape function derivatives
DERIV(1,1) = -0.25*(1.-T) ! partial derivatives w.r.t. S
DERIV(1,2) = 0.25*(1.-T)
DERIV(1,3) = 0.25*(1.+T)
DERIV(1,4) = -0.25*(1.+T)
DERIV(2,1) = -0.25*(1.-S) ! partial derivatives w.r.t. T
DERIV(2,2) = -0.25*(1.+S)
DERIV(2,3) = 0.25*(1.+S)
DERIV(2,4) = 0.25*(1.-S)
ELSE
SHAPE(1)=(-1.0+ST+SS+TT-SST-STT)/4.0
SHAPE(2)=(1.0-T-SS+SST)/2.0
SHAPE(3)=(-1.0-ST+SS+TT-SST+STT)/4.0
SHAPE(4)=(1.0+S-TT-STT)/2.0
SHAPE(5)=(-1.0+ST+SS+TT+SST+STT)/4.0
SHAPE(6)=(1.0+T-SS-SST)/2.0
SHAPE(7)=(-1.0-ST+SS+TT+SST-STT)/4.0
SHAPE(8)=(1.0-S-TT+STT)/2.0
! shape function derivatives
```

```
DERIV(1,1)=(T+S2-ST2-TT)/4.0 ! partial derivatives w.r.t. S

DERIV(1,2)=-S+ST

DERIV(1,3)=(-T+S2-ST2+TT)/4.0

DERIV(1,4)=(1.0-TT)/2.0

DERIV(1,5)=(T+S2+ST2+TT)/4.0

DERIV(1,6)=-S-ST

DERIV(1,7)=(-T+S2+ST2-TT)/4.0

DERIV(1,8)=(-1.0+TT)/2.0

DERIV(2,1)=(S+T2-SS-ST2)/4.0 ! partial derivatives w.r.t. T

DERIV(2,2)=(-1.0+SS)/2.0

DERIV(2,3)=(-S+T2-SS+ST2)/4.0

DERIV(2,4)=-T-ST

DERIV(2,5)=(S+T2+SS+ST2)/4.0

DERIV(2,6)=(1.0-SS)/2.0

DERIV(2,7)=(-S+T2+SS-ST2)/4.0

DERIV(2,8)=-T+ST

END IF

RETURN

END SUBROUTINE SFR2

!---------------- RETURN GAUSSIAN POINT COORDINATES (JACOB)-
------------------------

SUBROUTINE JACOB(KGASP,IELEM,NNODE)

USE DATA, ONLY : ELCOD,GPCOD,SHAPE
```

```
INTEGER :: IDIME, INODE, NNODE
DO IDIME=1,2
DO INODE=1,NNODE
GPCOD(IELEM,IDIME,KGASP)=GPCOD(IELEM,IDIME,KGASP)+&
ELCOD(IDIME,INODE)*SHAPE(INODE)
END DO
END DO
RETURN
END SUBROUTINE JACOB

!------- calculates coordinates of guass points ----------------------------
!------- and the jacobian matrix and its determinant -----------------------
!------- and the inverse for 2D elements -----------------------------------
!------- calculate coordinates of sampling point ---------------------------
!------- create jacobian matrix XJACM --------------------------------------

!---------------- JACOB2 -------------------------
SUBROUTINE JACOB2(DJACB,NNODE,IELEM)
USE DATA, ONLY: CARTD,DERIV,ELCOD,SHAPE
INTEGER NNODE,IELEM,IDIME
REAL(KIND=8) :: DJACB,XJACM(2,2),XJACI(2,2)
XJACM=0.0D0
DO IDIME=1,2
DO JDIME=1,2
DO INODE=1,NNODE
```

```
XJACM(IDIME,JDIME)=XJACM(IDIME,JDIME)+&
DERIV(IDIME,INODE)*ELCOD(JDIME,INODE)
END DO
END DO
END DO
! calculate determinant and inverse of Jacobian matrix
DJACB=XJACM(1,1)*XJACM(2,2)-XJACM(1,2)*XJACM(2,1)
! determinant of the Jacobian matrix
IF (DJACB.GT.0.0) GOTO 30
WRITE(6,900) IELEM
STOP
30 XJACI(1,1)= XJACM(2,2)/DJACB
XJACI(2,2)= XJACM(1,1)/DJACB
XJACI(1,2)=-XJACM(1,2)/DJACB
XJACI(2,1)=-XJACM(2,1)/DJACB
! calculate Cartesian derivatives
CARTD=0.0D0
DO IDIME=1,2
DO INODE=1,NNODE
DO JDIME=1,2
CARTD(IDIME,INODE)=CARTD(IDIME,INODE)+&
XJACI(IDIME,JDIME)*DERIV(JDIME,INODE)
END DO
END DO
```

```
END DO
900 FORMAT(//,24HPROGRAM HALTED IN JACOB2,&
/,11X,22H ZERO OR NEGATIVE AREA,/,&
10X,16H ELEMENT NUMBER ,I5)
RETURN
END SUBROUTINE JACOB2

!------------------------ DETERMINE STRAIN MATRIX ------------------------

SUBROUTINE BMATPB(NNODE,B,V)
USE DATA, ONLY: CARTD,SHAPE,BMATX
INTEGER NNODE,B,V
! calculates strain matrix B for plate bending element
BMATX = 0.0D0
JGASH = 0
IF (B == 1 .AND. V == 0) GOTO 10
IF (B == 0 .AND. V == 1) GOTO 20
!------------------------
! calculates B matrix for bending
!------------------------
10 DO INODE=1,NNODE
IGASH=JGASH+1
BMATX(4,IGASH)=0.0D0
BMATX(5,IGASH)=0.0D0
IGASH=IGASH+1
```

```
JGASH=IGASH+1
BMATX(1,IGASH)=-CARTD(1,INODE)
BMATX(3,IGASH)=-CARTD(2,INODE)
BMATX(4,IGASH)=0.0D0
BMATX(2,JGASH)=-CARTD(2,INODE)
BMATX(3,JGASH)=-CARTD(1,INODE)
BMATX(5,JGASH)=0.0D0
END DO
RETURN
!-----------------------
! calculates B matrix for shear
!-----------------------
20 DO INODE=1,NNODE
IGASH=JGASH+1
BMATX(4,IGASH)=CARTD(1,INODE)
BMATX(5,IGASH)=CARTD(2,INODE)
IGASH=IGASH+1
JGASH=IGASH+1
BMATX(1,IGASH)=0.0D0
BMATX(3,IGASH)=0.0D0
BMATX(4,IGASH)=-SHAPE(INODE)
BMATX(2,JGASH)=0.0D0
BMATX(3,JGASH)=0.0D0
BMATX(5,JGASH)=-SHAPE(INODE)
```

```
END DO
RETURN
END SUBROUTINE BMATPB

!-------------------------- B D Matrix -------------------------------------

SUBROUTINE DBE(DMATX,NSTRE,NEVAB,NNODE,N)
USE DATA, ONLY: DBMAT,BMATX
INTEGER NSTRE,NEVAB,NNODE,N
REAL(KIND=8) :: DMATX(5,5)
! calculates DxB
DBMAT=0.0D0
DO ISTRE=1,NSTRE
DO IEVAB=1,NEVAB
DO JSTRE=1,NSTRE
DBMAT(ISTRE,IEVAB)=DBMAT(ISTRE,IEVAB)+DMATX(ISTRE,JSTRE)&
*BMATX(JSTRE,IEVAB)
END DO
END DO
END DO
RETURN
END SUBROUTINE DBE

!------------ assemble the element stiffness matrices ----------------------

SUBROUTINE ASSEMK(NNODE,IELEM)
```

```
USE DATA, ONLY : GSTIF,ESTIF,LNODS,COUPLING

INTEGER ::
NNODE,INODE,IDOFN,NROWS,NROWE,JNODE,NODEJ,NCOLS,NCOLE,IELEM

IF (COUPLING == "N") GO TO 10

DO INODE=1,NNODE

NODEI=LNODS(IELEM,INODE)

DO IDOFN=1,3

NROWS=(NODEI-1)*3+IDOFN

NROWE=(INODE-1)*3+IDOFN

DO JNODE=1,NNODE

NODEJ=LNODS(IELEM,JNODE)

DO JDOFN=1,3

NCOLS=(NODEJ-1)*3+JDOFN

NCOLE=(JNODE-1)*3+JDOFN

GSTIF(NROWS,NCOLS)=GSTIF(NROWS,NCOLS)&

+ESTIF(IELEM,NROWE,NCOLE)

END DO

END DO

END DO

END DO

RETURN

10 DO I = 1,3*NNODE ! full stiffness matrix

DO J = 1,I
```

```
INODE=(I-1)/3+1 ; IDOFN=I-3*(INODE-1) ; INODE=LNODS(IELEM,INODE)

JNODE=(J-1)/3+1 ; JDOFN=J-3*(JNODE-1) ;
JNODE=LNODS(IELEM,JNODE)

NROWS=3*(INODE-1)+IDOFN ; NCOLS=3*(JNODE-1)+JDOFN

IF (NROWS > NCOLS) THEN

NODEJ=NROWS ; NROWS=NCOLS ; NCOLS=NODEJ

ENDIF

GSTIF(NROWS,NCOLS-NROWS+1)= GSTIF(NROWS,NCOLS-NROWS+1)+
&

ESTIF(IELEM,J,I)

END DO

END DO

END SUBROUTINE ASSEMK

!---------------- GRILLAGE ------------------------------------------

!------------ assemble the grillage element to global stiffness matrices -------

SUBROUTINE GRILLAGE

USE DATA, ONLY : GSTIF,NPOIN,COUPLING,GNODE,MCON,TORCON, &

& SMA,OLEN,JCT,NM,GSTIF1,GCORD

INTEGER :: J1,J2,I,J,K

REAL(KIND=8) :: ASAT(6,6),A(6,3),S(3,3),SAT(3,6)

IF (COUPLING == "N") GO TO 10

GSTIF1=0.

DO I = 1,NM

CALL SATMAK (I,J1,J2,A,S,SAT)
```

```
DO J = 1,6

DO K = J,6

ASAT(J,K) = 0.0

DO L =1,3

ASAT(J,K) = ASAT(J,K) + A(J,L) * SAT(L,K)

END DO

ASAT(K,J)=ASAT(J,K) ! form lower triangle

END DO

END DO

DO J = 1,3

DO K = 1,3

GSTIF1(((J1-1)*3+J),((J1-1)*3+K))=GSTIF1(((J1-1)*3+J),((J1-1)*3+K)) + ASAT(J,K)

GSTIF1(((J1-1)*3+J),((J2-1)*3+K))=GSTIF1(((J1-1)*3+J),((J2-1)*3+K)) + ASAT(J,(K+3))

GSTIF1(((J2-1)*3+J),((J1-1)*3+K))=GSTIF1(((J2-1)*3+J),((J1-1)*3+K)) + ASAT((J+3),K)

GSTIF1(((J2-1)*3+J),((J2-1)*3+K))=GSTIF1(((J2-1)*3+J),((J2-1)*3+K)) + ASAT((J+3),(K+3))

END DO

END DO

END DO

DO I = 1,NPOIN

DO J = 1,NPOIN

GSTIF(I,J)=GSTIF(I,J) + GSTIF1(I,J)

END DO
```

```
END DO

10 RETURN
END SUBROUTINE GRILLAGE

! ---------- SAMAK ( TO BUILD ARRAY [A], [S] AND [S].[A]T) ---------------------

SUBROUTINE SATMAK (I,J1,J2,A,S,SAT)
! IMPLICIT DOUBLE PRECISION (A-H,O-Z)
! --------------------------------------------------------------------
USE DATA, ONLY : JCT,NM,GCORD,MCON,TORCON,SMA,OLEN
REAL(KIND=8) A(6,3),S(3,3),SAT(3,6)
INTEGER :: J1,J2

! --------------------------------------------------------------------
! Build the transformed stiffness matrix for grillage analysis
! --------------------------------------------------------------------
J1 = IABS (MCON(I,1))
J2 = IABS (MCON(I,2))
ALEN = OLEN(I)
XO = GCORD(J2,1) - GCORD(J1,1)
YO = GCORD(J2,2) - GCORD(J1,2)
SA = YO/ALEN
CA = XO/ALEN

S(1,1) = TORCON(I)/ALEN
V4 = 4.0 * SMA(I)/ALEN
```

```
V2 = 0.5 * V4
V3 = 0.75 * V4
S(2,2) = V4
S(2,3) = V2
S(3,2) = V2
S(3,3) = V4
S(1,2) = 0.0
S(1,3) = 0.0
S(2,1) = 0.0
S(3,1) = 0.0
IF (MCON(I,1) .LE. 0) THEN
S(3,3) = V3
S(2,2) = 0.0
S(2,3) = 0.0
S(3,2) = 0.0
END IF
IF (MCON(I,2) .LE. 0) THEN
S(2,2) = V3
S(3,3) = 0.0
S(2,3) = 0.0
S(3,2) = 0.0
IF (MCON(I,1) * MCON(I,2) .GT. 0) S(2,2) = 0.0
END IF
A = 0.0
```

```
A(2,1) = CA
A(5,1) = -CA
A(1,2) = -1.0/ALEN
A(1,3) = A(1,2)
A(4,2) = -A(1,2)
A(4,3) = A(4,2)
A(2,2) = -SA
A(5,3) = -SA
A(3,1) = SA
A(3,2) = CA
A(6,1) = -SA
A(6,3) = CA
DO J = 1,3
DO K = 1,6
SAT(J,K) = 0.0
DO L = 1,3
SAT(J,K) = SAT(J,K) + S(J,L) * A(K,L)
END DO
END DO
END DO
RETURN
END SUBROUTINE SATMAK

!---------------------------- INPUT NODAL SPRING ---------------------------
```

```
SUBROUTINE INSPRG(NN,NSPRG)

USE DATA, ONLY: GSTIF,SPRING,NSPRING,COUPLING

INTEGER NN,NSPRG

! put SPRING constants into global stiffness matrix

DO ISPRG=1,NSPRG

I=NSPRING(ISPRG)

IF (COUPLING == "N") THEN

GSTIF(3*(I-1)+1,1)=GSTIF(3*(I-1)+1,1)+SPRING(ISPRG)

ELSE

GSTIF(3*(I-1)+1,3*(i-1)+1)=GSTIF(3*(I-1)+1,3*(i-1)+1)+SPRING(ISPRG)

ENDIF

END DO

RETURN

END SUBROUTINE INSPRG

!------------------------ FORM NODAL FORCE VECTOR FROM UDL -
------------------------

SUBROUTINE LOADPB(NELEM,PROPS,NEVAB,NGAUB,&

POSGP,WEIGP,NNODE,NPOIN,NN)

USE STRPAK

USE FIOPAK

USE DATA, ONLY:
ELOAD,SHAPE,DERIV,CARTD,GPCOD,COORD,LNODS,&

ELCOD,GLOAD,GLOAD1,IFIX,UDL,REACT

INTEGER NELEM,NGAUB,LNODE,NPOSN,NPOIN,KGASP,IELEM,&
```

```
IGAUS,JGAUS

REAL(KIND=8) ::
PROPS(4),POSGP(3),EXISP,ETASP,WEIGP(3),DAREA,DJACB

!! calculates nodal forces for plate element

!! read data controlling loading types to be input

60 CALL GAUSSQ(NGAUB,POSGP,WEIGP)

DO IELEM=1,NELEM

!! evaluate the coordinates of the element nodal points

ELOAD = 0.0D0

IF (DABS(UDL(IELEM)) < .000001) CYCLE

DO INODE=1,NNODE

LNODE=LNODS(IELEM,INODE)

ELCOD(1:2,INODE)=COORD(LNODE,1:2)

END DO

KGASP=0

!! enter loops for numerical integration

DO IGAUS=1,NGAUB

EXISP=POSGP(IGAUS)

K=1.

IF (IGAUS.EQ.2) THEN

K=-1.

END IF

DO JGAUS=1,NGAUB

ETASP=POSGP(JGAUS)*K
```

```
KGASP=KGASP+1

!! evaluate the shape functions at the sampling

!! points and elemental area

CALL SFR2(EXISP,ETASP,NNODE)

CALL JACOB2(DJACB,NNODE,IELEM)

DAREA=DJACB*WEIGP(IGAUS)*WEIGP(JGAUS)

!! calculate loads and associate with element nodal points

DO INODE=1,NNODE

NPOSN=(INODE-1)*3+1

ELOAD(NPOSN)=ELOAD(NPOSN)+SHAPE(INODE)*UDL(IELEM)*DAREA

END DO

END DO

END DO

!! assemble the element load matrices

DO INODE=1,NNODE

NODEI=LNODS(IELEM,INODE)

DO IDOFN=1,3

I=(INODE-1)*3+IDOFN

J=(NODEI-1)*3+IDOFN

IF (IFIX(J) == 0) GLOAD(J)=GLOAD(J)+ELOAD(I)

IF (IFIX(J) == 1) REACT(J)=REACT(J)-ELOAD(I)

END DO

END DO

END DO
```

```
GLOAD1=GLOAD

RETURN

END SUBROUTINE LOADPB

!----------------- PILE GROUP SUBROUTINE -----------------------------------

SUBROUTINE PILEG(PS,PF)

USE STRPAK

USE FIOPAK

USE DATA, ONLY:
NNODE,GSTIF,NPILE,PIPROP,COORD,EP,GS,GB,VS,PA,NPOIN, &

PILEN

REAL(KIND=8) ::L,R0,RB,AP,RM,K1,K2,U,R,KK1,KK2,RG

REAL(KIND=8) :: PS(NPILE,NPILE),PF(NPILE,NPILE),PCOORD(NPILE,2)

!-------------------

! input control data

!-------------------

IF(NPILE==0) RETURN

RG=DSQRT(PA/3.14159265359)

!-----------------

! pile coordinates

!-----------------

DO IPILE=1,NPILE

DO IDIME=1,2

PCOORD(IPILE,IDIME)=PIPROP(IPILE,IDIME)
```

```
END DO
END DO
!------------------------
! pile flexibility matrix
!------------------------
DO IPILE=1,NPILE
DO JPILE=1,NPILE
L =PIPROP(JPILE,3)
R0=PIPROP(JPILE,4)/2
RB=PIPROP(JPILE,5)/2
AP=PIPROP(JPILE,6)
RM=2.5*L*(1-VS)+RG
K1=2*3.14159265359*GS/DLOG(RM/R0)
K2=2*RB*GB/(1-VS)
IF (IPILE==JPILE) THEN
U=DSQRT(K1/AP/EP)
PF(IPILE,JPILE)=(1+(K2/AP/EP)*(DTANH(U*L)/U))/ &
(K2+K1*DTANH(U*L)/U)
ELSE
R=DSQRT((PCOORD(IPILE,1)-PCOORD(JPILE,1))**2+ &
(PCOORD(IPILE,2)-PCOORD(JPILE,2))**2)
KK1=K1*DLOG(RM/R0)/DLOG(RM/R)
KK2=K2*3.14159265359*R/2/RB
U=DSQRT(KK1/AP/EP)
```

```
PF(IPILE,JPILE)=(1+(KK2/AP/EP)*(DTANH(U*L)/U))/&
(KK2+KK1*DTANH(U*L)/U)
END IF
END DO
! WRITE(*,730) (PF(IPILE,JPILE),JPILE=1,NPILE)
END DO

CALL GJINV(NPILE,PF,PS) ! INVERSION OF PILE FLEXIBILITY MATRIX
!-----------------------------------------------------
! put pile stiffness matrix into global stiffness matrix
!-----------------------------------------------------

DO IPILE=1,NPILE
I=PILEN(IPILE)*3-2
DO JPILE=1,NPILE
J=PILEN(JPILE)*3-2
GSTIF(I,J)=GSTIF(I,J)+PS(IPILE,JPILE)
END DO
! WRITE(*,730) (PS(IPILE,JPILE),JPILE=1,NPILE)
END DO
RETURN
END SUBROUTINE PILEG

!----------------------- MATRIX INVERSE -----------------------------------

SUBROUTINE GJINV(N,PF,PS)
```

```
INTEGER N,P
REAL(KIND=8) :: PF(N,N),PS(N,N),M
!--------------
! initialise PS
!--------------
PS=0.0D0
DO K=1,N
PS(K,K)=1
END DO
!-----------------------------------------------
! calculate matrix inverse by Gauss-Jordon method
!-----------------------------------------------
DO K=1,N-1
!------------------------------
! eliminate PF to upper triangle
!------------------------------
DO J=K+1,N
M=PF(J,K)/PF(K,K)
DO P=1,N
PF(J,P)=PF(J,P)-M*PF(K,P)
PS(J,P)=PS(J,P)-M*PS(K,P)
END DO
END DO
END DO
```

```
!---------------------------------------
! reduce all diagonal entity to 1
!---------------------------------------
DO K=1,N
M=PF(K,K)
DO J=1,N
PF(K,J)=PF(K,J)/M
PS(K,J)=PS(K,J)/M
END DO
END DO
!---------------------------------------
! further eliminate PF to identity matrix
!---------------------------------------
DO K=1,N-1
DO J=1,N-K
M=PF(J,N-K+1)
DO P=1,N
PF(J,P)=PF(J,P)-M*PF(N-K+1,P)
PS(J,P)=PS(J,P)-M*PS(N-K+1,P)
END DO
END DO
END DO
RETURN
END SUBROUTINE GJINV
```

```
!------------------------ INPUT BOUNDARY CONDITION -------------------------

SUBROUTINE BOUND
USE DATA, ONLY: GLOAD,GSTIF,IFIX,BAND,COUPLING,NN,GLOAD1
INTEGER :: I,K,KV,J
LOOP1: DO K=1,NN
IF (IFIX(K) == 0) CYCLE LOOP1
IF (COUPLING == "N") THEN
DO J=2,BAND
KV=K+J-1
IF (NN-KV >= 0) THEN
GLOAD(KV)=GLOAD(KV)-GSTIF(K,J)*GLOAD1(K)
GSTIF(K,J)=0.D0
ENDIF
KV=K-J+1
IF (KV > 0) THEN
GLOAD(KV)=GLOAD(KV)-GSTIF(KV,J)*GLOAD1(K)
GSTIF(KV,J)=0.D0
ENDIF
GSTIF(K,1)=1.D0 ; GLOAD(K)=GLOAD1(K)
END DO
ELSE
DO I = 1,NN
GLOAD(I)=GLOAD(I)-GSTIF(I,K)*GLOAD1(K)
```

```
GSTIF(K,I)=0.D0
GSTIF(I,K)=0.D0
END DO
GSTIF(K,K)=1.D0 ; GLOAD(K)=GLOAD1(K)
ENDIF
END DO LOOP1
RETURN
END SUBROUTINE BOUND

!---------- Gauss Jordon Elimination ---------------------------------------

SUBROUTINE GAUSSJ(NJ2,AK,P)
INTEGER :: NJ2
REAL(KIND=8) AK(nj2,nj2),C,p(nj2)
DO 82 K=1,NJ2-1
K1=K+1
DO 82 I=K1,NJ2
C=AK(I,K)/AK(K,K)
P(I)=P(I)-C*P(K)
DO 82 J=1,NJ2
82 AK(I,J)=AK(I,J)-C*AK(K,J)
P(NJ2)=P(NJ2)/AK(NJ2,NJ2)
DO 84 I=NJ2-1,1,-1
I1=I+1
DO 83 J=I1,NJ2
```

```
83 P(I)=P(I)-AK(I,J)*P(J)
84 P(I)=P(I)/AK(I,I)
RETURN
END SUBROUTINE GAUSSJ

!----------------------------------------------------------------------
SUBROUTINE GAUSSB(GK,B,NT,NW)
INTEGER :: NT,NW,K,IK,I,J,JK,II,JJ
REAL(KIND=8) :: GK(NT,NW),B(NT),X,CN
DO 30 I=1,NT
X=GK(I,1)
IF(DABS(X)<= 0.0000000000001) THEN
WRITE(6,2) I
2 FORMAT(' **** SINGULARITY IN ROW',I5)
STOP
ENDIF
IK=I
DO 20 J=2,NW
IK=IK+1
CN=GK(I,J)/X
JK=0
DO 10 K=J,NW
JK=JK+1
10 GK(IK,JK)=GK(IK,JK)-CN*GK(I,K)
```

```
GK(I,J)=CN
20 B(IK)=B(IK)-CN*B(I)
30 B(I)=B(I)/X
DO 50 J=2,NT
II=NT-J+1
DO 40 K=2,NW
JJ=II+K-1
40 B(II)=B(II)-GK(II,K)*B(JJ)
50 CONTINUE
RETURN
END SUBROUTINE GAUSSB

!---------------------- WRITE NODAL DISPLACEMENT -----------------------

SUBROUTINE DISPT(NN,NPOIN,B)
REAL(KIND=8) B(NN)
INTEGER NPOIN,NN
WRITE (6,11)
WRITE(6,10)
DO I=1,NPOIN
J=(I-1)*3+1
WRITE(6,20) I,B(J),B(J+1),B(J+2)
END DO
10 FORMAT(//' DISPLACEMENT',/' ------------'/,1X,'NODE',8X,'DISP.'&
,5X,'XZ-ROT.',5X,'YZ-ROT.')
```

```
11 FORMAT(//' XZ-ROTATION IS THE ROTATION ABOUT YY AXIS',/ &
' YZ-ROTATION IS THE ROTATION ABOUT XX AXIS')
20 FORMAT(1X,I4,4X,3E12.4)
RETURN
END SUBROUTINE DISPT

!--------- calculate the support reactions ---------------------------------

SUBROUTINE REACTION(NEVAB,NNODE,NELEM,NPOIN)
USE DATA, ONLY: ESTIF,GLOAD,LNODS,REACT,GLOAD1,NN,IFIX,NSPRG, &
NSPRING,SPRING
INTEGER NEVAB,NNODE,NELEM,NPOIN
DO IELEM=1,NELEM
II=0
DO INODE=1,NNODE
DO IDOFN=1,3
J=(LNODS(IELEM,INODE)-1)*3+IDOFN
II=II+1
DO JJ=1,NNODE
DO I=1,3
K=(LNODS(IELEM,JJ)-1)*3+I
REACT(J)=REACT(J)+ESTIF(IELEM,II,(JJ-1)*3+I)&
*GLOAD(K)
END DO
```

```
END DO

END DO

END DO

END DO

DO I=1,NN

IF (IFIX(I) == 0) REACT(I)=0

END DO

WRITE(6,910)

DO I=1,NPOIN

IF (IFIX(3*(I-1)+1) == 0 .AND. IFIX(3*(I-1)+2) == 0 &

.AND. IFIX(3*I) == 0) CYCLE

WRITE(6,900) I,REACT((I-1)*3+1),REACT((I-1)*3+2),REACT((I-1)*3+3)

END DO

IF (NSPRG == 0) RETURN

WRITE (6,902)

DO I = 1,NSPRG

WRITE (6,900) NSPRING(I),SPRING(I)*GLOAD(3*(NSPRING(I)-1)+1)

END DO

902 FORMAT(//' SPRING FORCE',/' ---------------'/,1X,'NODE',5X,'FORCE')

900 FORMAT(1X,I4,1X,3E13.5)

910 FORMAT(//' NODAL REACTION',/' -
--------------'/,1X,'NODE',5X,'FORCE',&

6X,'XY-MOMENT',5X,'XZ-MOMENT')

RETURN
```

```
END SUBROUTINE REACTION

!---------------------------- CALCULATE STRESSES AFTER ANALYSIS-
--------------------------

SUBROUTINE
STREPB(NELEM,NGAUV,NGAUB,NSTRE,NNODE,NEVAB,&

NGAUS2)

USE DATA, ONLY:
SMATX,GPCOD,LNODS,GLOAD,NPOIN,COORD,SHAPE

INTEGER
NELEM,NPOSN,LNODE,KGASP,NGAUV,NGAUB,NSTRE,KGASH,INODE

REAL(KIND=8) :: ELDIS(3,8),NMOME(8,5),S,T,KK(NPOIN)

REAL(KIND=8) :: STRSG(13,5),STRSQ(13,5),NSTRES(NPOIN,5)

REAL(KIND=8) :: CENTROID(2)

! calculates stress resultants at gauss points

! for plate bending element

WRITE (6,895)

WRITE(6,900)

WRITE(6,905)

NSTRES=0. ; KK=0. ; STRSG=0.D0 ; STRSQ=0.D0

! loop over each element

DO IELEM=1,NELEM

! read the stress matrix, sampling point

! coordinates for the element

WRITE(6,910) IELEM

! identify the displacements of the element nodal points
```

```
DO INODE=1,NNODE
LNODE=LNODS(IELEM,INODE)
NPOSN=(LNODE-1)*3
DO IDOFN=1,3
NPOSN=NPOSN+1
ELDIS(IDOFN,INODE)=GLOAD(NPOSN)
END DO
END DO
KGASP=0
!---------------------------------------------
! enter loops over each sampling point - bending
!---------------------------------------------
DO IGAUS=1,NGAUB
DO JGAUS=1,NGAUB
KGASP=KGASP+1
! initialize stress values
STRSG(KGASP,:)=0.0
DO ISTRE=1,3
KGASH=0
! compute the stress resultants
DO INODE=1,NNODE
DO IDOFN=1,3
KGASH=KGASH+1
STRSG(KGASP,ISTRE)=STRSG(KGASP,ISTRE)+SMATX(IELEM,ISTRE,KG
```

```
ASH,KGASP)&
*ELDIS(IDOFN,INODE)
END DO
END DO
END DO
! output the stress resultants
END DO
END DO
IF (NNODE == 4) THEN
DO ISTRE = 1,3
STRSG(1,ISTRE)=(STRSG(1,ISTRE)+STRSG(2,ISTRE)+ &
STRSG(3,ISTRE)+STRSG(4,ISTRE))/4.
END DO
CENTROID=0.
DO ISTRE = 1,4
CENTROID(1)=CENTROID(1)+GPCOD(IELEM,1,ISTRE)/4.
CENTROID(2)=CENTROID(2)+GPCOD(IELEM,2,ISTRE)/4.
END DO
ENDIF
!-----------------------------------------------
! enter loops over each sampling point - shear
!-----------------------------------------------
DO IGAUS=1,NGAUV
DO JGAUS=1,NGAUV
```

```
KGASP=KGASP+1

! initialize stress values

STRSQ(KGASP,:)=0.0D0

DO ISTRE=4,5

KGASH=0

! compute the stress resultants

DO INODE=1,NNODE

DO IDOFN=1,3

KGASH=KGASH+1

STRSQ(KGASP,ISTRE)=STRSQ(KGASP,ISTRE)+SMATX(IELEM,ISTRE,KG
ASH,KGASP)&

*ELDIS(IDOFN,INODE)

END DO

END DO

END DO

! output the stress resultants

IF (NNODE == 8) THEN

WRITE(6,915) KGASP-4,(GPCOD(IELEM,IDIME,KGASP-4),IDIME=1,2),&

(STRSG(KGASP-4,ISTRE),ISTRE=1,3), &

(STRSQ(KGASP,ISTRE),ISTRE=4,NSTRE)

ENDIF

END DO

END DO

IF (NNODE == 4) THEN
```

```
WRITE(6,917) (CENTROID(IDIME),IDIME=1,2), &
(STRSG(1,ISTRE),ISTRE=1,3), &
(STRSQ(KGASP,ISTRE),ISTRE=4,NSTRE)
ENDIF
!--------------------------------------------------------------------
! enter loops for nodal moment & shear extrapolation for 8-node element
!--------------------------------------------------------------------
NMOME=0. !NMOME - nodal moment & shear
IF (NNODE.EQ.8) THEN

DO INODE=1,NNODE
IF (INODE.EQ.1) THEN
S=-1/0.577350269189626;T=-1/0.577350269189626
END IF
IF (INODE.EQ.2) THEN
S= 0. ;T=-1/0.577350269189626
END IF
IF (INODE.EQ.3) THEN
S= 1/0.577350269189626;T=-1/0.577350269189626
END IF
IF (INODE.EQ.4) THEN
S= 1/0.577350269189626;T= 0.
END IF
IF (INODE.EQ.5) THEN
```

```
S= 1/0.577350269189626;T= 1/0.577350269189626

END IF

IF (INODE.EQ.6) THEN

S= 0. ;T= 1/0.577350269189626

END IF

IF (INODE.EQ.7) THEN

S=-1/0.577350269189626;T= 1/0.577350269189626

END IF

IF (INODE.EQ.8) THEN

S=-1/0.577350269189626;T= 0.

END IF

CALL SFR2(S,T,4)

DO IGAUS=1,4 !for 2x2 guass points only

NCOOR(INODE,1)=NCOOR(INODE,1)+SHAPE(IGAUS)*GPCOD(IELEM,1,I
GAUS)

NCOOR(INODE,2)=NCOOR(INODE,2)+SHAPE(IGAUS)*GPCOD(IELEM,2,I
GAUS)

NMOME(INODE,1)=NMOME(INODE,1)+SHAPE(IGAUS)*STRSG(IGAUS,1)

NMOME(INODE,2)=NMOME(INODE,2)+SHAPE(IGAUS)*STRSG(IGAUS,2)

NMOME(INODE,3)=NMOME(INODE,3)+SHAPE(IGAUS)*STRSG(IGAUS,3)

! extrapolate from Gaussian points to get nodal bending moment

NMOME(INODE,4)=NMOME(INODE,4)+SHAPE(IGAUS)*STRSQ((IGAUS+
4),4) ! extrapolate from Gaussian points to get nodal shear

NMOME(INODE,5)=NMOME(INODE,5)+SHAPE(IGAUS)*STRSQ((IGAUS+
4),5)

END DO
```

```
LNODE=LNODS(IELEM,INODE)

NSTRES(LNODE,1)=NSTRES(LNODE,1)+NMOME(INODE,1)

NSTRES(LNODE,2)=NSTRES(LNODE,2)+NMOME(INODE,2)

NSTRES(LNODE,3)=NSTRES(LNODE,3)+NMOME(INODE,3)

NSTRES(LNODE,4)=NSTRES(LNODE,4)+NMOME(INODE,4)

NSTRES(LNODE,5)=NSTRES(LNODE,5)+NMOME(INODE,5)

KK(LNODE)=KK(LNODE)+1.

END DO

END IF

!-----------------------------------------------------------------------

! enter loops for nodal moment & shear extrapolation for 4-node element

!-----------------------------------------------------------------------

IF (NNODE.EQ.4) THEN

DO INODE=1,NNODE

DO IGAUS=1,4 !for 2x2 guass points only

NCOOR(INODE,1)=NCOOR(INODE,1)+SHAPE(IGAUS)*GPCOD(IELEM,1,I
GAUS)

NCOOR(INODE,2)=NCOOR(INODE,2)+SHAPE(IGAUS)*GPCOD(IELEM,2,I
GAUS)

NMOME(INODE,1)=NMOME(INODE,1)+STRSG(IGAUS,1)

NMOME(INODE,2)=NMOME(INODE,2)+STRSG(IGAUS,2)

NMOME(INODE,3)=NMOME(INODE,3)+STRSG(IGAUS,3)        !interpolate
bending and coor

NMOME(INODE,4)=NMOME(INODE,4)+STRSQ((IGAUS+4),4) ! interpolate
for shear

NMOME(INODE,5)=NMOME(INODE,5)+STRSQ((IGAUS+4),5)
```

```
END DO

LNODE=LNODS(IELEM,INODE)

NSTRES(LNODE,1)=NSTRES(LNODE,1)+NMOME(INODE,1)/4

NSTRES(LNODE,2)=NSTRES(LNODE,2)+NMOME(INODE,2)/4

NSTRES(LNODE,3)=NSTRES(LNODE,3)+NMOME(INODE,3)/4

NSTRES(LNODE,4)=NSTRES(LNODE,4)+NMOME(INODE,4)

NSTRES(LNODE,5)=NSTRES(LNODE,5)+NMOME(INODE,5)

KK(LNODE)=KK(LNODE)+1.

END DO

END IF

END DO

WRITE (6,904)

WRITE (6,906)

DO I=1,NPOIN

NSTRES(I,1:5)=NSTRES(I,1:5)/KK(I)

WRITE (6,916) I,COORD(I,1:2),NSTRES(I,1:5)

END DO

895 FORMAT(/' X-MOMENT IS THE MOMENT ABOUT ABOUT YY AXIS. Y-MOMENT IS THE ', 'MOMENT ABOUT XX'/' AXIS. XY-MOMENT IS THE TORSION. ','XZ-S FORCE IS THE', ' SHEAR FORCE ABOUT YY AXIS')

900 FORMAT(//' INTERNAL FORCES AT CENTROID',/' ---------------------------')

904 FORMAT(/'BE CAREFUL ! THE INTERNAL NODAL FORCES ARE INTERPOLATED FROM THE', /'INTERNAL FORCES AT THE CENTROIDS OF THE PLATE ELEMENTS. IF NODAL FORCES',/'ARE PRESENT AT ANY PARTICULAR NODE, THE INTERPOLATED NODAL FORCES ARE',
```

```
/'PRACTICALLY MEANINGLESS'/)

905 FORMAT(7X,8HX-COORD.,2X,8HY-COORD.,3X,8HX-MOMENT,4X,&

8HY-MOMENT,5X,10HXY-MOMENT,3X,10HXZ-S,FORCE,2X,10-
YZ-S,FORCE)

906 FORMAT(1X,4H PT.,2X,8HX-COORD.,2X,8HY-COORD.,3X,8-
X-MOMENT,4X,&

8HY-MOMENT,5X,10HXY-MOMENT,3X,10HXZ-S,FORCE,2X,10-
YZ-S,FORCE)

910 FORMAT(/,5X,12HELEMENT NO.=,I5)

915 FORMAT(1X,I4,2F10.4,5(1X,E12.5))

916 FORMAT(1X,I4,2F10.4,5(1X,E12.5))

917 FORMAT(5X,2F10.4,5(1X,E12.5))

920 FORMAT(I3,1X,2F11.4,36X,2E12.5)

RETURN

END SUBROUTINE STREPB

!------------------------ PILE LOAD-----------------------------------------

SUBROUTINE PREACT(B,PD)

USE DATA, ONLY: NPILE,GSTIF,PIPROP,PSTIF,NN,NPOIN,COORD,PILEN

REAL(KIND=8) :: B(NN),PD(NPILE),PLD(NPILE)

IF(NPILE==0) RETURN

!--------------------------

! retrieve pile displacement

!--------------------------

DO IPILE=1,NPILE
```

```
I=PILEN(IPILE)*3-2
PD(IPILE)=B(I)
END DO
!---------------------
! calculate pile load
!---------------------
PLD=0.0D0
WRITE(6,850)
DO IPILE=1,NPILE
DO JPILE=1,NPILE
PLD(IPILE)=PLD(IPILE)+PSTIF(IPILE,JPILE)*PD(JPILE)
END DO
WRITE (6,855) PIPROP(IPILE,1),PIPROP(IPILE,2),-1*PLD(IPILE)
END DO
850 FORMAT(//' PILE REACTION',/ '--- --- --- --'/ &
'PILE AT X-COORD Y-COORD ','REACTION')
855 FORMAT(8X,F7.3,7X,F7.3,9X,F10.3)
RETURN
END SUBROUTINE PREACT

!------------------ drawmesh -------------------------------------

subroutine drawmesh(ndim,nn,nod,nels,count)
use dislin
USE DATA, ONLY :
```

```
NELEM,NNODE,COORD,LNODS,IFIX,NPOIN,GLOAD,fildat,filout
implicit none
character(len=60) :: command
real :: x_scale1,y_scale1,x_scale2,y_scale2
logical :: alive
integer,intent(in out):: ndim,nn,nod,nels,count
integer::i,j,N,COMPARE
real,dimension(2) :: u,v
real,dimension(2):: X,Y
real,dimension(5):: XX,YY
real :: RANGE,M1,M2
real (KIND=4) ::
ZMAT(NPOIN,NPOIN),XRAY(NPOIN),YRAY(NPOIN),ZLEV(12)
!-----draw mesh--------------------------------------------------------

! u1 IS MAX.X, u2 IS MAX.Y
! v1 IS MIN.X, v2 IS MIN.Y
u=0. ; v=0.
N=NPOIN
DO i=1,nn
IF (coord(i,1).GE.u(1)) u(1)=coord(i,1)
IF (coord(i,1).LE.v(1)) v(1)=coord(i,1)
IF (coord(i,2).GE.u(2)) u(2)=coord(i,2)
IF (coord(i,2).LE.v(2)) v(2)=coord(i,2)
END DO
```

```
x_scale1=INT(v(1)-1) ! MIN.X - 1
x_scale2=INT(u(1)+1) ! MAX.X + 1
y_scale1=INT(v(2)-1) ! MIN.Y - 1
y_scale2=INT(u(2)+1) ! MAX.Y + 1

compare=x_scale2-x_scale1
if (y_scale2-y_scale1 > compare) compare=y_scale2-y_scale1
call scrmod('REVERS')
if (count == 3) then
call metafl('pscl') ! output postscript file
endif
if (count /= 3) then
CALL METAFL('cons') ! grahpic output on the screen
INQUIRE(FILE='DISLIN.PSC',EXIST=ALIVE)
IF (ALIVE) CALL SYSTEM('DEL DISLIN.PSC')
ENDIF
call setpag('da4l')
CALL DISINI()
call setvlt('rain')
CALL PAGERA()
CALL COMPLX()
CALL AXSPOS(350,1900)
CALL AXSLEN(2300,1600)
CALL NAME('Scale in Metre (m)','X')
```

```
CALL NAME('Scale in Metre (m)','Y')
CALL TICKS(2,'XY')
if (compare > 0.1 .and. compare < 1.0) then
range=0.1
elseif (compare > 1 .and. compare <= 10) then
range=1
elseif (compare > 10 .and. compare <= 30) then
range=2
elseif (compare > 30 .and. compare <= 100) then
range = 5
elseif (compare > 100) then
range = 10
endif
command(1:35)='Finite element mesh for input file '
command(36:)=fildat
CALL MESSAG(command,1000,75)
CALL
GRAF(x_scale1,x_scale2,x_scale1,range,y_scale1,y_scale2,y_scale1,range)
call height(55)
CALL COLOR('white') ! draw mesh
DO j=1,nelem
DO i=1,nnode
X(1)=coord(lnods(j,i),1)
Y(1)=coord(lnods(j,i),2)
```

```
if (i < nnode) then
X(2)=coord(lnods(j,i+1),1)
Y(2)=coord(lnods(j,i+1),2)
endif
if (i == nnode) then
X(2)=coord(lnods(j,1),1)
Y(2)=coord(lnods(j,1),2)
endif
CALL CURVE(X,Y,2)
IF (IFIX((LNODS(j,i)-1)*3+1) == 1) THEN
XX(1)=X(1)-0.03 ; YY(1)=Y(1)-0.03
XX(2)=X(1)+0.03 ; YY(2)=Y(1)-0.03
XX(3)=X(1)+0.03 ; YY(3)=Y(1)+0.03
XX(4)=X(1)-0.03 ; YY(4)=Y(1)+0.03
XX(5)=X(1)-0.03 ; YY(5)=Y(1)-0.03
CALL CURVE(XX,YY,5)
END IF
IF (IFIX((LNODS(j,i)-1)*3+2) == 1) THEN
XX(1)=X(1)-0.03 ; YY(1)=Y(1)-0.03
XX(2)=X(1)+0.03 ; YY(2)=Y(1)-0.03
XX(3)=X(1) ; YY(3)=Y(1)+0.03
XX(4)=X(1)-0.03 ; YY(4)=Y(1)-0.03
CALL CURVE(XX,YY,4)
END IF
```

```
IF (IFIX((LNODS(j,i)-1)*3+3) == 1) THEN
XX(1)=X(1)-0.03 ; YY(1)=Y(1)+0.03
XX(2)=X(1)+0.03 ; YY(2)=Y(1)+0.03
XX(3)=X(1) ; YY(3)=Y(1)-0.03
XX(4)=X(1)-0.03 ; YY(4)=Y(1)+0.03
CALL CURVE(XX,YY,4)
END IF
end do
END DO

call disfin() ; call system('cls')
if (count == 3) then
COMMAND=''
count=index(filout,'.')
if (count > 0) then
filout(count+1:count+3)='psc'
else
count=len_trim(filout)
filout(count+1:count+4)='.psc'
endif
inquire (file=filout,exist=alive)
if (alive) then
command(1:4)='del '
command(5:)=filout
```

```
call system(command)

endif

command(1:18)='rename dislin.psc '

count=len_trim(filout)

command(19:18+count)=filout

call system(command)

endif

end subroutine drawmesh
```

A sample input for a raft foundation problem is given below. Readers should find it easy to define a problem by reading the code and this sample input.

```
E_PLATE 10000000 POISSON_RATIO_PLATE 0.3

THICKNESS_PLATE=1.0

GLOBAL_PRESSURE=100 ! UDL of 100 acting on the whole plate area

Winkler_spring 20000

! Winkler spring 20000 acting over the whole plate area

NODAL_LOADS

15.95 15.95 100 0 0 ! xy-ordinates, v, moment about yy, moment about xx

RECTANGULAR_GRID Y

GRID_X 0 8 20 28 32

GRID_Y 0 8 20 28 32

LOCAL_PRESSURE 0.4

-0.01 -0.01 32.01 -0.01 32.01 32.01 -0.01 32.01

nodal_spring
```

```
12 12 1000000

24 24 1000000

12 24 1000000

24 12 1000000 ! xy-ordinates, spring values

V_FIX 0.

-0.01 -0.01 32.01 -0.01 32.01 0.01

0.01 0.01 0.01 32.01 -0.01 32.01

V_FIX 0.

31.99 -0.01 32.01 -0.01 32.01 32.01

-0.01 32.01 -0.01 31.99 31.99 31.99

rx_FIX 0.0001

-0.01 -0.01 32.01 -0.01 32.01 0.01

0.01 0.01 0.01 32.01 -0.01 32.01

rx_FIX 0.0001

31.99 -0.01 32.01 -0.01 32.01 32.01

-0.01 32.01 -0.01 31.99 31.99 31.99

END
```

APPENDIX 2-5. ONE-DIMENSIONAL CONSOLIDATION

There are many reclamation works undergoing in Hong Kong, and in general, pore water pressure and consolidation settlement are monitored during the consolidation process. In many cases, many Hong Kong engineers found that the use of Terzaghi one-dimensional consolidation theory and the the use of volume compressibility concept cannot give a good prediction to the consolidation settlement. This phenomenon can be explained by the fact that in many reclamation works in Hong Kong, there is a very large change in the vertical stresses, and the effective vertical stress can change by several times during the consolidation. Under such condition, the volume compressibility coefficient cannot be assumed to be a constant. Refer to the Terzaghi consolidation equation as governed by eq.(A2-1)

$$\frac{\partial u}{\partial t} = C_v \frac{\partial^2 u}{\partial z^2} \qquad \text{where} \qquad C_v = \frac{K}{m_v \gamma_w} \tag{A2-1}$$

Classically, to solve this problem, we can consider the boundary conditions for 1 way drainage as

$$\begin{aligned} &t > 0,\ z = 0 && u(0,t) = 0 \\ &t > 0,\ z = H && \frac{\partial u(z,t)}{\partial z} = 0 \\ &\text{Initial condition, } t = 0 && u(z,0) = u_0(z) \end{aligned} \tag{A2-2}$$

The solution of eq. (A2-1) can be obtained by using Fourier Series as

$$u(z,t) = \frac{2}{H}\left(\sum_{N=1}^{\infty}\int_0^H u_o(z)\sin\left(\frac{nz}{H}\right)dz\right)\cdot\sin\frac{nz}{H}e^{-n^2T_v}$$

$$\text{where} \quad n = \frac{(2N-1)}{2}\pi, T_v = \frac{C_v t}{H^2}, N = 1,2,3...... \tag{A2-3}$$

If $u_0(z)$ is constant

$$u(z,t) = u_o \sum_{N=1}^{\infty} \frac{2}{n} \sin\frac{nz}{H} e^{-n^2T_v} \tag{A2-4}$$

Ground settlement at time S(t) can be evaluated as

$$S(t)=\int m_v\left[u_o(z)-u(z,t)\right]dz \tag{A2-5}$$

$$=m_v u_o H\left[1-\sum_{N=1}^{\infty}\frac{2}{n^2}e^{-n^2T_v}\right] \tag{A2-6}$$

Degree of consolidation $u_v(t)$ can be defined as

$$u_v(t)=\frac{S(t)}{S(\infty)}=\left[1-\sum_{N=1}^{\infty}\frac{2}{n^2}e^{-n^2T_v}\right] \tag{A2-7}$$

These consolidation equations can be found in virtually all textbook in soil mechanics and foundation engineering, but are found to be not reflecting the true consolidation phenomenon from many field measurement results in Hong Kong. Actually, in eq.(A2-1), both m_v and K can change during the consolidation process, particularly for the volume compressibility m_v. K will also change due to the decrease in void ratio, but this term is in general less sensitive than the volume compressibility. Based on the compression index, the consolidation settlement can also be expressed in term of eq.(A2-8)

$$\Delta H=\frac{C_c H}{1+e_o}\log_{10}\frac{p_o+\Delta p}{p_o} \tag{A2-8}$$

Eq.(A2-8) should be compared with the use of the simple consolidation settlement equation

$$\Delta H=m_v\Delta\sigma_v' H \tag{A2-9}$$

Eq.(A2-8) is not commonly used by engineers, as no equation or design chart are available for the engineers to compute the time of consolidation. This equation is however a better equation to be adopted, as the compression index appears to be constant over a wide stress range, as compared with the volume compressibility. Actually, from eq.(A2-8), if the stress change is small, it can be approximated as

$$\Delta H = \frac{C_c H}{1+e_0} \log \frac{p_0' + \Delta p}{p_0'} \approx \frac{C_c H}{2.303(1+e_o)} \frac{\Delta p}{p_0'} \quad \textbf{(A2-10)}$$

The volume compressibility can hence be approximated as

$$m_v = \frac{C_c}{2.303(1+e_o)p_0'} \quad \textbf{(A2-11)}$$

Since the effective overburden stress can change by several times during the reclamation consolidation process, the volume compressibility will also change by several times. Even K has also decreased during the consolidation, the overall result is that Cv will also change significantly during the consolidation process. The poor prediction from eq.(A2-7) is hence not surprising.

To overcome this problem, eq.(A2-1) can be rewritten as

$$\frac{\partial u}{\partial t} = C_v \frac{\partial^2 u}{\partial z^2} = \frac{K}{\gamma_w} \frac{2.303 p_o'(1+e_o)}{C_c} \frac{\partial^2 u}{\partial z^2} \quad \textbf{(A2-12)}$$

To solve for eq.(A2-12) and hence the consolidation as well as the degree of consolidation, the authors have adopted the Galerkin finite element formulation, where eq.(A2-12) is written as

$$[C]\{\dot{u}\} + [K]\{u\} - \{F\} = 0 \quad \textbf{(A2-13)}$$

Equation (A2-13) can be solved by implicit finite difference scheme as

$$\{[C] + \theta \times \Delta t[K]\}\{u\}_b = \{[C] - (1-\theta)\Delta t[K]\}\{u\}_a + \Delta t \times [(1-\theta)\{F\}_a + \theta\{F\}_b] \quad \textbf{(A2-14)}$$

where [C] = global capacitance matrix

[K] = global stiffness matrix

{F} is the equivalent force vector

$0 < \theta < 1$ specified by user

i. $\theta = 0$, forward difference method. The accuracy is low and a small time step is required but the steps in computation is simple. Unstable if Δt is large.
ii. $\theta = 0.5$, central difference method which is commonly used. Numerical oscillation can occur if Δt is large.
iii. $\theta = 2/3$, Galerkin difference method. Numerical oscillation can occur if Δt is large.
iv. $\theta = 1$, backward difference method. Unconditionally stable.

The time-step must be controlled carefully, otherwise, numerical oscillation may occur in the computation as shown. Based on eq.(A2-14), a computer program has been developed where the compression index instead of the volume compressibility or coefficient of consolidation is used. In the following program, the permeability is assumed to be constant, but the readers can modify this easily if the relation between the permeability and the void ratio is known (this is not implemented into the present code, as such relation is generally not known to Hong Kong soil). A finite element code in Fortran 77 (Lahey Fortran compiler) is given here, and the code be modified easily to Fortran 90/95 easily. In general, this code is easy to read, and a sample input and output is attached herewith for easy reference. The readers should take care about two major parameters : FINITE and PROCES. The parameter FINITE control the numerical integration as discussed above. In most cases, the authors choose a value of 1.0 or 0.5 for application. For the parameter PROCES, it fine-tunes the computation with the options:

Consolidation process:

1. Constant clay thickness and constant effective stress, and the resulting effect will be similar to the use of volume compressibility.
2. Constant clay thickness and the effective stress will change during the computation.
3. The clay thickness will change during consolidation, with constant effective stress.
4. The clay thickness as well as the effective stress will change during consolidation.

```
C PROGRAM ODC

C Date : 15th April, 1994
```

```
C
C ****************************************
C * Argument Declarations of main program *
C ****************************************
C FILEOP - the option whether create or read a file;
C FINP - the name for input data file;
C FFPP - the name for output data file;
C NODNUM - the number of nodes;
C ELENUM - the number of elements;
C FORNUM - the option of the formulation;
C BOUND - the choice of the boundary condition;
C ROW - the no. of row in the element matrix;
C COL - the no. of column in the element matrix;
C NN - the no. of row in the UCAP and USTF arrays;
C DEGCON - the specified degree of consolidation
C for termination of the program;
C STEP - the no. of iteration;
C TIMFAC - the time factor;
C YCOOR - the y-coordinates of the nodes (meter);
C START - the first node of the element;
C END - the second node of the element;
C ELELEN - the element length (meter);
C CC - the compression index;
C STRESS - the effective stress (mega-newtons per square meter);
```

```
C K - the permeability (meter per year);
C E - the void ratio;
C CV - the coefficient of consolidation;
C (square meter per second)
C DELTAT - the time step (second);
C H - the thickness of the whole strata (meter);
C AVECV - the average value of CV (square meter per second);
C TIMFIN - the total time taken when finish (second);
C ASU - the vector for the external load;
C EC - the element capacitance matrix;
C UCAP - the banded matrix of the upper part of the symmetric
C global capacitance matrix;
C USTF - the banded matrix of the upper part of the symmetric
C global stiffness matrix;
C RCEXU - the rate of change of excess pore water pressure;
C TIME1 - starting time getting from PC;
C TIME2 - ending time getting from PC;
C COMTIME- total time taken by PC;
C FINITE - the method of finite difference;
C PROCES - the method of consolidation process;
C STEP - the number of iteration;
C TV - the time factor;
C TIME - the time for consolidation used at that STEP;
C FIN,U0,EXU,UA,UB,MV,AV are the auxiliary vectors and numbers.
```

```
C
C PRINTS - time for printing of results
INTEGER :: NODNUM,ELENUM,FORNUM,ROW,COL,NN, &
& FILEOP,BOUND,FINITE,PROCES,STEP,PRINTS
REAL :: DELTAT,DEGCON,H,AVECV,TIMFIN,TIME1,TIME2, &
& COMTIME,FIN,STRMPD,+TIMDAY,TV,TIME
CHARACTER :: FINP*12,FFPP*12
INTEGER, allocatable :: START(:), END(:)
REAL, allocatable :: YCOOR(:), ELELEN(:), CC(:), K(:), &
& STRESS(:),E(:), CV(:),EC(:,:,:), UCAP(:,:),USTF(:,:), &
& ASU(:),RCEXU(:), EXU(:),U0(:),UA(:),UB(:), MV(:)
LOGICAL BEGIN,SINGU
C
C Creat or Read Data File
CALL INPUT (FINP,FFPP,FILEOP)
PRINT *, 'THE PROGRAM EXECUTION BEGINS.'
PRINT *
C
C Read the data from Data File and find out the values of ROW,COL,NN
CALL KEY1 (FINP,NODNUM,ELENUM,FORNUM,ROW,COL,NN)
C
C Allocate the dynamic arrays
ALLOCATE (YCOOR(NODNUM),START(ELENUM),END(ELENUM), &
& CC(ELENUM),K(ELENUM),E(ELENUM),ELELEN(ELENUM), &
```

```
& CV(ELENUM),EC(ELENUM,ROW,COL),UCAP(NN,COL), &

& USTF(NN,COL),ASU(NN),RCEXU(NN),U0(NN),EXU(NN), &

& UA(NN),UB(NN),MV(ELENUM),STRESS(ELENUM))

C

C Read the remaining part of the data file

CALL KEY2 (NODNUM,ELENUM,DEGCON,YCOOR,START,END, &

&                    CC,STRESS,K,E,STRMPD,TIMDAY,BOUND,STEP,FINITE,
PROCES,PRINTS)

BEGIN=.TRUE.

DO WHILE (BEGIN)

C

C Mathematical model is made

CALL GRDMOD (NODNUM,ELENUM,FORNUM,YCOOR,

START,END,CC,STRESS,K,E,ELELEN,CV,DELTAT,H,AVECV,STEP,TIMFI
N,EXU,UA,FINITE,MV,NN,FIN,COL,UCAP,USTF,BOUND,UB,BEGIN,SING
U,TIMDAY,TIME,PROCES)

C

C Formation and Assemble the matrice: [UCAP], Capacitance matrix

C [USTF], Stiffness matrix

CALL CAPMAT (NODNUM,FORNUM,ROW,COL,ELENUM,CV,

ELELEN,NN,UCAP,EC)

CALL STFMAT (NODNUM,FORNUM,ROW,COL,ELENUM,ELELEN,

NN,USTF)

IF (STEP .EQ. 1) THEN

PRINT 10, 'THE SIZE OF THE GLOBAL CAPACITANCE MATRIX
=',NN,'x',COL
```

```
10 FORMAT (1X,A,1X,I5,3X,A1,1X,I4)
PRINT 20, 'THE SIZE OF THE GLOBAL STIFFNESS MATRIX =',NN,'x',COL
20 FORMAT (1X,A,1X,I5,3X,A1,1X,I4)
PRINT *
PRINT *, 'THE SIMULTANEOUS EQUATIONS ARE BEING SOLVED.'
PRINT *, 'IT IS THE MOST TIME CONSUMING PROCESS, PLEASE WAIT.'
ENDIF
C
C Formation of the Surcharge Vector
CALL SURCHG (ELENUM,FORNUM,NN,ASU,STRMPD,EC,ROW,COL)
C
C Solving the equations
IF (PROCES .EQ. 1) THEN
DO WHILE (BEGIN)
CALL COMBIN (FFPP,FORNUM,ELENUM,NN,COL,UCAP,USTF,ASU,DELTAT,
+TIMFIN,DEGCON,BOUND,H,AVECV,ELELEN,STEP,EXU,U0,UA,FIN,BEGIN,
+SINGU,TV,TIME,PROCES,PRINTS)
END DO
GOTO 30
ENDIF
CALL COMBIN (FFPP,FORNUM,ELENUM,NN,COL,UCAP,USTF,ASU,DELTAT,
+TIMFIN,DEGCON,BOUND,H,AVECV,ELELEN,STEP,EXU,U0,UA,FIN,BEGIN,
```

```
+SINGU,TV,TIME,PROCES,PRINTS)
END DO
30 CONTINUE
C
C Write down the result into Output File
PRINT *, 'WRITE DOWN THE RESULT INTO OUTPUT FILE.'
CALL RESULT (TIME,FORNUM,NN,EXU,NODNUM,SINGU)
C
PRINT *, 'OK!'
PRINT *
C
DEALLOCATE
(CV,ASU,UCAP,USTF,RCEXU,START,END,YCOOR,ELELEN,CC,
+STRESS,K,E,EC,U0,EXU,UA,UB,MV)
STOP
END
C-------------------------------------------------------------------
C-------------------------------------------------------------------
SUBROUTINE INPUT (FINP,FFPP,FILEOP)
C - Create or Read a data file
C
C Argument Declarations
CHARACTER NAME*8,FINP*12,FFPP*12
C Local Declarations
```

```
C STRMPD - the increasing rate of external loading in MPD per day;
C TIMDAY - the total time for addition of external load;
INTEGER NODNUM,ELENUM,FORNUM,BOUND,I,J,PROCES,FINITE
REAL STRMPD,TIMDAY,DEGCON
INTEGER, allocatable :: ST(:), EN(:)
REAL, allocatable :: YC(:), TE(:), TCC(:), TSS(:),TK(:)
C
C <<< Direct Input >>>
! CALL CLEARSCREEN ($GCLEARNSCREEN)
PRINT *, 'Select the options :-'
PRINT '(1X,A,//,1X,A)', '1. Create new file.',
+'2. Read existing file.'
10 PRINT *, 'Enter the option: 1, 2'
READ (*,*) FILEOP
IF ((FILEOP .LT. 1) .OR. (FILEOP .GT. 2)) THEN
PRINT *, '********* Illegal Option **********'
GOTO 10
ELSE
PRINT *, 'Enter the name of file '
READ (*,*) NAME
FINP=NAME
FFPP='out'
ENDIF
IF (FILEOP .EQ. 2) RETURN
```

```
C Open file for data

OPEN
(UNIT=10,FILE=FINP,FORM='FORMATTED',ACCESS='SEQUENTIAL',

+STATUS='NEW')

C

C <<< Step 1 >>>

PRINT *, 'STEP 1: INPUT THE PARAMETERS'

PRINT *

PRINT *, 'Enter the number of nodes.'

READ *, NODNUM

PRINT *

PRINT *, 'Enter the number of elements.'

READ *, ELENUM

PRINT *

PRINT *, 'There are three kinds of formulations:'

PRINT '(1X,A,/,1X,A,/,1X,A)', '1. Consistent Formulation',

+'2. Lumped Formulation','3. Cubic Formulation'

20 PRINT *, 'Enter the option: 1, 2, 3'

READ *, FORNUM

IF ((FORNUM .LT. 1) .OR. (FORNUM .GT. 3)) THEN

PRINT *, '********* Illegal Option **********'

GOTO 20

ENDIF

PRINT *
```

```
30 PRINT *, 'Enter the degree of consolidation for termination.'
READ *, DEGCON
IF (DEGCON .GT. 1) THEN
PRINT *, 'It should not be greater than 1 !'
GOTO 30
ENDIF
C
C Allocation
ALLOCATE (ST(ELENUM),EN(ELENUM),YC(NODNUM),TE(ELENUM),
+TCC(ELENUM),TK(ELENUM),TSS(ELENUM))
C
C <<< Step 2 >>>
PRINT *, 'STEP 2: DEFINE THE MATHEMATICAL MODEL'
PRINT *
DO 40 I=1,NODNUM
PRINT '(1X,A,1X,I4)', 'Enter the y-coordinates for node ', I
READ *, YC(I)
40 CONTINUE
PRINT *
PRINT *, 'Enter the connectivities of the elements'
DO 50 J = 1,ELENUM
PRINT '(1X,A,1X,I4)', 'Enter the node nos. of the 1st and 2nd
+ends of the element ', J
READ *, ST(J),EN(J)
```

```
50 CONTINUE
C
C <<< Step 3 >>>
PRINT *, 'STEP 3: INPUT THE SOIL PROPERTIES'
PRINT *
PRINT '(1X,A,1X,A)','Enter the compression index, effective stress
+, permeability,','and void ratio.'
DO 60 J=1,ELENUM
PRINT '(1X,A,1X,I4)', 'For the soil-layer number',J
READ *, TCC(J),TSS(J),TK(J),TE(J)
60 CONTINUE
C
C <<< Step 4 >>>
PRINT *, 'STEP 4: INPUT THE LOADING CASE'
PRINT *
PRINT *, 'Enter the surcharge increasing rate in MPa per day'
READ *, STRMPD
PRINT *
PRINT *,'Enter the total time taken for adding surcharge in days'
READ *, TIMDAY
C
C <<< Step 5 >>>
PRINT *, 'STEP 5: DEFINE THE BOUNDARY CONDITION'
PRINT *
```

```
PRINT 80, 'Input the boundary condition :','1. Free draining at
+both ends','2. Free draining at upper end','3. Free draining a
+t lower end'
70 PRINT *,'Enter the option : 1, 2, 3'
80 FORMAT (1X,A,/,1X,A,/,1X,A,/,1X,A)
READ *, BOUND
IF ((BOUND .LT. 1) .OR. (BOUND .GT. 3)) THEN
PRINT *, '********* Illegal Option **********'
GOTO 70
ENDIF
C
C <<< Step 6 >>>
PRINT *, 'STEP 6: DEFINE THE FINITE DIFFERENCE METHOD'
PRINT *
PRINT 90, 'Input the finite difference method :','1. Forward dif
+ference method','2. Central difference method','3. Galerkin`s
+difference method','4. Backward difference method'
90 FORMAT (1X,A,/,1X,A,/,1X,A,/,1X,A,/1X,A)
100 PRINT *,'Enter the option : 1, 2, 3, 4'
READ *, FINITE
IF ((FINITE .LT. 1) .OR. (FINITE .GT. 4)) THEN
PRINT *, '********* Illegal Option **********'
GOTO 100
ENDIF
```

```
C
C <<< Step 7 >>>
PRINT *, 'STEP 7: DEFINE THE METHOD OF CONSOLIDATION PROCESS'
PRINT *
PRINT 110, 'Input the method of consolidation process :','1. Con
+stant clay thickness and constant effective stress','2. Constant
+ clay thickness and changeable the effective stress','3. Changea
+ble clay thickness and constant effective stress','4. Changeable
+ clay thickness and changeable effective stress'
110 FORMAT (1X,A,/,1X,A,/,1X,A,/,1X,A,/1X,A)
120 PRINT *,'Enter the option : 1, 2, 3, 4'
READ *, PROCES
IF ((PROCES .LT. 1) .OR. (PROCES .GT. 4)) THEN
PRINT *, '********* Illegal Option **********'
GOTO 120
ENDIF
C
C Write data onto file
WRITE (10,300) 'Name of File (excluding extension)'
300 FORMAT (A)
WRITE (10,310) NAME
310 FORMAT (A8)
WRITE (10,320) 'Number of Nodes'
320 FORMAT (A)
```

```
WRITE (10,330) NODNUM
330 FORMAT (I4)
WRITE (10,340) 'Number of Elements'
340 FORMAT (A)
WRITE (10,350) ELENUM
350 FORMAT (I4)
WRITE (10,360) 'Method of Formulation : 1=Consistent, 2=Lumped, 3=
+Cubic'
360 FORMAT (A)
WRITE (10,370) FORNUM
370 FORMAT (I1)
WRITE (10,380) 'Degree of Consolidation for Termination'
380 FORMAT (A)
WRITE (10,390) DEGCON
390 FORMAT (F5.3)
WRITE (10,400) 'Y-Coordinate of Nodes (from top to bottom)'
400 FORMAT (A)
WRITE (10,410) (YC(I),I=1,NODNUM)
410 FORMAT (F8.3)
WRITE (10,420) 'Connectivities of the Elements (from Element 1)'
420 FORMAT (A)
WRITE (10,430) (ST(J),EN(J),J=1,ELENUM)
430 FORMAT (I4,1X,I4)
WRITE (10,440) 'Soil Properties: Compression index, Effective stre
```

```
+ss Permeability, Void Ratio'
440 FORMAT (A)
WRITE (10,450) (TCC(J),TSS(J),TK(J),TE(J),J=1,ELENUM)
450 FORMAT (F10.8,1X,F11.8,1X,F12.10,1X,F5.3)
WRITE (10,460) 'Increasing Rate of External Load Per Day'
460 FORMAT (A)
WRITE (10,470) STRMPD
470 FORMAT (F10.6)
WRITE (10,480) 'Total Time Taken in Days'
480 FORMAT (A)
WRITE (10,490) TIMDAY
490 FORMAT (F7.2)
WRITE (10,500) 'Boundary Condition: Free Drain at: 1=Both ends, 2=
+Upper end, 3=Lower end'
500 FORMAT (A)
WRITE (10,510) BOUND
510 FORMAT (I1)
WRITE (10,520) 'Finite difference method: 1=Forward, 2=Central,
+3=Galerkin`s, 4=Backward'
520 FORMAT (A)
WRITE (10,530) FINITE
530 FORMAT (I1)
WRITE (10,540) 'Consolidation process:','1. Constant clay thickn
+ess and constant effective stress','2. Constant clay thickness a
```

```
+nd changeable the effective stress','3. Changeable clay thicknes
+s and constant effective stress','4. Changeable clay thickness a
+nd changeable effective stress'
540 FORMAT (1X,A,/,1X,A,/,1X,A,/,1X,A,/1X,A)
WRITE (10,550) PROCES
550 FORMAT (I1)
C
C Exit
DEALLOCATE (ST,EN,YC,TCC,TSS,TK,TE)
ENDFILE (UNIT=10)
CLOSE (UNIT=10)
RETURN
END
C-----------------------------------------------------------------------
SUBROUTINE KEY1 (FINP,NODNUM,ELENUM,FORNUM,ROW,COL,NN)
C - Read the basic parameters such as the no.s of node and element,
C the method of formulation.
C - Find out the values of ROW, COL, NN
C
C Argument Declaration
INTEGER NODNUM,ELENUM,FORNUM,ROW,COL,NN
CHARACTER FINP*12
C Open file
OPEN
```

```
(UNIT=10,FILE=FINP,FORM='FORMATTED',ACCESS='SEQUENTIAL', &
& STATUS='OLD')
C READ data from file FINP
READ (10,*)
READ (10,*) NODNUM
20 FORMAT (/,I4)
READ (10,*)
READ (10,*) ELENUM
30 FORMAT (/,I4)
READ (10,*)
READ (10,*) FORNUM
40 FORMAT (/,I1)
C
IF ((FORNUM .EQ. 1) .OR. (FORNUM .EQ. 2)) THEN
ROW=2
COL=2
NN=NODNUM
ELSE
ROW=4
COL=4
NN=NODNUM*2
ENDIF
C
C Exit
```

```
RETURN
END
C-------------------------------------------------------------------
SUBROUTINE KEY2 (NODNUM,ELENUM,DEGCON,YCOOR,START, &
& END,CC,STRESS,K,E,STRMPD,TIMDAY,BOUND,STEP,FINITE, &
& PROCES,PRINTS)
C - Read the remaining part of the data file.
C
C Argument Declaration
INTEGER NODNUM,ELENUM,BOUND,FINITE,PROCES,STEP,PRINTS
REAL :: DEGCON,STRMPD,TIMDAY
INTEGER :: START(ELENUM),END(ELENUM)
REAL
YCOOR(NODNUM),CC(ELENUM),STRESS(ELENUM),K(ELENUM),E(ELE
NUM)
C Local Declaration
INTEGER I,J
CHARACTER UNUSED*73
C
C READ data from file FINP
READ (10,*)
READ (10,*) DEGCON
10 FORMAT (/,F5.3)
READ (10,*) UNUSED
20 FORMAT (A)
```

```
READ (10,*) (YCOOR(I),I=1,NODNUM)
30 FORMAT (F8.3)
DO J = 1,ELENUM
START(J)=J ; END(J)=J+1
END DO
READ (10,*) UNUSED
60 FORMAT (A)
READ (10,*) (CC(J),STRESS(J),K(J),E(J),J=1,ELENUM)
70 FORMAT (F10.8,1X,F11.8,1X,F12.10,1X,F5.3)
READ (10,*)
READ (10,*) STRMPD
80 FORMAT (/,F10.6)
READ (10,*)
READ (10,*) TIMDAY
90 FORMAT (/,F7.2)
READ (10,*)
READ (10,*) BOUND
100 FORMAT (/,I1)
READ (10,*)
READ (10,*) FINITE
110 FORMAT (/,I1)
READ (10,*)
READ (10,*)
READ (10,*)
```

```
READ (10,*)
READ (10,*)
READ (10,*) PROCES
120 FORMAT (/////,I1)
READ (10,*) PRINTS
C
C Set STEP=1
STEP=1
C
C Exit
CLOSE (UNIT=10)
RETURN
END
C-----------------------------------------------------------------------
SUBROUTINE GRDMOD (NODNUM,ELENUM,FORNUM,YCOOR,START, &
& END,CC,STRESS,K,E,ELELEN,CV,DELTAT,H,AVECV,STEP, &
& TIMFIN,EXU,UA,FINITE,MV,NN,FIN,COL,UCAP,USTF,BOUND, &
& UB,BEGIN,SINGU,TIMDAY,TIME,PROCES)
C - Define the model;
C - Calculate the ELELEN, CV, H, AVECV, DELTAT, TIMFIN
C
C Argument Declaration
INTEGER NODNUM,ELENUM,FORNUM,FINITE,STEP,NN,COL,BOUND,PROCES
```

```
REAL DELTAT,H,AVECV,TIMFIN,FIN,TIMDAY,TIME

INTEGER START(ELENUM),END(ELENUM)

REAL
YCOOR(NODNUM),ELELEN(ELENUM),CC(ELENUM),STRESS(ELENUM)

REAL K(ELENUM),E(ELENUM),CV(ELENUM),MV(ELENUM)

REAL UA(NN),EXU(NN),UB(NN),UCAP(NN,COL),USTF(NN,COL)

LOGICAL BEGIN,SINGU

C Local Declaration

INTEGER :: J

REAL :: A,B,C,AVEK,AVEMV,MINH,MAXCV

REAL, allocatable :: SETTLE(:), AV(:)

ALLOCATE (SETTLE(ELENUM),AV(ELENUM))

C

C <<< Calculate the element length and soil properties >>>

IF (STEP .EQ. 1) THEN

DO 10 J=1,ELENUM

ELELEN(J)=ABS(YCOOR(START(J))-YCOOR(END(J)))

K(J)=K(J)/(365.25*24*3600)

10 CONTINUE

C TIMFIN calculation

TIMFIN=TIMDAY*24.0*3600.0

TIME=0.0

DELTAT=0.0

ENDIF
```

```
C
C <<< For adding loading cases >>>
IF (TIME-TIMFIN .LT. DELTAT) THEN
DO 20 J=1,ELENUM
AV(J)=CC(J)/2.303/STRESS(J)
MV(J)=(AV(J)/1000)/(1.0+E(J))
CV(J)=K(J)/(10.0*MV(J))
20 CONTINUE
C Calculate the average values to evaluate the average Cv
A=0.0
B=0.0
C=0.0
DO 30 J=1,ELENUM
A=A+ELELEN(J)
B=B+ELELEN(J)/K(J)
C=C+MV(J)*ELELEN(J)
30 CONTINUE
H=A/2
AVEK=A/B
AVEMV=C/A
AVECV=AVEK/(AVEMV*10)
C Calculate the change of length, effective stress and void ratio
IF ((STEP .NE. 1) .AND. (TIME-TIMFIN .LT. 0)) THEN
CALL COM2 (FORNUM,NN,COL,UCAP,USTF,DELTAT, &
```

```
& BOUND,EXU,FIN,UB,BEGIN,SINGU)
DO 40 J=1,ELENUM
IF ((FORNUM .EQ. 1) .OR. (FORNUM .EQ. 2)) THEN
C Calculate the settlement
SETTLE(J)=MV(J)*ELELEN(J) &
& *(EXU(J)+EXU(J+1)-UB(J)-UB(J+1))/2
C Calculate the effective stress
IF ((PROCES .EQ. 2) .OR. (PROCES .EQ. 4)) THEN
STRESS(J)=STRESS(J) &
& +(EXU(J)+EXU(J+1)-UB(J)-UB(J+1))/2/1000
ENDIF
ELSE
SETTLE(J)=MV(J)*ELELEN(J) &
& *(EXU(2*J-1)+EXU(2*J+1)-UB(2*J-1)-UB(2*J+1))/2
IF ((PROCES .EQ. 2) .OR. (PROCES .EQ. 4)) THEN
STRESS(J)=STRESS(J) &
& +(EXU(2*J-1)+EXU(2*J+1)-UB(2*J-1)-UB(2*J+1))/2/1000
ENDIF
ENDIF
40 CONTINUE
C
IF ((PROCES .EQ. 3) .OR. (PROCES .EQ. 4)) THEN
DO 50 J=1,ELENUM
C Calculate the void ratio
```

```
E(J)=E(J)-SETTLE(J)*(1+E(J))/ELELEN(J)
ELELEN(J)=ELELEN(J)-SETTLE(J)
50 CONTINUE
ENDIF
ENDIF
ENDIF
C
C <<< For constant loading cases >>>
C Calculate the change of length, effective stress and void ratio
IF (TIME-TIMFIN .GE. DELTAT) THEN
DO 60 J=1,ELENUM
IF ((FORNUM .EQ. 1) .OR. (FORNUM .EQ. 2)) THEN
SETTLE(J)=MV(J)*ELELEN(J) &
& *(UA(J)+UA(J+1)-EXU(J)-EXU(J+1))/2
IF ((PROCES .EQ. 2) .OR. (PROCES .EQ. 4)) THEN
STRESS(J)=STRESS(J) &
& +(UA(J)+UA(J+1)-EXU(J)-EXU(J+1))/2/1000
ENDIF
ELSE
SETTLE(J)=MV(J)*ELELEN(J) &
& *(UA(2*J-1)+UA(2*J+1)-EXU(2*J-1)-EXU(2*J+1))/2
IF ((PROCES .EQ. 2) .OR. (PROCES .EQ. 4)) THEN
STRESS(J)=STRESS(J) &
& +(UA(2*J-1)+UA(2*J+1)-EXU(2*J-1)-EXU(2*J+1))/2/1000
```

```
ENDIF
ENDIF
60 CONTINUE
C Calculate the element length and soil properties
DO 70 J=1,ELENUM
IF ((PROCES .EQ. 3) .OR. (PROCES .EQ. 4)) THEN
E(J)=E(J)-SETTLE(J)*(1+E(J))/ELELEN(J)
ELELEN(J)=ELELEN(J)-SETTLE(J)
ENDIF
AV(J)=CC(J)/2.303/STRESS(J)
MV(J)=(AV(J)/1000)/(1.0+E(J))
CV(J)=K(J)/(10.0*MV(J))
70 CONTINUE
C Calculate the average values to evaluate the average Cv
A=0.0
B=0.0
C=0.0
DO 80 J=1,ELENUM
A=A+ELELEN(J)
B=B+ELELEN(J)/K(J)
C=C+MV(J)*ELELEN(J)
80 CONTINUE
H=A/2
AVEK=A/B
```

```
AVEMV=C/A
AVECV=AVEK/(AVEMV*10)
ENDIF
C Define value for FIN
IF (STEP .EQ. 1) THEN
IF (FINITE .EQ. 1) FIN=0.0
IF (FINITE .EQ. 2) FIN=0.5
IF (FINITE .EQ. 3) FIN=2./3.
IF (FINITE .EQ. 4) FIN=1.0
ENDIF
C <<< Calculate the time step >>>
MINH=ELELEN(1)
MAXCV=CV(1)
DO 90 J=2,ELENUM
IF (ELELEN(J) .LT. MINH) MINH=ELELEN(J)
IF (CV(J) .GT. MAXCV) MAXCV=CV(J)
90 CONTINUE
C * * * * * (Section 5.3.2) * * * * *
IF (FORNUM .EQ. 1) THEN
IF (FINITE .NE. 4) THEN
DELTAT=(MINH**2)/(12*MAXCV)/(1-FIN)
ELSE
DELTAT=(MINH**2)/(6*MAXCV*FIN)
ENDIF
```

```
ELSE IF (FORNUM .EQ. 2) THEN
IF (FINITE .NE. 4) THEN
DELTAT=(MINH**2)/(4*MAXCV)/(1-FIN)
ELSE
DELTAT=(MINH**2)/(4*MAXCV)/(1-0.7)
ENDIF
ELSE
DELTAT=(MINH**2)/(50*MAXCV)
ENDIF
C
C Exit subroutine
IF (STEP .EQ. 1) THEN
PRINT '(1X,A,1X,F15.5,2X,A7)', 'TIME STEP = ',DELTAT,'seconds'
PRINT '(1X,A,1X,F15.12,1X,A6)', 'AVERAGE VALUE OF COEFFICIENT OF
+CONSOLIDATION =',AVECV,'m^2/kN'
ENDIF
DEALLOCATE (SETTLE,AV)
RETURN
END
C------------------------------------------------------------------------
SUBROUTINE CAPMAT
(NODNUM,FORNUM,ROW,COL,ELENUM,CV,ELELEN,NN,UCAP,
+EC)
C - Calculate the components of the element capacitence matrice
```

```
C - Assemble to the banded global capacitence matrix, [UCAP]
C
C Argument Declaration
INTEGER NODNUM,ELENUM,FORNUM,ROW,COL,NN
REAL
CV(ELENUM),ELELEN(ELENUM),UCAP(NN,COL),EC(ELENUM,ROW,CO
L)
C Local Declarations
INTEGER I,I1,L,J,K
C
C Initialise the [EC] array
DO 10 I=1,ELENUM
DO 20 J=1,ROW
DO 30 K=1,COL
EC(I,J,K)=0.0
30 CONTINUE
20 CONTINUE
10 CONTINUE
C 1st : Calculate the element matrices:
C <<< Local Capacitance Matrices for the Consistent Formulation >>>
IF (FORNUM .EQ. 1) THEN
DO 40 J=1,ELENUM
EC(J,1,1)=ELELEN(J)/(3.0*CV(J))
EC(J,2,2)=EC(J,1,1)
EC(J,1,2)=ELELEN(J)/(6.0*CV(J))
```

```
EC(J,2,1)=EC(J,1,2)
40 CONTINUE
C <<< Local Capacitance Matrices for the Lumped Formulation >>>
ELSE IF (FORNUM .EQ. 2) THEN
DO 50 J=1,ELENUM
EC(J,1,1)=ELELEN(J)/(2*CV(J))
EC(J,2,2)=EC(J,1,1)
EC(J,1,2)=0.0
EC(J,2,1)=0.0
50 CONTINUE
C <<< Local Capacitance Matrices for the Cubic Formulation >>>
ELSE
DO 60 J=1,ELENUM
EC(J,1,1)=ELELEN(J)*156.0/(420.0*CV(J))
EC(J,2,2)=(ELELEN(J)**3)*4.0/(420.0*CV(J))
EC(J,3,3)=EC(J,1,1)
EC(J,4,4)=EC(J,2,2)
EC(J,1,2)=(ELELEN(J)**2)*22.0/(420.0*CV(J))
EC(J,1,3)=ELELEN(J)*54.0/(420.0*CV(J))
EC(J,1,4)=(ELELEN(J)**2)*(-13.0)/(420.0*CV(J))
EC(J,2,3)=(ELELEN(J)**2)*13.0/(420.0*CV(J))
EC(J,2,4)=(ELELEN(J)**3)*(-3.0)/(420.0*CV(J))
EC(J,3,4)=(ELELEN(J)**2)*(-22.0)/(420.0*CV(J))
EC(J,2,1)=EC(J,1,2)
```

```
EC(J,3,1)=EC(J,1,3)
EC(J,4,1)=EC(J,1,4)
EC(J,3,2)=EC(J,2,3)
EC(J,4,2)=EC(J,2,4)
EC(J,4,3)=EC(J,3,4)
60 CONTINUE
ENDIF
C
C 2nd : Assemble
C Initialise the [UCAP] array
DO 70 I=1,NN
DO 80 L=1,COL
UCAP(I,L)=0
80 CONTINUE
70 CONTINUE
C
C <<< Assembly Process >>>
C <<< For Consistent and Lumped Formulations >>>
IF ((FORNUM .EQ. 1) .OR. (FORNUM .EQ. 2)) THEN
I=1
DO 90 I1=1,NN
DO 100 L=1,2
UCAP(I1,L)=UCAP(I1,L)+EC(I,1,L)
100 CONTINUE
```

```
UCAP(I1+1,1)=UCAP(I1+1,1)+EC(I,2,2)
I=I+1
IF (I .EQ. NODNUM) GOTO 150
90 CONTINUE
C <<< For Cubic Formulation >>>
ELSE
I=1
DO 110 I1=1,NN,2
DO 120 L=1,4
UCAP(I1,L)=UCAP(I1,L)+EC(I,1,L)
120 CONTINUE
DO 130 L=2,4
UCAP(I1+1,L-1)=UCAP(I1+1,L-1)+EC(I,2,L)
130 CONTINUE
DO 140 L=3,4
UCAP(I1+2,L-2)=UCAP(I1+2,L-2)+EC(I,3,L)
140 CONTINUE
UCAP(I1+3,1)=UCAP(I1+3,1)+EC(I,4,4)
I=I+1
IF (I .EQ. NODNUM) GOTO 150
110 CONTINUE
ENDIF
C
C Exit
```

```
150 CONTINUE
RETURN
END
C----------------------------------------------------------------------
SUBROUTINE STFMAT (NODNUM,FORNUM,ROW,COL,ELENUM, &
& ELELEN,NN,USTF)
C - Calculate the components of the element stiffness matrix
C - Assemble to the banded global stiffness matrix, [USTF]
C
C Argument declaration
INTEGER NODNUM,FORNUM,ELENUM,ROW,COL,NN
REAL ELELEN(ELENUM),USTF(NN,COL)
C Local Declarations
INTEGER :: I,I1,L,J,K
REAL, allocatable :: ES(:,:,:)
ALLOCATE (ES(ELENUM,ROW,COL))
C
C Initialise the [ES] array
DO 10 I=1,ELENUM
DO 20 J=1,ROW
DO 30 K=1,COL
ES(I,J,K)=0.0
30 CONTINUE
20 CONTINUE
```

```
10 CONTINUE
C 1st: Calculate the element matrices:
C <<< Local Stiffness Matrices for Consistent & Lumped Formulations >>>
IF ((FORNUM .EQ. 1) .OR. (FORNUM .EQ. 2)) THEN
DO 40 J=1,ELENUM
ES(J,1,1)=1.0/ELELEN(J)
ES(J,2,2)=ES(J,1,1)
ES(J,1,2)=-1.0/ELELEN(J)
ES(J,2,1)=ES(J,1,2)
40 CONTINUE
C <<< Local Stiffness Matrices for Cubic Formulation >>>
ELSE
DO 50 J=1,ELENUM
ES(J,1,1)=36.0/(ELELEN(J)*30.0)
ES(J,2,2)=ELELEN(J)*4.0/30.0
ES(J,3,3)=ES(J,1,1)
ES(J,4,4)=ES(J,2,2)
ES(J,1,2)=3.0/30.0
ES(J,1,3)=(-36.0)/(ELELEN(J)*30.0)
ES(J,1,4)=3.0/30.0
ES(J,2,3)=(-3.0)/30.0
ES(J,2,4)=ELELEN(J)*(-1.0)/30.0
ES(J,3,4)=(-3.0)/30.0
ES(J,2,1)=ES(J,1,2)
```

```
ES(J,3,1)=ES(J,1,3)
ES(J,4,1)=ES(J,1,4)
ES(J,3,2)=ES(J,2,3)
ES(J,4,2)=ES(J,2,4)
ES(J,4,3)=ES(J,3,4)
50 CONTINUE
ENDIF
C
C 2nd: Assemble
C Initialise the array
DO 60 I=1,NN
DO 70 L=1,COL
USTF(I,L)=0
70 CONTINUE
60 CONTINUE
C
C <<< Assembly Process >>>
C <<< For the Consistent and Lumped Formulation >>>
IF ((FORNUM .EQ. 1) .OR. (FORNUM .EQ. 2)) THEN
I=1
DO 80 I1=1,NN
DO 90 L=1,2
USTF(I1,L)=USTF(I1,L)+ES(I,1,L)
90 CONTINUE
```

```
USTF(I1+1,1)=USTF(I1+1,1)+ES(I,2,2)
I=I+1
IF (I .EQ. NODNUM) GOTO 140
80 CONTINUE
C <<< For the Cubic Formulation >>>
ELSE
I=1
DO 100 I1=1,NN,2
DO 110 L=1,4
USTF(I1,L)=USTF(I1,L)+ES(I,1,L)
110 CONTINUE
DO 120 L=2,4
USTF(I1+1,L-1)=USTF(I1+1,L-1)+ES(I,2,L)
120 CONTINUE
DO 130 L=3,4
USTF(I1+2,L-2)=USTF(I1+2,L-2)+ES(I,3,L)
130 CONTINUE
USTF(I1+3,1)=USTF(I1+3,1)+ES(I,4,4)
I=I+1
IF (I .EQ. NODNUM) GOTO 140
100 CONTINUE
ENDIF
C
140 CONTINUE
```

```
C Exit
DEALLOCATE (ES)
RETURN
END
C-----------------------------------------------------------------------
SUBROUTINE SURCHG (ELENUM,FORNUM,NN,ASU,STRMPD, &
& EC,ROW,COL)
C Pre: -enter the stress increasing rate in MPa per day
C -enter the total time taken for adding surcharge in days
C Pos: -output the finish time for stress increasing, TIMFIN, in second
C -calculate the component in the surcharge vector, [ASU]
C
C Argument Declarations
INTEGER :: ELENUM,FORNUM,NN,ROW,COL
REAL :: STRMPD
REAL :: ASU(NN),EC(ELENUM,ROW,COL)
C Local Declarations
INTEGER :: J,K
REAL :: STRATE
REAL, allocatable :: ESU(:,:)
ALLOCATE (ESU(ELENUM,COL))
C
C STRATE calculation
STRATE=STRMPD*1000.0/(24.0*3600.0)
```

```
C
C Initialise the [ESU] and [ASU] array
DO 10 J=1,ELENUM
DO 20 K=1,ROW
ESU(J,K)=0.0
20 CONTINUE
10 CONTINUE
c
DO 30 J=1,NN
ASU(J)=0.0
30 CONTINUE
C <<< For Consistent and Lumped Formulations >>>
C <<< Evaluate the Element Load Vector >>>
IF ((FORNUM .EQ. 1) .OR. (FORNUM .EQ. 2)) THEN
DO 40 J=1,ELENUM
DO 50 K=1,ROW
ESU(J,K)=STRATE*(EC(J,K,1)+EC(J,K,2))
50 CONTINUE
40 CONTINUE
C
C <<< Assembly Process >>>
ASU(1)=ESU(1,1)
DO 60 J=1,ELENUM-1
ASU(J+1)=ESU(J,2)+ESU(J+1,1)
```

```
60 CONTINUE
ASU(NN)=ESU(ELENUM,2)
C <<< For Cubic Formation >>>
C <<< Evaluate the Element Load Vector >>>
ELSE
DO 70 J=1,ELENUM
DO 80 K=1,ROW
ESU(J,K)=STRATE*(EC(J,K,1)+EC(J,K,3))
80 CONTINUE
70 CONTINUE
C <<< Assembly Process >>>
ASU(1)=ESU(1,1)
ASU(2)=ESU(1,2)
DO 90 J=1,ELENUM-1
ASU(2*J+1)=ESU(J,3)+ESU(J+1,1)
ASU(2*J+2)=ESU(J,4)+ESU(J+1,2)
90 CONTINUE
ASU(NN-1)=ESU(ELENUM,3)
ASU(NN)=ESU(ELENUM,4)
ENDIF
C Exit
DEALLOCATE (ESU)
RETURN
END
```

```
C----------------------------------------------------------------------
SUBROUTINE COMBIN (FFPP,FORNUM,ELENUM,NN,COL, &
& UCAP,USTF,ASU,DELTAT,TIMFIN,DEGCON,BOUND, &
& H,AVECV,ELELEN,STEP,EXU,U0,UA,FIN,BEGIN,SINGU, &
& TV,TIME,PROCES,PRINTS)
C Solving the first-order differential equation to find out the final
C pore pressure
C
C Argument Declarations
INTEGER FORNUM,ELENUM,NN,COL,BOUND,STEP,PROCES,PRINTS
REAL :: DELTAT,TIMFIN,DEGCON,H,AVECV,FIN,TV,TIME
REAL :: EXU(NN),UA(NN),U0(NN)
REAL :: UCAP(NN,COL),USTF(NN,COL),ASU(NN),ELELEN(ELENUM)
CHARACTER FFPP*12
LOGICAL :: BEGIN,SINGU
C Local Declarations
INTEGER I,J,COUNT
REAL AVEU0,DT
REAL, allocatable :: A(:,:),B(:)
REAL, allocatable :: F(:),X(:,:)
C Allocation
ALLOCATE (A(NN,COL),B(NN),F(NN),X(NN,COL))
C
IF (STEP .EQ. 1) THEN
```

```
C Open file
OPEN
(UNIT=16,FILE=FFPP,FORM='FORMATTED',ACCESS='SEQUENTIAL', &
& STATUS='UNKNOWN')
C
COUNT=1
DO 10 I=1,NN
EXU(I)=0
U0(I)=0
10 CONTINUE
ENDIF
C <<< Calculate the time being used for this STEP >>>
TIME=TIME+DELTAT
C
C <<< Check the finish loading time >>>
IF (TIME-TIMFIN .GE. DELTAT) THEN
DO 20 J=1,NN
ASU(J)=0.0
20 CONTINUE
ELSE
IF (TIME .GT. TIMFIN) THEN
DT=TIME-TIMFIN
DO 30 J=1,NN
ASU(J)=ASU(J)*(DELTAT-DT)/DELTAT
```

```
30 CONTINUE
ENDIF
ENDIF
C
C <<< Calculate the LHS the first-order differential equation >>>
C * * * * * (Section 6.1.8) * * * * *
C Compute {C}+FIN*DELTAT{K}
DO 40 I=1,NN
DO 50 J=1,COL
A(I,J)=UCAP(I,J)+FIN*DELTAT*USTF(I,J)
50 CONTINUE
40 CONTINUE
C
C <<< Boundary Conditions >>>
IF ((FORNUM .EQ. 1) .OR. (FORNUM .EQ. 2)) THEN
IF (BOUND .EQ. 1) THEN
EXU(1)=0.0
EXU(NN)=0.0
A(1,1)=A(1,1)*1E9
A(NN,1)=A(NN,1)*1E9
ELSE IF (BOUND .EQ. 2) THEN
EXU(1)=0.0
A(1,1)=A(1,1)*1E9
ELSE
```

```
EXU(NN)=0.0
A(NN,1)=A(NN,1)*1E9
ENDIF
ELSE
IF (BOUND .EQ. 1) THEN
EXU(1)=0.0
EXU(NN-1)=0.0
A(1,1) = A(1,1)*1E9
A(NN-1,1) = A(NN-1,1)*1E9
ELSE IF (BOUND .EQ. 2) THEN
EXU(1)=0.0
EXU(NN)=0.0
A(1,1)=A(1,1)*1E9
ELSE
EXU(2)=0.0
EXU(NN-1)=0.0
A(NN-1,1)=A(NN-1,1)*1E9
ENDIF
ENDIF
C <<< Calculate the RHS the first-order differential equation >>>
C 1st: Compute {C}-(1-FIN)*DELTAT*{K}
DO 60 I=1,NN
DO 70 J=1,COL
X(I,J)=UCAP(I,J)-(1-FIN)*DELTAT*USTF(I,J)
```

```
70 CONTINUE

60 CONTINUE

C 2nd: Compute the product of X and EXU, store the result in vector F

C <<< For Consistent and Lumped Formulations >>>

IF ((FORNUM .EQ. 1) .OR. (FORNUM .EQ. 2)) THEN

F(1)=X(1,1)*EXU(1)+X(1,2)*EXU(2)

DO 80 I=2,NN-1

F(I)=X(I-1,2)*EXU(I-1)+X(I,1)*EXU(I)+X(I,2)*EXU(I+1)

80 CONTINUE

F(NN)=X(NN-1,2)*EXU(NN-1)+X(NN,1)*EXU(NN)

ELSE

C <<< For Cubic Formulation >>>

F(1)=X(1,1)*EXU(1)+X(1,2)*EXU(2)+X(1,3)*EXU(3) &

& +X(1,4)*EXU(4)

F(2)=X(1,2)*EXU(1)+X(2,1)*EXU(2)+X(2,2)*EXU(3) &

& +X(2,3)*EXU(4)

DO 90 I=3,NN-3,2

F(I)=X(I-2,3)*EXU(I-2)+X(I-1,2)*EXU(I-1)+ &

& X(I,1)*EXU(I)+X(I,2)*EXU(I+1)+ &

& X(I,3)*EXU(I+2)+X(I,4)*EXU(I+3)

F(I+1)=X(I-2,4)*EXU(I-2)+X(I-1,3)*EXU(I-1)+ &

& X(I,2)*EXU(I)+X(I+1,1)*EXU(I+1)+ &

& X(I+1,2)*EXU(I+2)+X(I+1,3)*EXU(I+3)

90 CONTINUE
```

```
F(NN-1)=X(NN-3,3)*EXU(NN-3)+X(NN-2,2)*EXU(NN-2)+ &
& X(NN-1,1)*EXU(NN-1)+X(NN-1,2)*EXU(NN)
F(NN)=X(NN-3,4)*EXU(NN-3)+X(NN-2,3)*EXU(NN-2)+ &
& X(NN-1,2)*EXU(NN-1)+X(NN,1)*EXU(NN)
ENDIF
C
C 3rd: Sum up the vectors F and DELTAT*ASU, store the result in vector B
DO 100 I=1,NN
B(I)=F(I)+DELTAT*ASU(I)
100 CONTINUE
C
C <<< Check the finish loading time >>>
IF (TIME-TIMFIN .GE. DELTAT) THEN
IF (COUNT .EQ. 1) THEN
DO 110 I=1,NN
U0(I)=EXU(I)
110 CONTINUE
COUNT=COUNT+1
AVEU0=0.0
IF ((FORNUM .EQ. 1) .OR. (FORNUM .EQ. 2)) THEN
DO 120 I=1,ELENUM
AVEU0=AVEU0+(U0(I)+U0(I+1))*ELELEN(I)
120 CONTINUE
ELSE
```

```
DO 130 I=1,ELENUM
AVEU0=AVEU0+(U0(2*I-1)+U0(2*I+1))*ELELEN(I)
130 CONTINUE
ENDIF
AVEU0=AVEU0/(4.0*H)
ENDIF
ENDIF
C
CALL SUBCOM (FORNUM,ELENUM,NN,COL,DELTAT,DEGCON, &
& BOUND,H,AVECV,ELELEN,STEP,EXU,U0,UA,BEGIN, &
& SINGU,COUNT,AVEU0,A,B,TV,TIME,PROCES,PRINTS)
C Exit
DEALLOCATE (A,B,F,X)
RETURN
END
C-------------------------------------------------------------------
SUBROUTINE SUBCOM (FORNUM,ELENUM,NN,COL,DELTAT, &
& DEGCON,BOUND,H,AVECV,ELELEN,STEP,EXU,U0,UA,BEGIN, &
& SINGU,COUNT,AVEU0,A,B,TV,TIME,PROCES,PRINTS)
C
C Argument Declarations
INTEGER :: FORNUM,ELENUM,NN,COL,BOUND,STEP,COUNT, &
& PROCES,PRINTS
REAL :: DELTAT,DEGCON,H,AVECV,AVEU0,TV,TIME
```

```
REAL :: EXU(NN),UA(NN),U0(NN),ELELEN(ELENUM)
REAL :: A(NN,COL),B(NN)
LOGICAL :: BEGIN,SINGU
C Local Declarations
INTEGER :: I,J,L,N1,NI,K,K1,K2,K3
REAL :: C,AVEXU,AVEDEG,TIMCON
REAL, allocatable :: D(:),U(:)
C Allocation
ALLOCATE (D(NN),U(NN))
C
DO 10 I=1,NN
U(I)=0.0
10 CONTINUE
C
C <<< Calculate the Average Tv or time is used >>>
IF (PROCES .EQ. 1) THEN
IF (STEP .EQ. 1) TV=0.0
TV=TV+AVECV*DELTAT/H**2
ELSE
TIMCON=TIME/(24*3600)
ENDIF
C
C <<< Store the pervious excess pore pressure in UA >>>
IF (STEP .EQ. 1) THEN
```

```
SINGU=.TRUE.
DO 20 I=1,NN
UA(I)=0
20 CONTINUE
ELSE
DO 30 I=1,NN
UA(I)=EXU(I)
30 CONTINUE
ENDIF
C <<< Elimination to find the Excess Pressure >>>
N1=NN-1
DO 40 K=1,N1
C=A(K,1)
K1=K+1
IF ((ABS(C)-0.000001) .LT. 0) THEN
GOTO 1000
ELSE
C
C Divide row by diagonal coefficient
NI=K1+COL-2
L=MIN(NI,NN)
DO 50 J=2,COL
D(J)=A(K,J)
50 CONTINUE
```

```
DO 60 J=K1,L
K2=J-K+1
A(K,K2)=A(K,K2)/C
60 CONTINUE
B(K)=B(K)/C
C
C Eliminate unknown X(K) from ROW I
DO 70 I=K1,L
K2=I-K1+2
C=D(K2)
DO 80 J=I,L
K2=J-I+1
K3=J-K+1
A(I,K2)=A(I,K2)-C*A(K,K3)
80 CONTINUE
B(I)=B(I)-C*B(K)
70 CONTINUE
ENDIF
40 CONTINUE
C
C Compute last unknown
IF ((ABS(A(NN,1))-0.000001) .LT. 0) GOTO 1000
B(NN)=B(NN)/A(NN,1)
C
```

```
C Apply back-substitution process to compute remaining unknowns
DO 90 I=1,N1
K=NN-I
K1=K+1
NI=K1+COL-2
L=MIN(NI,NN)
DO 100 J=K1,L
K2=J-K+1
B(K)=B(K)-A(K,K2)*B(J)
100 CONTINUE
90 CONTINUE
C
DO 110 I=1,NN
EXU(I)=B(I)
110 CONTINUE
C
C <<< Calculate the Degree of Consolidation >>>
IF (COUNT .GT. 1) THEN
C For the Consistent and Lumped Formulations
IF ((FORNUM .EQ. 1) .OR. (FORNUM .EQ. 2)) THEN
IF (BOUND .EQ. 1) THEN
DO 120 I=2,NN-1
U(I)=1.0-EXU(I)/U0(I)
120 CONTINUE
```

```
ELSE IF (BOUND .EQ. 2) THEN
DO 130 I=2,NN
U(I)=1.0-EXU(I)/U0(I)
130 CONTINUE
ELSE
DO 140 I=1,NN-1
U(I)=1.0-EXU(I)/U0(I)
140 CONTINUE
ENDIF
C
AVEXU=0.0
DO 150 I=1,ELENUM
AVEXU=AVEXU+(EXU(I)+EXU(I+1))*ELELEN(I)
150 CONTINUE
AVEXU=AVEXU/(4.0*H)
AVEDEG=1.0-AVEXU/AVEU0
ELSE
C For Cubic Formulation
IF (BOUND .EQ. 1) THEN
DO 160 I=3,NN-3,2
U(I)=1.0-EXU(I)/U0(I)
160 CONTINUE
ELSE IF (BOUND .EQ. 2) THEN
DO 170 I=3,NN-1,2
```

```
U(I)=1.0-EXU(I)/U0(I)
170 CONTINUE
ELSE
DO 180 I=1,NN-3,2
U(I)=1.0-EXU(I)/U0(I)
180 CONTINUE
ENDIF
C
AVEXU=0.0
DO 190 I=1,ELENUM
AVEXU=AVEXU+(EXU(2*I-1)+EXU(2*I+1))*ELELEN(I)
190 CONTINUE
AVEXU=AVEXU/(4.0*H)
AVEDEG=1.0-AVEXU/AVEU0
ENDIF
ENDIF
C
C <<< Write down into the Output File >>>
IF (PROCES .EQ. 1) THEN
WRITE (16,200) 'STEP =',STEP,'TIME FACTOR =',TV
200 FORMAT (1X,A7,1X,I6,7X,A13,1X,F10.5)
WRITE (16,210) '---------------------------------------------
+----------------------'
210 FORMAT (A)
```

```
ELSE

IF (STEP/PRINTS*PRINTS-STEP == 0) THEN

WRITE (16,220) 'STEP =',STEP,'TIME IS USED =',TIMCON,'DAYS'

220 FORMAT (1X,A7,1X,I6,7X,A14,1X,F10.3,1X,5A)

WRITE (16,230) '---------------------------------------------

+-----------------------'

ENDIF

ENDIF

230 FORMAT (A)

IF (STEP/PRINTS*PRINTS-STEP == 0) THEN

WRITE (16,240) 'NODE','EXCESS PORE PRESSURE','DEGREE OF
CONSOLIDATION'

240 FORMAT (1X,A5,2X,A22,5X,A25)

IF ((FORNUM .EQ. 1) .OR. (FORNUM .EQ. 2)) THEN

WRITE (16,250) (I,EXU(I),U(I),I=1,NN)

ELSE

WRITE (16,250) ((I+1)/2,EXU(I),U(I),I=1,NN-1,2)

ENDIF

ENDIF

250 FORMAT (1X,I4,3X,F22.6,5X,F25.6)

IF (STEP/PRINTS*PRINTS-STEP == 0) THEN

WRITE (16,260) '-----------------------------------------------

+--------------------'

260 FORMAT (A)
```

```
WRITE (16,270) 'AVERAGE DEGREE OF CONSOLIDATION =',AVEDEG
270 FORMAT (1X,A,1X,F10.6)
WRITE (16,280)
'==================================================
+===================='
ENDIF
280 FORMAT (A)
WRITE (16,290)
290 FORMAT (/)
C
C <<< Increasing STEP by 1 >>>
STEP=STEP+1
C
C <<< Check the Degree of Consolidation >>>
IF (AVEDEG .GE. DEGCON) BEGIN=.FALSE.
DO 300 I=1,NN
B(I)=0.0
300 CONTINUE
GOTO 1100
1000 CONTINUE
WRITE(16,1010) K
1010 FORMAT (' **** SINGULARITY IN ROW',I5)
BEGIN=.FALSE.
SINGU=.FALSE.
```

```
1100 CONTINUE
C Exit
DEALLOCATE (D,U)
RETURN
END
C-----------------------------------------------------------------------
SUBROUTINE COM2 (FORNUM,NN,COL,UCAP,USTF,DELTAT, &
& BOUND,EXU,FIN,UB,BEGIN,SINGU)
C Solving the first-order differential equations
C
C Argument Declarations
INTEGER :: FORNUM,NN,COL,BOUND
REAL :: DELTAT,FIN
REAL :: EXU(NN),UB(NN)
REAL :: UCAP(NN,COL),USTF(NN,COL)
LOGICAL :: BEGIN,SINGU
C Local Declarations
INTEGER I,J,L,N1,NI,K,K1,K2,K3
REAL :: C
REAL, allocatable :: A(:,:),B(:), D(:),X(:,:)
C Allocation
ALLOCATE (A(NN,COL),B(NN),X(NN,COL),D(NN))
C
C <<< Calculate the LHS the first-order differential equation >>>
```

```
C Compute {C}+FIN*DELTAT{K}
DO 10 I=1,NN
DO 20 J=1,COL
A(I,J)=UCAP(I,J)+FIN*DELTAT*USTF(I,J)
20 CONTINUE
10 CONTINUE
C
C <<< Boundary Conditions >>>
IF ((FORNUM .EQ. 1) .OR. (FORNUM .EQ. 2)) THEN
IF (BOUND .EQ. 1) THEN
EXU(1)=0.0
EXU(NN)=0.0
A(1,1)=A(1,1)*1E9
A(NN,1)=A(NN,1)*1E9
ELSE IF (BOUND .EQ. 2) THEN
EXU(1)=0.0
A(1,1)=A(1,1)*1E9
ELSE
EXU(NN)=0.0
A(NN,1)=A(NN,1)*1E9
ENDIF
ELSE
IF (BOUND .EQ. 1) THEN
EXU(1)=0.0
```

```
EXU(NN-1)=0.0
A(1,1) = A(1,1)*1E9
A(NN-1,1) = A(NN-1,1)*1E9
ELSE IF (BOUND .EQ. 2) THEN
EXU(1)=0.0
EXU(NN)=0.0
A(1,1)=A(1,1)*1E9
ELSE
EXU(2)=0.0
EXU(NN-1)=0.0
A(NN-1,1)=A(NN-1,1)*1E9
ENDIF
ENDIF
C <<< Calculate the RHS the first-order differential equation >>>
C 1st: Compute {C}-(1-FIN)*DELTAT*{K}
DO 30 I=1,NN
DO 40 J=1,COL
X(I,J)=UCAP(I,J)-(1-FIN)*DELTAT*USTF(I,J)
40 CONTINUE
30 CONTINUE
C 2nd: Compute the product of X and EXU, store the result in vector B
C <<< For Consistent and Lumped Formulations >>>
IF ((FORNUM .EQ. 1) .OR. (FORNUM .EQ. 2)) THEN
B(1)=X(1,1)*EXU(1)+X(1,2)*EXU(2)
```

```
DO 50 I=2,NN-1
B(I)=X(I-1,2)*EXU(I-1)+X(I,1)*EXU(I)+X(I,2)*EXU(I+1)
50 CONTINUE
B(NN)=X(NN-1,2)*EXU(NN-1)+X(NN,1)*EXU(NN)
ELSE
C <<< For Cubic Formulation >>>
B(1)=X(1,1)*EXU(1)+X(1,2)*EXU(2)+X(1,3)*EXU(3) &
& +X(1,4)*EXU(4)
B(2)=X(1,2)*EXU(1)+X(2,1)*EXU(2)+X(2,2)*EXU(3) &
& +X(2,3)*EXU(4)
DO 60 I=3,NN-3,2
B(I)=X(I-2,3)*EXU(I-2)+X(I-1,2)*EXU(I-1)+ &
& X(I,1)*EXU(I)+X(I,2)*EXU(I+1)+ &
& X(I,3)*EXU(I+2)+X(I,4)*EXU(I+3)
B(I+1)=X(I-2,4)*EXU(I-2)+X(I-1,3)*EXU(I-1)+ &
& X(I,2)*EXU(I)+X(I+1,1)*EXU(I+1)+ &
& X(I+1,2)*EXU(I+2)+X(I+1,3)*EXU(I+3)
60 CONTINUE
B(NN-1)=X(NN-3,3)*EXU(NN-3)+X(NN-2,2)*EXU(NN-2)+ &
& X(NN-1,1)*EXU(NN-1)+X(NN-1,2)*EXU(NN)
B(NN)=X(NN-3,4)*EXU(NN-3)+X(NN-2,3)*EXU(NN-2)+ &
& X(NN-1,2)*EXU(NN-1)+X(NN,1)*EXU(NN)
ENDIF
C
```

```
C <<< Elimination to find the Excess Pressure >>>
N1=NN-1
DO 70 K=1,N1
C=A(K,1)
K1=K+1
IF ((ABS(C)-0.000001) .LT. 0) THEN
GOTO 1000
ELSE
C
C Divide row by diagonal coefficient
NI=K1+COL-2
L=MIN(NI,NN)
DO 80 J=2,COL
D(J)=A(K,J)
80 CONTINUE
DO 90 J=K1,L
K2=J-K+1
A(K,K2)=A(K,K2)/C
90 CONTINUE
B(K)=B(K)/C
C
C Eliminate unknown X(K) from ROW I
DO 100 I=K1,L
K2=I-K1+2
```

```
C=D(K2)
DO 110 J=I,L
K2=J-I+1
K3=J-K+1
A(I,K2)=A(I,K2)-C*A(K,K3)
110 CONTINUE
B(I)=B(I)-C*B(K)
100 CONTINUE
ENDIF
70 CONTINUE
C
C Compute last unknown
IF ((ABS(A(NN,1))-0.000001) .LT. 0) GOTO 1000
B(NN)=B(NN)/A(NN,1)
C
C Apply back-substitution process to compute remaining unknowns
DO 140 I=1,N1
K=NN-I
K1=K+1
NI=K1+COL-2
L=MIN(NI,NN)
DO 150 J=K1,L
K2=J-K+1
B(K)=B(K)-A(K,K2)*B(J)
```

```
150 CONTINUE
140 CONTINUE
C
DO 160 I=1,NN
UB(I)=B(I)
160 CONTINUE
C
GOTO 1100
1000 CONTINUE
WRITE(16,1010) K
1010 FORMAT (' **** SINGULARITY IN ROW',I5)
BEGIN=.FALSE.
SINGU=.FALSE.
1100 CONTINUE
C Exit
DEALLOCATE (A,B,D,X)
RETURN
END
C-----------------------------------------------------------------------
SUBROUTINE RESULT (TIME,FORNUM,NN,EXU,NODNUM,SINGU)
C Find out the time for consolidation
C Write down the result into output file
C
C Argument Declarations
```

```
INTEGER :: FORNUM,NN,NODNUM
REAL :: TIME
REAL :: EXU(NN)
LOGICAL :: SINGU
C Local Declarations
INTEGER :: I
REAL :: TIMCON
REAL, allocatable :: FIPP(:)
C Allocation
ALLOCATE (FIPP(NN))
C
DO 10 I=1,NN
FIPP(I)=0
10 CONTINUE
IF (SINGU) THEN
C <<< Find out the Time for Consolidation and the Final Nodal Excess
C Pore Pressure >>>
TIMCON=TIME/(24*3600)
IF ((FORNUM .EQ. 1) .OR. (FORNUM .EQ. 2)) THEN
DO 20 I=1,NN
FIPP(I)=FIPP(I)+EXU(I)
20 CONTINUE
ELSE
DO 30 I=1,NN,2
```

```
FIPP((I+1)/2)=FIPP((I+1)/2)+EXU(I)
30 CONTINUE
ENDIF
C <<< Write down into Output File >>>
WRITE(16,40) '*****************************************************
+*********************'
40 FORMAT (A)
WRITE (16,50) 'TOTAL TIME TAKEN OF CONSOLIDATION = ',TIMCON,
+'DAYS'
50 FORMAT (1X,A,1X,F10.3,1X,A4)
WRITE (16,60) '---------------------------------------------------
+--------------------'
60 FORMAT (A)
ENDIF
C
WRITE (16,70) 'NODE','FINAL PORE PRESSURE'
70 FORMAT (1X,A4,10X,A19)
WRITE (16,80) (I,FIPP(I),I=1,NODNUM)
80 FORMAT (1X,I4,9X,F20.6)
C
C Close file and exit
DEALLOCATE (FIPP)
ENDFILE (UNIT=16)
CLOSE (UNIT=16)
```

```
RETURN
END
C-------------------------------------------------------------------
```

Sample input file for consolidation analysis

```
Number of Nodes
17
Number of Elements
16
Method of Formulation : 1=Consistent, 2=Lumped, 3=Cubic
1
Degree of Consolidation for Termination
.950
Y-Coordinate of Nodes (from top to bottom)
.000
.250
.500
.750
1.000
1.250
1.500
1.750
2.000
2.250
2.500
```

2.750

3.000

3.250

3.500

3.750

4.000

Soil Properties: Compression index, Effective stress Permeability, Void Ratio

.50000000 .15507720 .0200000000 .500

.50000000 .15507720 .0200000000 .500

.50000000 .15507720 .0200000000 .500

.50000000 .15507720 .0200000000 .500

.50000000 .15507720 .0200000000 .500

.50000000 .15507720 .0200000000 .500

.50000000 .15507720 .0200000000 .500

.50000000 .15507720 .0200000000 .500

.50000000 .15507720 .0200000000 .500

.50000000 .15507720 .0200000000 .500

.50000000 .15507720 .0200000000 .500

.50000000 .15507720 .0200000000 .500

.50000000 .15507720 .0200000000 .500

.50000000 .15507720 .0200000000 .500

.50000000 .15507720 .0200000000 .500

.50000000 .15507720 .0200000000 .500

Increasing Rate of External Load Per Day

.001000

Total Time Taken in Days

10.00

Boundary Condition: Free Drain at: 1=Both ends, 2=Upper end, 3=Lower end

1

Finite difference method: 1=Forward, 2=Central, 3=Galerkin`s, 4=Backward

1

Consolidation process:

1. Constant clay thickness and constant effective stress

2. Constant clay thickness and change the effective stress

3. Change clay thickness and constant effective stress

4. Change clay thickness and change effective stress

4

5

The output file for this input is long, but some selected output is attached for discussion.

STEP = 30 TIME IS USED = 26.050 DAYS

-

NODE	EXCESS PORE PRESSURE	DEGREE OF CONSOLIDATION
1	0.000000	0.000000
2	3.796853	0.469873
3	6.828033	0.298791
4	8.739305	0.129695

5	9.647088	0.036979
6	9.949143	0.004651
7	10.005071	**-0.000547**
8	10.003174	**-0.000288**
9	10.000105	**-0.000013**
10	10.003176	**-0.000289**
11	10.005074	**-0.000547**
12	9.949146	0.004651
13	9.647091	0.036978
14	8.739305	0.129695
15	6.828034	0.298791
16	3.796853	0.469873
17	0.000000	0.000000

AVERAGE DEGREE OF CONSOLIDATION = 0.110646

STEP = 65 TIME IS USED = 55.664 DAYS

-

NODE EXCESS PORE PRESSURE DEGREE OF CONSOLIDATION

1	0.000000	0.000000
2	2.501854	0.650684
3	4.770308	0.510111
4	6.644112	0.338345
5	8.036170	0.197789

6	8.962826	0.103326
7	9.508471	0.049115
8	9.779602	0.022068
9	9.859620	0.014036
10	9.779603	0.022068
11	9.508471	0.049115
12	8.962826	0.103326
13	8.036169	0.197789
14	6.644113	0.338345
15	4.770308	0.510111
16	2.501854	0.650684
17	0.000000	0.000000

AVERAGE DEGREE OF CONSOLIDATION = 0.233295

At time step 30, the pore pressure actually increases at some locations inside the clay, even though this is a one-dimensional consolidation problem. The Mandel Cryer effect is hence not limited to 2 or 3 dimensional problems. Based on the actual observations for some reclamation works, the present program can actually perform better than the classical consolidation equations.

APPENDIX 2-6. EXTENSION TO TWO AND THREE – DIMENSIONAL BIOT CONSOLIDATION

For a raft foundation on saturated clay, the use of 1-D consolidation is far from realistic as there is a spread of stress due to three-dimensional effect. The use of 2-D or 3-D consolidation will be required. Unlike 1-D consolidation, seepage can occur in different directions, and the stresses changes include all the three normal and shear stresses. The treatment by Biot which is given below.

From elasticity, the equilibrium equation of an infinitesimal element can be expressed as (equivalent to $\sum Fx=0$, $\sum Fy=0$, $\sum Fz=0$)

$$\frac{\partial \sigma_{xx}'}{\partial x} + \frac{\partial \tau_{xy}}{\partial y} + \frac{\partial \tau_{xx}}{\partial z} = fx$$

$$\frac{\partial \tau_{xy}}{\partial x} + \frac{\partial \sigma_{yy}'}{\partial y} + \frac{\partial \tau_{yz}}{\partial z} = fy \qquad \textbf{(A2-15)}$$

$$\frac{\partial \tau_{xz}}{\partial x} + \frac{\partial \tau_{yz}}{\partial y} + \frac{\partial \sigma_{zz}'}{\partial z} = fz$$

where compression is taken as +ve and fx, fy, fz are body forces.

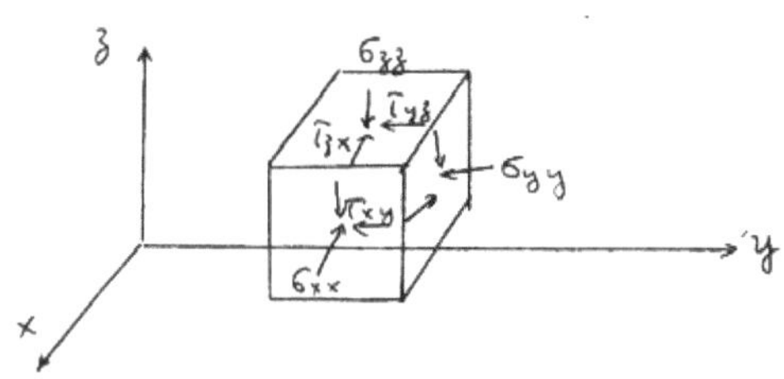

Fig. (E1). Stresses on a three-dimensional element.

Strain compatibility equations are given by

$$\xi_x = -\frac{\partial u_x}{\partial x}, \quad \xi_y = -\frac{\partial u_y}{\partial y}, \quad \xi_z = -\frac{\partial u_z}{\partial z}$$

(A2-16)

$$\gamma_{xy} = -\left(\frac{\partial u_x}{\partial y} + \frac{\partial u_y}{\partial x}\right), \quad \gamma_{yz} = -\left(\frac{\partial u_z}{\partial y} + \frac{\partial u_y}{\partial z}\right), \quad \gamma_{zx} = -\left(\frac{\partial u_x}{\partial z} + \frac{\partial u_z}{\partial x}\right)$$

Darcy's law can be expressed as

$$v_x = -K_x \frac{\partial h}{\partial x}, \; v_y = -K_y \frac{\partial h}{\partial y}, \; v_z = -K_z \frac{\partial h}{\partial z} \qquad \textbf{(A2-17)}$$

where h = total head

Similar to 1-D condition, the difference of q_{in} and q_{out} will be equal to the change of volumetric strain, or

$$-\frac{\partial}{\partial x}\left(K_x\frac{\partial h}{\partial x}\right)-\frac{\partial}{\partial y}\left(K_y\frac{\partial h}{\partial y}\right)-\frac{\partial}{\partial z}\left(K_z\frac{\partial h}{\partial z}\right)=\frac{\partial \xi v}{\partial t} \qquad \textbf{(A2-18)}$$

For isotropic soil where $K_x = K_y = K_z = K$. Since

$$v_x=\frac{-K}{\gamma_w}\frac{\partial u}{\partial x},\ v_y=\frac{-K}{\gamma_w}\frac{\partial u}{\partial y},\ v_z=\frac{-K}{\gamma_w}\frac{\partial u}{\partial z} \qquad \textbf{(A2-19)}$$

$$\therefore \quad -\frac{K}{\gamma_w}\left(\frac{\partial^2 u}{\partial x^2}+\frac{\partial^2 u}{\partial y^2}+\frac{\partial^2 u}{\partial z^2}\right)=\frac{\partial \xi v}{\partial t} \qquad \textbf{(A2-20)}$$

Based on elasticity,

$$\begin{aligned}&\sigma_x'=2G\left(\frac{v}{1-2v}\xi v+\xi x\right),\qquad \sigma_y'=2G\left(\frac{v}{1-2v}\xi v+\xi y\right)\\&\sigma_z'=2G\left(\frac{v}{1-2v}\xi v+\xi z\right),\qquad \tau_{xy}=G\gamma_{xy},\qquad \ldots\ldots\end{aligned} \qquad \textbf{(A2-21)}$$

Put eq. (A2-16) into (A2-21) and (A2-15), we have

$$\begin{aligned}&-G\nabla^2 u_x+\frac{G}{1-2v}\frac{\partial}{\partial x}\left(\frac{\partial u_x}{\partial x}+\frac{\partial u_y}{\partial y}+\frac{\partial u_z}{\partial z}\right)+\frac{\partial u}{\partial x}=0\\&-G\nabla^2 u_y+\frac{G}{1-2v}\frac{\partial}{\partial y}\left(\frac{\partial u_x}{\partial x}+\frac{\partial u_y}{\partial y}+\frac{\partial u_z}{\partial z}\right)+\frac{\partial u}{\partial y}=0\\&-G\nabla^2 u_z+\frac{G}{1-2v}\frac{\partial}{\partial z}\left(\frac{\partial u_x}{\partial x}+\frac{\partial u_y}{\partial y}+\frac{\partial u_z}{\partial z}\right)+\frac{\partial u}{\partial z}=-\gamma\end{aligned} \qquad \textbf{(A2-22)}$$

$$\xi v = \xi x + \xi y + \xi z = -\left(\frac{\partial u_x}{\partial x} + \frac{\partial u_y}{\partial y} + \frac{\partial u_z}{\partial z}\right) \quad \textbf{(A2-23)}$$

since

$\therefore$ eq. (A2-20) can be re-written as (continuity equation)

$$\frac{K}{\gamma_w}\nabla^2 u = \frac{\partial}{\partial t}\left(\frac{\partial u_x}{\partial x} + \frac{\partial u_y}{\partial y} + \frac{\partial u_z}{\partial z}\right) \quad \textbf{(A2-24)}$$

(A2-22) and (A2-24) form the governing equations and is termed as couple analysis. For 2-D problem where $\xi y = 0$, eq. (A2-22) and (A2-24) can be simplified as:

$$\begin{aligned} -G\nabla^2 u_x + \frac{G}{1-2v}\frac{\partial}{\partial x}\left(\frac{\partial u_x}{\partial x} + \frac{\partial u_z}{\partial z}\right) + \frac{\partial u}{\partial x} &= 0 \\ -G\nabla^2 u_z + \frac{G}{1-2v}\frac{\partial}{\partial z}\left(\frac{\partial u_x}{\partial x} + \frac{\partial u_z}{\partial z}\right) + \frac{\partial u}{\partial z} &= -\gamma \end{aligned} \quad \textbf{(A2-25)}$$

$$\frac{K}{\gamma_w}\nabla^2 u = \frac{\partial}{\partial t}\left(\frac{\partial u_x}{\partial x} + \frac{\partial u_z}{\partial z}\right) \quad \textbf{(A2-26)}$$

The above formulation is called Biot consolidation which coupled the dissipation of pore water pressure with soil displacement. The Biot equation is very difficult to be solved in general, and only complicated expressions can be obtained for simple cases. Even when analytical expressions exist for some simple cases, these expressions are difficult to be evaluated by hand calculation, and computer analysis is necessary. Under general condition, the solution of Biot equation can only be solved by numerical method, and finite element method is the best method for this problem.

Terzaghi-Redulic assumes the mean pressure $(\sigma_x + \sigma_y + \sigma_z) / 3$ remains unchanged during consolidation and simplified the Biot equation to

$$\xi v = \frac{p'}{K'} = \frac{1}{K'}(p-u)$$

where $p = (\sigma_x + \sigma_y + \sigma_z)/3$ $K' = \frac{E}{3(1-2v)}$ **(A2-29)**

= bulk modulus of volume compressibility

$$\frac{\partial \xi v}{\partial t} = \frac{1}{K'}\frac{\partial p'}{\partial t} = \frac{1}{K'}\left(\frac{\partial p}{\partial t} - \frac{\partial u}{\partial t}\right) \quad \textbf{(A2-30)}$$

Assume $\frac{\partial p}{\partial t} = 0, \quad \frac{\partial \xi v}{\partial t} = -\frac{1}{K'}\frac{\partial u}{\partial t}$ **(A2-31)**

Put eq. (A2-31) in eq. (A2-20)

$$\frac{K}{\gamma_w}\nabla^2 u = \frac{1}{K'}\frac{\partial u}{\partial t} \quad \text{or} \quad C_{v3}\nabla^2 u = \frac{\partial u}{\partial t} \quad \textbf{(A2-32)}$$

where $C_{v3} = \frac{K}{\gamma_w}K' = \frac{K}{\gamma_w}\frac{E}{3(1-2v)}$

For 2-D problem, $\xi y = 0$, eq. (A2-32) reduce to

$$C_{v2}\nabla^2 u = \frac{\partial u}{\partial t} \quad \text{where} \quad C_{v2} = \frac{K}{\gamma_w}\frac{E}{2(1+v)(1-2v)} \quad \textbf{(A2-33)}$$

There are some noticeable differences between the Biot consolidation and Terzaghi Redulic consolidation in the early stage of consolidation. In general, the pore water pressure can increases with time initially and then decrease afterward for Biot consolidation for some points (but not all locations). This phenomenon is called Mandel-Cryer effect which does not occur for the Terzaghi's approach.

Biot consolidation is best analyzed by finite element analysis as the analytical solution is very complicated to be used. A Windows based program written in Fortran 90 and Winteracter is shown below for teaching and research purpose. This program uses 8/9 nodes elements for greater accuracy, and is applicable for general nonhomegeneous non-regular problem with elastic and Cam-clay model. The complete program code, including the Window interface, is too long to be included here. Interested readers can request for the computer code by sending an email to the authors.

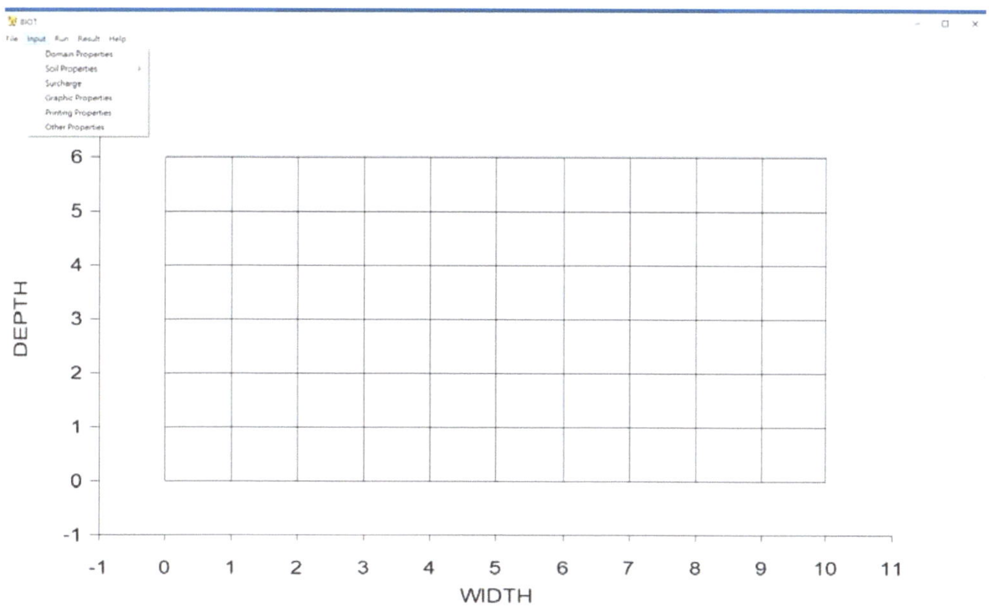

Fig. (E2). Basic illustration of the program BIOT.

CHAPTER 3

Plasticity, Limit Equilibrium and Limit Analysis Methods in Geotechnical Engineering

Abstract: In this chapter, the ultimate limit state of a system is considered by means of limit equilibrium, plasticity slip line method, limit analysis and DLO methods. The basic plasticity formulation for the slip line method is given, which is applied to some classical geotechnical problems. Following this, the three major geotechnical problems are unified under the extremum principle by the plasticity formulation. There is also discussion on the basic formulation for the DLO method and the limitations of the method or the commercial program.

Keyword: Axi-symmetric, Bearing capacity, DLO, Extremum, Lateral earth pressure, Limit equilibrium, Limit analysis, Plasticity, Slip line, Slope stability.

3.1. INTRODUCTION TO ULTIMATE LIMIT STATE ANALYSIS

Due to the difficulties in defining the *in situ* stress, the complications in many constitutive models as well as the difficulty in determining the various parameters required for a constitutive model, many geotechnical analysis and design works are still based on the ultimate limit state consideration, despite many finite element programs with various capabilities developed over the years. This is not surprising for the engineers and researchers, particularly for the engineers. The authors are greatly interested in a case in Hong Kong, where a highly theoretical soil constitutive model was developed and calibrated for the various required parameters in laboratory. This model was used for the interpretation of a plate load test in Hong Kong, and the percentage error of the prediction was found to be around 400%. On the other hand, the model can predict extremely well for the laboratory test results. It is not surprising that many engineers have various hesitations on the use of many modern and sophisticated soil constitutive models. In Hong Kong and many other countries, these advanced constitutive models are not commonly used for practical purposes. Unless this critical limitation can be overcome, the very large gap between the theoretical development and the actual applications will remain there. Actually, Cheng has written a very complicated nonlinear large strain elasto-plastic geotechnical program with 13 constitutive models, where different combinations of yield functions and plastic potential functions can be combined and used. So far, Cheng seldom adopts this program or other commercial programs using advanced constitutive model for real engineering design. The limited site investigations and laboratory/field test results for most projects cannot justify the choice of a highly sophisticated constitutive model, and many surprising results (usually local effects) can be obtained from every commercial program that the authors have tried when those advanced constitutive models are used. In fact, some engineers in Hong Kong will simply turn the complicated constitutive soil model to simple elastic/elasto-plastic model when numerical problems occur. Interestingly, many engineers find that a very famous geotechnical analysis program (the latest version already) can run into problems easily with the more advanced constitutive models for many excavation/lateral support problems. From the authors' view, a constitutive model which is not accurate enough for real problems or can run into numerical problems easily is not a good model, no matter how good is the theoretical background behind the model.

On the other hand, the use of the ultimate limit state for design is well-received and used by the engineers, with the application of a suitable factor of safety. With the experience accumulated over many years, engineers tend to rely more on the ultimate limit state analysis and design than the use of modern constitutive models for some types of problems. It is not surprising that some geotechnical designs are still based on the use of the limit state analysis up to the present. There many different references and research papers associated with this, and some additional references are given by Liu *et al.* (2019), Baars (2018), Nedderman (1992), Yu (2006), Davis and Selvadurai (2002), Rees (2006) as well as the classical works by Hill (1950), Sokolovskii (1965) and Chen (1975).

For stability analysis, there are various methods available to the engineers, and the choice of the method depends on the complexity of the geometry and the convenience in the solution. In this chapter, the slip line method, limit equilibrium method and limit analysis will be introduced for the lateral earth pressure, ultimate bearing capacity and slope stability problems. It is interesting to note that these three topics are usually considered separately in most of the books or research studies, and different methods of analyses have been proposed for individual problem even though they are governed by the same requirements for the ultimate conditions. Since the governing equations and boundary conditions for these problems are actually the same, Cheng and Li (2017) view that each problem can be viewed as the inverse of the other problems which will also be demonstrated in the present chapter. After the introduction of the three basic stability analysis methods, the unification of the three most important stability problems will be discussed.

The three stability methods together with the corresponding numerical solution techniques will be discussed with the use of different computer programs developed by the authors. The limit equilibrium methods as discussed in this chapter are available in the program SLOPE 2000 developed by Cheng, which can be obtained from the authors at natureymc@yahoo.com.hk. SLOPE2000 is also one of the analysis modules in the large scale geotechnical analysis and design package GEOCalc 1.0/2.0/4.0.

3.2. SLIP-LINE METHOD

At the ultimate condition, both equilibrium and yield conditions must be satisfied. Combining the Mohr-Coulomb yield criterion (which is generally adequate for soil) and the equilibrium equations, a set of hyperbolic partial differential equations of plastic equilibrium can be developed. In order to solve the governing partial differential equation, it is more convenient to transform the governing equations to curvilinear coordinates along the directions of the failure planes for mathematical convenience. Once the equations are solved, the failure modes with the corresponding systems of stresses will be automatically determined. The slip directions or slip lines constitute a network which is called slip-line field. The governing equations can be solved with adequate boundary conditions to investigate the stresses at the ultimate condition, and the solution of the problem is commonly taken as the rigorous solution, as the solutions are either similar to those from other methods or are better. Since the governing equations are written along the slip lines, the slip line fields corresponding to the solutions are commonly considered as the failure mechanism of the governing problem. For example, the bearing capacity of footing and the lateral earth pressure behind a retaining wall are commonly analyzed by the slip line analysis, but not for the slope stability problem.

Kötter (1903) was the first to derive the slip-line equations for two-dimensional ultimate problems, while Prandtl (1920) was the first to obtain an analytical solution for footing by assuming the weight of soil to be negligible. His results were then applied by Reissner (1924) and Novotortsev (1938) to different problems on the bearing capacity of footing on weightless soil. The inclusion of soil weight in the solution of the governing partial differential equation is analytically impossible, and Sokolovskii (1965) proposed a finite difference approximation of the slip-line equations for which the accuracy can be further improved by an iteration scheme (Cheng 2002, 2003), and such iteration to update the coordinates of the grid points on the slip line field has been demonstrated to be important for passive pressure evaluation. Sokolovskii (1965) solved many types of problems on the bearing capacity of footings, slopes as well as the lateral earth pressure on retaining walls. De Jong (1957) developed a graphical procedure for the solutions which appeared to be seldom used nowadays. There are other approximate solutions for the governing differential equations which include the applications of perturbation methods and series expansion methods, but these methods are not popular and versatile enough for more complicated problems and are seldom considered now. More recent results and numerical techniques are given by Cheng and Au (2005) for bearing capacity problem and Cheng (2003b) and Cheng *et al.* (2007b) for lateral earth pressure problem.

3.2.1. Slip-Line Method for Plane Strain Problem

Consider a general body force γ which may not align with XY axes as shown in Fig. (**3.1**) such consideration is commonly required when there is a rotation of axes applied to a problem or when earthquake load is considered, otherwise, ξ is usually 90°. Under the rotation of axes technique as proposed by Cheng (2003b), ξ may not be 90° for general case, if there is a horizontal earthquake.

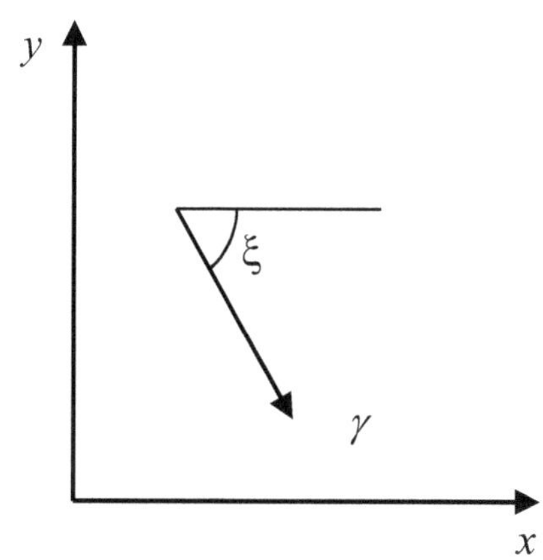

Fig. (3.1). A general two-dimensional body force system with non-vertical gravity.

Stress equilibrium of an infinitesimal element gives:

$$\frac{\partial \sigma_x}{\partial x} + \frac{\partial \tau_{xy}}{\partial y} = \gamma \cos \xi$$
$$\frac{\partial \tau_{xy}}{\partial x} + \frac{\partial \sigma_y}{\partial y} = -\gamma \sin \xi \qquad (3.1)$$

Let R and p represent the radius and the middle of the Mohr-circle of the stresses.

From Fig. (**3.2**)

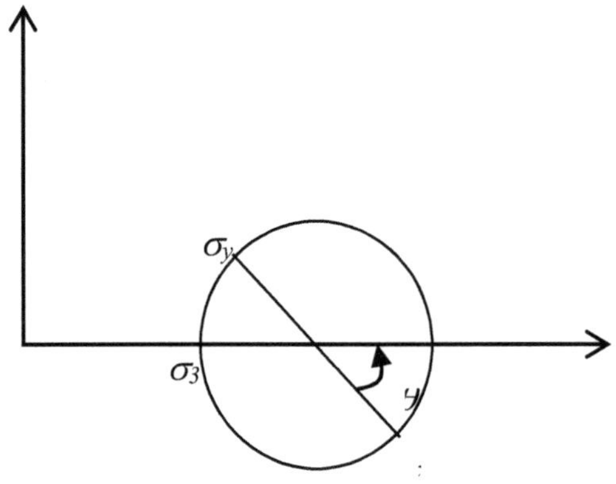

Fig. (3.2). Mohr transformation of stresses.

$$p = (\sigma_1 + \sigma_3)/\,2, \qquad R = (\sigma_1 - \sigma_3)/\,2 \tag{3.2}$$

At yield condition, from the Mohr-Coulomb relation:

$$\sigma_x = p + R\cos 2\theta \qquad \sigma_y = p - R\cos 2\theta \qquad \tau_{xy} = R\sin 2\theta \tag{3.3}$$

Put eq.(3.3) into (3.1) gives

$$\frac{\partial p}{\partial x}(1+\sin\phi\cos 2\theta) + \frac{\partial p}{\partial y}\sin\phi\sin 2\theta + 2R\left(-\frac{\partial \theta}{\partial x}\sin 2\theta + \frac{\partial \theta}{\partial y}\cos 2\theta\right) = \gamma\cos\xi \tag{3.4}$$

$$\frac{\partial p}{\partial x}\sin\phi\sin 2\theta + \frac{\partial p}{\partial y}(1-\sin\phi\cos 2\theta) + 2R\left(\frac{\partial \theta}{\partial x}\cos 2\theta + \frac{\partial \theta}{\partial y}\sin 2\theta\right) = -\gamma\sin\xi \tag{3.5}$$

where ϕ is the friction angle of soil. Multiply eq.(3.4) by $\sin(\theta \pm \mu)$ and eq. (3.5) by $-\cos(\theta \pm \mu)$, where $\mu = \frac{\pi}{4} - \frac{\phi}{2}$ and then add up eq.(3.4) and (3.5) gives

$$\begin{aligned}&\frac{\partial p}{\partial x}\left[\sin(\theta\pm\mu) + \sin\phi\sin(-\theta\pm\mu)\right] + \frac{\partial p}{\partial y}\left[-\cos(\theta\pm\mu) + \sin\phi\cos(-\theta\pm\mu)\right] \\ &+ 2R\left[-\frac{\partial \theta}{\partial x}\cos(-\theta\pm\mu) + \frac{\partial \theta}{\partial y}\sin(-\theta\pm\mu)\right] = \gamma\sin(\theta+\xi\pm\mu)\end{aligned} \tag{3.6}$$

Using the formula

$$\sin(\theta \pm \mu) = \pm\cos(\theta \pm \mu\,\mathrm{m}\frac{\pi}{2}) = \pm\cos(\theta\,\mathrm{m}\mu\,\mathrm{m}\phi) \qquad \textbf{(3.7)}$$

$$= \pm\cos(\theta\,\mathrm{\mu}\,\mu)\cos\phi + \sin\phi\sin(\theta\,\mathrm{\mu}\,\mu)$$

➔ $\sin(\theta \pm \mu) + \sin\phi\sin(-\theta \pm \mu) = \pm\cos(\theta\ \mathrm{\mu}\ \mu)\cos\phi$

➔ $$\cos(\theta \pm \mu) = \sin(\frac{\pi}{2} - \mu\,\mathrm{\mu}\,\theta) = \sin(\frac{\pi}{2} - 2\mu\,\mathrm{\mu}\,\theta + \mu) \qquad \textbf{(3.8)}$$

$$= \sin(\phi\,\mathrm{\mu}\,\theta + \mu) = \sin[\phi\,\mathrm{\mu}\,(\theta\,\mathrm{\mu}\,\mu)]$$

$$= \sin\phi\cos(\theta\,\mathrm{\mu}\,\mu)\,\mathrm{\mu}\,\cos\phi\sin(\theta\,\mathrm{\mu}\,\mu) \qquad \textbf{(3.9)}$$

∴ $$\sin\phi\cos(-\theta \pm \mu) - \cos(\theta \pm \mu) = \pm\sin(\theta\,\mathrm{\mu}\,\mu)\cos\phi \qquad \textbf{(3.10)}$$

and $$\sin(\theta + \xi \pm \mu) = \sin(\theta\ \mathrm{\mu}\,\mu + \xi \pm 2\mu) \qquad \textbf{(3.11)}$$

$$= \sin(\theta\,\mathrm{\mu}\,\mu)\cos(\xi \pm 2\mu) + \cos(\theta\,\mathrm{\mu}\,\mu)\sin(\xi \pm 2\mu)$$

Therefore, eq. (3.6) can be written as

$$\pm\cos(\theta\,\mathrm{\mu}\,\mu)\cos\phi\frac{\partial p}{\partial x} \pm \sin(\theta\,\mathrm{\mu}\,\mu)\cos\phi\frac{\partial p}{\partial y} - 2R\left[\cos(\theta\,\mathrm{\mu}\,\mu)\frac{\partial \theta}{\partial x} + \sin(\theta\,\mathrm{\mu}\,\mu)\frac{\partial \theta}{\partial y}\right] \qquad \textbf{(3.12)}$$

$$= \gamma\left[\sin(\theta\,\mathrm{\mu}\,\mu)\cos(\xi \pm 2\mu) + \cos(\theta\,\mathrm{\mu}\,\mu)\sin(\xi \pm 2\mu)\right]$$

Now a pair of line defined by α line and β line can be defined as the failure plane direction as shown in Fig. **(3.3)**

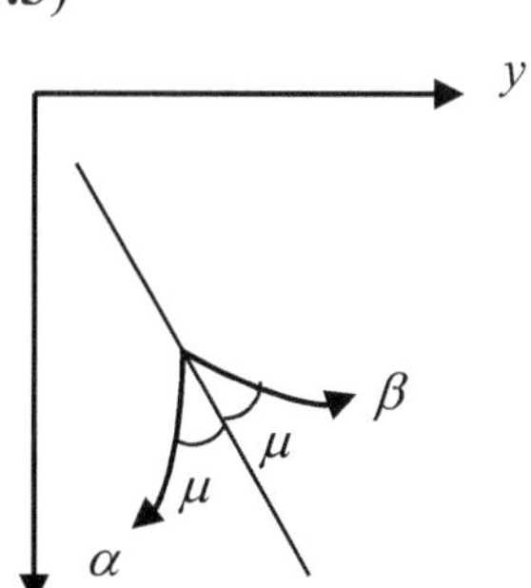

Fig. (3.3). The α and β characteristic lines.

The slopes of these two lines are:

$$\frac{dy}{dx} = \tan(\theta - \mu), \qquad \frac{dy}{dx} = \tan(\theta + \mu) \tag{3.13}$$

The directional derivatives along these two lines are given by

$$\frac{\partial}{\partial S_\alpha} = \cos(\theta - \mu)\frac{\partial}{\partial x} + \sin(\theta - \mu)\frac{\partial}{\partial y}$$

$$\therefore \qquad \frac{\partial x}{\partial S_\alpha} = \cos(\theta - \mu), \qquad \frac{\partial y}{\partial S_\alpha} = \sin(\theta - \mu) \tag{3.14}$$

and

$$\frac{\partial}{\partial S_\beta} = \cos(\theta + \mu)\frac{\partial}{\partial x} + \sin(\theta + \mu)\frac{\partial}{\partial y} \tag{3.15}$$

$$\therefore \qquad \frac{\partial x}{\partial S_\beta} = \cos(\theta + \mu), \qquad \frac{\partial y}{\partial S_\beta} = \sin(\theta + \mu) \tag{3.16}$$

where S_α, S_β are the arc length. Since $\cos\phi = \sin 2\mu$, from (3.12) and (3.15), we can transform the two equations to axes along the failure surfaces as:

$$-\sin 2\mu \frac{\partial p}{\partial S_\alpha} + 2R\frac{\partial \theta}{\partial S_\alpha} + \gamma\left[\sin(\xi + 2\mu)\frac{\partial x}{\partial S_\alpha} + \cos(\xi + 2\mu)\frac{\partial y}{\partial S_\alpha}\right] = 0 \tag{3.17}$$

$$\sin 2\mu \frac{\partial p}{\partial S_\beta} + 2R\frac{\partial \theta}{\partial S_\beta} + \gamma\left[\sin(\xi - 2\mu)\frac{\partial x}{\partial S_\beta} + \cos(\xi - 2\mu)\frac{\partial y}{\partial S_\beta}\right] = 0 \tag{3.18}$$

Eqs.(3.4) and (3.5) are now transformed to eq. (3.17) and (3.18) which are the characteristic equations along the failure surfaces. Analytical solutions to eqs.(3.17) and (3.18) can be evaluated if the unit weight of soil is 0, otherwise, numerical method has to be adopted which are complicated for complicated geometry with non-homogeneous conditions. There are three possible boundary conditions for the present problem: Cauchy type, Riemann type and mixed type. To solve the slip line equations, iterative finite difference method is commonly used.

Cauchy type boundary condition: Starting from two points A and B with known solutions as shown in Fig. (**3.4**) the results at point P along the intersection of α and β lines are given by the following relations.

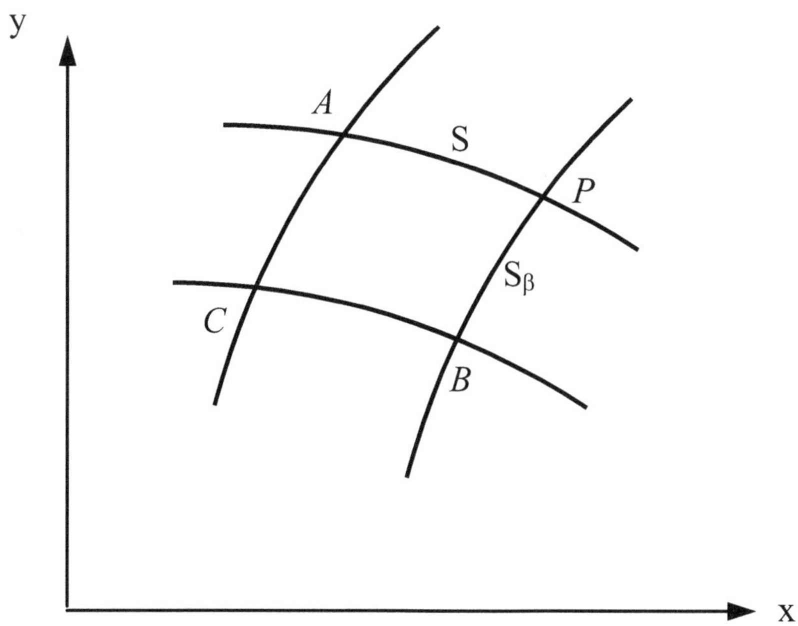

Fig. (3.4). α and β lines and finite difference solution of slip line problem.

Along α characteristic line :

$$R=\frac{R_P+R_A}{2} \;;\; \theta=\frac{\theta_P+\theta_A}{2}$$
$$\frac{\partial p}{\partial S_\alpha}=p_P-p_A \;;\; \frac{\partial \theta}{\partial S_\alpha}=\theta_P-\theta_A \;;\; \frac{\partial y}{\partial S_\alpha}=y_P-y_A \;;\; \frac{\partial x}{\partial S_\alpha}=x_P-x_A \qquad \textbf{(3.19)}$$

Along β characteristic line :

$$R=\frac{R_P+R_B}{2} \;;\; \theta=\frac{\theta_P+\theta_B}{2}$$
$$\frac{\partial p}{\partial S_\beta}=p_P-p_B \;;\; \frac{\partial \theta}{\partial S_\beta}=\theta_P-\theta_B \;;\; \frac{\partial y}{\partial S_\beta}=y_P-y_B \;;\; \frac{\partial x}{\partial S_\beta}=x_P-x_B \qquad \textbf{(3.20)}$$

If point P in Fig. (**3.4**) is close to A and B, eqs.(3.17) and (3.18) can be rewritten as:

$$-(p_P - p_A)\sin 2\mu + (R_P + R_A)(\theta_P - \theta_A) \approx -\gamma(y_P - y_A)\sin 2\mu - \gamma(x_P - x_A)\cos 2\mu \quad \textbf{(3.21)}$$

$$(p_P - p_B)\sin 2\mu + (R_P + R_B)(\theta_P - \theta_B) \approx -\gamma(y_P - y_B)\sin 2\mu - \gamma(x_P - x_B)\cos 2\mu \quad \textbf{(3.22)}$$

where (x_p - x_A) and (x_p - x_B) are given by Sokolovskii (1965) as

$$x_P - x_A \approx \tan(\frac{\theta_P + \theta_A}{2} - \mu)(y_P - y_A) \approx (\frac{dx}{dy})_1(y_P - y_A) \quad \textbf{(3.23)}$$

$$x_P - x_B \approx \tan(\frac{\theta_P + \theta_B}{2} + \mu)(y_P - y_B) \approx (\frac{dx}{dy})_2(y_P - y_B) \quad \textbf{(3.24)}$$

To solve for equations (3.21) and (3.22), boundary conditions of known values of x, y, p and θ must be available on each of the slip lines. Such boundary condition can usually be specified for specific problem.

Riemann type problem: For the two segments CA and CB in Fig. (**3.4**) on two corresponding α and β lines, the point P at the intersection of the slip lines passing through A and B can be initially estimated as:

$$x_P \approx x_A + x_B - x_C \;;\; y_P \approx y_A + y_B - y_C \;;\; p_P \approx p_A + p_B - p_C \;;\theta_P \approx \theta_A + \theta_B - \theta_C \quad \textbf{(3.25)}$$

In eq.(3.25), the results for point C can be estimated as the average values of the corresponding results from points A and B. After obtaining the initial estimates for x, y and p, equations (3.26) and (3.27) are used to determine the first iteration values of p_p and θ_p by substituting the initial estimate of x_p, y_p and R_p (=$p_p\sin\phi + c\cos\phi$) into equations (3.21) and (3.22) which gives:

$$\theta_P = \frac{1}{R_A + R_B + 2R_P}\{[\gamma(y_A - y_B)\sin 2\mu - \gamma(2x_P - x_A - x_B)\cos 2\mu + (p_B - p_A)\sin 2\mu + (R_P + R_A)\theta_A + (R_P + R_B)\theta_B\} \quad \textbf{(3.26)}$$

$$p_P = \frac{1}{\sin 2\mu + (\theta_P - \theta_B)\sin\phi}\{[\sin 2\mu - (\theta_P - \theta_B)\sin\phi]p_B - 2c(\theta_P - \theta_B)\cos\phi - \gamma(y_P - y_B)\sin 2\mu - \gamma(x_P + x_B)\cos 2\mu\} \quad \textbf{(3.27)}$$

$(\frac{dx}{dy})_1$ and $(\frac{dx}{dy})_2$ can be determined from eqs.(3.23) and (3.24) while x_p and y_p can be determined from eqs.(3.28) and (3.29).

$$y_P = \frac{x_B - x_A + (\frac{dx}{dy})_1 y_A - (\frac{dx}{dy})_2 y_B}{(\frac{dx}{dy})_1 - (\frac{dx}{dy})_2} \quad \textbf{(3.28)}$$

$$x_P = x_B + (\frac{dx}{dy})_2 (y_P - y_B) \quad \textbf{(3.29)}$$

where x_P, y_P, θ_p and p_P obtained in the second step will be used in a refined analysis by using eqs(3.23), (3.24), (3.28) and (3.29) to produce a more accurate solution. The differences between x_P, y_P, θ_p and p_P before and after each iteration are checked. If the changes are less than 0.1% or any small value, convergence is assumed to be achieved. If the accuracy for a specific point is achieved, then the values for other points within the soil can be calculated in a similar manner. The importance of iteration in the solution of the finite difference equations will be illustrated in the later section.

A finite difference grid for numerical analysis is shown in Fig. (**3.6**) It should be noted that due to the effect of soil weight, the slip lines are not straight lines. In Fig. (**3.6**) j=1 starts from the edge of the footing and increases towards the right. i starts from the edge of the footing as well and increases in a radial direction. The calculations are divided into three parts: (1) the points at the second radial slip line where i=2; (2) the points on the slip lines just below the boundary where j=2; and (3) other points on the slip lines. The centre of the radial slip line has the serial numbers i=1, j=1. Starting from this point, the next outer radial slip line is i=2 and so on. The maximum number value of i or the number of β lines is denoted by N_i. The number j increases as the slip line approaches the base of the footing. The maximum number of j in each radial slip line is denoted by N_k. The maximum number of points at the transition zone in each radial slip line is denoted by N_j. The authors find that if N_j and N_k are set to be greater than 50, the results of analysis are practically unaffected by the values of N_j and N_k.

For the points where i>2 and j>2, the trial values of x, y, p at point P can be estimated from the corresponding values from points A, B and C on the corresponding α and β lines by eq.(3.25). For the points at the second radial slip line where i=2, the average value of previous two points A and B is used to estimate the y value as:

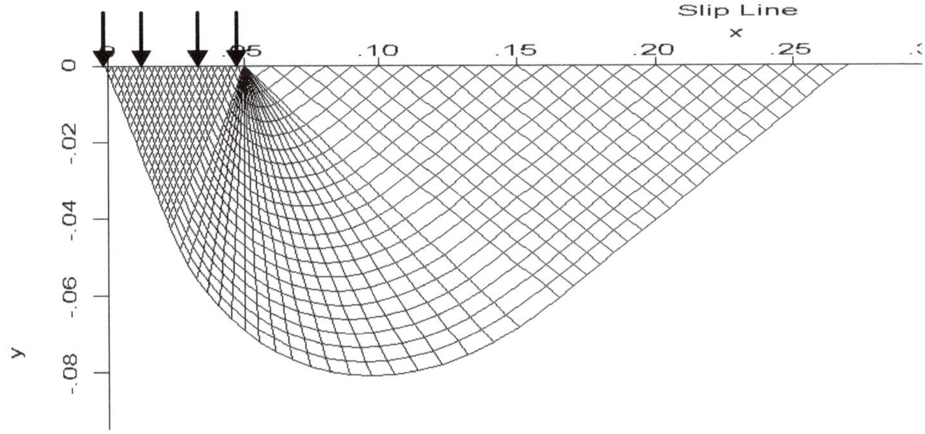

Fig. (3.5). Typical solution for slip lines fields for N_c and N_q factors (uniform foundation load).

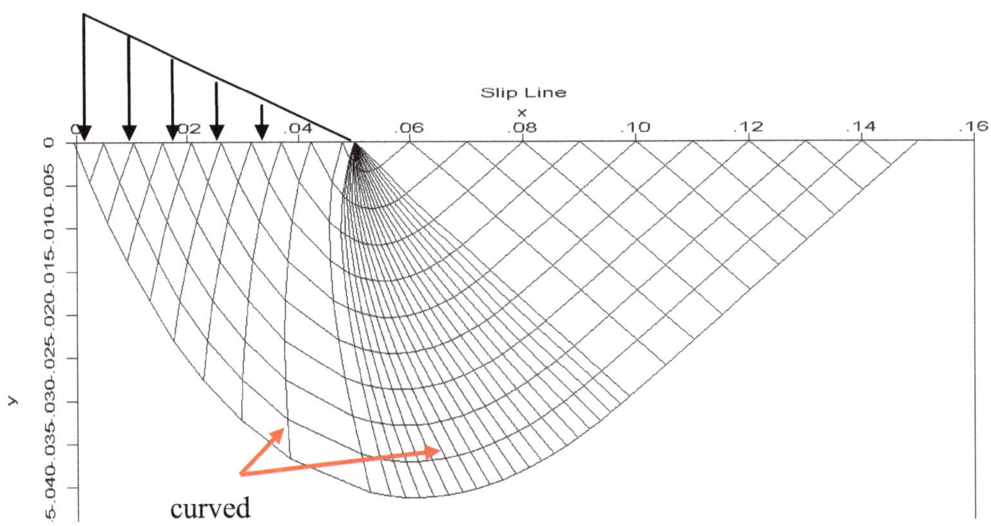

Fig. (3.6). Typical Slip lines for N_γ (triangular foundation load), ϕ=30°.

$$x_P=\frac{1}{2}(x_A+x_B)\ ;\ y_P=\frac{1}{2}(y_A+y_B)\ ;\ \theta_P=\frac{1}{2}(\theta_A+\theta_B)\ ;\ p_P=\frac{1}{2}(p_A+p_B) \qquad \textbf{(3.30)}$$

For the points on the slip lines just below the boundary where j=2, the initial estimates of the values at point P are given by the average values of the corresponding values at point A and B as:

$$\theta_P \approx \theta_B - \frac{\theta_{i=2,j=1}}{N_j + 1} \tag{3.31}$$

$$(\frac{dx}{dy})_1 = \tan[\frac{1}{2}(\theta_P + \theta_A) - \mu] \tag{3.32}$$

$$(\frac{dx}{dy})_2 = \tan[\frac{1}{2}(\theta_P + \theta_B) + \mu] \tag{3.33}$$

$$y_P \approx step * \tan(\pi/4 + \phi/2) + \frac{1}{2}(y_A + y_B) \tag{3.34}$$

$$x_P \approx x_B + (\frac{dx}{dy})_2 (y_A - y_B) \tag{3.35}$$

$$p_P \approx \frac{1}{2}(p_A + p_B) \tag{3.36}$$

After obtaining the initial estimates for x,y and p, equations (3.30) to (3.36) are used to determine the first iteration values of p_p and θ_p by substituting the initial guess of x_p, y_p R_p ($=p_p\sin\phi + c\cos\phi$) into the equations eqs.(3.35) and (3.36) which gives:

$$\theta_P = \frac{1}{R_A + R_B + 2R_P}\{[-\gamma\sin(\varepsilon - 2\mu)(y_A - y_B) - \gamma\cos(\varepsilon - 2\mu)(2x_P - x_A - x_B) \\ + \sin 2\mu(p_B - p_A) + (R_P + R_A)\theta_A + (R_P + R_B)\theta_B\} \tag{3.37}$$

$$p_P = \frac{1}{\sin 2\mu + (\theta_P - \theta_B)\sin\phi}\{[\sin 2\mu - (\theta_P - \theta_B)\sin\phi]p_B - 2c(\theta_P - \theta_B)\cos\phi \\ - \gamma\sin(\varepsilon - 2\mu)(y_P - y_B) - \gamma\cos(\varepsilon - 2\mu)(x_P + x_B)\} \tag{3.38}$$

The refined values of x_p and y_p in the first iteration are then given by:

$$y_P = \frac{x_B - x_A + (\frac{dx}{dy})_1 y_A - (\frac{dx}{dy})_2 y_B}{(\frac{dx}{dy})_1 - (\frac{dx}{dy})_2} \tag{3.39}$$

$$x_P = x_B + (\frac{dx}{dy})_2 (y_P - y_B) \quad \textbf{(3.40)}$$

The differences of x_P, y_P, θ_P, p_P, before and after each iteration analysis are checked. If the change is less than 0.1% (as used in the present study), convergence is assumed to be achieved. Based on trial and error analysis, it is found that a tolerance of 1% is actually sufficiently accurate for most cases but a smaller tolerance is adopted in the present study. If the accuracy for a specific point is achieved, then the values for other points within the soil can be calculated in a similar way. Otherwise, the newly obtained iteration values will be substituted into equations (3.21) and (3.22) to determine the next iteration values. This process continues until accurate values satisfying the convergence criterion is achieved.

In general, for a bearing capacity problem, the pressure outside the foundation is assumed to be known, and the slip line equations are then solved towards the foundation. To determine the bearing capacity factors Nc and Nγ, a very small surcharge is applied outside the foundation, while for the bearing capacity factor Nq, a unit surcharge is usually applied. If the pressure beneath the foundation is obtained as q_{ult}, then it can be equated to cNc (with a practically zero q) or qNq to give the bearing capacity factors. For Nγ, the bearing pressure beneath the foundation will be obtained as triangular. For the interpretation, some researchers assume symmetric failure mode while some others assume asymmetric failure mode, and the bearing capacity for the asymmetric failure mode will then be only one half of that for the symmetric failure mode.

3.2.1.1. Boundary Conditions in a Bearing Capacity Problem

In the case of a smooth horizontal footing base, the values y and θ are set to zero along this boundary. For the points below the footing, the values required for solution can be estimated from the corresponding values at points B, C and D at any two slip lines as:

At β characteristics line:

$$x_P - x_B = (\frac{dx}{dy})_2 (y_P - y_B) \quad \textbf{(3.41)}$$

$$\frac{x_P - x_C}{x_D - x_C} = \frac{y_P - y_C}{y_D - y_C} \quad \textbf{(3.42)}$$

x_p and y_p can then be determined by the following equations,

$$y_P = \frac{y_D - y_C}{x_D - x_C}[x_B - x_C + (\frac{dx}{dy})_2(y_P - y_B)] + y_C \quad \textbf{(3.43)}$$

$$x_P = (\frac{dx}{dy})_2(y_P - y_B) + x_B \quad \textbf{(3.44)}$$

From the x,y obtained and the known value of q, p_p can be determined by eq. (3.27).

3.2.1.2. Boundary Conditions for a Lateral Earth Pressure Problem

After the solution of the slip line equations for the whole domain, for retaining wall problem with wall friction, the equivalent stress σ along the retaining wall can be determined from *p* by eq.(3.45) given by Sokolovskii (1965).

$$\sigma = p\frac{\sin(\Delta \pm \delta)}{\sin\Delta} \quad ; \quad \sin\Delta = \frac{\sin\delta}{\sin\phi} \quad \textbf{(3.45)}$$

The actual active or passive pressure acting along the retaining wall can be recovered from eq. (3.46) as

$$\sigma_{a,p} = \sqrt{(\sigma\cos\delta - c\cot\phi)^2 + (\sigma\sin\delta)^2} \quad \textbf{(3.46)}$$

In eq. (3.45), the positive sign refers to the passive case while the negative sign refers to the active case. After the determination of the lateral earth pressure or bearing stress, the active/passive pressure coefficients due to γ, c and q or the bearing capacity factor can be back determined then. Sokolovskii (1965) and Kumar (2001, 2001) have neglected the cohesive strength c in formulating eq.(3.46) which is included by Cheng (2003b). Cheng (2003b) shown that the formulation as described above can be applied to a general c-ϕ material behind a retaining wall under seismic loading condition simply by a rotation of axes approach.

Based on the solution of the governing partial differential equations as given by eqs.(3.12) and (3.22), Cheng has developed program SLIP for the solution of bearing capacity problem and KA/KP for solution of general lateral earthquake problem with consideration of earthquake coefficient. The programs have been validated with other programs and published results, and extensions of these programs have been done so that these programs are more versatile for more complicated cases. Typical slip line fields for bearing capacity factors N_c, N_q and N_γ are given in Figs. (**3.5** and **3.6**). Based on such known boundary condition (set q to practically 0 as the initial boundary condition for N_c and N_γ determination and 1 unit for N_q determination), the slip line fields are evaluated from right to left. For N_c and N_q determination, it is assumed that the weight of soil is negligible, and a uniform distributed load (UDL) will be obtained at the base of the footing. Based on such condition, the classical Prandtl's failure mechanism will be automatically obtained from the iterative finite difference analysis for N_c and N_q determination with unit weight of soil being 0, and the failure mechanisms are the same for both bearing capacity factors. The failure mechanism as shown in (Fig. **3.5**) is also exactly the same as the analytical solutions from eqs.(3.17) and (3.18) if the unit weight of soil is 0 (Prandtl's mechanism). On the other hand, if the effect of the weight of soil is considered alone, a triangular pressure will be obtained at the base of the foundation which is shown in (Fig. **3.6**), and N_γ is determined by the average stress of this triangular pressure. The failure mechanism for this case deviates from the classical solution in two aspects: (1) the log-spiral zone will be distorted and is no longer a true log-spiral zone; (2) the triangular zone underneath the foundation as shown in (Fig. **3.5**) will also be distorted to that as shown in (Fig. **3.6**). Interestingly, even though the classical failure mechanism as shown in (Fig. **3.5**) is different from that as shown in (Fig. **3.6**) for the effect of soil weight, the N_γ factor obtained by limit analysis using the mechanism in (Fig. **3.5**) (Chen 1975) is still close to that as obtained by slip line analysis with the failure mechanism as that in (Fig. **3.6**). As shown in Table. **3.1**, **3.2**, **3.3**, **3.4** and **3.5**, the N_γ factors obtained by the slip line method and limit analysis are relatively similar, but the differences between these factors and those by Meyerhof (1963) , Hansen (1970) and Vesic (1973) are noticeable. While it is well known that N_c, N_q are the same among all methods for level ground, if the ground outside the foundation is sloping at an angle β, this understanding (using correction factors as compared with slip line solutions) will no longer be true as illustrated in Tables **3.2** and **3.3**. While the iterative finite difference solutions by program SLIP agree exactly with the analytical solution by Cheng and Au (2005), minor differences in these factors with those by Meyerhof (1963), Hansen (1970) and Vesic (1973) are noticed. It appears that the correction factors by Meyerhof, Hansen and Vesic are still acceptable for practical use even though they are not the exact solutions. For the N_γ factor, the classical results by Vesic (1973) and Hansen (1970) have however over-estimated the bearing capacity and are not safe to be used. This is important for engineering design in Hong Kong, as there are many engineering problems in Hong Kong which rely on this factor on sloping ground. Cheng and Au (2005) further investigated the N_γ factor under earthquake condition (by an equivalent pseudo-static earthquake coefficient) which is given in Table **3.5**, while analytical solutions for the N_c and N_q factors are also provided by Cheng and Au (2005). Classically, the effects due to surcharge, cohesive strength and friction angle are directly superimposed in the bearing capacity and lateral earth pressure problems. Cheng (2002) has demonstrated that such direct superposition approach is usually good enough for engineering use except for some very special case. In program SLIP or KA/KP, the global analysis or the analysis of individual effect can be considered, but there will be no corresponding coefficients in bearing capacity or lateral earth pressure then, as every factor is coupled with other factors for such case. It should also be noted that the use of a logspiral curve as the transition is not rigorous if the self weight of soil is not negligible (Li *et al.* 2018). Nevertheless, the use of logspiral curve can be a good approximation for normal engineering design works.

Table 3.1. Comparison of N_γ on level ground by various methods of analysis (N_c, N_q are the same among all methods).

$\phi \backslash N_\gamma$	SLIP	Sokolovskii	Hansen	Meyerhof	Vesic	Chen
10	0.56	0.56	0.4	0.4	1.2	0.72
20	3.16	3.16	2.9	2.9	5.4	3.45
30	15.31	15.3	15.1	15.7	22.4	15.2
40	86.6	86.5	79.4	93.6	109.3	81.79

Note: The results by Chen (1975) is based on limit analysis using associated flow rule

Table 3.2. Comparison of N_c for sloping ground with ground slope angle β with various methods of analysis (ϕ=30°).

β\Nc	SLIP	Eq.	Vesic	Hansen
10	24.33	24.32	24.66	28.09
20	19.57	19.57	20.15	26.04
30	15.68	15.68	16.48	24
40	12.5	12.5	13.46	21.94

Table 3.3. Comparison of N_q for sloping ground with various methods of analysis (ϕ=30°).

$\beta \backslash N_q$	SLIP	Eq.	Vestic	Hansen
10	12.94	12.94	12.94	11.6
20	7.96	7.95	7.44	6.74
30	2.74	2.74	3.29	3.35
40	5.67	5.67	5.24	4.88

Table 3.4. Comparison of N_γ for sloping ground for different methods of analysis (ϕ=35°).

$\beta \backslash N_\gamma$	SLIP	Vesic	Hansen
10	22.5	32.59	21.38
20	14.9	19.43	12.42
30	9.7	8.58	6.18
35	0	4.32	3.9

Table 3.5. N_γ under different earthquake coefficient.

$\phi \backslash k_h$	0	0.05	0.1	0.15	0.2
25	6.86	6.43	6.03	5.54	5.04
30	15.31	14.53	13.22	12.99	12.15
35	35.15	33.85	32.43	30.86	29.36
40	86.6	83.95	80.8	78.32	74.86

For the solution of lateral earth pressure problem, the stress condition on the ground surface behind the retaining wall is usually prescribed (usually prescribed to be a very small value), and the slip line equations are solved to give the lateral earth pressure on the retaining wall. Typical slip line fields for vertical rough retaining wall with level back are shown in (Fig. **3.7**) Based on the consideration on the effect of cohesive strength, surcharge and unit weight of soil, three lateral earth pressure coefficients K_{ac}, K_{aq} and $K_{a\gamma}$ (or K_{pc}, K_{pq} and $K_{p\gamma}$) are also obtained for either the active or passive condition.

The importance of iteration analysis as discussed previously are considered with respect to active and passive pressures, and the results are illustrated in Tables **3.5**, **3.6a** and **3.6b** with comparisons with the results by Kerisel and Absi (1990). From these two tables by Cheng (2003b) , it is observed that the use of multiple iterations results in more accurate values for the passive pressure coefficients (alternatively, a very fine grid can be used to improve the solution). For active pressure coefficients, use of multiple iterations offers no significant advantage as the effects can only be observed in the 4th or 5th significant figure. The use of multiple iterations in active and passive pressure determination have great difference from the analysis, and this phenomenon can be explained by reference to Coulomb equations for active and passive pressure. It is well-known that Coulomb active pressure is close to that by slip line solution which is equivalent to active pressure determination will not be sensitive to the precise location of the failure mechanism or equivalently the use of multiple iterations. With no iteration analysis in slip line analysis, minor errors in the precise location of the slip lines have a small effect on the computation. On the other hand, Coulomb passive pressure can be a poor passive pressure determination, which is equivalent to passive pressure determination being sensitive to the precise location of the failure mechanism and hence the use of multiple iterations. For earthquake load, the classical the Mononobe-Okabe (1929) equation appears to be popular among the engineers, but Cheng (2003b) demonstrated that this equation have over-estimated the passive pressure coefficients when the wall friction is high which is not favourable, which are illustrated in Tables **3.7** and **3.8**.

Table 3.6a. Active pressure coefficients for δ=ϕ and ω/ϕ=0.8 (δ and ω are the wall friction and the inclination of ground beyond the retaining wall with respect to horizontal direction, ε is wall inclination).

	Kerisel and Absi			No Iteration			2 Iterations		
ε\ϕ	20	30	40	20	30	40	20	30	40
80	0.743	0.634	0.533	0.754	0.644	0.541	0.754	0.644	0.541
90	0.633	0.488	0.364	0.631	0.486	0.362	0.631	0.486	0.362
100	0.517	0.357	0.233	0.524	0.361	0.235	0.524	0.361	0.235

Table 3.6b. Passive pressure coefficients for δ=ϕ and ω/ϕ=0.8.

	Kerisel and Absi			No Iteration			2 Iterations		
ε\ϕ	20	30	40	20	30	40	20	30	40
80	3.900	10.800	47.000	3.800	10.122	39.318	3.939	11.072	47.985
90	4.500	13.900	69.000	4.280	12.431	54.431	4.508	13.984	69.245
100	5.100	17.700	100.000	4.871	15.571	77.575	5.243	18.086	101.837

Table 3.7. A comparison of $K_{p\gamma}$ values for a vertical wall (β = 90°, δ = 0, ω=0) with different earthquake coefficient K..

ϕ	K	Slip Line (Cheng, 2003)	Kumar and Chitikela (2002)	Kumar (2001)	Soubra (2000)	Morrison & Ebeling (1995)	Mononobe-Okabe (1929)
30°	0	3.000	3.000	3.000	3.000	3.000	3.000
	0.1	2.821	2.817	2.821	2.820	2.892	2.821
	0.2	2.621	2.610	2.618	2.630	2.646	2.629
	0.3	2.394	2.374	2.393	2.420	2.461	2.418
	0.4	2.130	2.096	2.128	-	2.185	2.175
	0.5	1.792	1.738	1.787	-	1.866	1.866
40°	0	4.599	4.590	4.599	4.600	-	4.599
	0.1	4.381	4.377	4.400	4.380	-	4.380
	0.2	4.146	4.138	4.148	4.150	-	4.152
	0.3	3.897	3.880	3.894	3.910	-	3.912
	0.4	3.627	3.598	3.624	-	-	3.657
	0.5	3.333	3.286	3.328	-	-	3.381

Table 3.8. A comparison of $K_{p\gamma}$ values for a vertical wall (β = 90°, δ = 0, ω=0) with different earthquake coefficient K.

φ	K	Slip Line (Cheng 2003)	Kumar and Chitikela (2002)	Kumar (2001)	Soubra (2000)	Morrison & Ebeling (1995)	Mononobe-Okabe (1929)
30°	0	6.556	6.563	6.677	6.860	7.077	10.095
	0.1	6.084	6.083	6.187	6.350	6.661	9.020
	0.2	5.566	5.562	5.655	5.790	6.154	7.921
	0.3	4.996	4.986	5.065	5.170	5.538	6.784
	0.4	4.343	4.327	4.390	-	4.846	5.577
	0.5	3.526	3.501	3.545	-	3.923	4.208
40°	0	18.150	18.186	18.863	19.620	-	92.586
	0.1	17.106	17.131	17.736	18.460	-	83.254
	0.2	16.011	16.002	16.607	17.230	-	73.902
	0.3	14.853	14.852	15.385	15.940	-	64.522
	0.4	13.619	13.606	14.082	-	-	55.097
	0.5	12.287	12.260	12.677	-	-	45.600

3.2.2. Slip Line Analysis for Axi-Symmetric Problem

Besides the plane strain problem, Cheng and Hu (2005), Cheng *et al.* (2007b) and Cheng *et al.* (2008b) also extended the slip line analysis to axi-symmetric problem for a more general case. The equilibrium equations for a toroidal element (Fig. **3.8**) can be written in cylindrical coordinates system r, θ, z as:

$$\frac{\partial \sigma_r}{\partial r}+\frac{\partial \tau_{rz}}{\partial z}+\frac{\sigma_r-\sigma_\theta}{r}=0, \qquad \frac{\partial \tau_{rz}}{\partial r}+\frac{\partial \sigma_z}{\partial z}+\frac{\tau_{rz}}{r}=\gamma \tag{3.47}$$

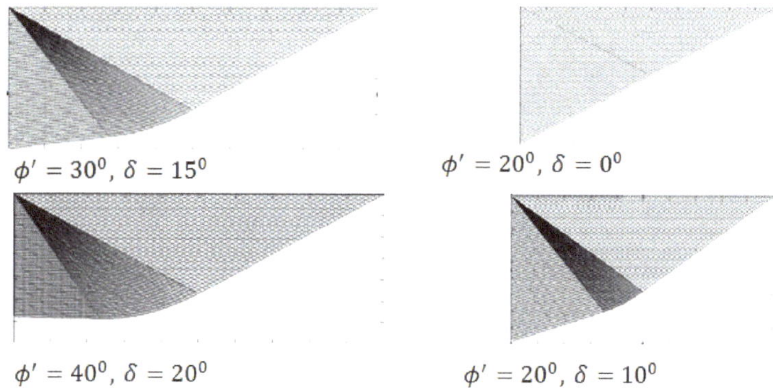

Fig. (3.7). Slip network for passive earth pressure arising from unit weight of soil from slip line method.

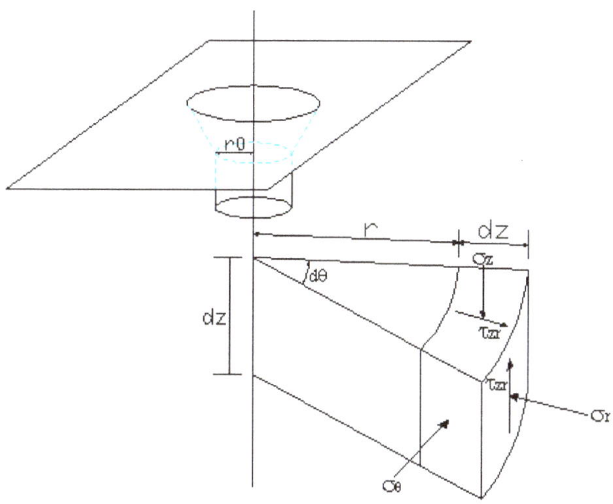

Fig. (3.8). Cylindrical coordinate system and stress components for axi-symmetric problem (z is the vertical direction).

Hill and Cox (2000) assumed the condition of stable hole at great depth so that eq. (3.47) can be simplified by taking σ_z as constant. The four stress components can be expressed in terms of the mean stress σ and the inclination angle ψ which is formed by extending the major principal stress to the r axis (Fig. **3.9**) as:

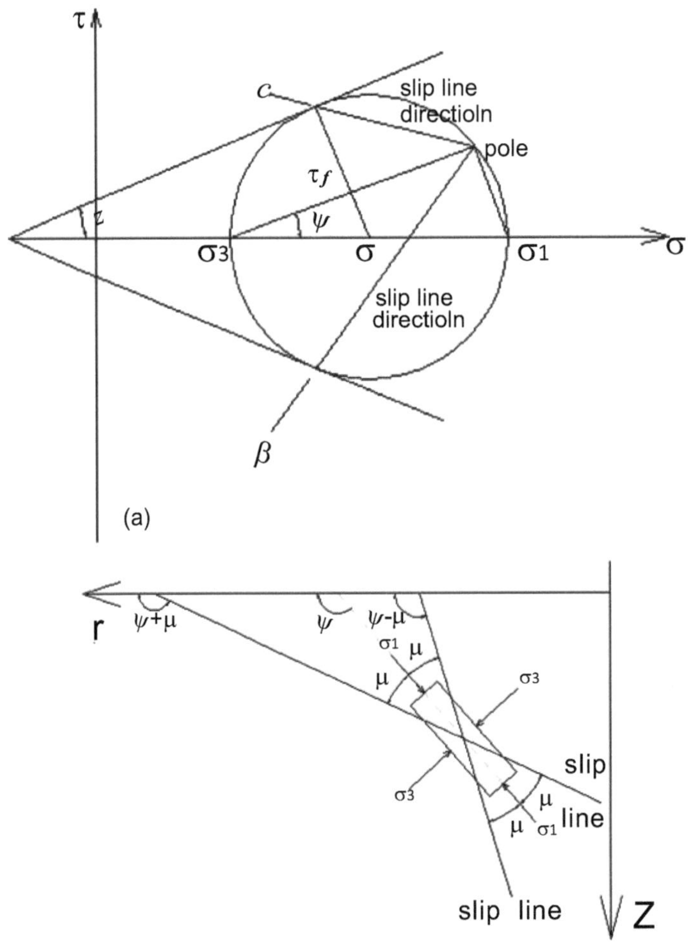

Fig (3.9). (**a**) Mohr circle under failure condition. (**b**) Sign convention and notation.

$$\sigma_r = \sigma(1+\sin\phi\cos 2\psi) - c\cdot\cot\phi \qquad \sigma_z = \sigma(1-\sin\phi\cos 2\psi) - c\cdot\cot\phi \quad \textbf{(3.48)}$$

$$\tau_{rz} = \sigma\sin\phi\sin 2\psi \qquad \sigma_\theta = k\sigma_1 = k\sigma(1+\sin\phi) - kc\cdot\cot\phi \quad \textbf{(3.49)}$$

where $\sigma = (\sigma_1+\sigma_3)/2 + c\cdot\cot\phi$, c is cohesive strength and k is a ratio between σ_θ and σ_1.

k is taken to be 1.0 by most of the researchers in the past for simplicity which is known as the Haar-von Karman hypothesis. This ratio must lie somewhere between 1.0 and the active pressure coefficient k_a theoretically, and more precisely should be taken as a value between k_0 (at-rest) and 1.0 in general (depending on the mobilization of active pressure). Prater (1977) considered both $k=k_0$ and $k=k_a$ in the analysis with limit equilibrium method, where k_0 is the at-rest earth pressure condition. Cheng *et al.* (2008b) pointed out the problem in adopting $k=k_a$ in general. Berezantzev (1958) adopted the Harr-von Karman hypothesis in his simplified slip line solution of axi-symmetric active pressure equation which is actually a special case of that by Cheng *et al.* (2008b). Jenike and Yen's approach (1962) implicitly implied that the intermediate principal stress is equal to the minor principal stress while Hill and Cox (2000) further considered the case of k=1.0. Hill and Cox also mentioned that their solutions (2000) are also not applicable to those plastic regimes where k is not 1.0. During shaft excavation, the radial stress will decrease while the tangential stress will increase. The assumption of intermediate principal stress equal to the minor principal stress by Jenike and Yen (1962) or Hill and Cox (2000) is not a reasonable assumption which tends to over-estimate the ultimate active pressure. Drescher (1986) also obtained the solution using Harr-von Karman hypothesis while Cheng *et al.* (2007b) demonstrated that the Harr-von Karman hypothesis is a good assumption in general.

For the characteristic α and β lines, the geometry dictates the slopes of the slip lines which can be written as

$$\frac{dz}{dr} = \tan(\psi + m\mu) \quad \text{where} \quad \mu = \frac{\pi}{4} - \frac{\phi}{2} \tag{3.50}$$

and m takes the value -1 for an α line and +1 for a β line (shown in Fig. **3.9b**). Equilibrium and yield equations from eqs.(3.47), (3.48) and (3.50) give a set of hyperbolic partial differential equations that will reduce to two ordinary differential equations expressing the changes in stress along each characteristic line in terms of the changing inclination ψ and position (r, z) as

$$\begin{aligned} d\sigma + m2\sigma\tan\phi d\psi + \frac{(1-k-k\sin\phi)\sigma - c(1-k)\cot\phi}{r}dr \\ + m\frac{k(1+\sin\phi)\sigma + c(1-k)\cot\phi}{r}\tan\phi dz = \gamma(m\tan\phi dr + dz) \end{aligned} \tag{3.51}$$

To simplify the analysis, the variables will be normalized with the radius r_0 which take the form

$$\Omega = \frac{\sigma}{r_0\gamma}, \qquad R = \frac{r}{r_0}, \qquad Z = \frac{z}{r_0}, \quad C = \frac{c}{r_0\gamma}, \quad Q_0 = \frac{q_0}{r_0\gamma} \tag{3.52}$$

where r_0 is the radius of the circular excavation, z is the vertical depth under consideration, γ is the unit weight of soil, q_0 is the external surcharge on ground surface (which can be inclined) behind the circular excavation. Substituting eq.(3.52) into eqs.(3.50) and (3.51), we have:

$$\frac{dZ}{dR} = \tan(\psi + m\mu) \qquad \text{where } \mu = \frac{\pi}{4} - \frac{\varphi}{2} \tag{3.53}$$

$$d\Omega + m2\Omega\tan\phi d\psi + \frac{(1-k-k\sin\phi)\Omega - C(1-k)\cot\phi}{R}dR$$

$$+m\frac{k(1+\sin\phi)\Omega + C(1-k)\cot\phi}{R}\tan\phi dZ = m\tan\phi dR + dZ \tag{3.54}$$

Approximate analytical solution to eq.(3.54) based on the Berezantzev's assumption on the failure mechanism (1958) has been provided by Cheng *et al.* (2008b), while the numerical solution to eq.(3.54) will be discussed here. It should be noted that the simplified approximate analytical solution can provide a relatively good estimation of the axi-symmetric active pressure but not the failure mechanism. Houlsby and Wroth (1982) and Drescher (1983, 1986) adopted method of characteristics for the solution of simple axi-symmetric problem, but the effects of inclined ground and wall friction are not considered. The limitations of the previous works by various researchers can be overcome by the solution of eqs. (3.53) and (3.54) which can be written in finite difference form for iterative analysis. For a point $W(R_W, Z_W, \Omega_W, \psi_W)$ as shown in (Fig. **3.10**) (a typical Riemann type problem) which is the intersection point of an α line which passed through a known "point" P $(R_P, Z_P, \Omega_P, \psi_P)$ and a β line which passed through another known "point' Q $(R_Q, Z_Q, \Omega_Q, \psi_Q)$, the finite difference solution can be derived:

$$A = (\psi_P - \psi_Q)\tan\phi + [\frac{k(\sin\phi+1)\tan\phi(Z_W - Z_Q) + (1-k-k\sin\phi)(R_W - R_Q)}{R_W + R_Q}$$

$$-\frac{k(\sin\phi+1)\tan\phi(Z_W - Z_P) - (1-k-k\sin\phi)(R_W - R_P)}{R_W + R_P}] \tag{3.55}$$

$$B = A*(\Omega_P + \Omega_Q) + \tan\phi(R_Q - R_P) - (2Z_W - Z_P - Z_Q) - 2C(1-k)[\frac{R_W - R_P}{R_W + R_P}\cot\phi +$$

$$+\frac{Z_W - Z_P}{R_W + R_P} - \frac{R_W - R_Q}{R_W + R_Q}\cot\phi + \frac{Z_W - Z_Q}{R_W + R_Q}] \tag{3.56}$$

$$C_1 = (A-2)*\Omega_P*\Omega_Q + \Omega_Q[\tan\phi(R_W - R_P) - (Z_w - Z_P) - 2C(1-k)(\frac{R_W - R_P}{R_W + R_P}\cot\phi$$

$$+\frac{Z_W - Z_P}{R_W + R_P})] - \Omega_P[\tan\phi(R_W - R_Q) + (Z_W - Z_Q) + 2C(1-k)(\frac{R_W - R_Q}{R_W + R_Q}\cot\phi - \frac{Z_W - Z_Q}{R_W + R_Q})] \tag{3.57}$$

$$D = \psi_P + \psi_Q - \frac{k(\sin\phi+1)\tan\phi(Z_W - Z_Q) + (1-k-k\sin\phi)(R_W - R_Q)}{(R_W + R_Q)\tan\phi}$$

$$-\frac{k(\sin\phi+1)\tan\phi(Z_W - Z_P) - (1-k-k\sin\phi)(R_W - R_P)}{(R_W + R_P)\tan\phi}] \tag{3.58}$$

$$E = Z_P - Z_Q + \tan\phi(2R_W - R_P - R_Q) - 2C(1-k)[\frac{(R_W - R_P)\cot\phi + Z_W - Z_P}{R_W + R_P}$$

$$+\frac{(R_W - R_Q)\cot\phi - Z_W + Z_Q}{R_W + R_Q}] \tag{3.59}$$

The finite difference equations for point W can be written as

$$R_W = \frac{Z_Q - Z_P + R_P\tan(\frac{\psi_W + \psi_P}{2} - \mu) - R_Q\tan(\frac{\psi_W + \psi_Q}{2} + \mu)}{\tan(\frac{\psi_W + \psi_P}{2} - \mu) - \tan(\frac{\psi_W + \psi_Q}{2} + \mu)} \tag{3.60}$$

$$Z_W = \frac{Z_Q + Z_P + (R_W - R_P)\tan(\frac{\psi_W + \psi_P}{2} - \mu) + (R_W - R_Q)\tan(\frac{\psi_W + \psi_Q}{2} + \mu)}{2} \tag{3.61}$$

$$\Omega_W = \frac{B}{2A+4} + \frac{\sqrt{B^2 - 4(A+2)C_1}}{2|A+2|} \tag{3.62}$$

$$\psi_W = \frac{D}{2} + \frac{(\Omega_P - \Omega_Q)(A-2) + 2E}{2(2\Omega_W + \Omega_P + \Omega_Q)\tan\phi} \tag{3.63}$$

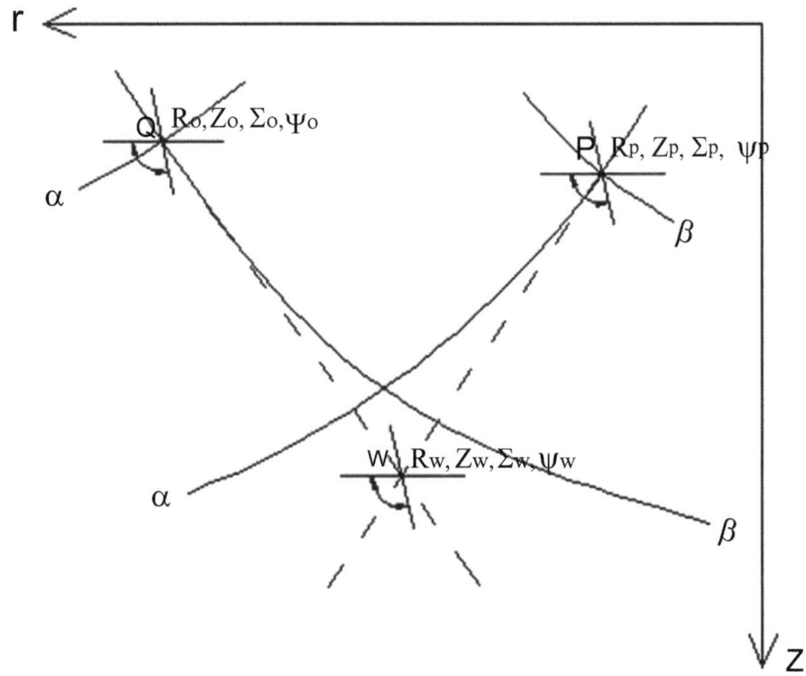

Fig. (3.10). Computation scheme of new point W from known points P and Q.

Refer to (Fig. **3.10**) where results for points P and Q are known and results for W are to be determined. R_W and Z_W can be determined from eqs. (3.60) and (3.61). Ω_W and ψ_W can then be calculated from eqs. (3.62) and (3.63) by using an initial trial of $2.33\psi_W = (\psi_P+\psi_Q)/2$. Since the characteristics lines are curved in general, the solution can be improved by updating ψ_W from eq.(3.47) and repeating the iteration analysis until convergence is achieved (within a specified tolerance of 0.1% by Cheng *et al.* 2007b).

For the general problem as shown in (Fig. **3.11**), the ground line KS makes an angle α with the horizontal direction. Proceeding from the known boundary condition (external surcharge q_0) on the free surface KS (Fig. **3.11**), the entire stress field within KLMT can be determined. Firstly, the boundary KS is subdivided with to a set of equally spaced points, and a fine grid is suggested for analysis. The solution then marches in towards the wall boundary. The whole calculation iterates until an acceptable accuracy (0.1% or any other value) is achieved. A very coarse mesh applied to a circular excavation for calculating active pressure case is llustrated in (Fig. **3.11**) which is used to illustrate the marching scheme used on the computer program WELLKA developed by

Cheng *et al*. (2007b), and a typical slip line field is shown in (Fig. **3.12**).

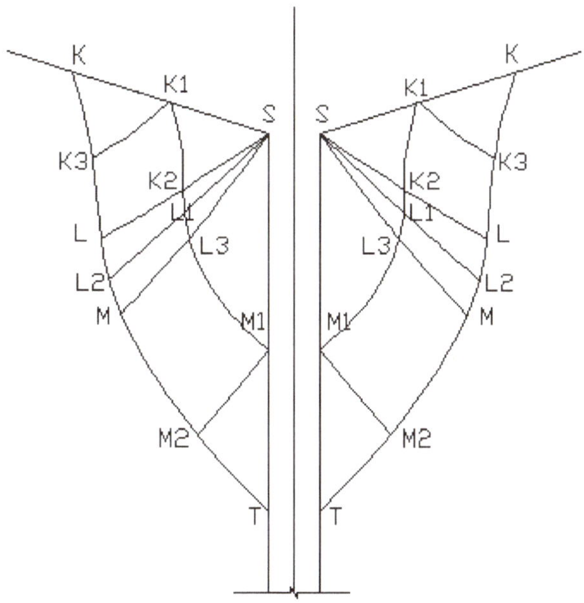

Fig. (3.11). Typical stress characteristic mesh for axi-symmetric problem.

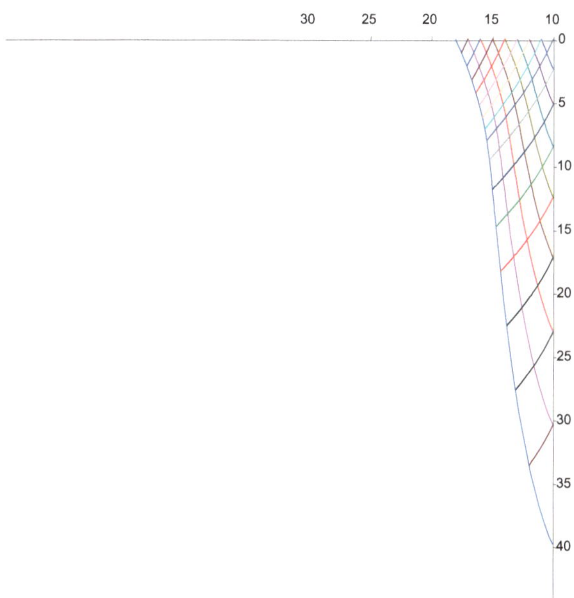

Fig. (3.12). Typical slip lines field for ϕ=40° and k=1.

3.2.2.1. Steps for Solution

(1) Active zone—This is a typical Cauchy type boundary condition and the boundary conditions at S, K1 and K are known. K2 is found from S and K1, K3 is found from K1 and K; L is found from K3 and K2. S, K2 and L become the known boundary conditions for the fan zone.

(2) Fan zone—Node S can be viewed as a degenerate β characteristic with unique R and Z, but varying ψ and Ω. The first two terms of eq. (3.51). show that the stress increases exponentially with rotation ▲ψ of the β line from K2 to L1 in the fan zone. For example, L1 is found from S and K2. Similarly, L2 is found from L1 and L. S, L3 and M now become the known boundary conditions for the passive zone.

(3) Passive zone—Adjacent to the side of the circular wall, R is known and the value of ψ can be determined from the wall condition. The mobilization of the wall friction is assumed to be known. The solutions for Ω and Z at M1 can therefore be determined from the boundary condition at L3. M2 can then be found from the results at M1 and M. Finally, Ω and Z can be found at T using M2 together with the known values of R and ψ at T. The active pressure at excavation wall can then be determined according to following formulae:

$$\Delta = \arcsin\frac{\sin\delta}{\sin\phi} \quad \text{and} \quad P = \Omega\frac{\sin(\Delta-\delta)}{\sin\Delta} \tag{3.64}$$

where δ is the wall friction angle wall, Ω is the mean stress at the wall. The dimensionless active earth pressure P_a and the actual active earth pressure p_a are given by

$$P_a = \sqrt{(P\cos\delta - C\cot\phi)^2 + (P\sin\delta)^2} \quad \text{and} \quad p_a = r_0\gamma P_a \tag{3.65}$$

Based on the above procedures, the effect of friction angle, wall friction and cohesive strength, backfill angle as well as external surcharge can all be determined for general problems which is a major extension of the works by Houlsby and Wroth (1982).

It is found that the failure surface from iterative finite difference solution has an interesting shape different from the corresponding plane strain problem. The failure surface is similar to plane strain at top (R is greater than z and is close to plane strain condition), but arch action will have noticeable effect at greater depth.

Even more interesting is that the angle of ground inclination α behind the wall affects only the active pressure coefficient at the top of the excavation. If the depth is great enough, the arch action (circular effect) will mask out this effect so that the ground inclination angle will only play a very small role on the active pressure. This new result has demonstrated that the arch action is more important than the ground inclination angle when the depth is great. The active pressure coefficient due to self-weight of soil and surcharge is found to decrease with depth. The active pressure due to self-weight of soil may increase indefinitely with depth (at a reduced rate with depth) or attains a maximum value which depends on the lateral stress coefficient k and the friction angle of soil. The actual active pressure due to surcharge will decrease with depth, which means that the circular effect (arch action) has masked the effect of surcharge if the depth is great. The behavior of the active pressure due to self-weight of soil and external surcharge under axi-symmetric condition is drastically different from the corresponding plane strain results. For the cohesive strength of soil, its effect on the active pressure increases with depth and attains a maximum value if the depth is great. The maximum value can be attained by any value of lateral stress coefficient k which is also different from the effect of self-weight of soil. Cheng *et al.* (2008b) also demonstrated that the plane strain relation $k_{ac}=[k_{aq}-1]\cot\phi$ is also generally applicable for axi-symmetric problem. Based on the results by Cheng *et al.* (2008b), it can be concluded that the arch action is actually due to the effects of the intermediate stress plays which is absent from the plane strain condition.

Cheng *et al.* (2008) used the Berezantzev's assumption and derived an analytical solution to the general axi-symmetric lateral earth pressure problem. In this solution, the failure surface is assumed to be straight line which is different from the slip line solution as given in this section. If the depth of excavation is relative small, there is no major difference the two approaches. This is however not true if the depth is great, as the failure surface will be curved in the slip line approach, which means that the volume of failure mass and the lateral earth pressure from the slip line approach will smaller than that by Cheng *et al.* (2008). Interested readers can use the program in Appendix 3-3 and the equations by Cheng *et al.* (2008) for a comparison.

3.2.3. Discussion on Slip Line Analysis

One of the limitations of the slip-line method is the neglect of the stress-strain relationship of the soil, and only the equilibrium and yield conditions are used in the formulation of the stability/ultimate condition. Such idealization may however be acceptable and even necessary for many types of geotechnical problems, as the initial stresses and the constitutive models are usually not known with good accuracy. For the case of plane deformations, the characteristic equations with the given stress boundary conditions have the same number of equations as the unknown stresses so that the problem is statically determinate. If the boundary conditions involve stress and rates of displacement, then the constitutive relation of the soil must be specified in order to obtain a solution.

In general, in a slip-line solution, only part of the soil mass is assumed to be in the state of plastic equilibrium. The solution consists of constructing a slip-line field in the region, which satisfies all the stress boundary conditions that directly concern the region, as well as the equilibrium and yield conditions at every point inside the region. The stress field so obtained has been termed as partial stress field. The stress distribution outside this partial stress field region is not defined or determined. For the solution to be valid, there should exist an associated stress distribution in the non-yielded region, which is in equilibrium with the partial stress field and nowhere violates the yield criterion. The extended stress field defines the stress distribution over the whole body concerned. Even if such an extended stress field can be specified, there is no guarantee that this extended stress solution will be correct.

If the partial stress distribution in the slip-line solution is acceptable as a lower-bound solution, the stress–strain relations should be used to determine whether the given stress and displacement states correspond. If the solution is the correct one, the associated displacement mode will be compatible with a continuous distortion satisfying the displacement boundary conditions and everywhere the rate of plastic work will be positive. Provided that this is the case, then, the solutions so obtained is also an *upper* bound to the collapse value according to the upper-bound rules of limit analysis, and hence is identical with the correct value.

It must be kept in mind that a partial stress field obtained from the slip-line equations is not necessarily the correct solution nor is it known when it is an upper-bound or a lower-bound solution. If a compatible displacement or velocity field can be associated with the specified partial stress field through a given stress–strain relation, the slip-line solution is an upper-bound solution. If, in addition, the partial stress distribution in the plastic zone can be extended to the entire body, satisfying the equilibrium equations, the yield criterion and the stress boundary conditions, the slip-line solution will also a lower bound and is hence the correct solution.

3.3. INTRODUCTION TO LIMIT EQUILIBRIUM METHOD

Classically, the limit equilibrium method (LEM) is commonly adopted by the engineers for the simplified analysis of many geotechnical problems, which include the lateral earth pressure, bearing capacity and slope stability problem. This is due to the fact that limit equilibrium method employs only force and moment equilibrium without tedious mathematics, and such approach is well-received for normal engineering use. More importantly, the use of slip line or other similar methods (variational principle or others) are usually limited to problems with regular geometry and single material. In fact, many computer programs have been developed for the solution of various stability problems by LEM, but no general purpose program based on the slip line method has been developed so far, due to the inherent limitations of the method as discussed previously. Interestingly, for many well-known standard problems where the rigorous solutions are known, LEM will be able to give good approximate solutions with ease. In views of the versatile application of the LEM, it is not surprising that this is the most popular method to the engineers. LEM is more commonly adopted in lateral earth pressure and slope stability problem. The well-known Coulomb active and passive earth pressures as well as the trial wedge method are actually highly simplified limit equilibrium method for lateral earth pressure problem. Under such condition, the factor of safety of the system is actually taken to be 1.0 in the lateral earth pressure determination. On the other hand, when LEM is used for slope stability problem, the factor of safety is commonly determined. Other than this requirement, there is no major difference between LEM for lateral earth pressure and slope stability problem, and Cheng *et al.* (2013a, 2013b) demonstrated that by varying the lateral pressure until critical factor of safety is 1.0 in slope stability methods, the lateral earth pressure as determined will be similar to that based on the Rankine and Coulomb's method. The major difference between LEM for slope stability and lateral earth pressure problems are the requirement on the internal force distribution. LEM for lateral earth pressure analysis usually neglects this assumption and the whole soil mass is considered as a rigid body but neglect the moment equilibrium requirement, while refined internal force distribution is usually allowed in slope stability analysis so that moment equilibrium can also be satisfied.

3.3.1. Definition of the Factor of Safety for Slope Stability Analysis

There are different definitions to the term factor of safety. In general, the factor of safety for slope stability analysis is commonly defined as the ratio of the ultimate shear strength divided by the mobilized shear stress at failure state. The precise definition of factor of safety is however governed by the force or moment equilibrium equation as:

1. ***Moment Equilibrium:*** The factor of safety F_m is defined with respect to moment as:

$$F_m = \frac{M_r}{M_d} \tag{3.66}$$

where M_r is the sum of the resisting moments and M_d is the sum of the driving moment. For a circular failure surface, the centre of the circle is usually taken as the moment point for convenience, but other point of rotation can actually be used if necessary (very uncommon). For a non-circular failure surface, any arbitrary point can be used for the moment consideration, and as long as the system truly obey force and moment equilibrium, then the factor of safety will not depend on the location of the moment point. Interestingly, a moment high above the soil mass can usually result in better convergence in iteration analysis. For Bishop or similar method which do not satisfy horizontal force equilibrium, the factor of safety will depend on the choice of the moment point as 'true' moment equilibrium that requires force equilibrium. Actually, the use of the moment equilibrium equation alone without enforcing force equilibrium cannot guarantee 'true' moment equilibrium.

2. ***Force Equilibrium*:** The factor of safety F_f is defined with respect to force as:

$$F_f = \frac{F_r}{F_d} \tag{3.67}$$

where F_r is the sum of the resisting forces and F_d is the sum of the driving forces.

For "simplified methods" which can only fulfill either the force or moment equilibrium, these two definitions will be slightly different in the values and meaning, however, most of the design codes do not have a clear requirement on the differentiation of these two factors of safety, and a single factor of safety with no clear about the precise definition is actually specified in many design codes. In this respect, a slope may actually possess more than one factors of safety. It should also be pointed out that for soil nail or similar, the loading can be specified as active or passive in some computer programs, and the corresponding forces or moment from the nails are either deducted from the bottom of eqs. (3.66) and (3.67) or added to the top of these two equations. Such operation will result in different factor of safety, particularly when the nail load is large.

3.3.2. Formulation of Limit Equilibrium Methods

Slope stability problem by classical limit equilibrium method is actually a statically indeterminate problem, and there are different assumptions and methods of analysis available to the engineers. Most of the limit equilibrium methods are based on the techniques of slices which can be vertical (most common due to the vertical weight being the most critical factor in most cases), horizontal or inclined (more suitable when there are horizontal/inclined layers with soil parameters critically different from the other soil layers). Various assumptions and methods have been proposed in the past by Bishop (1955), Janbu (1957), Lowe and Karafiath (1960), Morgenstern and Price (1965), Spencer (1967), Janbu (1973) and many others. The two-dimensional slice methods have been well summarized by Fredlund and Krahn (1984), Nash (1987), Morgenstern (1992), Duncan (1996) and Cheng and Lau (2014). The common features of the methods of slices have been summarized by Zhu *et al.* (2003) as:

(a) The sliding mass defined by the failure surface is divided into a finite number of slices. The slices are usually cut vertically, but horizontal as well as inclined cut have also been used by various researchers. In general, the differences between different methods of cutting are not major, and vertical cut is preferred by most engineers at present as gravity is the major source of instability.

(b) There is only a single factor of safety which is applied throughout the whole failure mass.

(c) Assumptions regarding inter-slice forces/thrust line are used to render the problem determinate.

(d) The factor of safety is determined from force or/and moment equilibrium equations based on the assumptions.

The classical limit equilibrium analysis considers only the ultimate limit state of the system without any consideration on the development of strain, and this feature is similar to that by the slip line method or limit analysis. Based upon the static equilibrium conditions and the concept of limit equilibrium, the number of equations and unknown variables of LEM are summarized in Tables **3.9** and **3.10**.

Table 3.9. A comparison of $K_{p\gamma}$ values for a vertical wall ($\beta = 90°$, $\delta = \phi$, ω=0).

Equations	Condition
n	Moment equilibrium for each slice
2n	Force equilibrium in X and Y directions for each slice
n	Mohr-Coulomb failure criterion
4n	Total number of equations

Table 3.10. Summary of unknown variables.

Unknowns	Description
1	Factor of safety
n	Normal force at base of slice
n	Location of normal force at base of slice
n	Shear force at base of slice
n-1	Interslice horizontal force
n-1	Interslice tangential force
n-1	Location of interslice force (thrust line)
6n-2	Total number of unknowns

From the above analysis, it is clear that the classical limit equilibrium problem is statically indeterminate in the order of 6n-2-4n = 2n-2. Hence, additional (2n-2) assumptions have to be introduced in order to solve the problem. The locations of the base normal forces are usually assumed to be at the middle of the slice (n unknowns are eliminated), and this is usually a reasonable assumption unless the width of the slice is wide. In this respect, a limited width of slice (or adequate number of slices) is necessary for a good analysis. This assumption will reduce the number of unknowns to n-2. The most common additional assumptions are either the location of the interslice normal forces or the angle of inclination of the internal forces (equivalent the relation between the interslice normal and shear forces). Such assumption will provide additional n-1 equations so that the problem will now become over-specified by 1. There is not any classical LEM method which can avoid this limitation and can satisfy all the global and local force/moment equilibrium automatically under all cases. At present, there are more than 20 different types of slope stability analysis methods developed by different researchers with different assumptions. The extremum principle by Cheng *et al.* (2010) which is practically equivalent to the lower bound method is an advancement as compared with the classical LEM in that all the local and global force/moment equilibrium can be satisfied without any assumption on the interslice force function, but the amount of computations required for the extremum determination is much more than that based on the classical LEM methods.

By considering the overall moment equilibrium and overall force equilibrium separately, two different expressions are obtained for the factor of safety, F_m and F_f with a given internal force distribution function. Classically, slope stability analysis is commonly formulated in terms of forces instead of stresses for convenience, and we will keep this approach in this chapter.

With respect to Fig. (**3.13**):

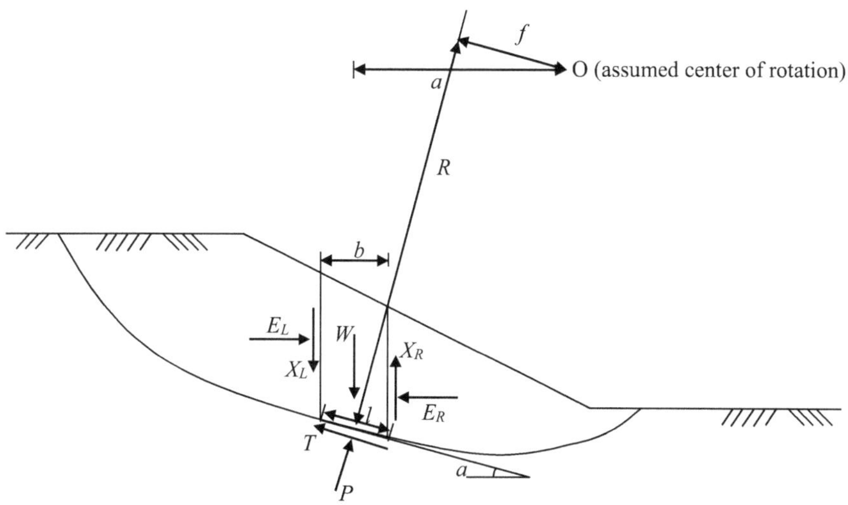

Fig. (3.13). Internal forces for limit equilibrium analysis.

At the base of slice-base normal stress=σ, base shear stress=τ, pore pressure=u

Failure criterion: ultimate shear strength $s = c' + (\sigma - u)\tan\phi'$

Mobilized shear strength $\tau = s/F$, where F is the factor of safety. For the normal and shear forces P and T,

Now

$$P = \sigma l \qquad T = \tau l \quad \text{so } T = \frac{1}{F}(c'l + (P - ul)\tan\phi') \tag{3.68}$$

Resolve vertically:

$$P\cos\alpha + T\sin\alpha = W - (X_R - X_L) \tag{3.69}$$

Rearranging and substituting for T gives

$$P = [W - (X_R - X_L) - \frac{1}{F}(c'l\sin\alpha - ul\tan\phi'\sin\alpha)]/m_\alpha \tag{3.70}$$

$$\text{where } m_\alpha = \cos\alpha(1 + \tan\alpha\frac{\tan\phi'}{F})$$

Resolve horizontally:

$$T\cos\alpha - P\sin\alpha + E_R - E_L = 0 \tag{3.71}$$

Rearranging and substituting for T gives

$$E_R - E_L = P\sin\alpha - \frac{1}{F}\left[c'l + (P - ul)\tan\phi'\right]\cos\alpha \tag{3.72}$$

Overall MOMENT equilibrium (about O):

$$\Sigma Wd = \Sigma TR + \Sigma Pf \tag{3.73}$$

Rearranging and substituting for T gives

$$F_m = \frac{\Sigma\left[c'l + (P - ul)\tan\phi'\right]R}{\Sigma(Wd - Pf)} \tag{3.74}$$

For circular slip surfaces $f = 0$; $d = R\sin\alpha$; $R =$constant

$$\text{so}\ \ F_m = \frac{\Sigma\left[c'l + (P - ul)\tan\phi'\right]}{\Sigma W\sin\alpha} \tag{3.75}$$

Overall FORCE equilibrium

In the absence of surface loading

$$\Sigma(E_R - E_L) = 0 \tag{3.76}$$

$$\Sigma(X_R - X_L) = 0 \tag{3.77}$$

From eq.(3.76), $\Sigma(E_R - E_L) = \Sigma P\sin\alpha - \Sigma\frac{1}{F_f}(c'l + (P - ul)\tan\phi')\cos\alpha = 0$

So

$$\text{so } F_f = \frac{\Sigma\big(c'l + (P - ul)\tan\phi'\big)\cos\alpha}{\Sigma P \sin\alpha} \quad \textbf{(3.78)}$$

In order to solve for F_m and F_F, P must be evaluated, and this requires the evaluation of X_R, P_L the interslice shear forces. As the problem is indeterminate an assumption must be made. Some common assumptions are:

$$X_R - X_L = 0 \qquad \text{Bishop (1955)}$$

$$\frac{X}{E} = \text{constant}=\lambda \qquad \text{Spencer (1967)} \quad \textbf{(3.79)}$$

$$\frac{X}{E} = \lambda f(x) \qquad \text{Morgenstern and Price (1965)}$$

Bishop (1955) showed that F_m is much less sensitive to the assumption about interslice forces than F_f. As the expression of P contains the interslice shear forces X_R and X_L, assumptions are required to render the problem determinate for solution. It is these assumptions and whether overall/local force equilibrium, moment equilibrium or both are considered that distinguish the different methods of limit equilibrium analysis in literature.

3.3.3. Interslice Force Function

The interslice shear force X is assumed to be related to the interslice normal force E by the relation, and $\lambda f(x)$ can be considered as the inclination of the internal force. With reference to Fig. **(3.14)** L is the overall horizontal length of the slip surface while x is the horizontal distance of the section under consideration with respect to the left exit end of the slip surface. x_i is dimensionless and is defined as x/L so that $0.\leq x \leq 1.0$. The angle of inclination of the internal force is taken as θ_i, and $\lambda_i f(x_i)$ is defined as $\tan\theta_i$. λ for a slip surface is then defined as the maximum of λ_i of all sections. For $f(x)$, there is actually not theoretical basis to determine it for a general case under the classical formulation, and there are 7 types of $f(x)$ commonly used form are implemented in SLOPE2000:

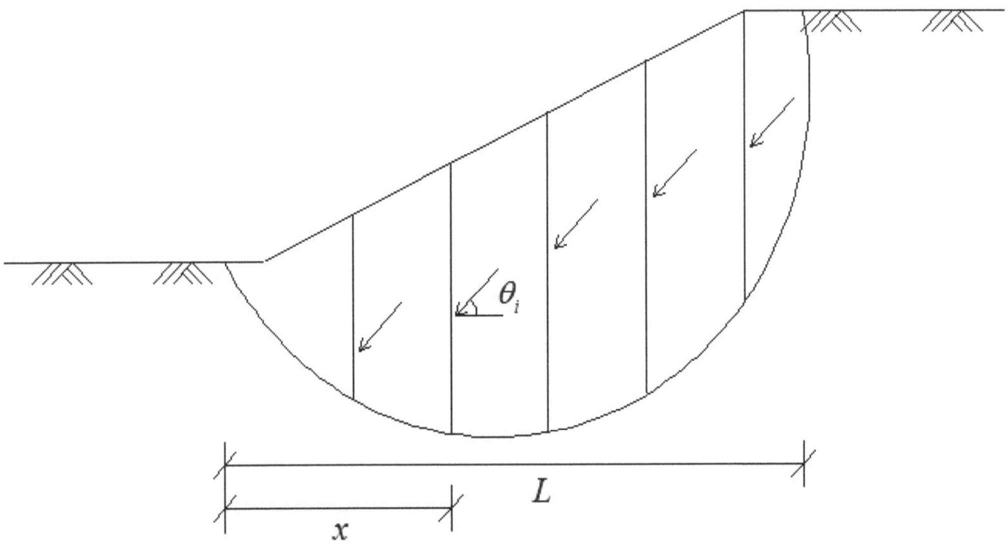

Fig. (3.14). Definition of tanθ for defining λ and *f(x)* for limit equilibrium analysis.

Type 1: *f(x)* = 1.0. This case is equivalent to the Spencer method and is commonly adopted by many engineers. It will be demonstrated later that *f(x)*=1.0, which is the most popular option to the engineers, may not be realistic and good enough in some special cases, though it may be acceptable for normal engineering problems. This option is popular simply because it is simple and there is no particular good choice for *f(x)* for general case.

Type 2: *f(x)* = sin(x). This is a relatively popular alternative to *f(x)*=1.0. This function is adopted simply purely because of its simplicity, but some engineers also report that convergence is usually better with this function than assuming *f(x)* = 1.0.

Type 3: *f(x)* = trapezoidal shape as shown in Fig. (**3.15**). Type 3 *f(x)* can reduce to type 1 as a special case, but this option is seldom adopted in practice due to the difficulty in defining the parameters for this function. In fact, the authors never used this option for real projects.

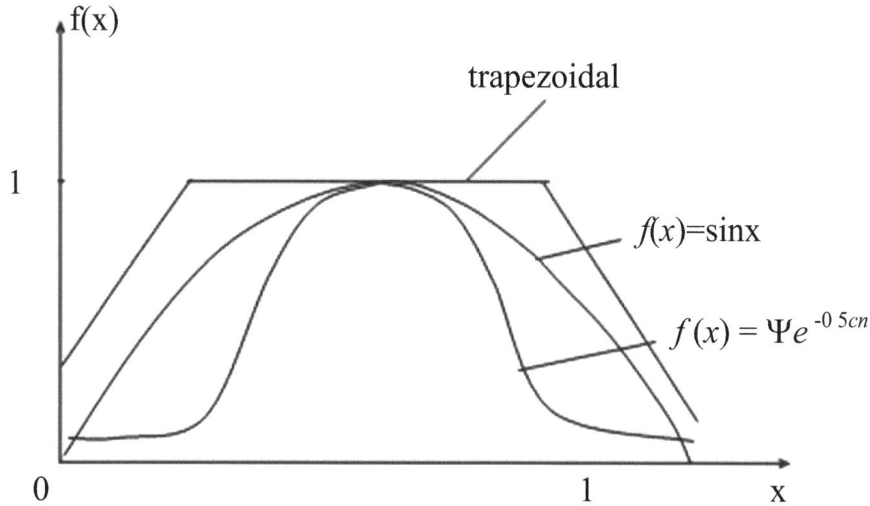

Fig. (3.15). Different types of f(x) commonly implemented into computer programs.

Type 4: *f(x)* = error function or the Fredlund-Wilson-Fan force function (1986) which is defined by $f(x) = \Psi \exp(-0.5c^n\eta^n)$, where Ψ, c and n have to be defined by the users. η is a normalized dimensional factor which has a value of -0.5 at left exit end (x=0) and 0.5 at right exit end of the failure surface (x=1). η varies linearly with the x-ordinates of the failure surface. The factor ψ appears originally in the work by Fan *et al.* (1986) is now absorbed into the parameter λ so that this parameter can be removed in actual computation. This error function is actually based on the curve fitting solution from an elastic finite element stress analysis by Fan *et al.* (1986). Since the stress state in the LEM is the ultimate condition instead of an elastic stress state, the suitability of this interslice force function cannot be justified in general. It is also difficult to define the suitable parameters for a general problem with soil nails, water table and external loads, but it appears that this function is also easy to converge in the stability calculation.

Type 5: Corps of Engineers interslice force function. λ*f(x)* is assumed to be constant and is equal to the slope defined by the two extreme ends of the failure surface. Moment equilibrium is not considered if this function is adopted.

Type 6: Lowe-Karafiath interslice force function (1960). λ*f(x)* is assumed to be the average of the slope of the ground profile and the failure surface at the section under consideration. Similar to the Corps of Engineers method, moment equilibrium cannot be achieved by this function as λ*f(x)* is prescribed to be known.

Type 7: *f(x)* is defined as the tangent of the base slope angle at the current section, and this assumption is used in the Load Factor method in China. Similar to the previous two cases, this method cannot satisfy moment equilibrium.

Type 8: In the extremum option in Slope2000, f(x) is taken to be a variable to be determined. That means, the user needs not to specify a f(x). In theory, this option is free from the assumption of f(x) which can be important for some special cases, at the expense of computer time.

For the above first four types of interslice force functions, they are commonly adopted in the Morgenstern-Price and GLE methods (Sarma method by some researchers, but appear to be not rare), and both the moment (globally) and force equilibrium (locally and globally) can be satisfied simultaneously. The back-computed thrust line may however lie outside the soil mass which means that local moment equilibrium is actually not achieved. An interslice force function should satisfy all local and global equilibrium, but there is no simple way or theoretical background in defining this function except for the extremum principle by Cheng *et al.* (2010).

For type 5 to type 7 interslice force functions, only force equilibrium is utilized in the formulation. The factors of safety from these methods are however usually close to those by the "rigorous" methods, and are usually better than the results by the Janbu simplified method (1957). In fact, the Janbu simplified method (1957) is given by the case of λ=0 for the Corps of Engineers method, Lowe-Karafiath method (1960) and the Load factor method, and results from the Janbu simplified analysis (1957) can also be taken as the first approximation trial in the Morgenstern-Price analysis (1965) if iteration method is used.

Based on a Mohr circle transformation analysis, Chen and Morgenstern (1983) established that $\lambda f(x)$ at the two ends of a slip surface should be equal to the ground slope angle. Other than this requirement, there is no simple way to establish *f(x)* for a general problem. Although there is no simple way to define *f(x)*, Morgenstern (1992) pointed out that for normal problems, factor of safety from different methods of analyses are similar so that the assumptions on the internal force distributions are not major issues for practical use, except for some isolated cases. In views of the difficulty in prescribing a suitable *f(x)* for general case, most engineers will choose *f(x)*=1.0 for simplicity which appears to be satisfactory for most cases. Cheng *et al.* (2010) have however established the upper and lower bounds of the factor of safety and the corresponding *f(x)* based on the extremum principle. The true factor of safety of any system must lie in between these two limits. Cheng *et al.* (2010) also demonstrated that the results from maximum extermum principle can satisfy the requirement as stipulated by

Chen and Morgenstern (1983).

Besides the interslice force assumption, the locations of the interslice internal force (thrust line) or the base normal forces can also be taken as the basic assumption in formulating the limit equilibrium method. Janbu rigorous method (1973) appears to be appealing in that the local moment equilibrium is used explicitly while only global instead of local moment equilibrium is enforced with the interslice force function (hence the back-computed line of thrust can lie outside the soil mass which is not possible). As suggested by Janbu (1973), the line of thrust ratio is usually taken as 1/3 of the interslice height, which is basically compatible with the classical lateral earth pressure distribution. It should be noted that the equilibrium of the last slice is actually not checked in the Janbu rigorous method (1973), as the problem is over specified by 1 by nature. The compliance of local moment equilibrium from the Janbu rigorous method (1973) is hence not strictly rigorous. A practical limitation of this method is the relatively poor convergence in the iteration analysis, particularly when the failure surface is highly irregular or there are external loads. This is due to the fact that the line of thrust ratio is pre-determined, and a small change in the line of thrust can give rise to a large change in the internal force, hence convergence is usually poorer. This method is hence seldom implemented into commercial program, but is available as an option in SLOPE2000 with some minor revisions to improve the convergence. If the method is modified by assuming ht/h=$\lambda f(x)$, where ht=height of line of thrust above slice base and h=length of the vertical interslice, the convergence of this method may be be improved for suitable $f(x)$. There is however difficulty in defining $f(x)$ for the line of thrust for general case, hence this approach is practically not used. Cheng *et al.* (2013b) developed another version of Janbu rigorous method with better convergence which is also implemented in the program SLOPE 2000.

With reference to (Fig. **3.16**), eqs.(3.68) to (3.71) which are the Mohr-Coulomb relation and vertical force equilibrium will hold for every method of analysis, including the Janbu rigorous method.

Resolve parallel to base of slice: $T+(E_R-E_L)\cos\alpha=(W-(X_R-X_L))\sin\alpha$

$$E_R-E_L=(W-(X_R-X_L))\tan\alpha-\frac{1}{F}(c'l+(P-ul)\tan\phi')\sec\alpha \tag{3.80}$$

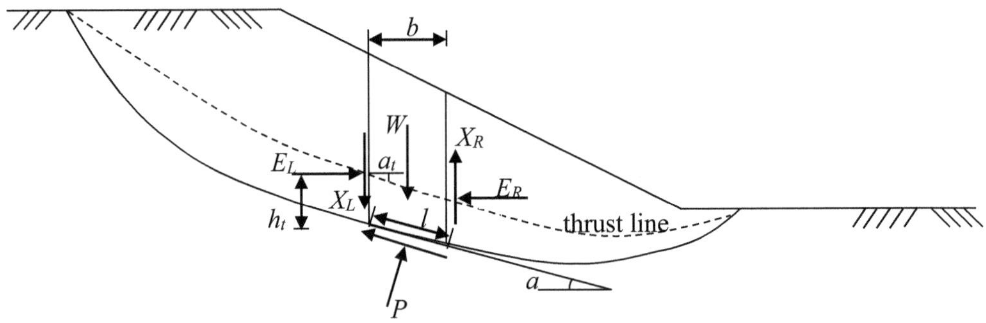

Fig. (3.16). Line of thrust in the Janbu rigorous formulation.

Take moments about centre of base of slice (for thin slice):

$$E_R b \tan\alpha_t - X_R b - (E_R - E_L)h_t = 0$$

or $$X_R = E_R \tan\alpha_t - (E_R - E_L)\frac{h_t}{b} \quad \textbf{(3.81)}$$

From eq.(3.76)

$$\Sigma(E_R - E_L) = \Sigma(W - (X_R - X_L))\tan\alpha - \frac{1}{F_f}\Sigma(c'l + (P - ul)\tan\phi')\sec\alpha = 0 \quad \textbf{(3.82)}$$

hence $$F_f = \frac{\Sigma(c'l + (P - ul)\tan\phi')\sec\alpha}{\Sigma(W - (X_R - X_L))\tan\alpha} \quad \textbf{(3.83)}$$

The solution is usually achieved by an iterative method. First it is assumed that X_R-X_L=0. Then values of E and X are calculated using eq. (3.80) and (3.81) above, and the values of shear force X lag by one iteration as compared with E. As moment equilibrium is satisfied by eq. (3.81), hence $F_f = F_m$. In Janbu's original formulation (1973), P is eliminated and the following expression for F is obtained:

$$F = \frac{\Sigma[c'b + (W - (X_R - X_L) - ub)\tan\phi']/n_\alpha}{\Sigma(W - (X_R - X_L))\tan\alpha} \quad \text{in which } n_\alpha = \cos\alpha . m_\alpha \quad \textbf{(3.84)}$$

Factor of safety is sometimes sensitive to line of thrust. X obtained should be checked with E to ensure the correct F.O.S. If violated, the thrust line should be adjusted. This approach is however not possible if an automatic search for global minimum is progressing.

For the determination of the factor of safety using the interslice force function, λ and F are the two unknowns which can be determined by different methods. In most of the commercial programs, the factor of safety is obtained by iteration method with an initial trial factor of safety (usually 1.0) which is effective for most cases. The use of the iteration method is actually equivalent to expressing the complicated factor of safety polynomial in a functional form as (fixed point theorem):

$$F=f(F) \tag{3.85}$$

Chen and Morgenstern (1983) and Zhu *et al.* (2001) proposed the gradient type Newton-Rhapson technique in the evaluation of the factor of safety F and λ. The gradient type methods are more complicated in the formulation but are fast in solution. Chen and Morgenstern (1983) have also provided the expressions for the derivatives of the moment and shear required in the Newton-Rhapson analysis. Zhu *et al.* (2001) admitted that the initial trials of F and λ can greatly affect the efficiency of the computation. In some cases, poor initial trials can even lead to divergence in analysis. Zhu *et al.* (2001) proposed a variant of the gradient type method which appears to work fine for smooth failure surface, but this method can also break down and diverges. The authors's experience is that for non-smooth or deep-seated failure surface, it is not easy to estimate a good initial trial value, and Zhu's proposal may not work well for these cases even though the answer exist (Cheng *et al.* 2008).

The interslice forces at any interface are the same for the slice to the left and to the right of that interface, an overall equation can hence be assembled in a way similar to that in the stiffness method which will result in a matrix equation (Cheng, 2003a) or equivalently a polynomial equation. The complete solution of all the real positive factors of safety from the matrix can be determined by the double QR method by Cheng (2003a) (double QR method computes all the positive roots while single QR method cannot), which is a useful numerical method to calculate all the positive roots associated with the Hessenberg matrix arising from force equilibrium. It should be noted that imaginary numbers or negative numbers may also satisfy the factor of safety polynomial, so the double QR method instead of the classical QR method is necessary to determine the real positive factors of safety. As an alternative, Cheng (2003a) and Cheng and Zhu (2004) proposed that the factor of safety based on the force equilibrium is determined directly from the polynomial arising from the force equilibrium with a given λ, and these approaches can avoid the problems that may be encountered using the Newton-Rhapson method or iteration method. These approaches can still work under difficult problems while Chen and Morgenstern's (1983) or Zhu's (2001) methods may fail to give the answer. To solve for the factor of safety, the following steps can be used:

1. λ is increased gradually. From slice 1 to n, based on λ and *f(x)*, the factors of safety using force equilibrium can be determined from the polynomial by the Gauss-Newton method with a line search step selection or the double QR method (direct computation with the expense of time). The internal forces *E,X,P* and *T* can be then be determined directly without using any iteration analysis.
2. Based on the internal forces as determined in step 1 for a positive and physically acceptable factor of safety (extremely small value may occasionally appear but should be removed), moment equilibrium is then back checked (net moment of the system). If the moment equilibrium is not satisfied with that specific factor of safety, increase λ and repeat step 1.
3. In the actual implementation, the signs of the unbalanced system moments associated with different λ are monitored. If the sign of the unbalanced moment changes between λ and λ+δλ, then the result will be refined by bisection method until the unbalanced moment is small enough.

There are many cases where iteration analysis (as adopted in many commercial programs) may fail to converge, but the factors of safety actually exist and can be determined (Cheng *et al.* 2008c). Using the polynomial form by Cheng and Zhu (2004) or the matrix form and the double QR method by Cheng (2003a), it is possible to determine the factor of safety without iteration analysis. Zheng *et al.* (2009) also suggested procedures to solve for the factor of safety which is similar to the procedure as suggested above. The advantage of the methods by Cheng is that the factor of safety and the internal forces with respect to force equilibrium are obtained directly without any iteration analysis. Cheng (2003a) also demonstrated that there can be at most n possible factor of safety (including negative value and imaginary number) from the double QR analysis for a failure mass with n slices, which is actually the famous theorem for a nth order polynomial with n roots.

Since all the possible factors of safety for each λ are examined, this method is the ultimate method in the determination of the factor of safety. If other methods of analysis fail to determine the factor of safety, this method may still work. On the other hand, if no physically acceptable solution is found from the double QR method, the problem under consideration has no solution by nature of the specified *f(x)*. Failure to convergence is thus equivalence to the specification of an unsuitable *f(x)*.

3.3.4. Discussion on the Interslice Force Function

From the authors's experience on the application of LEM in many complicated and large scale slope stability problems in Hong Kong, China, and Europe, some of the important questions about fundamental problems of the LEM as raised by the engineers include:

1. The meaning of failure to converge during stability calculation – this is particularly important for slopes with external loads and soil reinforcement which may create local stress concentration so that the problem is difficult to be defined by a simple interslice force function as 1.0.
2. $f(x)$ can be important for some special problems but there is no way to define it.
3. For cases where $f(x)$ is important, there will be a wide range of results based on different $f(x)$, and the choice of a suitable result is important to the engineers.

These questions are important to both researchers and engineers for certain difficult problems, but there are very few previous studies devoted to these three questions. These questions also review the fundamental problem of the basic formulation of the classical LEM methods.

Failure to converge with the "rigorous" methods is well-known to many engineers, in particular, for complicated problems with heavy external loads or soil reinforcement. From the experience of the authors, Janbu rigorous is less satisfactory while Sarma's method appears to be the best in terms of convergence. Cheng *et al.* (2008c) carried out a detailed study on the fundamental problem in convergence and have concluded that there are two reasons for failure to converge with the rigorous methods. First, the iteration method that is commonly used to determine the factor of safety may fail to converge because the interslice shear force is commonly assumed to be zero in the first step of the iterative analysis for simplicity. Cheng (2003a) developed the double QR method which can evaluate the factor of safety and internal forces directly from the solution of a Hessenberg matrix. Based on this method, failure to converge in stability calculations due to the search path (first reason) can be totally eliminated. There are, however, many cases for which the double QR method determines that a physically acceptable answer does not exist for a given $f(x)$ (usually for non-smooth slip surface with soil nails), which means that no meaningful factor of safety will be available unless $f(x)$ can be varied. Actually, some engineers have already questioned the meaning of "no factor of safety" for a given slip surface, as such a concept does not appear in the corresponding structural engineering discipline. So far, there is little previous study on this type of failure to converge, and no rigorous method can guarantee convergence for the general case and provide complete satisfaction of the equilibrium requirements.

Since a critical slip surface may not necessarily converge according to the existing "rigorous" methods of analysis (Cheng *et al.* 2008c), there is always a chance that the critical slip surface may be missed during optimization analysis. Cheng *et al.* (2008c) found a practical case where failure to converge by using commercial program can lead to an error of over 30%, though this is not common in general. It is also interesting to note that it has never been proven that a slip surface that fails to converge in stability analysis is not a critical slip surface, but all commercial programs will simply neglect those slip surfaces that fail to converge, as nothing can be done if no answer can be found. Although this problem may not be critical in practice except for very limited case, a failure surface with no FOS is still physically surprising. A system without factor of safety is not the real behavior, and it is just a human deficiency in making the wrong assumption in $f(x)$ or the stability calculation, and this situation never appears in other engineering disciplines. Factor of safety always exist for a problem, but it is possible that we are not able to determine by classical computation simply because of the use of wrong assumption or mathematical technique, and this is supported by the study by Cheng *et al.* (2008c) that many smooth slip surfaces can still fail to converge using the popular Spencer's method.

Cheng has solved the above three problems raised by engineers using an approach which is equivalent to the variational principle (Cheng *et al.* 2010, Cheng *et al.* 2011). The upper and lower bounds to the factor of safety for any problem can be evaluated within a reasonable time for any complicated problem by using modern heuristic optimization method. The present approach can also be viewed (has also been proved by Cheng *et al.* 2013a) as a lower bound approach, based on which $f(x)$ can be determined, and failure to converge will be virtually eliminated in the stability analysis.

Baker and Garber (1978), Baker (1980) and Revilla and Castillo (1977) applied calculus of variation to determine the factor of safety of a slope which does not require any assumption on the internal force distribution. The variational formulation by Baker (1980) was criticized by De Jong (1980, 1981) as the stationary value may have an indefinite character rather than a minimum. The global minimum is not necessarily given by the gradient of the function as zero if the global minimum lies at the boundary of the solution domain (Cheng 2003). This conclusion was also supported by Castilo and Luenco (1980, 1982) which was based on a series of counter-examples. Baker (2003) later incorporated some additional physical restrictions into the basic limiting equilibrium framework to guarantee that the slope stability problem has a well-defined minimum solution. Although the variational principle requires very few assumptions with no convergence problems during the solution, it is difficult to adopt when the geometry or the ground/loading conditions are complicated. This approach is also never adopted in any commercial program or used in any practical problem.

Pan (1980) and Cheng *et al.* (2008a) stated that slope stability problem is actually a dual optimization problem which appears to be not well-known outside China. On one hand, the soil mass should redistribute the internal forces and utilize its strength to resist the failure, which will result in a maximum factor of safety for any given slip surface. This is called the maximum extremum principle, and the factor of safety will then be compatible with that for reinforced concrete/stell design, and such value should be considered as accurate instead of approximate. On the other hand, the slip surface with the minimum factor of safety is the most possible slip surface, which is called the minimum extremum principle. The maximum and minimum extremum principles are actually equivalent to the lower and upper bound methods. Mathematically, the solution from the use of variational principles is an extremum of a function, and this is also equal to the global maximum/minimum of the function. The "present proposal" can hence be viewed as a form of the discretized variational principle.

The maximum extremum principle is not new in engineering, and at the ultimate limit state of a reinforced concrete beam, the compressive zone of the concrete beam will start to yield and the yield zone will propagate until a failure mechanism is formed. The ultimate limit state design of a reinforced concrete beam under application of a moment is hence equivalent to the maximum extremum principle. For any prescribed failure surface, the maximum "strength" of the system will be mobilized when a continuous yield zone is formed which is similar to a concrete beam. Pan's extremum principle (1980) can provide a practical guideline for slope stability analysis, and it is equivalent to the calculus of variation method used by Baker and Garber (1978), Baker (1980) and Revilla and Castillo (1977). This dual extremum principle is proved by Chen (1998) based on the lower and upper bound analyses, and it is further elaborated upon with applications to rock slope problems by Chen *et al.* (2001). The maximum extremum is actually the lower bound solution, and the present approach is actually a lower bound approach as well as a variational principle approach.

To determine the bounds to the factor of safety and $f(x)$ as suggested by Cheng *et al.* (2010, 2011), the slope as shown in Fig. (**3.17**) can be considered. For a failure surface with n slices, there will be (n-1) interfaces and hence (n-1) $f(x_i)$. All $f(x)$ must lie within the range of 0 to 1.0, while the mobilization factor λ and the objective function FOS based on Morgenstern-Price's method (1965) will be determined for each set of $f(x_i)$. The maximum and minimum factors of safety of a prescribed failure surface satisfying force and moment equilibrium will then be given by the various possible $f(x_i)$ satisfying eq. (3.86) and the physical constraints eq. (3.87).

$$\text{Maximize (or minimize) FOS} \quad \text{subject to} \quad 0 \leq f(x_i) \leq 1.0 \text{ for all } i \qquad \textbf{(3.86)}$$

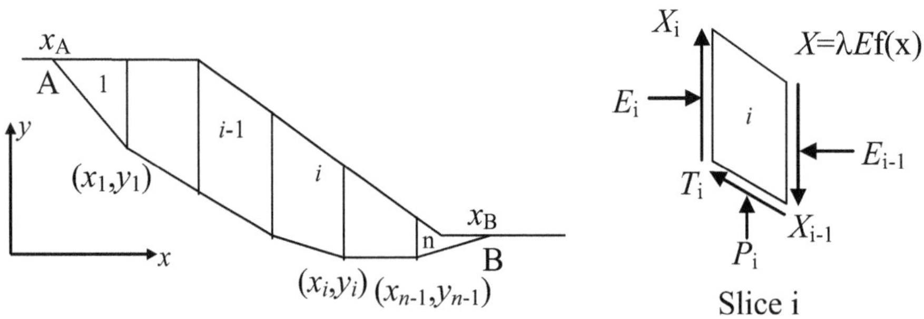

Fig. (3.17). Slip surface and slices.

In carrying out the optimization analysis as given by eq. (3.86), the constraints from the Mohr-Coulomb relation along the interfaces between slices as given by eq. (3.87) should be considered.

$$X \leq E \tan(\phi') + c'L \qquad \textbf{(3.87)}$$

where L is the vertical length of the interface between slices. The constraint given by eq.(3.71) should also satisfy the requirement that the thrust line of the internal forces lies within the soil mass. eq. (3.87) can have a major impact on the FOS in some cases, which will be illustrated by numerical examples in the following section. Since other than the $f(x)$, the Morgenstern-Price's method (1965) is totally governed by force and moment equilibrium, the maximum and minimum factors of safety found from varying $f(x)$ with the constraints (3.87) and the acceptability of the thrust line will provide the upper and lower bounds to the true factor of safety of the slope that are useful for some difficult problems.

The difficulties of the present global optimization analysis are:

1. The number of control variables for the present non-polynomial (NP) type optimization problem is not small.
2. No solution may be obtained for many combinations of $f(x_i)$. This can be due to two possible reasons: (1) no physically acceptable FOS can be found, or (2) eq. (3.87) or thrust line location is violated. It is found that violation of eq. (3.87)

can be a major problem, and the objective function can be highly discontinuous over the solution domain. This is an important and difficult problem for global optimization analysis.

3. There are multiple local minima for the objective function, and the objective function is not necessarily a convex function.

The present approach relies on the use of a stable and robust global optimization method which can be effective for the present highly discontinuous problem. Various optimization methods for slope stability analysis have been used by different researchers: the simplex method by Nguyen (1985) and Chen and Shao (1983), alternating variable methods by Celestino and Duncan (1981), the conjugate-gradient method by Arai and Tagyo (1985), and the Monte Carlo method by Greco (1996) and Malkawi *et al.* (2001). These methods are applicable for simple problems but can be trapped by the presence of a local minimum for general cases. The simulated annealing method, particle swarm optimization (PSO), harmony search (HM), ant-colony, Tabu search, and fish swarm methods are first adopted by Cheng (2003a), Cheng (2007), Cheng *et al.* (2015b), Cheng and Li (2007), Cheng *et al.* (2007a, 2007b, 2008d, 2008e), Cheng and Lau (2014), while genetic algorithms have been adopted by Zolfaghari *et al.* (2005) and Cheng *et al.* (2007a), and the leap-frog algorithm was adopted by Bolton *et al.* (2003). Cheng *et al.* (2012) also proposed the use of coupled optimization method which will be more suitable for complicated difficult problem. Cheng *et al.* (2007a) carried out detailed comparisons between six major types of heuristic global optimization methods, and the sensitivity of the analysis with different optimization parameters has been investigated.

Since the objective function is highly discontinuous by nature, the factor of safety is obtained by the double QR method by Cheng (2003a). The simulated annealing method which is more stable but less efficient is used to determine the extrema (or the maximum factor of safety) with any given slip surface according to eq. (3.86). To evaluate the global minimum factor of safety of a slope, a global optimization analysis (inner loop) should be carried out for the factor of safety, which is an outer loop of the global optimization analysis. To ensure that "false" failure-t--converge due to iteration analysis (Cheng *et al.* 2010) is not encountered so as to reduce the discontinuity of the objective function, the factor of safety is determined by the more time-consuming but robust double QR method. Since discontinuity of the objective function is improved, the more efficient harmony search method (or similar) can be used for locating the critical failure surface. The complete process is computationally intensive, but the use of modern global optimization processes can make this process practical on a personal computer within an acceptable computation time.

Cheng has adopted and modified the simulated annealing method (Cheng 2003a, 2007) complying with eqs. (3.86) and (3.87) to evaluate the two extrema of the factor of safety, and the method is coded into a general purpose program SLOPE 2000 by Cheng (2003a). To determine the maximum and minimum extrema during the simulated annealing analysis, a tolerance of 0.0001 is used to control the optimization search. The authors have adopted 15 slices in the computation, hence there are 14 $f(x_i)$ unknowns in the analysis, and the number of trials required to evaluate the two extrema ranges from 25000 to 32000, which is controlled by the tolerance during the optimization search. Based on this study, it was found that about 30-80% of the trials can converge when $f(x)$ is varied, and those trials that fail to converge are controlled by either eq. (3.87) or the thrust line location, or no physically acceptable answer can be found from the double QR analysis. The number of trials which fail to comply with eq. (3.87) is about 3-5 times that where no physically acceptable answer can be found by the double QR method, so the compliance with eq. (3.87) (and the thrust line location), which has been neglected in the Morgenstern-Price's method (1965), is actually important if an arbitrary $f(x)$ is defined.

For the slope as shown in Fig. (**3.18**) with the soil parameters given in Table **3.11**, the various factors of safety are given in Tables **3.12a** and **3.12b**. If only the lower nail is applied, the differences between the two extrema as compared with Spencer's result (1967) are about 5.9% and 4.4% when eq. (3.87) is used or not used respectively, and such a result is not negligible. The corresponding results/differences when the two soil nails are present are 10.5% and 4.1%. It can be observed that when the soil nail or external load is applied, the choice of $f(x)$ has a noticeable influence on the results, and the compliance with eq. (3.87) is a critical issue which must be considered in the determination of the extrema.

Table 3.11. Soil parameters for Fig. (3.18).

Soil	Unit weight (kN/m^3)	Saturated Unit Weight (kN/m^3)	c' (kPa)	ϕ' (°)
Top	18	20	5	36
Second Layer	15	17	3	30

Table 3.12a. Factors of safety from lower bound approach and Spencer's analysis for (Fig. 3.18).

Case	Max. FOS no eq.3.87	Max FOS with eq.3.87	Min. FOS no eq. (3.87)	Min. FOS with eq. (3.87)	Spencer
Bottom nail	1.856	1.841	1.750	1.763	1.790
2 nails	2.661	2.600	2.398	2.498	2.515

Table 3.12b. λ from lower bound approach and Spencer's analysis for Fig. (3.18).

Case	Max. FOS no eq.3.87	Max FOS with eq.3.87	Min. FOS no eq. (3.87)	Min. FOS with eq. (3.87)	Spencer
Bottom nail	1.149	0.944	0.924	1.902	0.488
2 nails	1.435	1.281	1.149	2.011	0.547

Consider the slope with a steep failure surface (near to the right hadn side) as shown in Fig. (**3.19**). The factors of safety are 1.542, 1.570, 1.526, and 1.550 based on Bishop's method, Janbu's simplified method (without the correction factor), the Swedish method, and Sarma's method. The extrema are 1.602 and 1.547 if eq. (3.87) is not enforced and are 1.564 and 1.559 if eq. (3.87) is considered. No physically acceptable result can be found for Spencer's method (1967) using the double QR method, and "failure to converge" is the fundamental problem in assuming $f(x)$=1.0 as there is no answer by nature for this problem using the Spencer's method (1967).

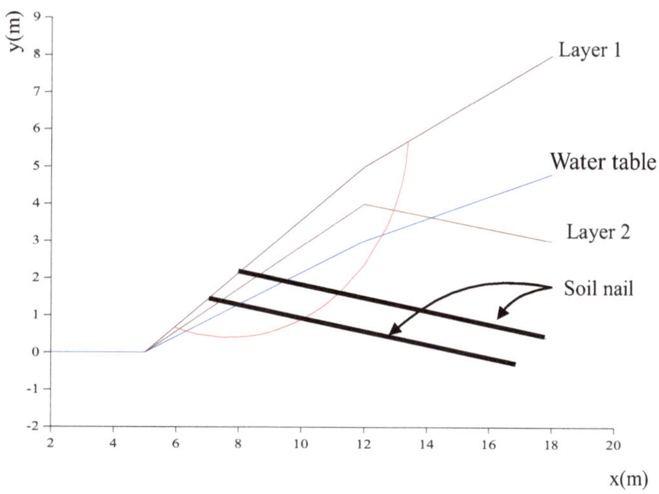

Fig. (3.18). A problem with 2 soil nails.

For the two extrema from the extermum principle, the authors view that the maximum extremum should be taken as the factor of safety for the prescribed failure surface, as the internal forces within the soil mass should redistribute until the maximum resistance capacity of the soil mass is fully mobilized (definition of ultimate condition). Beyond that limit, the soil mass will start to fail. The "extremum proposal" also possesses an advantage in that it is independent of the choice of $f(x)$ or the unknown variables to define the problem. It is well known that there are also cases where $f(x)$ may have a noticeable influence on the factor of safety. There is no clear guideline on the acceptance of a suitable FOS for different *f*(*x*). The use of the maximum extremum can avoid this dilemma, which is an advantage over the classical definition. The present approach is a typical lower bound approach, as statically admissible forces associated with a prescribed *f*(*x*) are considered, and this understanding will be further illustrated in later section. The selection of the maximum FOS is hence justified from the lower bound theorem.

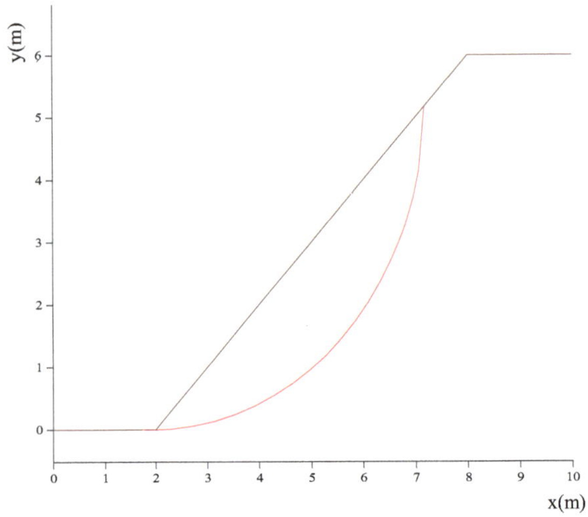

Fig. (3.19). A slope with steep slip surface.

The extremum principle as discussed above can be further demonstrated by a simple bearing capacity problem for clay soil, where the bearing capacity factor N_c is well-known to be π+2 and is given by Prandtl (1920), Sokolovskii (1965) and Chen (1975). So far, there is not any reported success in using the classical "rigorous" slope stability method to evaluate this factor, and an approximate solution is provided by Rocscience (2006). Consider a smooth shallow foundation on clay as shown in Fig. (**3.20**) where the failure mechanism can be obtained analytically. The stresses within the failure mass can be determined by the iterative finite difference solution of the slip line equations as illustrated by Cheng and Au (2005), and a bearing capacity factor Nc of 5.14 (or π+2) is determined by the authors from iterative finite difference analysis of the slip line equations. Consider the vertical plane AB as shown in Fig. (**3.20**), the normal and shear stresses on this plane are determined from slip line analysis, and the integration of the normal and shear stresses along AB will give the total normal and shear forces E and X on this surface. The inclination of X/E for different dimensionless distance x are hence obtained. Denote the largest ratio of X/E as λ, *f(x)* is determined by the ratio between X/E and λ, and the corresponding *f(x)* is shown in

Fig. (**3.21**) while λ is obtained as 0.388. For the left half of triangle CDO and the right half of triangle GFO in Fig. (**3.20**), the principal stresses are along the horizontal and vertical directions respectively so that *f(x)* is equal 0. Within zone ODF, there is a rapid change of the principal stresses directions and *f(x)* will no longer be zero. From D to O in Fig. (**3.21**), *f(x)* increases rapidly to 1.0 at x=0.5. For this problem, the interslice force function is determinate from slip line analysis, but this kind of function appears to be never used in the past. If the cohesive strength and unit weight of the clay are 20 kPa and zero respectively, based on an applied pressure q_u of 102.83 kPa (20N_c), the factor of safety of this system should be 1.0. A factor of safety 0.941 is however given by the Spencer's method (*f(x)*=1.0) (Rocscience 2006) which is clearly not satisfactory, and clearly the result is a lower bound to the rigorous solution. This result is not good as the simple interslice force function 1.0 is far from the actual *f(x)* as shown in Fig. (**3.21**). Based on the slip line solution, a *f(x)* close to that as shown in Fig. (**3.21**) is obtained with a factor of safety 0.995 and λ equals to 0.391 using 15 slices in the LEM analysis, and the result is very close to the theoretical factor of safety 1.0. This example has clearly demonstrated that a correct *f(x)* can be important in some cases, and the applicability of the extremum principle is not limited by the assumption of *f(x)*, as there is no need to make this assumption explicitly. Furthermore, the result has also demonstrated that the maximum of the factor of safety as adopted from the extremum principle is correct, as other *f(x)* will give a smaller factor of safety different from the rigorous solution.

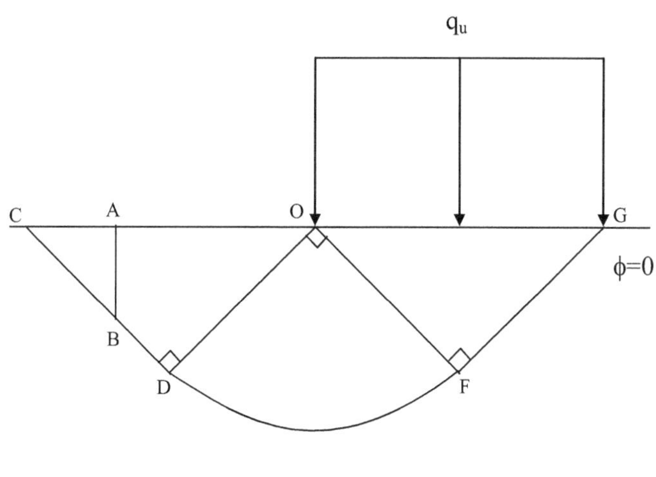

Fig. (3.20). A simple smooth footing on clay.

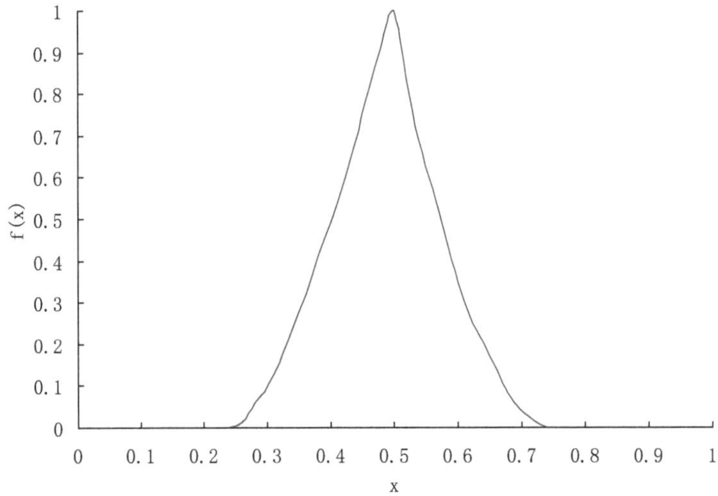

Fig. (3.21). *f(x)* for the simple footing on clay as shown in Fig. (**3.20**) based on slip line solution.

The interslice force function as shown in Fig. (**3.21**) can be derived from slip line analysis, but such a shape cannot be described by the commonly used interslice force function, and it also appears to be never adopted in the past. This example has also illustrated that *f(x)* can be important for some cases, and the maximum extremum which does not require *f(x)* explicitly can overcome this limitation.

While an arbitrary choice of $f(x)$ may be in conflict with eq. (3.87) or no solution can be evaluated, the extremum principle always provides results that are physically consistent and acceptable. The present approach possesses the following advantages: (1) it avoids the assumption of $f(x)$/thrust line or other similar relation, (2) it exhibits virtually no failure to converge for arbitrary problem, (3) consistent and acceptable internal forces complying with force and moment equilibrium are always determined, (4) it provides the bounds to the true factor of safety, (5) it determines $f(x)$ during the evaluation of the FOS, and (6) the problem of the variational principle as discussed by De Jong (1980, 1981) and Castilo and Luenco (1980, 1982) is automatically eliminated by using the global optimization analysis.

Currently, there are more than 20 slope stability analysis methods available (but only very few are used for practice), or equivalently, more than 20 sets of FOS under different assumptions for a problem. Although the differences between these factors of safety are usually small, there are cases where $f(x)$ can be important, and it is not easy to select a suitable FOS for these difficult cases. The authors view that the maximum extremum FOS can be taken as the FOS of the slope under consideration. Under this case, $f(x)$ can be determined from the present approach. The question of *f(x)* (or equivalently, the thrust line in Janbu's rigorous method, 1973) and convergence can then be considered as settled under the present formulation. Based on the use of mixed optimization methods, which can be considered as a discretized variational principle, the "extremum principle" can also be a practical tool for application to complicated problems that cannot be solved by the classical variational principle.

3.4. UNIFICATION OF BEARING CAPACITY, LATERAL EARTH PRESSURE AND SLOPE STABILITY PROBLEMS

Classically, lateral earth pressure, bearing capacity and slope stability problems are considered separately, though these stability problems are all related to the ultimate limit state of a system. The authors view that these stability problems are governed by similar ultimate requirements and should be equivalent in terms of the failure mechanism, and these problems can be considered as different view of the ultimate limit state of the system. In this section, based on the method of characteristics, the authors will determine the bearing capacity factors N_c, N_q and N_γ, for shallow foundation and the lateral earth pressure coefficients K_{ac}, K_{aq} and $K_{p\gamma}$ for retaining wall. Through the method of characteristics, the bearing capacity and lateral earth pressure coefficients are demonstrated to be closely related, and these two problems can be viewed as the inverse of the other problem. The authors will also apply the extremum principle to demonstrate the equivalence between the lateral active pressure, bearing capacity and slope stability problems. Overall, these three problems can be considered as equivalent in the basic principles, and the classification of a problem is simply a matter of convenience (Cheng and Li, 2017).

For bearing capacity and lateral earth pressure problems, the use of slip line method is more common, as the geometry under consideration is usually more regular in nature. For slope stability problem where the geometry is usually irregular with complicated soil reinforcements and external loads, the use of the slip line method is practically not possible. Engineers usually adopt the limit equilibrium method (an approximate slip line form) with different assumptions on the internal force distribution (*f(x)*) for the solution of the problems. Cheng *et al.* (2010) and Cheng *et al.* (2011) developed the extremum principle and have demonstrated the equivalence of the maximum extremum from the limit equilibrium method and the classical plasticity solution by a simple footing on clay. Cheng *et al.* (2010) treated *f(x)* as a variable to be determined, and complete equilibrium is enforced during the search for the maximum extremum of a system. Cheng *et al.* (2010) also pointed out that as long as a *f(x)* is prescribed, the limit equilibrium solution will be a lower bound to the ultimate limit state which is the lower bound theorem. Under the ultimate condition where the strength of a system is fully mobilized, *f(x)* is actually determinate by this requirement which is a boundary condition which has not been used in the past.

If the self-weight of soil is neglected (i.e. γ=0), equations (3.17) and (3.18) can be simplified as follows:

$$-\sin 2\mu \frac{\partial p}{\partial S_\alpha} + 2R\frac{\partial \theta}{\partial S_\alpha} = 0 \tag{3.88}$$

$$\sin 2\mu \frac{\partial p}{\partial S_\beta} + 2R\frac{\partial \theta}{\partial S_\beta} = 0 \tag{3.89}$$

Integrates the above equations along the α and β lines gives:

$$\ln \bar{p} - 2\theta \cot 2\mu = C_\alpha \Rightarrow \ln \bar{p} - 2\theta \tan\varphi = C_\alpha \text{ (along } \alpha \text{ line)} \tag{3.90}$$

$$\ln \bar{p} + 2\theta \cot 2\mu = C_\beta \Rightarrow \ln \bar{p} + 2\theta \tan\varphi = C_\beta \text{ (along } \beta \text{ line)} \tag{3.91}$$

where $\bar{p} = p + c\cot\phi$, c is cohesive strength of soil and C_α and C_β are constants. The sign convention is that compressive stress being positive, while the angles θ and others are taken as positive if measured in a counter-clockwise direction. eqs. (3.90) and (3.91) will be used to derive the expressions for N_q, N_c, K_{aq} and K_{ac} in the following sections.

For a more general condition, the friction along the interface between the soil and the footing or the retaining wall is considered. If a pressure p' act on one-point M on the soil surface and the angle between p' and the normal line to boundary is δ *i.e.* friction angle, (Fig. **3.22a**), p' can be described as:

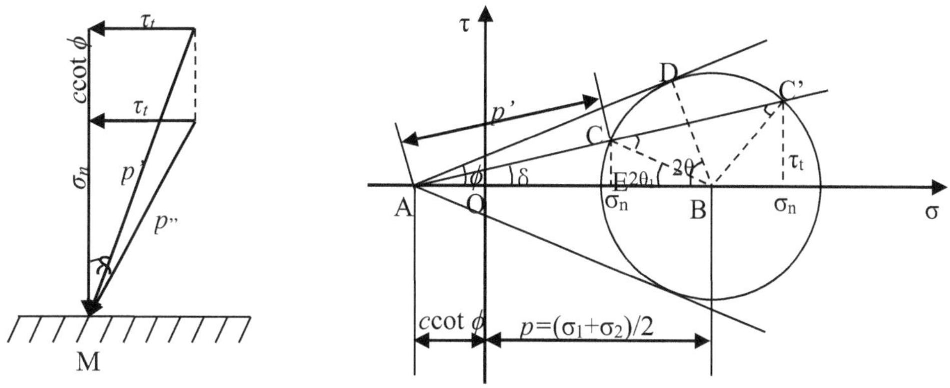

Fig. (3.22). The stress state of point M on a rough surface.

$$p' = \sqrt{(\sigma_n + c\cot\phi)^2 + \tau_t^2} \tag{3.92}$$

where σ_n and τ_t are normal and shear stress acting on the point M respectively. The stress state at the point M can be represented by the Mohr circle as shown in Fig. (**3.22b**). AD is the Mohr-Coulomb line and the angle between AD and horizontal line AB is ϕ. AC represents stress p', and the angle between AC and horizontal line is δ ($\angle CAB$), and δ is taken as positive in Fig. (**3.22a**) in a clockwise direction, which corresponds to an active pressure condition. The line AC intersects the Mohr circle at point C, and the abscissa and ordinate of point C are σ_n and τ_t respectively. Line OB denotes the characteristic stress p, $\angle ABD = 2\theta$, $\angle ABC = 2\theta_1$, and angle θ_1 is inclination angle between the major principal stress and boundary. It can easily be seen from the geometrical relationship in the Fig. (**3.22b**) that:

$$\tau_t = p'\sin\delta \tag{3.93}$$

$$\overline{\sigma_n} = \sigma_n + c\cot\phi = p'\cos\delta \tag{3.94}$$

Therefore, eq. (3.93) divided by eq. (3.94) gives,

$$\frac{\tau_t}{\sigma_n + c\cot\phi} = \frac{\sin\delta}{\cos\delta} = \tan\delta \tag{3.95}$$

For convenience, δ in eq. (3.95) is a boundary condition which is defined in terms of τ_f, σ_n, c and φ instead of purely τ_f, σ_n for simplicity. This boundary condition is usually the footing or the back of a retaining wall (surface AO in Fig. **3.23**) where σ_n is constant (γ is neglected here). The true friction can be transformed to the apparent friction δ easily. According to the geometrical relationship in Fig. **(3.23b)**,

$$\overline{\sigma_n} = \sigma_n + c\cot\varphi = \overline{AB} - \overline{BE} = \overline{AB} - \overline{BC}\cos 2\theta_1 = \overline{AB} - \overline{BD}\cos 2\theta_1 \tag{3.96}$$

$$\tau_t = \overline{CE} = \overline{BC}\cos 2\theta_1 = \overline{BD}\sin 2\theta_1 \tag{3.97}$$

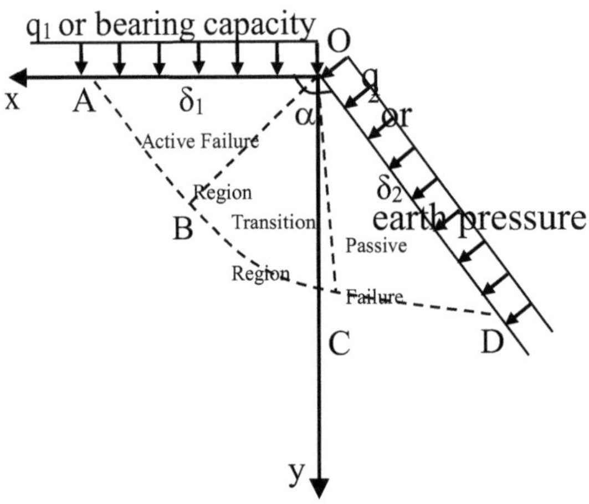

Fig. (3.23). Unified model of bearing capacity and lateral earth pressure problem.

From triangle ABD, it is known that,

$$\overline{AB} = p + c\cot\phi \tag{3.98}$$

$$\overline{BD} = \overline{AB}\sin\phi = (p + c\cot\phi)\sin\phi \tag{3.99}$$

Substitute the above equations into eqs. (3.96) and (3.97),

$$\overline{\sigma_n} = \sigma_n + c\cot\phi = (p + c\cot\phi) - (p + c\cot\phi)\sin\phi\cos 2\theta_1 \tag{3.100}$$

$$\tau_t == (p + c\cot\phi)\sin\phi\sin 2\theta_1 \tag{3.101}$$

So, eq. (3.101) divided by eq. (3.100) gives:

$$\frac{\tau_t}{\sigma_n + c\cot\phi} = \frac{(p + c\cot\phi)\sin\phi\sin 2\theta_1}{(p + c\cot\phi) - (p + c\cot\phi)\sin\phi\cos 2\theta_1} = \frac{\sin\phi\sin 2\theta_1}{1 - \sin\phi\cos 2\theta_1} \tag{3.102}$$

Solve eq. (3.95) and (3.102) simultaneously,

$$\frac{\sin\delta}{\cos\delta} = \frac{\sin\phi\sin 2\theta_1}{1 - \sin\phi\cos 2\theta_1} \tag{3.103}$$

Rearrange eq. (3.103) give,

$$\frac{\sin\delta}{\sin\phi} = \sin(2\theta_1 + \delta) \tag{3.104}$$

The general solution to eq. (3.104) is,

$$2\theta_1 + \delta = 2m\pi + \Delta \tag{3.105}$$

$$\text{and} \quad 2\theta_1 + \delta = (2m+1)\pi - \Delta \tag{3.106}$$

where $\Delta = \arcsin\frac{\sin\delta}{\sin\phi}$. Combine the above two solutions,

$$2\theta_1 + \delta = 2m\pi + k\Delta + (1-k)\frac{\pi}{2} \tag{3.107}$$

$$\text{or} \quad \theta_1 = m\pi + \frac{1}{2}(k\Delta - \delta) + (1-k)\frac{\pi}{4} \tag{3.108}$$

where k=±1; m is an integer and generally m=0 or m=±1. According to the triangle ABC in Fig. (**3.22b**),

$$p + c\cot\phi = p'\frac{\sin\Delta}{\sin(\Delta - k\delta)} = \frac{(\sigma_n + c\cot\phi)}{\cos\delta}\frac{\sin\Delta}{\sin(\Delta - k\delta)} \qquad \textbf{(3.109)}$$

Based on eqs. (3.108) and (3.109), the pressure and friction angle on the soil boundary can be converted into the characteristic pressure p and θ_1. For the angle θ between the major principal stress σ_1 and boundary, if the boundary is horizontal, $\theta=\theta_1$. If the boundary inclines to the horizon at α, $\theta=\theta_1+\alpha$. It is noted here that the sign (±) of k is defined based on the direction of displacement of the boundary when the soil attains the limit equilibrium state. When the boundary of soil moves in the direction of p', k=-1; when the boundary of soil move in the opposite direction of p', k=1. In other words, if the soil is in active failure state, k=-1; otherwise, the soil is in the passive failure state and k=1.

For the case of a smooth soil surface ($\delta=0$) with tangential stress (shear stress) $\tau_t=0$, the characteristic pressure on the boundary OA can be simplified as,

$$p = \frac{(\sigma_n + c\cot\phi)}{1-\sin\phi} - \cot\phi = \frac{\sigma_n + c\cos\phi}{1-k\sin\phi} \qquad \textbf{(3.110)}$$

Equation (3.94) is obtained because of the relation $\lim_{\delta\to 0}\frac{\sin\Delta}{\sin(\Delta-\delta)} = \frac{1}{1-\sin\phi}$.

The ultimate limit state of a bearing capacity and lateral earth pressure problem is shown in Fig. (**3.23**). For the ultimate bearing capacity problem, q_2 is the surcharge on the ground surface, and q_1 is an unknown to be determined. By contrast, when active lateral earth pressure problem is considered, q_1 is the surcharge acting behind the retaining wall, and q_2 is the lateral earth pressure to be determined. Except for the known and unknown variables and the direction in solving the problem (from left to right or from right to left), the governing equations and conditions for the bearing capacity and lateral earth pressure problems are actually similar. Friction angle is considered in the present study, i.e. the friction angle at the interface OA is δ_1; and the friction angle at the interface OD is δ_2. The total slipping wedge can be divided into three limit equilibrium regions, i.e. active failure region OAB, transition region OBC and passive failure region OCD.

For a plane strain case, the ultimate bearing capacity of shallow foundation can be determined from the 'superposition' approach which has been shown to be an approximate but good assumption for normal problems by Michalowski (1997) and Cheng (2002). For a strip footing loaded vertically in the plane of symmetry, the ultimate bearing capacity pressure q_u is given by the bearing capacity factors N_γ, N_c and N_q as:

$$q_u = q_{u\gamma} + q_{uc} + q_{uq} = \frac{1}{2}\gamma B N_r + cN_c + qN_q \tag{3.111}$$

Similarly, the total active earth pressure is considered a combination of the effects due to the weight of the soil ($p_{a\gamma}$), the cohesive strength of soil (p_{ac}) and the surcharge loading (p_{aq}). The lateral active earth pressure coefficients $K_{a\gamma}$, K_{ac}, *and* K_{aq} and the total active earth pressure can be expressed as,

$$p_a = p_{a\gamma} + p_{ac} + p_{aq} = \gamma h K_{ar} - cK_{ac} + qK_{aq} \tag{3.112}$$

To derive N_q from the bearing capacity model, assume c=0 and γ=0 and a uniform surcharge q_2 is applied along interface OD. The boundary conditions along the slope surface OA (active boundary) are:

$$p = p' \frac{\sin\Delta}{\sin(\Delta - k\delta)} = \frac{q_{uq}}{\cos\delta_1} \frac{\sin\Delta_1}{\sin(\Delta_1 + \delta_1)} \tag{3.113}$$

$$\theta = \theta_1 = \pi - \frac{1}{2}(\Delta_1 + \delta_1) + \frac{\pi}{2} \tag{3.114}$$

The boundary conditions along the slope surface OD (passive boundary) are:

$$p = \frac{q_2}{\cos\delta_2} \frac{\sin\Delta_2}{\sin(\Delta_2 - \delta_2)} \tag{3.115}$$

$$\theta = \theta_1 + \alpha = \pi + \frac{1}{2}(\Delta_2 - \delta_2) + \alpha \tag{3.116}$$

According to equations (3.90) or (3.91), substitute the above boundary conditions into the equation (3.91) respectively and the following relation can be obtained,

$$\ln\left(\frac{q_{uq}}{\cos\delta_1}\frac{\sin\Delta_1}{\sin(\Delta_1+\delta_1)}\right)+2\left(\pi+\frac{\pi}{2}-\frac{1}{2}(\Delta_1+\delta_1)\right)\tan\phi=$$
$$\ln\left(\frac{q_2}{\cos\delta_2}\frac{\sin\Delta_2}{\sin(\Delta_2-\delta_2)}\right)+2\left(\pi+\frac{1}{2}(\Delta_2-\delta_2)+\alpha\right)\tan\phi \quad \textbf{(3.117)}$$

From the above equation, N_q can be derived,

$$N_q=\frac{\cos\delta_1}{\cos\delta_2}\frac{\sin(\Delta_1+\delta_1)\sin\Delta_2}{\sin(\Delta_2-\delta_2)\sin\Delta_1}\exp\left[(2\alpha-\pi+\Delta_1+\delta_1+\Delta_2-\delta_2)\tan\phi\right] \quad \textbf{(3.118)}$$

Similar procedure can be used to derive N_c. Consider the where q_2=0 and γ=0, the boundary conditions at the base of the footing become:

$$p+c\cot\phi=\frac{q_{uc}+c\cot\phi}{\cos\delta_1}\frac{\sin\Delta_1}{\sin(\Delta_1+\delta_1)} \quad \textbf{(3.119)}$$

θ is unchanged and is same as eq. (3.114). The boundary conditions along the slope surface OD become:

$$p+c\cot\phi=\frac{c\cot\phi}{\cos\delta_2}\frac{\sin\Delta_2}{\sin(\Delta_2-\delta_2)} \quad \textbf{(3.120)}$$

θ is unchanged and is same as eq. (3.116). Similar to the previous process in deriving N_q, substitute the above boundary conditions into eq. (3.91) gives N_c as:

$$N_c=\left\{\frac{\cos\delta_1}{\cos\delta_2}\frac{\sin(\Delta_1+\delta_1)\sin\Delta_2}{\sin(\Delta_2-\delta_2)\sin\Delta_1}\exp\left[(2\alpha-\pi+\Delta_1+\delta_1+\Delta_2-\delta_2)\tan\phi\right]-1\right\}\cot\phi \quad \textbf{(3.121)}$$

For the lateral earth pressure problem, K_{aq} is derived by assuming c=0 and γ=0. The boundary conditions on the surface OA are:

$$p=p'\frac{\sin\Delta}{\sin(\Delta-k\delta)}=\frac{q_1}{\cos\delta_1}\frac{\sin\Delta_1}{\sin(\Delta_1+\delta_1)} \quad \textbf{(3.122)}$$

θ is the same as eq. (3.114). The boundary conditions on the interface OD are:

$$p = \frac{p_{aq}}{\cos\delta_2}\frac{\sin\Delta_2}{\sin(\Delta_2 - \delta_2)} \quad \textbf{(3.123)}$$

θ is the same as eq. (3.116). Substitute the above boundary conditions into equation (3.91), and K_{aq} can be derived as:

$$K_{aq} = \frac{\cos\delta_2}{\cos\delta_1}\frac{\sin(\Delta_2 - \delta_2)\sin\Delta_1}{\sin(\Delta_1 + \delta_1)\sin\Delta_2}\exp\left[(\pi - 2\alpha - \Delta_1 - \delta_1 - \Delta_2 + \delta_2)\tan\phi\right] \quad \textbf{(3.124)}$$

Considering another case where q_1=0 and γ=0, the boundary conditions on the surface OA become,

$$p + c\cot\phi = \frac{c\cot\phi}{\cos\delta_1}\frac{\sin\Delta_1}{\sin(\Delta_1 + \delta_1)} \quad \textbf{(3.125)}$$

θ is the same as eq. (3.114). The boundary conditions along the slope surface OD become,

$$p + c\cot\phi = \frac{-p_{ac} + c\cot\phi}{\cos\delta_2}\frac{\sin\Delta_2}{\sin(\Delta_2 - \delta_2)} \quad \textbf{(3.126)}$$

It is noted that the lateral earth pressure induced by cohesive strength is negative, so a sign (-) is added to p_{ac}. θ is the same as eq. (3.116). Similar to the previous process in deriving K_{ac}, substitute the above boundary conditions into eq. (3.91) gives:

$$K_{ac} = \left\{1 - \frac{\cos\delta_2}{\cos\delta_1}\frac{\sin(\Delta_2 - \delta_2)\sin\Delta_1}{\sin(\Delta_1 + \delta_1)\sin\Delta_2}\exp\left[(\pi - 2\alpha - \Delta_1 - \delta_1 - \Delta_2 + \delta_2)\tan\phi\right]\right\}\cot\phi \quad \textbf{(3.127)}$$

By comparing eq. (3.118) with eq. (3.127), it is clear that the relation between N_q and K_{aq} for general rough interface is given as,

$$N_q \cdot K_{aq} = 1 \quad \textbf{(3.128)}$$

Combining eq. (3.121), (3.124) and (3.127) together, it is noted that the following relation among N_c, K_{aq} and K_{ac} can be developed,

$$N_c = K_{ac}/K_{aq} \tag{3.129}$$

eq. (3.129) can also be viewed from another point of view. The active pressure along OD induced from the uniform surcharge q_1 along AO will be negative for the cohesive strength. If q_1 is chosen such that the lateral earth pressure on OD is exactly 0, this condition will correspond to the ultimate bearing capacity induced by cohesive strength. Using this concept, eq. (3.113) is also determined from eqs. (3.121), (3.124) and (3.127). The bearing capacity factors N_c and N_q are hence demonstrated to be related to the active lateral earth pressure coefficients K_{ac} and K_{aq} by eqs. (3.128) and (3.129). For the terms N_γ and $K_{a\gamma}$ related to the self-weight of soil, analytical expression is not possible and numerical computation will be adopted for the comparisons which will be explained in the later section.

To verify the above results, the unification of the active lateral earth pressure and bearing capacity problems are further assessed by the use numerical computation. Cheng (2003b) developed an iterative finite difference program called KA for the slip line analysis of lateral earth pressure. For the numerical study in this section, another program ABC which is coded by Martin (2004) for bearing capacity problem based on the method of characteristic is used for comparison. First of all, consider the case without friction on the interface. As shown in Table **3.13**, for N_q and K_{aq}, the values using the formula or the finite difference programs are very close to each other, and the relation between N_q and K_{aq} is consistent with eq. (3.128). The numerical results have further supported the validity of the previous results and the unification of the lateral earth pressure and bearing capacity problem. Table **3.13** shows that the results from ABC and KA are very close to the solutions from the analytical equations, and the relation between K_{ac} and N_c as given in Table **3.14** meet the requirement from eq. (3.129). The authors have however noticed a requirement on the iterative finite difference analysis in the present study. Cheng (2003b) found that the active pressure coefficients are not sensitive to the precise location of the slip line, and no iteration in the finite difference scheme is required for practical purpose (passive pressure coefficient is very sensitive and at least two iterations are required for a good accuracy). When lateral earth pressure program KA is used for bearing capacity problem, the transition zone is much larger and the precision of analysis becomes important. For the present study, iteration in program KA is required and convergence is deemed to be satisfied when the change of the coordinates of the slip line is less than 0.1%. In general, several iterations will be required for the analysis, and the number of iterations increases with the friction angle which is as expected. With the use of iteration analysis, the numerical results from KA (active pressure program) agree well with the analytical expressions for bearing capacity problem as given in this paper. For shallow foundations with sloping ground or active lateral earth pressure problem, the inverse relation between the above two classical soil mechanical problems also holds well which is illustrated by the results as shown in Tables **3.15** and **3.16**.

Table 3.13. Verification of the relationship between K_{aq} and N_q on level ground (smooth), and K_{aq} x N_q is very close to 1.0 for all cases.

ϕ (°)	N_q		K_{aq}	
	Eq. (3.118)	**ABC**	**Eq. (3.124)**	**KA**
10	2.471	2.471	0.405	0.405
20	6.399	6.34	0.156	0.156
30	18.401	18.407	0.054	0.054
40	64.195	64.256	0.016	0.016
50	319.057	319.92	0.003	0.003

Note: c=0, γ=0, α=180.

Table 3.14. Verification of the relationship between K_{ac} and N_c on level ground (smooth), and $N_c = K_{ac} / K_{aq}$ is satisfied.

ϕ (°)	N_c		K_{ac}	
	Eq. (3.112)	**ABC**	**Eq. (3.127)**	**KA**
10	8.345	8.345	-3.377	-3.382
20	14.835	14.836	-2.318	-2.329
30	30.14	30.149	-1.638	-1.648
40	75.313	75.385	-1.173	-1.178
50	266.882	267.606	-0.836	-0.838

Note: γ=0, q=0, α=180.

Table 3.15. Verification of the relationship between K_{aq} and N_q on sloping ground (smooth condition), and K_{aq} x N_q is very close to 1.0 for all cases.

α (°)	N_q	K_{aq}	
	Eq. (3.118)	**Eq. (3.124)**	**KA**
170	15.042	0.066	0.066
160	12.297	0.081	0.081

(Table 3.15) cont.....

150	10.052	0.099	0.100
140	8.218	0.122	0.122
130	6.718	0.149	0.149

Note: c=0, ϕ=30 , γ=0.

Table 3.16. Verification of the relationship between K_{ac} and N_c on sloping ground (smooth), and $N_c = K_{ac} / K_{aq}$ is satisfied.

α (°)	N_c	K_{ac}	
	Eq. (3.121)	**Eq. (3.127)**	**KA**
170	24.322	-1.617	-1.626
160	19.567	-1.591	-1.599
150	15.679	-1.560	-1.566
140	12.501	-1.521	-1.526
130	9.903	-1.474	-1.477

Note: ϕ=30 , q=0, γ=0.

For the more general cases with friction on the interface between soil and wall/footing, the inverse relation between the bearing capacity and active earth pressure problems is also confirmed which is shown in Tables **3.17-3.20**. Through the above results, it is clear that each problem is the inverse of the other, and the choice of the problem depends on the way the problem is being viewed and solved. In other words, the variables of one problem are just the variables of the corresponding image problem.

Table 3.17. Verification of the relationship between K_{aq} and N_q on level ground (rough), and K_{aq} x N_q is very close to 1.0 for all cases.

δ_1 (°)	δ_2 (°)	N_q	K_{aq}	
		Eq. (3.118)	**Eq. (3.124)**	**KA**
0	5	19.285	0.0519	0.0518
0	10	20.093	0.0498	0.0497
0	15	20.834	0.0480	0.0479
0	20	21.505	0.0465	0.0464
0	25	22.083	0.0453	0.0452
0	30	22.456	0.0445	0.0444
5	0	21.167	0.0472	0.0472
10	0	23.838	0.0419	0.0419
15	0	26.299	0.0380	0.0379
20	0	28.416	0.0352	0.0351

(Table 3.17) cont.....

25	0	30.020	0.0333	0.0332
30	0	30.829	0.0324	0.0322
5	5	22.184	0.0451	0.045
10	10	26.030	0.0384	0.0383
15	15	29.776	0.0336	0.0335
20	20	33.209	0.0301	0.0300
25	25	36.027	0.0277	0.0276
30	30	37.622	0.0266	0.0263

Note: c=0, ϕ=30 , γ=0.

Table 3.18. Verification of the relationship between K_{ac} and $N_c$on level ground (rough), and $N_c = K_{ac} / K_{aq}$ is satisfied.

δ_1 (°)	δ_2 (°)	N_c	K_{ac}	
		Eq. (3.121)	**Eq. (3.127)**	**KA**
0	5	31.671	-1.642	-1.652
0	10	33.07	-1.646	-1.656
0	15	34.353	-1.649	-1.66
0	20	35.516	-1.652	-1.663
0	25	36.516	-1.654	-1.666
0	30	37.162	-1.655	-1.672
5	0	34.931	-1.650	-1.660
10	0	39.557	-1.659	-1.670
15	0	43.819	-1.666	-1.677
20	0	47.485	-1.671	-1.682
25	0	50.265	-1.674	-1.685
30	0	51.666	-1.676	-1.687
5	5	36.692	-1.654	-1.664
10	10	43.353	-1.666	-1.676
15	15	49.841	-1.674	-1.683
20	20	55.787	-1.680	-1.692
25	25	60.668	-1.684	-1.697
30	30	63.432	-1.686	-1.702

Note: ϕ=30 , q=0, γ=0.

Table 3.19. Verification of the relationship between K_{aq} and N_q on sloping ground (rough), and K_{aq} x N_q is very close to 1.0 for all cases.

α	N_q	K_{aq}	
(°)	**Eq. (3.118)**	**Eq. (3.124)**	**KA**
170	24.341	0.0411	0.041
160	19.898	0.0503	0.0502
150	16.266	0.0615	0.0614
140	13.297	0.0752	0.0751
130	10.87	0.0920	0.091

Note: c=0, ϕ=30 , γ=0.

Table 3.20. Verification of the relationship between K_{ac} and N_c on sloping ground (rough), and $N_c = K_{ac} / K_{aq}$ is satisfied.

α	N_c	K_{ac}	
(°)	**Eq. (3.121)**	**Eq. (3.127)**	**KA**
170	40.428	-1.661	-1.672
160	32.733	-1.645	-1.656
150	26.442	-1.626	-1.636
140	21.299	-1.602	-1.611
130	17.096	-1.573	-1.58

Note: ϕ=30 , q=0, γ=0.

For N_γ, Sokolovskii (1965) and Cheng and Au (2005) and others determined the value by solving equations (3.17) and (3.18) from right to left. For the ultimate pressure at the bottom of the footing on a level ground, the slip line program SLIP by Cheng and Au (2005) starts at with a uniform distributed load (very small and tends to zero) along OD (which is horizontal now) and the construction of the slip-line field begins from the right-hand side to the left-hand side. The ultimate bearing stress along OA due to the weight of soil and friction angle of the soil will be a triangular pressure along OA, which are given by Sokolovskii (1965), Booker and Zheng (2000) and Cheng and Au (2005). However, the image passive pressure problem can be determined by applying a small uniform distributed load (tends to zero) on line OA as shown in (Fig. **3.24**) and solving the slip line equations from left to right. The passive pressure along line OD from passive pressure determination is virtually the same as the bearing stress on line OA from bearing capacity determination from the program output from SLIP and KP. This approach is conceptually the reverse of the classical method of characteristic for solving the bearing capacity problem, but it is not explicitly considered in the past.

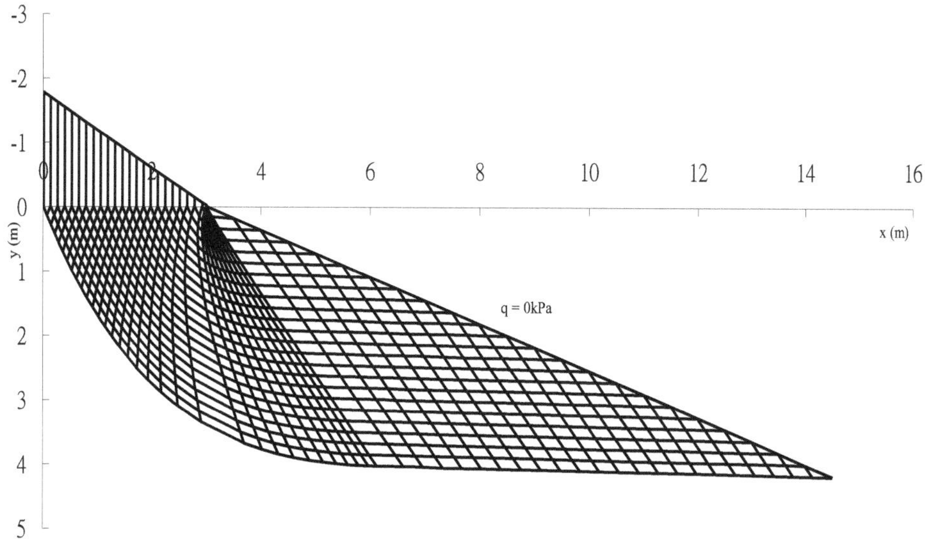

Fig. (3.24). Typical slip line pattern for lateral earth pressure problem with triangular surcharge.

A typical slip line system for the lateral pressure problem is shown in (Fig. **3.24**), which is actually the same as the slip lines obtained by Cheng and Au (2005) for bearing capacity problem. Just from (Fig. **3.24**) alone, it is actually not possible to tell whether it is a bearing capacity problem or a lateral earth pressure problem (even though the ways of solution are the inverse). The mean pressure acting on OA is the ultimate bearing pressure due to the self-weight of soil and N_γ is then determined accordingly. The present results are actually equal to that from the Prandtl's mechanism, and the results are compared with the classical solution by Sokolovskii (1966), Hansen (1970), Vesic (1973), Meyerhof (1963) and Chen (1975). Based on the slip line program ABC developed by Martin (2004) for bearing capacity determination and the slip line lateral earth pressure program KP developed by Cheng (2003b), the equivalent N_γ for level ground and sloping ground are given in Tables **3.21** and **3.22**. Since N_γ from SLIP is defined by the average value of the bearing stress along OA from bearing capacity determination (after Sokolovskii 1966) from the Prandtl's mechanism, numerically, it is the same as the passive pressure $K_{p\gamma}$ from KP, or

$$N_\gamma = K_{p\gamma} \tag{3.130}$$

N_γ can be considered as the corresponding passive earth pressure coefficient, and the classical bearing capacity equation $\gamma BN_\gamma/2$ can be considered as equivalent to $K_p\gamma H^2/2$. The classical definition of bearing capacity $\gamma BN_\gamma/2$ is a hence better definition as compared with using γBN_γ, as it actually reflects the fact that bearing capacity and passive pressure problems are just image problem of each other.

From Table **3.21** and **3.22**, it is clear that N_γ can also be determined from the lateral earth pressure program KP by the authors. It is interesting to note that when α equals 145°, N_γ should equal to 0 when c=0. This result is correctly reflected in the active pressure program KA or program ABC but is not correct in the Vesic's (1973) and Hansen's (1970) bearing capacity factors for purely cohesionless soil.

Table 3.21. Comparison of N_γ on level ground by various methods of analysis (Prandtl mechanism).

$\phi\backslash N_\gamma$	ABC	Sokolovskii	Hansen (1970)	Meyerhof (1963)	Vesic (1973)	Chen (1975)	KP
10	0.56	0.56	0.4	0.4	1.2	0.72	0.56
20	3.16	3.16	2.9	2.9	5.4	3.45	3.158
30	15.31	15.3	15.1	15.7	22.4	15.2	15.30

Table 3.22. Comparison of N_γ for different methods of analysis (ϕ=35) for sloping ground (correction factors by Vesic 1973 and Hansen 1970 are used).

$\alpha\backslash N_\gamma$	Vesic (1973)	Hansen	KP	ABC	SLIP
180 170	48.03 32.59	33.93 21.38	35.20 22.49	35.18 22.5	35.15 22.49
160	19.43	12.42	14.89	14.9	14.9
150	8.58	6.18	9.69	9.7	9.69
145	4.32	3.9	0	0	0

Cheng *et al.* (2010) demonstrated that the extremum principle provides a good approximation to the ultimate condition, and the results by extremum principle are very close to (usually equal to) the slip line solutions in general. In fact, Cheng *et al.* (2013a) demonstrated that the extremum principle is equivalent to the solution of partial differential equations, and the slope stability problem and bearing capacity problem can be viewed as equivalent problem under the extremum condition (Cheng *et al.* 2013b).

To determine N_c factor from slope stability analysis, a surcharge is applied underneath a foundation with a value equal to cN_c without any external surcharge while the self-weight is maintained at zero. For factor N_q, the self-weight and cohesive strength are maintained at zero while a surcharge of 1 unit is applied outside the foundation and the foundation load is maintained at N_q. If the slip surface based on the classical Prandtl's or Hill's mechanism (Chen 1975) is used, the factors of safety as determined for different ϕ are exactly 1.0 if the extremum principle is used and *f(x)* is kept on varying until the maximum factor of safety is found (Cheng 2013a). For the term N_r, a triangular pressure is applied underneath the foundation, and the maximum pressure is adjusted until the critical factor of safety from extremum limit equilibrium method is 1.0. The critical slip surface is allowed to be automatically determined from the extremum principle in this analysis, and the resulting critical slip surface with a factor of safety 1.0 from the extremum principle will be very close to that by slip line analysis. The bearing capacity factors N_r obtained from different approaches are given in Table **3.23**. For example, for ϕ=10°, the slip line system using a very fine grid for the case of N_r is given in (Fig. **3.25**). The failure zone at the right of (Fig. **3.25**) is a typical triangular zone underneath the footing with a curved narrow transition zone, and this slip surface is far from the Hill's or Prandtl's mechanism as adopted by Chen (1975). The slip surface from the extremum principle is close to that as shown in (Fig. **3.25**), and the N_r from extremum principle limit equilibrium slope stability analysis is also very close to that by the slip line method. For N_c and N_q, if the slip surfaces are not specified but is searched from the extremum principle in the way as suggested by Cheng *et al.* (2007a, 2010), the critical slip surfaces from extremum limit equilibrium solution are exactly equal to that from the slip line solution as shown by Cheng *et al.* (2013b).

Table 3.23. Bearing capacity N_r from different methods (solution from Slip is based on Cheng and Au (2005) using slip line solution, solution from Sokolovskii is also based on slip line solution, solution from Chen is based on limit analysis, present solution (automatic) is based on extremum principle with search for critical slip surface).

ϕ(°)	Program SLIP	Sokolovksii (1965)	Chen (1975)	Present (Automatic Search)
10	0.56	0.56	0.72	0.57
20	3.16	3.16	3.45	3.18
30	15.31	15.3	15.2	15.3
40	86.8	86.5	81.79	86.6

It is also interesting to note that the lower bound principle has been clearly illustrated in the present analysis. For the case of factor N_c using the Prandtl's mechanism, the authors have analyzed the intermediate results during the extremum analysis. *f(x)* is kept on changing during the analysis, with an initial value of 1.0 for all x_i during the simulated annealing optimization analysis. With *f(x)*=1.0, the factor of safety is actually 0.925 which is far from 1.0. As *f(x)* is changed, the factor of safety converges towards 1.0 during the simulated annealing analysis, but no factor of safety greater than 1.0 can be obtained. The same results are also obtained for the case of N_q as well. The results as shown in (Fig. **3.26**) are actually good illustration of the lower bound principle. It can hence be concluded that the LEM is practically a lower bound method.

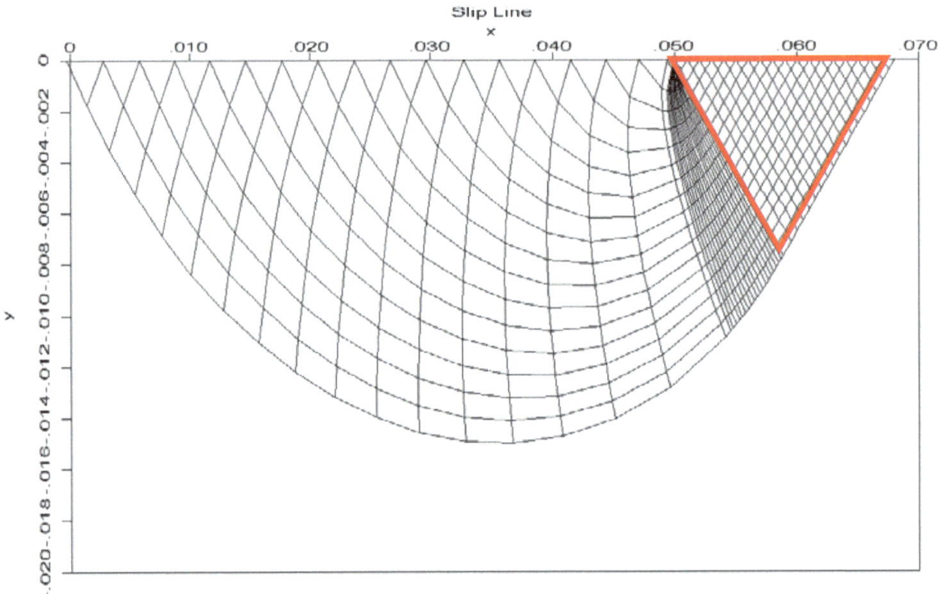

Fig. (3.25). Slip line for the bearing capacity problem when ϕ=10°.

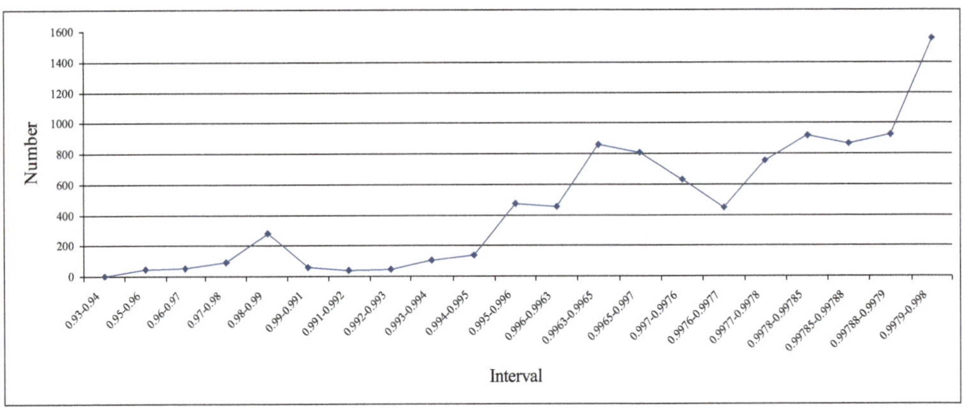

Fig. (3.26). Distribution of acceptable factor of safety during simulated annealing analysis for N_c (same results for N_q) with ϕ=30° using *f(x)* as the variable.

3.4.1. Discussion on Unification of Stability Analysis Methods

Cheng and Li (2017) applied the method of characteristics to study two classical and important geotechnical problems: active lateral earth pressure problem and ultimate bearing capacity problem. The active lateral earth pressure and bearing capacity problems are demonstrated to be equivalent, except for the ease of mathematically manipulation. Based on the slip line theory, the coefficients N_q, N_c, K_{aq} and K_{ac} are derived and the relations among them are established. Two finite difference programs ABC and KA are further used to confirm the validity of the results as derived. Other than the ways in solving eqs.(3.17) and (3.18), there are no practical difference between the two classical geotechnical problems, and the two problems can actually be viewed as equivalent problems under the ultimate condition. It is demonstrated that classical bearing capacity and active lateral earth pressure problems are actually the same, as they are controlled by both the yield and equilibrium equations. Based on the present study, for a normal shallow foundation problem, it can be viewed as a lateral earth pressure problem and vice versa. The surcharge (ultimate bearing capacity) behind an imaginary retaining wall which generate a net upward stresses outside the foundation (or equivalently the retaining wall) will be the ultimate limit state of the system. A bearing capacity problem can hence be viewed as a lateral earth pressure problem in this manner.

Cheng and Li (2017) also demonstrated the lower bound principle in (Fig. **3.26**) using the extremum analysis. Based on the minimum of the extremum from limit equilibrium method, the authors have also demonstrated that there are no practical differences between the slope stability problem and bearing capacity problems, provided that the extremum from the slope stability analysis is used in the comparisons. Based on the present study, it can be concluded that the three geotechnical problems are practically the same problem – the ultimate condition of the system where the maximum resistance of the system is fully mobilized. In this respect, the three geotechnical problems can be considered as equivalent under the ultimate condition. If the three methods are not compared on the same ground, for example, an inappropriate *f(x)* in slope stability analysis or simplified wedge analysis without the consideration of the internal stresses, then the results from the simplified methods for the three basic geotechnical stability problems will not be the same.

Sokolovskii (1965) hinted that the three problems are equivalent in nature, but has not actually demonstrated the equivalence of the problems. The classical plasticity slip line formulation is actually a lower bound problem where the maximum resistance of a system is fully mobilized. For classical slope stability problem where a *f(x)* is prescribed, the solution is a lower bound to the true collapse state which is clearly illustrated in (Fig. **3.26**) and is also discussed by Cheng *et al.* (2010) and Cheng *et al.* (2013a). By varying *f(x)* until the maximum resistance of the system is fully mobilized, the corresponding solution is practically equivalent to the plasticity slip line solution. The three classical geotechnical problems can then be unified, and such unification is not surprising because the fundamental principles behind the three problems are exactly the same: equilibrium and yield.

The classification of a problem is hence just a matter of convenience instead of the difference in the nature of the problem.

3.5. LIMIT ANALYSIS METHOD

Another popular stability formulation is the limit analysis which adopts an idealized stress-strain relation, and the soil is assumed to be a rigid perfectly plastic material with an associated flow rule (nonassociated flow rule can also be adopted). Without carrying out a tedious step-by-step elasto-plastic analysis, limit analysis can provide good (or even exact) solutions to many simple problems. Limit analysis is based on the bound theorems of classical plasticity theory (Drucker *et al.*, 1951; Drucker and Prager, 1952). The general procedure of limit analysis is to assume a kinematically admissible failure mechanism for an upper bound solution (most commonly adopted) or a statically admissible stress field for a lower bound solution (less commonly adopted), and the objective function will be optimized (maximize for lower bound analysis and minimize for upper bound analysis) with respect to the control variables. In the early stage of limit analysis development, direct algebraic method or analytical method is used to obtain the solutions for problems with simple geometry and soil profile (Chen, 1975). Since closed form solutions for most practical complicated nonhomogeneous problems are not available, later attention has been shifted to employing the slice techniques in traditional limit equilibrium to the upper bound limit analysis (Michalowski, 1997; Donald and Chen, 1997).

Limit analysis are based on two theorems: (a) the lower bound theorem, which states that any statically admissible stress field not violating the yield criterion will provide a lower bound estimate of the true ultimate load; and (b) the upper bound theorem, which states that when the energy dissipated by any kinematically admissible velocity field is equated with the energy dissipated by the external loads, then the external loads will be the upper bounds on the true ultimate load (Drucker and Prager, 1952).

A statically admissible stress field is one that satisfies the equilibrium equations, stress boundary conditions, and yield criterion. A kinematically admissible velocity field is one that satisfies strain and velocity compatibility equations, velocity boundary conditions, and the flow rule. When combined, the two theorems will provide a rigorous bound on the true ultimate load. Application of the lower bound theorem usually proceeds as: (a) First, a statically admissible stress field is constructed which is often a discontinuous field. (b) These unknown stresses are then adjusted so that the load on the soil is maximized but the soil remains unyielded. The resulting load becomes the lower bound estimate for the actual collapse load.

Stress fields used in lower bound approaches are often constructed without a clear relation to the real stress fields. Thus, the lower bound solutions for practical geotechnical problems are often difficult to find. Collapse mechanisms used in the upper bound calculations, however, have a distinct physical interpretation associated with actual failure patterns and thus have been extensively used in practice.

3.5.1. Lower Bound Approach

The application of the classical analytical limit analysis was usually limited to simple problems. Numerical methods therefore have been used to compute the lower and upper bound solutions for more complicated problems. The first lower bound formulation based on the finite element method was proposed by Lysmer (1970) for plane strain problems. The approach used the concept of finite element discretization and linear programming. The soil mass is subdivided into simple three-node triangular elements where the nodal normal and shear stresses were taken as the unknown variables. The stresses were assumed to vary linearly within an element, while stress discontinuities were permitted to occur at the interface between adjacent triangles. The statically admissible stress field was defined by the constraints of the equilibrium equations, stress boundary conditions, and linearized yield criterion. Each nonlinear yield criterion was approximated by a set of linear constraints on the stresses that lie inside the parent yield surface, thus ensuring that the solutions were strictly lower bounds. The ultimate load was maximized subjected to a set of linear constraints on the stresses using linear programming method. Other investigations have worked on similar algorithms (Anderheggen and Knopfel, 1972; Bottero *et al.*, 1980). The major disadvantage of these formulations was the linearization of the yield criterion which generated a large system of linear equations, and required excessive computational times, especially if the traditional simplex or revised simplex algorithms were used (Sloan, 1988a).

Efficient analyses for solving the lower bound problem by the finite element method and linear programming method have been developed recently (Bottero *et al.*, 1980; Sloan 1988 and 1989). The key concept of these analyses was the introduction of an active s*et al*gorithm (Sloan 1989) to solve the linear programming problem where the constraint matrix was sparse. Sloan (1989) shown that the active s*et al*gorithm was ideally suited to the numerical lower bound formulation and could solve a large-scale linear programming problem efficiently. A second problem associated with the numerical lower bound solutions occurred when dealing with statically admissible conditions for an infinite-half space. Assdi and Sloan (1990) solved this problem by adopting the concept of infinite elements, and hence obtained rigorous lower bound solutions for general problems.

Lyamin and Sloan (1997) proposed a new lower bound formulation which used linear stress finite elements, incorporating nonlinear yield conditions, and exploiting the underlying convexity of the corresponding optimization problem. They showed that the lower bound solution could be obtained efficiently by solving the system of nonlinear equations that define the Kuhn-Tucker optimality conditions directly. Recently, Zhang (1999) presented a lower bound limit analysis in conjunction with another numerical method - the rigid finite element method (RFEM) to assess the stability of slopes. This formulation satisfies both static and kinematic admissibility of a discretized soil mass without requiring any assumption. Nonlinear programming method has to be adopted to search for the critical solution surface.

3.5.2. Upper Bound Approach

Upper bound theorem is generally carried out as follows: (a) Firstly, a kinematically admissible velocity field is constructed. No separations or overlaps should occur anywhere in the soil mass. (b) Secondly, two rates are then calculated: the rate of internal energy dissipation along the slip surface and discontinuities that separate the various velocity regions, and the rate of work done by all the external forces, including gravity forces, surface tractions and pore water pressures. (c) Thirdly, the above two rates are set to be equal. The resulting equation, called energy-work balance equation, is solved for the applied load on the soil mass. This load would be equal to or greater than the true ultimate load.

The first application of upper bound limit analysis to slope stability problem was by Drucker and Prager (1952) for the critical height of a slope. A failure plane was assumed, and analyses were performed for isotropic and homogeneous slopes with various angles. In the case of a vertical slope, it was found that the critical height obtained by the upper bound theorem was identical with that obtained by the limit equilibrium method. Similar studies have been done by Chen and Giger (1971) and Chen (1975). However, their attention was mainly limited to a rigid body sliding along a circular or log-spiral slip surface which could be minimized using an analytical technique.

Karel (1977a, 1977b) developed an energy method for soil stability analysis. The failure mechanisms used in the method included: (a) A rigid zone with a planar or a log-spiral transition layer, (b) A soft zone confined by plane or log-spiral surfaces, and (c) A composed failure mechanism consisting of rigid and soft zones. The internal dissipation of energy occurred along the transition layer for rigid zone, within zone and along transition layer for the soft zone. However, no numerical technique was proposed to determine the least upper bound of the factor of safety. Izbicki (1981) presented an upper bound approach to slope stability analysis. A translational failure mechanism which was confined by a circular slip surface in the form of rigid blocks similar to the traditional slice method was used. The factor of safety was determined by an energy balance equation and the equilibrium conditions of the field of force associated with the assumed kinematically admissible failure mechanism. Similar to the works by Karel (1977a, 1977b), no numerical technique was provided to search for the least upper bound of the factor of safety.

Michalowski (1995) presented an upper bound (kinematical) approach of limit analysis in which the factor of safety for slopes is associated with a failure mechanism in the form of rigid blocks analogous to the vertical slices used in traditional limit equilibrium methods. A convenient way to include pore water pressure has also been presented and implemented in the analysis of both translational and rotational slope collapse. The strength of the soil between blocks was assumed explicitly, and it was taken as zero or its maximum value set by the Mohr-Coulomb yield criterion. Donald and Chen (1997) proposed another upper bound approach to evaluate the stability of slopes based on a multi-wedge failure mechanism. The sliding mass was divided into a small number of discrete blocks, with linear interfaces between the blocks and with either linear or curved bases to individual blocks. The factor of safety was iteratively calculated by equating the work done by external loads and body forces to the energy dissipated along the bases and interfaces of the blocks.

There are works related to limit analysis by the finite element method which will discussed in details in chapter 5. Many researchers (Mroz and Drescher, 1969; Collins, 1974; Chen, 1975; Michalowski, 1989; Drescher and Detournay, 1993; Donald and Chen, 1997; Yu *et al*., 1998) pointed out that an upper bound limit analysis solution may be regarded as a special limit equilibrium solution but not vice versa. The equivalence of the two approaches plays a key role in the derivations of the limit load or factor of safety for materials following the non-associated flow rule.

Classically, algebraic expressions for the upper bound method are determined for the simple problems for optimization analysis. Assuming a log-spiral failure mechanism for failure surface A shown in (Fig. **3.27**), the work done by the weight of the soil is equated to the energy dissipated along the failure surface based on the upper bound approach by Chen (1975) using an associated flow rule, and the height of the slope can be expressed:

$$H = \frac{c'}{\gamma} f(\phi', \alpha, \beta, \theta_h, \theta_o) \quad \textbf{(3.131)}$$

where

$$f = \frac{\sin\beta\{\exp[2(\theta_h - \theta_o)\tan\phi] - 1\}}{2\sin(\beta-\alpha)\tan\phi(f_1 - f_2 - f_3)}\{\sin(\theta_h+\alpha)\exp[(\theta_h-\theta_o)\tan\phi] - \sin(\theta_o+\alpha)\}$$

$$f_1 = \frac{1}{3(1+9\tan^2\phi)}\{(3\tan\phi\cos\theta_h + \sin\theta_h)\exp[3(\theta_h-\theta_o)\tan\phi] - (3\tan\phi\cos\theta_o + \sin\theta_o)\}$$

$$f_2(\theta_h,\theta_o) = \frac{1}{6}\frac{L}{r_o}\left(2\cos\theta_o - \frac{L}{r_o}\cos\alpha\right)\sin(\theta_o+\alpha)$$

$$f_3(\theta_h,\theta_o) = \frac{1}{6}\exp[(\theta_h-\theta_o)\tan\phi]\left[\sin(\theta_h-\theta_o) - \frac{L}{r_o}\sin(\theta_h+\alpha)\right] *$$

$$\left\{\cos\theta_o - \frac{L}{r_o}\cos\alpha + \cos\theta_h\exp[(\theta_h-\theta_o)\tan\phi]\right\}$$

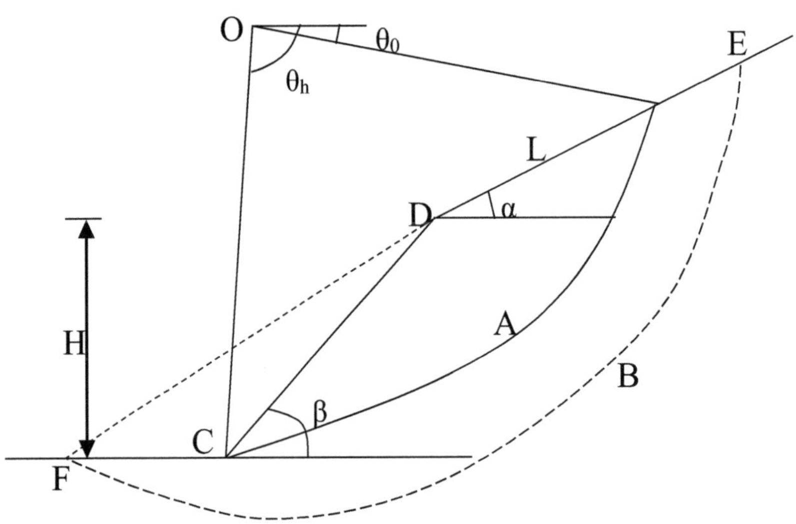

Fig. (3.27). Critical log-spiral failure surface by limit analysis for a simple homogeneous slope.

The critical height of the slope is obtained by minimizing eq. (3.131) (numerically) with respect to θ_0 and θ_h which has been obtained by Chen (1975). Chen has also found that failure surface A is the most critical log-spiral failure surface unless β is small. When β and ϕ' are small, a deep-seated failure shown by failure surface B in (Fig. **3.27**) may be more critical. The basic solution as given by eq. (3.131) can however be modified slightly for this case by joining points D and F. The virtual mass DCF can be deducted from the global mass OEF. The critical result of f(ϕ',α,β) as given by eq. (3.131) can be expressed as a dimensionless stability number N_s which is given eq. (3.132) as a replacement of eq. (3.131), and β' is angle DFC in (Fig. **3.27**). In general, the stability numbers by Chen (1975) are very close to that by Taylor (1948). Cheng has also determined similar normalized stability table as given in Table **3.24** based on the Bishop's method by using program SLOPE2000, and again, the results are also very similar to those based on the limit analysis by Chem (1975) as given in Table **3.25**.

Table 3.24. Stability Chart using 2D Bishop simplified analysis by SLOPE2000 (* means below toe failure).

ϕ(°)	β\α(°)	70	65	60	55	50	45	40	35	30	25	20	15
0	0	4.80	5.03	5.25	5.46	5.67	5.87	5.41*	5.43*	5.45*	5.43*	5.45*	5.46*
5	0	5.41	5.73	6.09	6.46	6.85	7.29	7.79	8.37	9.08*	9.97*	11.43*	14.38*
	5	5.30	5.63	5.96	6.32	6.70	7.11	7.59	8.14	8.77*	9.60*	10.96*	13.69*
10	0	6.05	6.52	7.09	7.71	8.40	9.21	10.22	11.54	13.45	16.62	23.14	45.57
	5	5.95	6.44	6.99	7.58	8.26	9.05	10.06	11.36	13.24	16.33	22.78	45.00
	10	5.84	6.33	6.86	7.41	8.07	8.83	9.78	11.11	12.84	15.79	21.90	42.86
15	0	6.94	7.58	8.37	9.36	10.50	11.94	13.90	16.79	21.69	32.14	69.23	-
	5	6.77	7.50	8.30	9.25	10.36	11.80	13.74	16.59	21.48	31.86	68.97	
	10	6.67	7.38	8.18	9.09	10.20	11.61	13.51	16.33	21.13	31.36	68.18	
	15	6.53	7.22	8.01	8.89	9.96	11.25	13.14	15.85	20.45	30.25	68.18	
20	0	7.97	9.01	10.14	11.61	13.51	16.07	20.00	26.67	41.38	94.74	-	
	5	8.04	8.91	10.04	11.50	13.38	15.93	19.82	26.55	41.10	94.74		
	10	7.69	8.82	9.92	11.35	13.22	15.76	19.61	26.28	40.72	93.75		
	15	7.60	8.66	9.75	11.16	12.97	15.49	19.25	25.79	40.18	92.78		
	20	7.59	8.44	9.55	10.91	12.66	15.06	18.71	25.00	38.54	88.24		
25	0	9.42	11.01	12.57	14.80	18.04	22.93	31.47	50.28	120.00	-		
	5	9.28	10.91	12.46	14.69	17.91	22.78	31.03	50.00	120.00			
	10	9.50	10.84	12.33	14.57	17.73	22.56	31.03	49.72	119.21			
	15	9.00	10.60	12.16	14.35	17.51	22.28	30.72	49.32	118.42			
	20	8.97	10.51	12.00	14.12	17.22	21.95	30.25	48.65	116.88			
	25	8.81	10.17	11.73	13.74	16.74	21.25	29.27	46.75	111.80			
30	0	11.89	13.79	16.07	19.65	25.53	35.64	58.63	144.00	-			

(Table 3.24) cont.....

	5	11.54	13.74	16.00	19.52	25.35	35.64	58.44	144.00				
	10	11.43	13.75	15.83	19.35	25.17	35.43	58.25	144.00				
	15	11.04	13.53	15.65	19.19	25.00	35.16	57.32	142.86				
	20	10.71	13.31	15.49	18.95	24.66	34.75	57.32	142.29				
	25	10.81	12.93	15.23	18.60	24.19	34.16	56.60	140.63				
	30	10.43	12.11	14.86	18.09	23.44	32.97	54.22	134.33				
35	0	14.83	18.00	21.25	27.48	39.30	65.93	166.67	-				
	5	14.94	17.82	21.18	27.40	39.30	65.69	166.67					
	10	14.25	17.65	21.08	27.27	39.13	65.45	165.90					
	15	14.04	17.54	20.93	27.03	38.79	65.45	165.14					
	20	13.85	17.65	20.69	26.87	38.63	64.98	165.14					
	25	13.19	16.32	20.55	26.55	38.30	64.29	165.14					
	30	12.82	15.76	20.18	26.01	37.50	63.16	163.64					
	35	12.54	15.67	19.35	25.14	36.14	61.02	155.17					
40	0	20.07	24.03	30.10	42.35	72.29	185.57	-					
	5	19.13	23.68	29.41	42.06	72.00	185.57						
	10	19.82	23.72	29.27	42.35	71.43	185.57						
	15	18.95	23.53	30.28	41.86	71.43	183.67						
	20	17.61	23.38	29.32	41.47	71.43	183.67						
	25	16.93	23.23	28.85	41.10	70.87	183.67						
	30	16.36	22.70	28.57	40.72	70.04	181.82						
	35	16.04	21.05	28.13	40.00	68.97	180.00						
	40	15.72	20.00	27.69	38.46	65.93	171.43						

Table 3.25. Comparison of results between limit analysis and other limit equilibrium methods for the sability number.

Slope	**Friction**	**Stability Factor (Ns) for α = 0**			
Angle	**Angle**	**Limit Equilibrium**			**Limit Analysis**
β	**ϕ**	**Slices**	**ϕ circle**	**Log-Spiral**	**Log-Spiral**
90	**0**	3.83	3.83	3.83	3.83
	5	4.19	4.19	4.19	4.19
	10	5.02	5.02	-	5.02
	25	6.06	6.06	6.06	6.06
75	**0**	4.57	4.57	4.57	4.56
	5	5.13	5.13	-	5.14
	15	6.49	6.52	-	6.57

(Table 3.25) cont.....

	25	8.48	8.54	-	8.58
60	**0**	5.24	5.24	5.24	5.25
	5	6.06	6.18	6.18	6.16
	15	8.33	8.63	8.63	8.63
	25	12.20	12.65	12.82	12.74
45	**0**	5.88	5.88*	5.88*	5.53*
	5	7.09	7.36	-	7.35
	15	11.77	12.04	-	12.05
	25	20.83	22.73	-	22.90
30	**0**	6.41*	6.41*	6.41*	5.53*
	5	8.77*	9.09*	-	9.13*
	15	20.84	21.74	-	21.69
	25	83.34	111.1	-	119.93
15	**0**	6.90*	6.90*	6.90*	5.53*
	5	13.89*	14.71*	14.71	14.38*
	10	-	43.62	-	45.49

* Critical failure surface passes below toe.

$$f\left(\theta_h,\theta_o,\beta'\right)=\frac{\sin\beta'\left\{\exp\left[2\left(\theta_h-\theta_o\right)\tan\phi\right]-1\right\}}{2\sin\left(\beta'-\alpha\right)\tan\alpha\left(f_1-f_2-f_3-f_4\right)}\left\{\sin\left(\theta_h+\alpha\right)\exp\left[\left(\theta_h-\theta_o\right)\tan\phi\right]-\sin\left(\theta_o+\alpha\right)\right\} \quad \textbf{(3.132)}$$

To determinate the lateral earth pressure of a fill on a retaining wall when frictional forces act on the back of the wall, upper-bound technique of limit analysis can be conveniently applied to this problem to obtain the solutions for the general lateral earth pressure problems including the effect of wall friction. Three failure mechanisms are selected for limit analysis to determine the active and passive soil pressures which are:

I. Two-Wedge Mechanism II. Log-Spiral Mechanism III. Arc sandwich Mechanism

The formulations as considered in (Fig. **3.28**) are similar to that by Chen and Liu (1990). For a Coulomb soil (c-[REMOVED 102 FIELD]) with uniform surcharge at the back of retaining wall, the lateral pressure can be expressed by eq. (3.112).

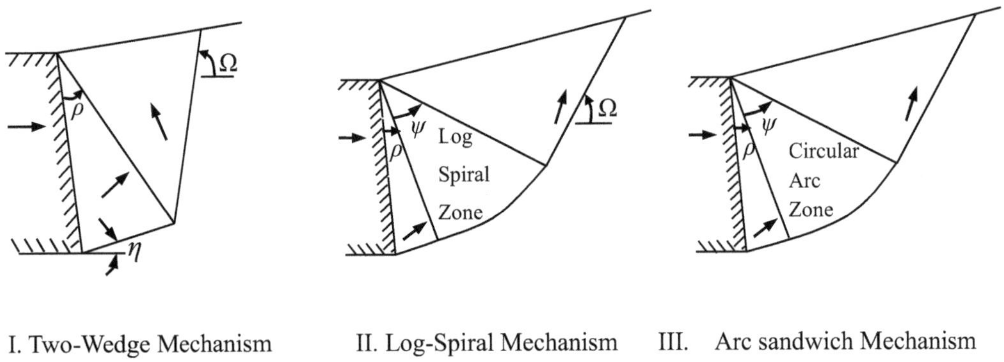

Fig. (3.28). Failure mechanism used in computer program "LEP".

where

Kaγ and Kpγ: for the case of cohesionless soil with no surcharge loading. (c=0, q=0)

K_{ac}, K_{pc}: for the case of cohesive weightless soil with no surcharge loading. (γ=0, q=0)

K_{aq}, K_{pq}: for the case of cohesionless weightless soil with surcharge loading (c=0, γ=0)

In finding the solution of a particular mechanism, two following work equations should be formed first:

i) the external rate of work due to the external applied loads and soil weight.

ii) the internal dissipation of energy in the plastically deformed region.

The solution can then be found by equating the above two work done equations, and the analytical solutions are obtained by the symbolic mathematics option in Mathcad. There are totally six long equations for the active and passive pressure coefficients for the cases in (Fig. **3.28**), and interested readers can obtain the long equations and detailed procedures for derivations from the authors (too long and tedious to include here). Since the equations are very long and tedious, to obtain the maximum active pressure and minimum passive pressure, a computer program as shown in the Appendix 3-1 of this chapter has been developed. The computer program will compute the three coefficients corresponding to active and passive pressure coefficients for the three mechanisms in (Fig. **3.28**) for wall with arbitrary inclination and wall friction. A more complete program including simple graphics support can be obtained from the authors at natureymc@yahoo.com.hk. After detailed study, it is found that mechanism III as shown in (Fig. **3.28c**) is always not the controlling mechanism (except for limited cases). In fact, when the wall friction is high, this mechanism will produce a very poor result which is shown in Table **3.26**. It should also be pointed out that according to the upper bound analysis, different failure mechanisms should be considered, and the critical solution from different mechanisms will then be a good estimate of the solution. In practice, it is impossible to derive the equations for arbitrary failure mechanism, and the finite element limit analysis has to be adopted for such purpose.

Table 3.26. Comparison between three mechanisms on passive pressure coefficients for zero ground slope behind the wall (Passive Case).

Angle of Internal Friction φ	Wall Friction Angle δ	Passive Pressure Coefficient ($K_{p\gamma}$)								
		α = 70			α = 90			α = 110		
		I	II	III	I	II	III	I	II	III
10	0	1.03601	1.3601	1.3601	1.4203	1.4203	1.4203	1.7406	1.7405	1.7401
	5	1.4547	1.4547	1.4547	1.5599	1.5599	1.5601	1.9620	1.9611	1.9549
	10	0.5380	1.5379	1.5373	1.6822	1.6815	1.6749	2.1677	2.1632	2.1402
20	0	0.7482	1.7482	1.7482	2.0396	2.0396	2.0396	2.9060	2.9053	2.9308
	10	2.0761	2.0761	2.0761	2.5830	2.5825	2.6076	3.9226	3.9108	3.9416
	20	2.4369	2.4356	2.4662	3.1812	3.1690	3.1860	5.1189	5.0373	4.9446
30	0	2.2743	2.2743	2.2743	3.0000	3.0000	3.0000	5.0876	5.0844	5.3229
	15	3.1597	3.1597	3.1616	4.7133	4.7083	4.9702	9.0294	8.9195	10.2399
	30	4.4258	4.4117	4.7562	7.2428	7.1021	8.3140	15.4732	14.3971	17.6223
40	0	3.0180	3.0180	3.0180	4.5989	4.5989	4.5989	9.7218	9.7045	10.6598
	20	5.3139	5.3137	5.3380	10.1156	10.0698	11.7720	26.6162	25.4523	68.9161
	40	10.0473	9.8821	15.5175	22.7146	20.9146	67.9121	73.5860	56.5992	560.7383
10	0	0.7148	0.7041	0.7041	1.4213	1.4203	1.4203	3.3385	3.3477	3.4152
	5	0.8305	0.8208	0.8208	1.7475	1.7472	1.7575	4.1555	4.0971	4.2084
	10	0.9785	0.9698	0.9698	2.1399	2.1364	2.1663	5.0775	4.9357	5.1074
20	0	0.8990	0.8830	0.8830	2.0415	2.0396	2.0396	5.7244	5.6759	6.2804
	10	1.3069	1.2931	1.2931	3.3384	3.3320	3.5129	9.4341	9.1039	10.8707
	20	2.0410	2.0396	2.0396	5.2905	5.1909	6.0054	14.9920	13.5283	17.5555
30	0	1.1229	1.0983	1.0983	3.0036	3.0000	3.0000	10.4710	10.3406	13.5474
	15	2.2262	2.2509	2.2509	7.3033	7.2352	8.7517	26.1481	24.0652	48.5650
	30	5.2976	5.2651	5.8777	16.4250	15.0533	30.2558	61.8636	47.0593	147.8946
40	0	1.3950	1.3554	1.3554	4.6066	4.5989	4.5989	21.6380	21.2036	32.6936

(Table 3.26) cont.....

	20	4.6053	4.5989	4.5989	19.9034	19.2927	32.6936	102.1064	84.4295	1469.7900
	40	17.8485	16.8482	41.5915	78.4460	59.3991	1420.2319	493.5247	258.1019	94912.523
10	**0**	1.3623	1.3601	1.3601	1.4204	1.4203	1.4203	1.7114	1.7094	1.7159
	5	1.4550	1.4547	1.4547	1.5469	1.5457	1.5507	1.8730	1.8603	1.8737
	10	1.5283	1.5254	1.5323	1.6353	1.6212	1.6350	1.9984	1.9636	1.9826
20	**0**	1.7544	1.7482	1.7482	2.0401	2.0396	2.0396	2.8058	2.7984	2.8884
	10	2.0762	2.0761	2.0761	2.5219	2.5153	2.5915	3.5164	3.4511	3.6963
	20	2.3870	2.3709	2.4626	2.9487	2.8724	3.1015	4.2156	3.9985	4.4243
30	**0**	2.2878	2.2743	2.2743	3.0012	3.0000	3.0000	4.8007	4.7773	5.3143
	15	3.1568	3.1568	3.1616	4.4705	4.4389	4.9679	7.3738	7.0687	9.6774
	30	4.2063	4.1274	4.7562	6.1881	5.8038	8.1812	10.8034	9.4999	15.9370
40	**0**	3.0463	3.0180	3.0180	4.6019	4.5989	4.5989	8.8706	8.7918	10.6598
	20	5.2830	5.2815	5.3380	9.0834	8.9157	11.7720	18.7419	17.0444	67.6366
	40	8.9657	8.5264	15.5175	16.8084	14.3934	67.7120	39.6483	29.0104	521.7122

Two-Wedge Case = I, Log-Spiral Case = II, Arc-Sandwich Case = III

3.5.3. Lateral Earth Pressure Coefficients by Limit Analysis

The lateral earth pressure coefficients as obtained for the failure mechanism in (Fig. **3.28**) are obtained by Mathcad. These coefficients are very long and tedious, but they are given in the appendix 3-2 for learning purposes. Any reader can program and obtain the critical values easily if the precise equations are available. The results from limit analysis of the lateral earth pressure problems are given in Table 3.26.

Although the true failure mechanism will deviate slightly from a log-spiral mechanism for the effect of self-weight which is illustrated in (Fig. **3.6**), the use of log-spiral mechanism can still provide good solution for normal engineering applications which is illustrated in Table **3.27**. In carrying out limit analysis or even slip line analysis, there is one important point which should be noted. In the evaluation of the individual lateral earth pressure coefficient, it is assumed that other effects are neglected in the computation (direct superposition). For the limit analysis program LEP as shown in the Appendix **3-1** or the slip line program SLIP/KA/KP by Cheng (2003), it is also possible to carry out a global analysis in that the effects due to self-weight of soil, cohesive strength and surcharge can be lumped together. By doing so, the concept of earth pressure coefficients will be absent. Cheng (2002) has demonstrated that the direct superposition of the three effects is a good approximation of the global analysis for normal engineering problems. The computer code as shown in the Appendix 3-1 actually carries out the global analysis besides the evaluation of the individual earth pressure coefficients. Besides slope stability and lateral earth pressure problem, limit analysis has also been used for bearing capacity problem by Chen (1975) which will not be repeated here. The results for the bearing capacity factors by limit analysis are found to be the same as that by limit equilibrium or slip line method for the factors N_c and N_q. For the terms N_γ, the differences between the results from slip line method and limit analysis are also small (Cheng and Au, 2005).

Table 3.27. Comparison of $K_{p\gamma}$ between limit analysis and slip line solutions for a vertical wall with no back slope.

$\phi=\delta$ (°)	LA	Slip Line
10	1.675	1.660
20	3.170	3.052
30	7.102	6.569
40	20.915	18.216

3.6. LIMIT ANALYSIS METHOD - DLO

The limit analysis adopts the concept of an idealized stress-strain relation, i.e., the soil is assumed as a rigid perfectly plastic material with an associated flow rule. Without carrying out the step-by-step elasto-plastic analysis, the limit analysis can provide solutions to many problems. Limit analysis is based on the bound theorems of classical plasticity theory (Drucker *et al.*, 1951; Drucker and Prager, 1952). The general procedure of limit analysis is to assume a kinematically admissible failure mechanism for an upper bound solution or a statically admissible stress field for a lower bound solution, and the objective function will be optimized with respect to the control variables. Early efforts of limit analysis were mainly made on using the direct algebraic method or analytical method to obtain the solutions for slope stability problems with simple geometry and soil profile (Chen, 1975). Since closed form solutions for most practical problems are not available, later attention has been shifted to employing the slice techniques in traditional limit equilibrium to the upper bound limit analysis (Michalowski, 1995; Donald and Chen, 1997).

Limit analysis are based on two theorems: (a) the lower bound theorem, which states that any statically admissible stress field will provide a lower bound estimate of the true collapse load; and (b) the upper bound theorem, which states that when the power dissipated by any kinematically admissible velocity field is equated with the power dissipated by the external loads, then the external loads are upper bounds on the true collapse load (Drucker and Prager, 1952).

A statically admissible stress field is one that satisfies the equilibrium equations, stress boundary conditions, and yield criterion. A kinematically admissible velocity field is one that satisfies strain and velocity compatibility equations, velocity boundary conditions, and the flow rule. When combined, the two theorems provide a rigorous bound on the true collapse load. Application of the lower bound theorem usually proceeds as stated next: (a) First, a statically admissible stress field is constructed. Often it will be a discontinuous field in the sense that we have a patchwork of regions of constant stress that together cover the whole soil mass. There will be one or more particular value of stress that is not fully specified by the conditions of equilibrium. (b) These unknown stresses are then adjusted so that the load on the soil is maximized but the soil remains unyielded. The resulting load becomes the lower bound estimate for the actual collapse load.

Stress fields used in lower bound approaches are often constructed without a clear relation to the real stress fields. Thus, the lower bound solutions for practical geotechnical problems are often difficult to find. Collapse mechanisms used in the upper bound calculations, however, have a distinct physical interpretation associated with actual failure patterns and thus have been extensively used in practice.

3.6.1. Discontinuity Layout Optimization

Discontinuity layout optimization (DLO) is a novel limit analysis based slope stability analysis method which is proposed by Smith and Gilbert (2007). DLO can identify a critical layout of lines of discontinuity, which form a failure, by using the rigorous mathematical optimization techniques. These lines of discontinuity are virtually the slip lines in two-dimensional geotechnical stability problems. DLO can be used to identify critical translational sliding block failure mechanisms which are familiar to the method of slices solution for slope stability problem. This traditional method, however, works only with failure soil mass in a few sliding blocks. DLO can overcome this limitation by the lines of discontinuity which define the boundaries between the rigid moving blocks of material being divided into a large number of sliding blocks. The failure mechanisms are actually the upper bound theory corresponding to a load factor or strength factor. The load factor or strength factor obtained from the trial process for any arbitrarily assumed mechanism is increased if necessary in order to find the exact solution which brings a slope to collapse. Based on this theory, DLO procedure, in essence, can be applied to a wide range of geotechnical stability problems by trial and error process.

In the DLO procedure, the discontinuities are assumed as variables that considers the relative displacement along the discontinuities. By using the relative displacement, compatibility can be checked at each node by applying the simple linear equation involving these variables. Finally, an objective function can be defined in accordance with the energy dissipation along all discontinuities when the failure occurs. A linear optimization problem can then be determined by a linear function of the slip displacement variables.

DLO procedure expresses the limit analysis problem entirely in terms of lines of discontinuity instead of elements as in the classical continuum problem (Smith and Gilbert 2007). Using DLO, a large number of potential discontinuities are set up at different orientations; while the continuum based element formulations, discontinuities are typically restricted to lie only at the edges of elements. With the use of modern optimization algorithms, an optimized solution can be achieved easily. After the initial success by Smith and Gilbert (2007), there are different works in DLO by Clarke *et al.* (2013), Smith and Gilbert (2013), Bauer and Lackner (2015), Al-Defae and Knappett (2015), Leshchinsky (2015), Vahedifard *et al.* (2014), Leshchinsky and Ambauen (2015). The original DLO formulation suffers from the limitation that only the translation mechanism can be considered. In view of such limitation, Gilbert *et al.* (2010) and later Smith and Gilbert (2013) extended the DLO formulation to cover the rotational formulation. Since DLO is actually a numerical form of limit analysis, and the basic limitation of limit analysis is similar to that for DLO.

As discussed above, the compatibility of the displacement of the nodes is important to achieve the potential critical mechanisms. There are three cases to demonstrate the equilibrium and compatibility conditions at a node in (Fig. **3.29**). (Fig. **3.29a**) consider an unloaded node in a truss:

$$\sum_{i=1}^{5} \alpha_i q_i = 0 \tag{3.133}$$

$$\sum_{i=1}^{5} \beta_i q_i = 0 \tag{3.134}$$

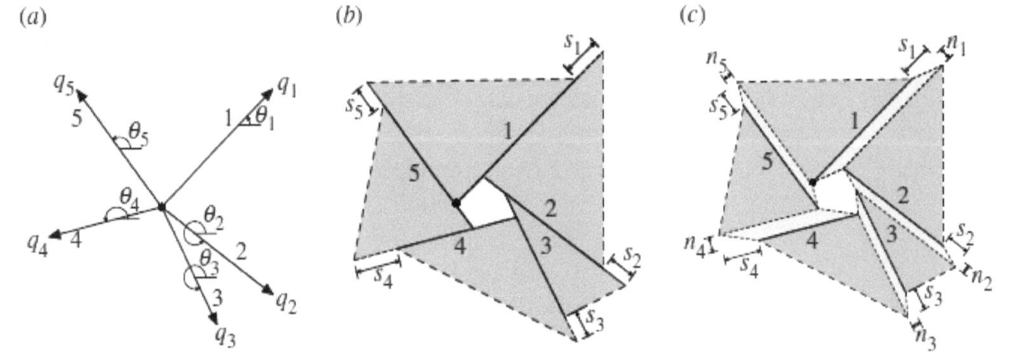

Fig. (3.29). Equilibrium and compatibility conditions at a node: (**a**) Equilibrium at (unloaded node in a truss; (**b**) compatibility at node in slip mechanism (translation only); (**c**) compatibility at node in slip mechanism (translation and dilation) (from Smith and Gilbert, 2007).

where $\alpha_i = \cos\theta_i$ and $\beta_i = \sin\theta_i$. Compatibility at node in slip is considered for translation only in Fig. 3.29b:

$$\sum_{i=1}^{5}\alpha_i s_i = 0 \qquad \textbf{(3.135)}$$

$$\sum_{i=1}^{5}\beta_i s_i = 0 \qquad \textbf{(3.136)}$$

where s_i is the shear displacement for i discontinuity. Compatibility at node in slip considers both the translation and dilation in (Fig. **3.29c**):

$$\sum_{i=1}^{5}\alpha_i s_i - \beta_i n_i = 0 \qquad \textbf{(3.137)}$$

$$\sum_{i=1}^{5}\beta_i s_i + \alpha_i n_i = 0 \qquad \textbf{(3.138)}$$

where n is normal displacement for i discontinuity.

3.6.2. Some Studies on Discontinuity Layout Optimization

The details of the DLO are covered by Smith and Gilbert, and this method is currently available in the program LimitState GEO. Interested readers should consult this commercial program about the operations and the underlying principle. It should be noted that there is only one commercial program for DLO up to the present, and 3D DLO is still not available for practical purposes, due to various technical problems and limitations. A typical output from DLO is given in (Fig. **3.30**) for reference. It should be noted that both the factor of safety and critical slip surface from DLO match very well with the results from LEM or SRM for this case, but the authors have also found that there are many cases for which noticeable differences between the results from DLO and LEM/SRM are found, particularly when some special soil parameters are defined.

With reference to the standard slope stability problem by Cheng *et al.* (2007) as shown in (Fig. **3.31**), 23 cases with different soil parameters are studied. It is noticed that the results from DLO and LEM as shown in Table. **3.28** match very well except for several cases (cases 6, 11, 16, 21, 22, 23) On the other hand, the results from SRM1 (zero dilation angle) and SRM2 (dilation angle=friction angle) match even better with that from the results using LEM. Actually, the authors have carried out many case studies with the DLO and SRM and has found that the DLO at present has more problem cases than SRM, particularly for some special combinations of soil parameters. For normal cases, the differences between DLO and other methods are actually small.

Table 3.28. Comparisons between the results from LEM, SRM and DLO.

Case	Cohesion c' (kPa)	Friction Angle Φ' (degree)	FOS (LEM)	FOS (DLO)	FOS by SRM (SRM1)	FOS by SRM (SRM2)
1	2	5	0.25	0.26	0.25	0.25
2	2	15	0.50	0.51	0.52	0.50
3	2	25	0.74	0.76	0.77	0.75
4	2	35	1.01	1.04	1.07	1.03
5	2	45	1.35	1.40	1.42	1.36
6	5	5	0.41	0.44	0.41	0.41
7	5	15	0.70	0.73	0.72	0.71
8	5	25	0.98	1.00	1.02	0.99
9	5	35	1.28	1.31	1.33	1.30
10	5	45	1.65	1.70	1.71	1.69
11	10	5	0.65	0.71	0.66	0.65
12	10	15	0.98	1.04	1.00	0.995
13	10	25	1.30	1.35	1.34	1.33
14	10	35	1.63	1.70	1.69	1.65

(Table 3.28) cont.....

15	10	45	2.04	2.09	2.09	2.07
16	20	5	1.06	1.22	1.12	1.12
17	20	15	1.48	1.61	1.51	1.50
18	20	25	1.85	1.96	1.87	1.87
19	20	35	2.24	2.34	2.26	2.27
20	20	45	2.69	2.79	2.72	2.74
21	5	0	0.20	0.24	0.22	0.22
22	10	0	0.40	0.47	0.44	0.44
23	20	0	0.80	0.95	0.88	0.88

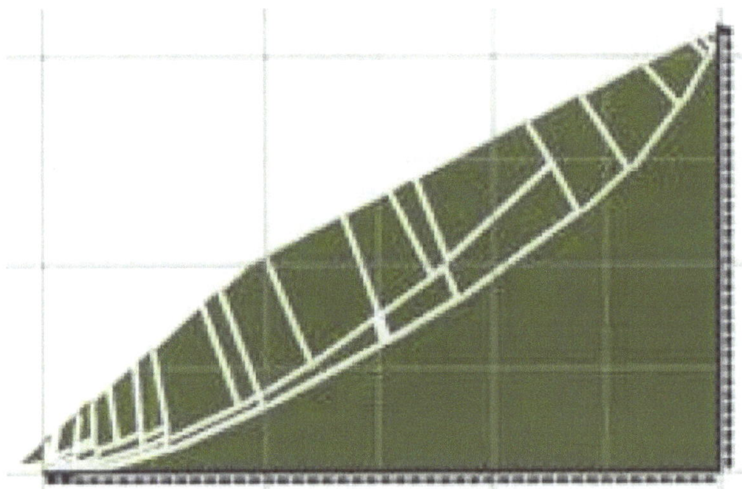

Fig. (3.30). An output from DLO analysis.

Yu *et al.* (1998) gave a very detailed comparison between the use of limit analysis and LEM, and it is found that the results from the two methods are similar and comparable in most cases for relatively simple problems. Recently, DLO has been adopted for slope stability analysis by Leshchinsky and Ambauen (2015), and it is found that the results by DLO and LEM are comparable in general. Leshchinsky and Ambauen (2015) have however found some cases for which there are noticeable differences between the DLO and LEM, and they have concluded that DLO requires less assumption on the location of collapse, and therefore may be more preferable than LEM, especially for complex, yet realistic geotechnical problems. After reviewing the examples by Leshchinsky and Ambauen (2015), the authors tend to disagree with the results and comments by Leshchinsky and Ambauen (2015). There are some limitations in the works by Leshchinsky and Ambauen (2015) which include: 1) use of classical LEM method which are greatly affected by convergence problem (Cheng *et al.* 2008, Cheng *et al.* 2010); 2) critical failure surface has not been determined (Fig. **12** from Leshchinsky and Ambauen (2015) has only considered 151 surfaces); 3) interslice force function can be critical in complex problems. As discussed by Cheng (2003), Cheng *et al.* (2010) and Cheng *et al.* (2013a), these three problems can lead to relatively poor solution in some cases by the classical LEM, and the extremum principle by Cheng *et al.* (2010), Cheng *et al.* (2011b) and Cheng *et al.* (2013a) can overcome these problems and can provide solutions similar to some classical plasticity problems which are not possible with the classical LEM. A fair comparison and commentary on these methods must be based on reliable and robust analyses that identify the differences between DLO and LEM.

With reference to (Fig. **3.32**) which is Fig. **5a** by Leshchinsky and Ambauen (2015), the soil parameters are unit weight=19 kN/m^3, c'=28 kPa and ϕ'=20°. The critical result by DLO pass below the toe of the slope at the right-hand side of (Fig. **3.32**) by Leshchinsky and Ambauen (2015). On the other hand, the critical result by the authors using the heuristic optimization method and Spencer method developed by Cheng *et al.* (2007a) and Cheng and Lau (2014) pass through the toe of the slope. The critical result by Baker (1980) also pass through the toe of the slope while the critical result by Krahn and Fredlund (1997) (not shown for clarity) is similar to that by DLO but extends further to the right of the toe. The result by Krahn and Fredlund (1997) is not determined by the use of advanced optimization algorithm, and the adequacy of the result has not been confirmed. The authors have tried several updated commercial programs and have obtained results similar to that by the authors as shown in (Fig. **3.32**). Since the friction angle of the soil is 20° which is not a small value, the critical result by limit analysis should pass through the toe of slope as demonstrated by Chen (1975) using limit analysis. In views of the above discussion, the authors will suggest that the result by DLO is not the critical solution for such a simple case which is surprising to the authors.

With reference to (Fig. **3.33**) where there is a 0.5m thickness of soft material for soil layer 2, the soil parameters are unit weight=19 kN/m^3, c'=28 kPa and ϕ'=20° for layer 1 and unit weight=19 kN/m^3, c'=0 kPa and ϕ'=10° for soil layer 2. The results by DLO and the authors are very similar except that the critical result by the authors lies at the bottom of the soft layer while the results by Leshchinsky and Ambauen (2015) lie at the top of the soft layer. The critical results by Baker (1980) and Krahn and Fredlund (1997) are also at the bottom of the soft band but are mistaken to be at the top of the soft band by Leshchinsky and Ambauen (2015). The authors have reduced the thickness of the soft layer to 1mm, and the factor of safety as well as the critical result will then be equal to that by Leshchinsky and Ambauen (2015), Baker (1980), and Krahn and Fredlund (1997).

When the authors increase the thickness of the soft layer to 1.5m, the critical result will still lie at the bottom of the soft layer with a factor of safety 1.14. Since the shear strength parameters at soil layer 2 are low, the weight of the soil tends to push the soft material to the right so that the critical slip surface should lie within the soft band, and the results by the authors are more reasonable as compared with other results. As discussed by Cheng (2007) and Cheng *et al.* (2012), the presence of a soft band is mathematically equivalent to a Dirac function, for which many optimization algorithms fail to work. The domain transformation technique by Cheng (2007) and the coupled optimization algorithm by Cheng *et al.* (2012) effectively overcome this problem without any special precaution required by the engineers in the analysis. In (Fig. **3.34**) which is same as that for (Fig. **3.33**) with a pore pressure ratio 0.25 (Fig. 5d by Leshchinsky and Ambauen, (2015), the critical results by Leshchinsky and Ambauen (2015) lies at the top of the soft band while the critical results by the authors, Baker (1980), and Krahn and Fredlund (1997) (mistaken to be at the top of the soft band by Leshchinsky and Ambauen (2015) lie at the bottom of the soft band, and the inability to locate the critical result for a soft band by DLO is clearly illustrated. In (Fig. **3.35**) which is same as that for (Fig. **3.33**) with a prescribed water table (Fig. **5f** by Leshchinsky and Ambauen, (2015), the critical result again lie at the bottom of the soft band which are different the critical result by Leshchinsky and Ambauen (2015) (again Baker 1980 and Krahn and Fredlund (1997) also get critical result at bottom of soft band but are mistaken to be top of soft band by Leshchinsky and Ambauen 2015).

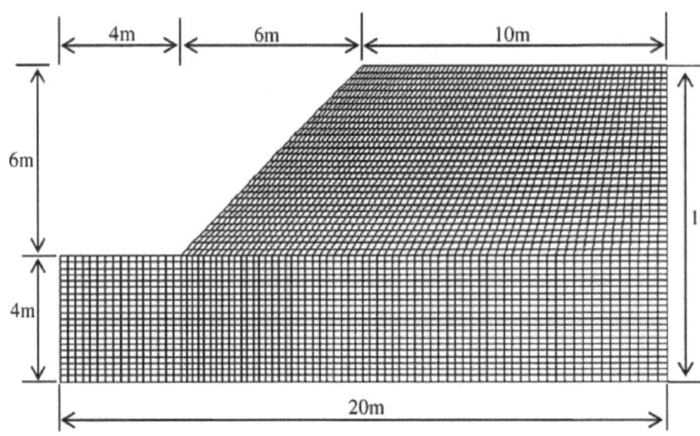

Fig. (3.31). Geometry of a standard problem by Cheng *et al.* (2007b).

In (Fig. **3.36**) (Fig. **12** by Leshchinsky and Ambauen 2015), there are great differences between the critical result by the authors and Leshchinsky and Ambauen (2015). The soil parameters are unit weight=20 kN/m^3, c'=0 kPa and ϕ'=30° for soil layer 1, unit weight=19 kN/m^3, c'=0 kPa and ϕ'=45° for soil layer

2 and unit weight=19 kN/m^3, c'=10 kPa and ϕ'=0° for soil layer 3. On the left-hand side of the critical slip surface by Leshchinsky and Ambauen (2015), there is a very sudden change in the slope of the critical failure surface which seems unlikely to happen. At the right-hand side of the critical slip surface by Leshchinsky and Ambauen (2015), the critical slip surface is nearly vertical, which is also highly unlikely, as the friction angle of soil layer 1 and 2 are 30° and 40° respectively with zero cohesive strength. When this same slip surface by Leshchinsky and Ambauen (2015) is considered with the M-P method using f(x)=sin(x), the authors actually gets a factor of safety of 1.05, which is significantly greater than the result of 0.95 by Leshchinsky and Ambauen (2015).

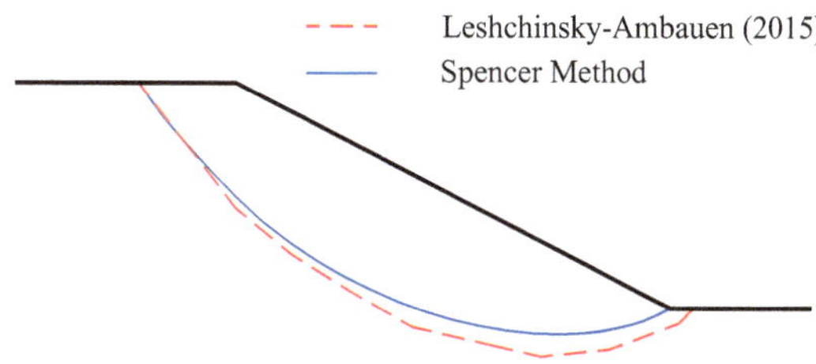

Fig. (3.32). Comparisons of DLO and LEM for Fig. **5a** by Leshchinsky and Ambauen (2015), FOS=2.03 by DLO and 1.94 by Spencer method.

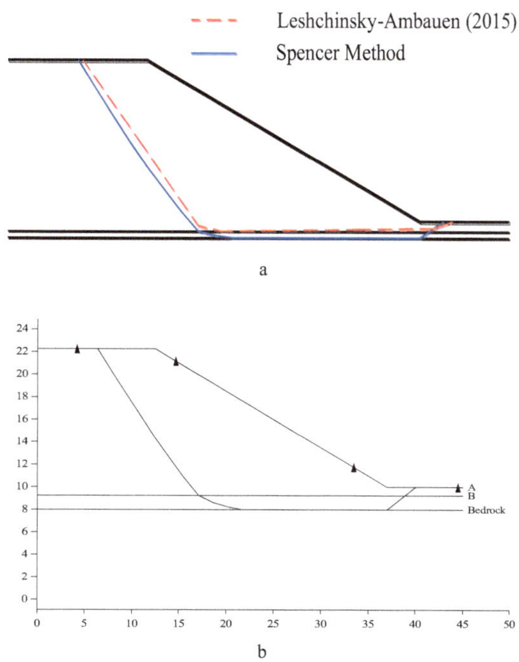

Fig. (3.33). Comparisons of DLO and LEM for Fig. **5b** by Leshchinsky and Ambauen (2015), FOS=1.28 by DLO and 1.21 by Spencer method. The thickness of the soft band has been increased to 1.25m for the result in Fig. **2b** with FOS=1.14.

This problem is then reanalyzed by the authors using LEM to locate the critical slip surface. For this problem, the use of f(x)=1 is poor in convergence, and the authors get a slightly different critical slip surface and a factor of safety of 0.97 by using f(x)=1.0. As mentioned by Cheng *et al.* (2008, 2010), f(x) can be critical in some cases which will affect the optimized solution. In this respect, the authors have also adopted the extremum principle (Cheng *et al.* 2010, 20110, 2013) and have obtained a critical solution 0.915 which is close to that by using f(x)=sin(x). Since the extremum principle, which is practically equivalent to the lower bound approach, satisfies all the force and moment equilibrium with acceptable internal forces and is free from convergence problem, this result can be considered as a very good estimation of the critical result for the present problem. The authors view that the critical result by Leshchinsky and Ambauen (2015) is possibly a local minimum instead of being the global minimum.

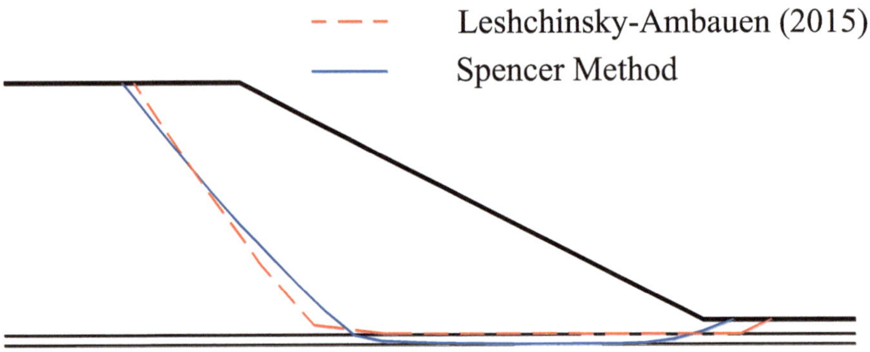

Fig. (3.34). Comparisons of DLO and LEM for Fig. **5d** by Leshchinsky and Ambauen (2015), FOS=1.01 by DLO and 0.99 by Spencer method.

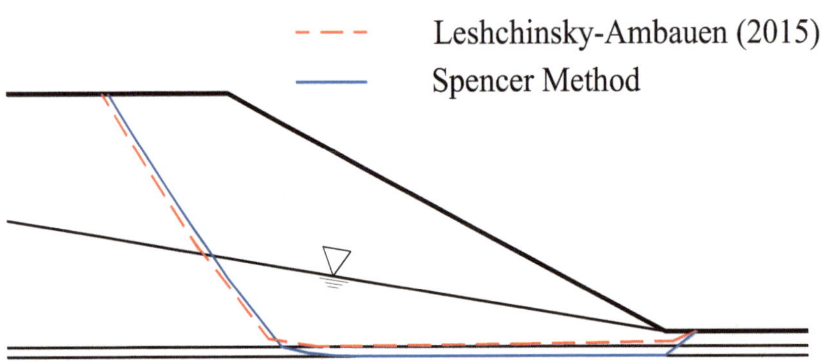

Fig. (3.35). Comparisons of DLO and LEM for Fig. **5f** by Leshchinsky and Ambauen (2015), FOS=1.14 by DLO and 1.07 by Spencer method.

The authors have adopted an accuracy of 0.001 in all the global optimization search in the present study, and the global minima of each example has been tested with different optimization algorithms for confirmation. Based on the above case studies, it can be concluded that some of the past reported results in literature which are not optimized with the modern optimization algorithms may not be reliable enough for comparisons. In particular, for the presence of a soft band which is a difficult problem, the present study and the works by Cheng (2007b), Cheng *et al.* (2012) demonstrated that great care must be taken in order to obtain a good result. Furthermore, as a relatively new computational method, DLO has been demonstrated to be affected by the soft band or local minima problem. Overall, the authors view that the problems presented in this section are not fundamental deficiencies of DLO. Instead, they highlight the limitations of the numerical technique in implementing the DLO up to the present moment. With refined and improved numerical technique coupled with DLO, the authors expect that better results will be produced by DLO in the future. On the other hand, it is dangerous to compare the advantages and limitations of different stability methods based on old results or computer programs which may have various limitations. Some of the comments in previous literature are possibly distorted by the limitations of the computational technique in computer programs instead of being the actual comparisons of different stability analysis methods.

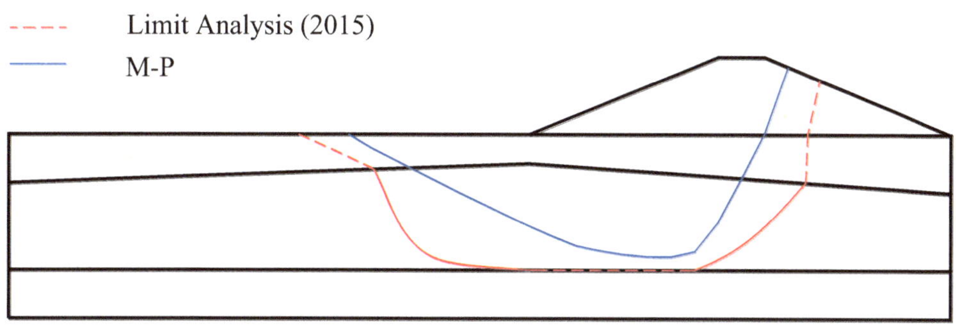

Fig. (3.36). Comparisons of DLO and LEM for Fig. **12** by Leshchinsky and Ambauen (2015), FOS=0.95 by DLO and 1.0 by Spencer method by Leshchinsky and Ambauen (2015) using 151 trial surface, 0.92 by M-P method using f(x)=sin(x) and 0.97 by f(x)=1.0 by the authors using about 20000 trials with simulated annealing optimization method.

For the problem with a soft band at soil layer 2 as discussed by Cheng *et al.* (2007b), surprising results are again obtained by DLO. The unit weight of the soils are 19 kN/m^3, and c'=20 kPa and ϕ'=35° for soil layer 1, c'=0 kPa and ϕ'=25° for soil layer 2 and c'=10 kPa and ϕ'=35° for soil layer 3. As discussed by Cheng *et al.* (2007a), it appears that some SRM programs are affected by the size of the solution domain. The factor of safety for LEM is obtained as 0.927 by the Spencer method by Cheng *et al.* (2007b), and this value lie within the SRM1 and SRM2 results by Plaxis and Phase 8.0. On the other hand, the factor of safety appears to be highly dependent on the nodal number adopted in the analysis. Even if 2000 nodal number is adopted, the factor of safety from DLO still appears to be unsatisfactory which is given in Table **3.29** The results by DLO are higher than those by LEM or SRM under all cases in Table. **3.29**, and the differences are not minor. Surprisingly, the critical failure surface from DLO as shown in (Fig. **3.38**) is similar to that by LEM or SRM (Cheng *et al.* 2007b). From Table **3.29**, it can be concluded that the most influential factor in a proper DLO analysis is the nodal number.

Table 3.29. FOS for different nodal number and tolerance of analysis (c'=0, ϕ'=25° for the soft band) for the problem in Fig. (3.37).

28m Domain Size, Solution Tolerance 0.01, Different Nodal Density				
Case	**Nodal No.**	**FOS by DLO**	**FOS by LEM**	**FOS Difference with LEM (DLO %)**
1	250	1.356	0.927	-46.28
2	500	1.069	0.927	-15.32
3	1000	1.082	0.927	-16.72
4	2000	1.055	0.927	-13.81
28m Domain Size, Nodal Density 500, Different Solution Tolerance				
Case	**Solution** Tolerance	**FOS by DLO**	**FOS by LEM**	**FOS Difference with LEM (DLO %)**
1	0.01	1.069	0.927	-15.32
2	0.001	1.069	0.927	-15.32
3	0.004	1.069	0.927	-15.32
4	0.005	1.069	0.927	-15.32
Solution Tolerance 0.01, Nodal Density 500, Different Domain Size				
Case	**Domain Size (m)**	**FOS by DLO**	**FOS by LEM**	**FOS Difference with LEM (DLO %)**
1	28	1.069	0.927	-15.32
2	20	1.093	0.927	-17.91
3	12	1.025	0.927	-10.57

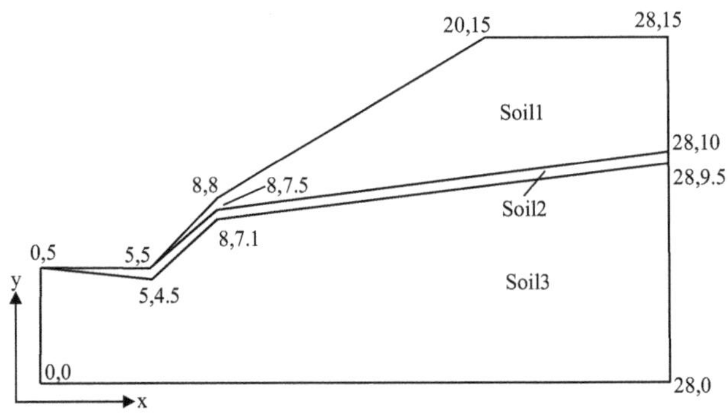

Fig. (3.37). A slope problem with a soft band as discussed by Cheng *et al.* (2007b).

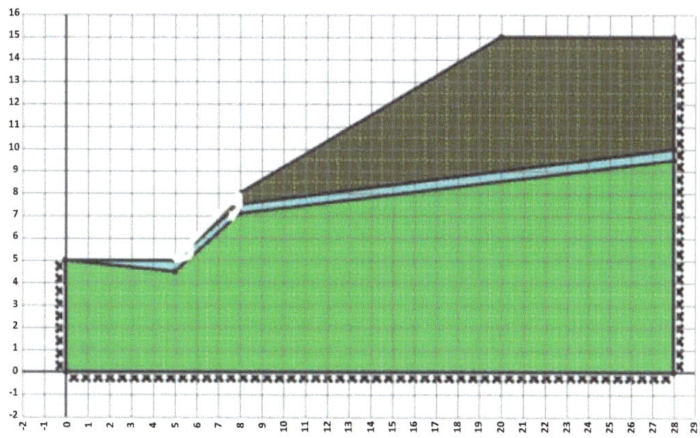

Fig. (3.38). Critical failure surface from DLO for the problem in Fig. (3.37).

If the third layer of soil instead of the second layer of soil is a soft material, the factor of safety has been established to be 1.29 from Spencer method, 1.27 from Extremum principle and 1.33 for SRM2 for all the SRM programs as discussed by Cheng *et al.* (2007b), and f(x) is relatively important for the present case (in general f(x) is not negligible if the friction angle is low). On the other hand, the result by DLO will approach the above factor of safety when the nodal number is large enough (Table **3.30**). However, while the critical failure surfaces from LEM and SRM agree quite well as shown in (Fig. **3.39a** and **3.39b**), the critical failure surface from DLO extends further to the right in (Fig. **3.39c**). To further examine these results, the authors have found another local minimum 1.29 with the Spencer method for the failure surface as shown in (Fig. **3.40**), which is very similar to that one by DLO as shown in (Fig. **3.39**). The authors view that a local minimum has been obtained from the DLO analysis. It should also be notes that the failure surfaces in (Fig. **3.39a** and **3.40**) bear virtually the same factors of safety, and the differences between the two values are small so that it can be viewed that there are two global minimum for this problem. LEM can analyzed such problem easily while it takes more effort for SRM to detect the result in (Fig. **3.40**). Actually, without the previous knowledge about the existence of another global minimum, engineers will miss the result in (Fig. **3.39**) easily. For DLO, the failure surface as given by (Fig. **3.39a** or **3.39b**) cannot be obtained even increasing nodal number as given in Table **3.30**. In this respect, there are some inherent limitation in the present development of DLO.

Table 3.30. DLO, SRM and LEM analysis for the problem in Fig. (3.37), when the third layer of soil has low shear strength.

DLO Analysis of Soft Soil Layer 3 with Different Nodal Density (28 Domain, Solution Tolerance 0.01)						
Case	**Nodal No.**	**FOS by DLO**	**FOS by SRM2**	**FOS by LEM**	**FOS Difference with LEM(DLO%)**	**FOS Difference with LEM(SRM2%)**
1	250	1.405	1.33	1.27	-8.91	-3.10
2	500	1.358	1.33	1.27	-5.27	-3.10
3	1000	1.35	1.33	1.27	-4.65	-3.10

3.7. OVERALL DISCUSSION

As demonstrated in this chapter, stability of a geotechnical system can be evaluated by means of slip line method, limit equilibrium method and limit analysis. There are no major differences between the results from the three different methods if the problem is properly considered. In fact, the authors have also illustrated that the three classical stability problems can be unified under the ultimate condition, and each problem can be viewed as the inverse of the other problem, depending on the angle of view of the engineer. Slip line method and limit analysis are however suitable only for simple problems with regular geometry and homogeneous soil condition. On the other hand, limit equilibrium method is an approximate lower bound method which relies on the assumption of an interslice force function. To evaluate this function, the extremum approach by Cheng *et al.* (2010) can be adopted which is a time-consuming computation. Luckily, for most of the practical problems, the classical limit equilibrium method can quickly provide a solution which is accurate and good enough for normal engineering applications. This method is also easily understood by most of the engineers and is well adapted to problems with irregular geometry and nonhomogeneous condition. It is not surprising that while no commercial program is available for slip line method or limit analysis, there are many commercial programs targeted towards different applications using the limit equilibrium method.

The most critical difference between the three different methods is the application in three-dimensional condition. Currently, there is no application of the slip line method (exclude axi-symmetric problem) and very few successful applications of the limit analysis for three-dimensional application (Cheng *et al.* 2015a). The extension of these two methods to three-dimensional applications is very difficult, and Cheng and Zhu (2002) proposed an improved three-dimensional Bishop's method (1955). On the other hand, the extension of limit equilibrium method to three-dimensional is relatively easy, and the most critical problem of three-dimensional limit equilibrium formulation is the sliding direction which is solved effectively by Cheng and Yip (2007). The general application of three-dimensional limit equilibrium method is mainly limited by the difficulty in defining a precise three-dimensional ground condition (expensive and time-consuming) and the search for critical three-dimensional solution. For the latter limitation, it is limited by the lack of a general equation which can approximate all the possible three-dimensional slip surfaces with limited control variables. Cheng *et al.* (2005) proposed the use of NURBS function for such purpose, but even this function will not be able to model all the possible real world three-dimensional failure surfaces. In this respect, the finite element method will be a simpler solution for three-dimensional stability analysis.

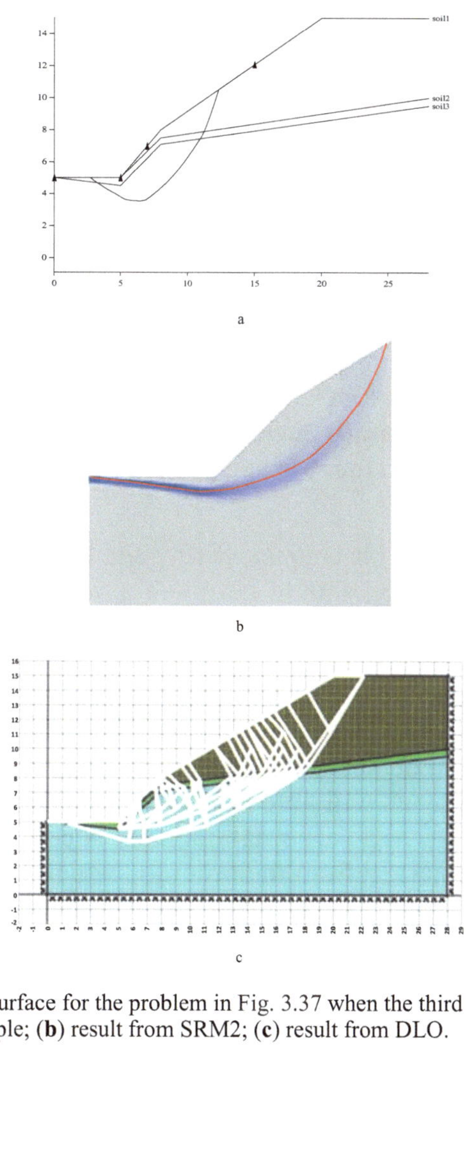

Fig. (3.39). Critical failure surface for the problem in Fig. 3.37 when the third layer of soil is soft material (**a**) result from Extremum principle; (**b**) result from SRM2; (**c**) result from DLO.

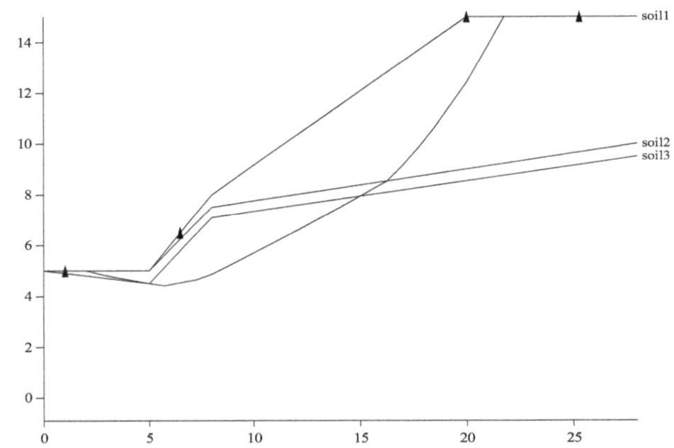

Fig. (3.40). Another LEM global minimum for the problem in Fig. 3.37, when the third layer of soil is soft material.

APPENDIX 3-1 - LATERAL EARTH PRESSURE SLIP LINE PROGRAM KA

The program KA is developed by Cheng (2003) for the analysis of general lateral earth pressure problem with earthquake. In this program, a very fine grid is defined for the ground level (which can be non-horizontal) behind the retaining wall. In general, the authors choose a value of 1mm to 0.1mm grid size, which is denoted by variable c2. The weight of soil can be arbitrary if only the lateral earth pressure coefficients are required. It should be noted that eq. (3.112) or eq. (3.133) assume direct superposition of the effects from weight of soil as well as that from cohesive strength and surcharge. Such direct superposition is strictly not exact, though the error is small is on the safe side for practical purpose.

In applying the programs in this chapter, if Kaγ and Kpγ are to be determined, the surcharge qload to be applied should be given a very small value to avoid the boundary stress to be 0 which may sometimes create numerical problem. As long as qload is kept to be very small, the results will be adequate. After defining the friction angle of soil phi, the wall friction delta, a5 (90° for a vertical retaining wall), set cohesive strength of soil cohe to be 0. The lateral earth pressure on the face of retaining wall will be determined by the solution as given previously. This lateral pressure divided by γh will give Kaγ. If Kac and Kpc are to be determined, set weigh = 0 and qload to be a very small number. A constant but negative lateral earth pressure will be obtained from the program, which should be equal to -cK_{ac}. For Kaq and Kpq, set qload to 1 or other value and weigh and cohe to 0. The constant lateral stress from the program will be equal to qKaq. The readers should note that this approach to determine the lateral earth pressure coefficients apply for both the slip line method and the limit analysis method. Furthermore, both the limit analysis and slip line programs can work for the case of global analysis, where qload, cohe and weigh are not 0. In such case, the lateral stress coefficients will be meaningless, as the coupled effect due to the weight of soil will distort the slip line field. Unless the weight of soil is very large (far beyond the normal unit weight of soil), the coupling effect is small, and the direct superposition can be used without sacrificing the accuracy of the solution. The readers can use the program to test for the coupling effect if interested.

```
! This program is used to evaluate active pressure coefficient on
! the retaining wall with the inclination of surcharge q and
! show the points along the boundary slipline
!
!
! INPUT:
!
! qload : uniformly distributed load q ... ...(kPa)
! weigh : unit weight of soil (gama) ... ...(kN/m3)
! phi : angle of friction of soil ... ...(degree)
! delta : angle of friction of retaining wall ... ...(degree)
! a5 : angle between horizontal line and the
! retaining wall ... ...(degree)
! a6 : no. of division in transition zone
! a7 : angle of inclination of backfill to the
! horizontal line ... ...(degree)
! c1 : no. of division on the boundary line of backfill
! c2 : separation of division in c1
```

```
! cohe : cohesion of soil ... ...(kN/m2)
!
!
! OUTPUT:
!
! ARRAY OF X,Y,M,N and P
!
! x and y are used to show the position of each point
! m is the stress measuring from centre of Mohr's Circle to
! the point that tangent of Mohr's circle intersect with
! the axis of abscissae
! n is angle between the major principal stress to the X axis
! p is the actual active pressure acting on the retaining wall
!
PROGRAM ACTIVE
implicit none
integer :: k1,k2,c1,r,i,ii,jj,j,c4,s,a6,count,counter
INTEGER :: ierror
real (kind=8) :: c2,qload,weigh,phi,delta,a5,a7,l1,l2,l3,d3,e3,f3,g3, &
& pi,cohe,ns,ss,nn,tt
real (kind=8) :: b1,b2,b3,b4,b5,d1,e1,f1,g1,d2,e2,f2,g2, &
h1,h2,h3,h4,h5,h6,h7,h8,beta,beta1
real (kind=8), allocatable :: x(:,:),y(:,:),sxx(:,:),sxy(:,:),syy(:,:),sv(:,:), &
m(:,:),n(:,:),p(:,:)
```

```
pi=3.1415926535
write (*,*) " Active earth pressure determination by solution of two partial "
write(*,*) "differential equations (slip lines). This program "
write (*,*) " can totally replace the 200 pages book of earth pressure"
write (*,*) " coefficients by Kerisel and Absi, Active and Passive Earth Pressure"
write (*,*) " Table, Balkema, 1990.
write (*,*)
write (*,*) " New feature : Earthquake loading is added on 30/7/2002"
write (*,*) " Dr. Y.M. Cheng, Department of Civil and Environmental Engineering"
write (*,*) " Hong Kong Polytechnic University, 30/7/2002"
write (*,*)
write (*,*)
write (*,*) 'Phi of soil = '
read (*,*) phi
write (*,*) 'Wall friction = '
read (*,*) delta
write (*,*) 'Angle of slope behind retaining wall = '
read (*,*) a7
write (*,*) 'Earthquake coefficient = '
read (*,*) beta
!!!!! set a5 to other angle if the wall is not vertical !!!!!
weigh=40. ; a5=90. ; a6=100 ; c1=200 ; c2=0.001d0
cohe = 0. ; counter=2 ; qload=0.01d0
```

```
! counter can set to more than 2 for more iterations and better accuracy
! cohe, weigh can be set to any values, weigh
! a6, c1 can be set to any value
! c2 denote the grid spacing. A fine spacing is suggested
! qload is the surcharge behind retaining wall, can be set to a large value to
! determine the lateral earth pressure coefficient due to surchage
! read (5,*) qload,weigh,phi,delta,a5,a6,a7,c1,c2,cohe,counter,beta
! Input File name : active
! Output file name : a_rwall
! : a_bline
!
k2 = c1 ; k1=a6+k2+c1 ! k1 and k2 are used for counting
b1=pi*phi/180. ; b2=pi*delta/180.
b3=pi*a5/180. ; b4=pi/4.-b1/2.
b5=pi*a7/180. ; beta1=beta
beta=(dsin(b5)+beta*dcos(b5))/(dcos(b5)-beta*dsin(b5))
beta=datan(beta)
!
! b1 : angle of friction of soil ... ... radian
! b2 : angle of friction of retaining wall ... ... radian
! b3 : angle between horizontal line and the retaining wall ... ... radian
! b4 : angle between slipline and major principal stress ... ... radian
! b5 : angle of inclination of backfill to the horizontal line ... ... radian
!
```

```
allocate (x(0:k1,0:k2),y(0:k1,0:k2),m(0:k1,0:k2),n(0:k1,0:k2),p(0:k1,0:k2))
allocate (sxx(0:k1,0:k2),syy(0:k1,0:k2),sxy(0:k1,0:k2),sv(0:k1,0:k2))
do 450 j = 0, k2
do 440 i = 0, k1
if (i > c1) go to 190
if (i < c1-j) go to 440
if (i == c1-j) go to 140
go to 420

140 x(i,j)=(c1-i)*c2
y(i,j)=0.
! Boundary condition if the ground behind retaining wall is not horizontal
if (cohe==0) then
ns=qload*dcos(b5)*dcos(b5)
else
ns=qload*dcos(b5)*dcos(b5)+cohe/dtan(b1)
endif
ss=qload*dsin(b5)*dcos(b5)
h1=((ns*dsin(b1))**2.)-((ss*dcos(b1))**2.)
if (h1 < 0.) then
m(i,j)=ns/(dcos(b1)**2.)
n(i,j)=(datan(ss/(ns-m(i,j))))/2.+pi/2.
else
m(i,j)=(ns-(h1**0.5))/(dcos(b1)**2.)
```

```
n(i,j)=(datan(ss/(ns-m(i,j))))/2.+pi/2.
ENDif
go to 440
!
! 140 ... ... calculate all the points on boundary line of backfill
!
! ns : normal stress acting on the plane of backfill due to q
! ss : shear stress acting on the plane of backfill due to q
!

190 c4=c1+a6
if (i > c4) go to 300
if (j > 0) go to 420
if (abs(phi-delta) < 0.001) then
l2=pi/2.
else
l1=dsin(b2)/dsin(b1)
l2=dasin(l1)
end if
l3=b3-(-l2+b2)/2.+b5-n(c1,0)
n(i,j)=n(c1,0)+(i-c1)*l3/a6
m(i,j)=m(c1,0)*dexp((2*n(c1,0)-2*n(i,j))*dtan(b1))
x(i,j)=0. ; y(i,j)=0.
go to 440
```

```
!
! 190 ... ... calculate m & n of point O (0,0) in each division in transition zone
!

300 if (i > c4+j) go to 440
if (i == c4+j) go to 330
go to 420

330 d3=x(i-1,j) ; e3=y(i-1,j)
f3=m(i-1,j) ; g3=n(i-1,j)
if (dabs(b3+b5-pi/2.) < 0.001) then
x(i,j)=0.
y(i,j)=e3-d3*dtan(g3+b4)
else
x(i,j)=(e3-d3*dtan(g3+b4))/(dtan(b3+b5)-dtan(g3+b4))
y(i,j)=x(i,j)*dtan(b3+b5)
endif
n(i,j)=b3+(l2-b2)/2.+b5
h6=weigh/dcos(b1)*(dsin(-beta+b1)*(x(i,j)-d3)+(y(i,j)-e3)*dcos(-beta+b1))
h7=f3-2.*f3*(n(i,j)-g3)*dtan(b1)
m(i,j)=h6+h7
go to 440
!
! 330 ... ... calculate all points on the retaining wall
!
```

```
420 r=j-1
d1=x(i,r) ; e1=y(i,r)
f1=m(i,r) ; g1=n(i,r) ; s=i-1
d2=x(s,j) ; e2=y(s,j) ; f2=m(s,j) ; g2=n(s,j)
count=0
11 h1=d1*dtan(g1-b4)-d2*dtan(g2+b4)+e2-e1
h2=dtan(g1-b4)-dtan(g2+b4)
x(i,j)=h1/h2
y(i,j)=(x(i,j)-d1)*dtan(g1-b4)+e1
h3=weigh/dcos(b1)*(dsin(-beta-b1)*(x(i,j)-d1)+(y(i,j)-e1)*dcos(-beta-b1))
h4=weigh/dcos(b1)*(dsin(-beta+b1)*(x(i,j)-d2)+(y(i,j)-e2)*dcos(-beta+b1))
h5=-m(i,r)+m(s,j)+2.*dtan(b1)*(f1*n(i,r)+f2*n(s,j))
h8=2.*(f1+f2)*dtan(b1)
n(i,j)=(h4-h3+h5)/h8
h6=weigh/dcos(b1)*(dsin(-beta-b1)*(x(i,j)-d1)+(y(i,j)-e1)*dcos(-beta-b1))
h7=m(i,r)+2.*f1*(n(i,j)-n(i,r))*dtan(b1)
m(i,j)=h6+h7
g1=(n(i,r)+n(i,j))/2. ; g2=(n(s,j)+n(i,j))/2.
f1=(m(i,r)+m(i,j))/2. ; f2=(m(s,j)+m(i,j))/2.
count=count+1
if (count > counter) go to 440
go to 11
!
! 420 ... ... calculate all points within boundary lines
```

```
!

440 continue
450 continue
!
! calculate the stress
!
do 491 j = 0, k2
do 481 i = 0, k1
sv(i,j)=m(i,j)*dsin(b1)+cohe*dcos(b1)
sxx(i,j)=m(i,j)+sv(i,j)*dcos(2*n(i,j))
syy(i,j)=m(i,j)-sv(i,j)*dcos(2*n(i,J))
sxy(i,j)=sv(i,j)*dsin(2*n(i,j))
481 continue
491 continue

!
! Remark: x, y axis are rotated with the plane of backfill in the above calculation
!
beta=dsqrt(1+beta1*beta1)
do j = 1,k2
i=a6+c1+j
nn=x(i,j)
tt=y(i,j)
x(i,j)=nn*dcos(b5)+tt*dsin(b5)
```

```
y(i,j)=tt*dcos(b5)-nn*dsin(b5)
n(i,j)=n(i,j)-b5
if (b2 < 0.001) then
h1=m(i,j)*(1-dsin(b1))
else
h1=m(i,j)*dsin(l2-b2)/dsin(l2)
end if
if (cohe==0) then
h2=h1*dcos(b2)
else
h2=h1*dcos(b2)-cohe/dtan(b1)
END if
h3=h1*dsin(b2)
p(i,j)=((h2**2.)+(h3**2.))**(0.5)
p(i,j)=p(i,j)*beta
end do
h1=p(a6+k2+c1,k2)/y(a6+k2+c1,k2)/40.
write (*,'(a,f5.3)') 'ka= ',h1
!
! Print out all the points on retaining wall in output file A_RWALL
!
open (UNIT=20 , FILE='pa.dat' , STATUS='new' , ACTION='write' ,
IOSTAT=ierror)
write (20,*) " list of relation between y and mean pressure and coefficient"
```

```
write (20,*) " y pa coefficient"
do j=1,k2
i=c4+j
if (b2 < 0.001) then
h1=m(i,j)*(1-dsin(b1))
else
h1=m(i,j)*dsin(l2-b2)/dsin(l2)
end if
if (cohe==0) then
h2=h1*dcos(b2)
else
h2=h1*dcos(b2)-cohe/dtan(b1)
END if
h3=h1*dsin(b2)
p(i,j)=((h2**2.)+(h3**2.))**(0.5)
h4=p((a6+k2+j),j)/y((a6+k2+j),j)/weigh
write (20,111) y(i,j) , p(i,j) , h4
111 FORMAT(1x, 3F10.3)
end do
close (UNIT=20)
!
! Print out all the points on retaining wall in output file A_RWALL
!
do 471 j = 0, k2
```

```
do 461 i = 0, k1

if (i < k2-j) then

ii=c1-j

x(i,j)=x(ii,j) ; y(i,j)=y(ii,j)

endif

if (i > c4+j) then

ii=c4+j

x(i,j)=x(ii,j) ; y(i,j)=y(ii,j)

endif

461 continue

471 continue

open (UNIT=21 , FILE='pa-bxy.dat' , STATUS='new' , ACTION='write' ,
IOSTAT=ierror)

write (21,*) " list of coordinates x and y and stress in x and y direction "

write (21,*) " along the b-slip line "

write (21,*) " ia ib x y sxx syy sxy"

do 472 i = 0, k1

do 462 j=0, k2,10

write (21,112) i, j, x(i,j),y(i,j), sxx(i,j), syy(i,j), sxy(i,j)

112 FORMAT(1x, I5,5x,I5,2X, 5f10.3)

462 continue

472 continue

close (UNIT=21)
```

```
do 475 j = 0, k2
do 465 i = 0, k1
if (i < c1-j) then
jj=c1-i
x(i,j)=x(i,jj) ; y(i,j)=y(i,jj)
endif
if (i > c4+j) then
jj=i-c4
x(i,j)=x(i,jj) ; y(i,j)=y(i,jj)
endif
465 continue
475 continue

open (UNIT=22 , FILE='pa-axy.dat' , STATUS='new' , ACTION='write' ,
IOSTAT=ierror)
write (22,*) " list of coordinates x and y and stress in x and y direction "
write (22,*) " along the a-slip line"
write (22,*) " ib ia x y sxx syy sxy"
do 473 j = k2, 0, -1
do 463 i=0, k1,10
write (22,113) j,i,x(i,j),y(i,j),sxx(i,j),syy(i,j),sxy(i,j)
113 FORMAT(1x,I5,5x,I5,2X,5f10.3)
463 continue
473 continue
close (UNIT=22)
```

```
open (UNIT=23, FILE='pa-xyst.dat' , STATUS='replace' , ACTION='write' ,
IOSTAT=ierror)

write (23,*) " list of coordinates x and y and mean pressure and angle of slipline"

write (23,*) " ia ib x y sigma Theta"

do 474 j = 0, k2

do 464 i = 0, k1

if (i < c1-j) go to 464

if (i > c4+j) go to 464

write (23,114) i,j,x(i,j),y(i,j),m(i,j),n(i,j)

114 FORMAT (1x,I5,5x,I5,2X,4f10.3)

464 continue

WRITE (23,115)

115 format (' ')

474 continue

close (UNIT=23)

open (UNIT=24 , FILE='pa-sxy.dat' , STATUS='replace' , ACTION='write' ,
IOSTAT=ierror)

write (24,*) " list of coordinates x and y and stres in the nodes"

write (24,*) " ia ib x y sxx syy sxy"

do 478 j = 0, k2

do 468 i = 0, k1

if (i < c1-j) go to 468

if (i > c4+j) go to 468

write (24,116) i,j,x(i,j),y(i,j),sxx(i,j),syy(i,j),sxy(i,j)

116 FORMAT (1x,I5,5x,I5,2X,5f10.3)
```

```
468 continue
WRITE(24,117)
117 format (' ')
478 continue
close (UNIT=24)
!
open (UNIT=25 , FILE='pa-face.dat' , STATUS='replace' , ACTION='write' ,
IOSTAT=ierror)
write (25,*) " list of coordinates x and y at the surface node"
write (25,*) " ia ib x y sxx syy sxy"
do 476 j = 0, k2
do 466 i = 0, k1
if (i == c4+j) then
write (25,118) i,j,x(i,j),y(i,j),sxx(i,j), syy(i,j), sxy(i,j)
118 FORMAT (1x,I5,5x,I5,2X,5f10.3)
ENDif
466 continue
476 continue
close (UNIT=25)
do i = 0,k1
j=k2
nn=x(i,j)
tt=y(i,j)
x(i,j)=nn*dcos(b5)+tt*dsin(b5)
```

```
y(i,j)=tt*dcos(b5)-nn*dsin(b5)

n(i,j)=n(i,j)-b5

!

! convert x,y back to the horizontal and vertical plane

!


end do

!

! Print out all the points on boundary slipline in outpout file a_BLINE

!

END PROGRAM ACTIVE
```

The typical output from KA is given below for illustration and study purpose.

Referring to the output along the retaining wall, the stresses and coordinates of the end points of the slip line are given below. Based on these stresses, the equivalent earth pressure coefficients can be computed easily.

list of coordinates x and y at the surface node

ia ib x y sxx syy sxy
300 0 0.000 0.000 0.004 0.011 -0.001
301 1 0.000 0.001 0.023 0.065 -0.008
302 2 0.000 0.003 0.044 0.123 -0.014
303 3 0.000 0.004 0.065 0.180 -0.021
304 4 0.000 0.005 0.086 0.238 -0.028
305 5 0.000 0.006 0.106 0.295 -0.034
306 6 0.000 0.008 0.127 0.352 -0.041
307 7 0.000 0.009 0.147 0.409 -0.048
308 8 0.000 0.010 0.168 0.466 -0.054
309 9 0.000 0.012 0.188 0.523 -0.061

Cont.....

310 10 0.000 0.013 0.209 0.580 -0.067
311 11 0.000 0.014 0.230 0.636 -0.074
312 12 0.000 0.015 0.250 0.693 -0.081
313 13 0.000 0.017 0.271 0.750 -0.087
314 14 0.000 0.018 0.291 0.807 -0.094
315 15 0.000 0.019 0.312 0.864 -0.101
316 16 0.000 0.021 0.332 0.921 -0.107
317 17 0.000 0.022 0.353 0.978 -0.114
318 18 0.000 0.023 0.373 1.035 -0.120
319 19 0.000 0.025 0.394 1.092 -0.127
320 20 0.000 0.026 0.414 1.148 -0.134
321 21 0.000 0.027 0.435 1.205 -0.140
322 22 0.000 0.028 0.455 1.262 -0.147
323 23 0.000 0.030 0.476 1.319 -0.153
324 24 0.000 0.031 0.496 1.376 -0.160
325 25 0.000 0.032 0.517 1.433 -0.167
326 26 0.000 0.034 0.537 1.490 -0.173
327 27 0.000 0.035 0.558 1.547 -0.180
328 28 0.000 0.036 0.578 1.603 -0.187
329 29 0.000 0.037 0.599 1.660 -0.193
330 30 0.000 0.039 0.619 1.717 -0.200
331 31 0.000 0.040 0.640 1.774 -0.206
332 32 0.000 0.041 0.660 1.831 -0.213
333 33 0.000 0.043 0.681 1.888 -0.220
334 34 0.000 0.044 0.701 1.945 -0.226
335 35 0.000 0.045 0.722 2.001 -0.233
336 36 0.000 0.046 0.742 2.058 -0.239
337 37 0.000 0.048 0.763 2.115 -0.246
338 38 0.000 0.049 0.783 2.172 -0.253

For p and θ along the slip lines, the output are given below.

list of coordinates x and y and mean pressure and angle of slipline

ia ib x y sigma Theta

Cont.....

200 0 0.000 0.000 0.007 1.835
201 0 0.000 0.000 0.007 1.835
202 0 0.000 0.000 0.007 1.834
203 0 0.000 0.000 0.007 1.833
204 0 0.000 0.000 0.007 1.832
205 0 0.000 0.000 0.007 1.831
206 0 0.000 0.000 0.007 1.830
207 0 0.000 0.000 0.007 1.829
208 0 0.000 0.000 0.007 1.828
209 0 0.000 0.000 0.007 1.827
210 0 0.000 0.000 0.007 1.826
211 0 0.000 0.000 0.007 1.825
212 0 0.000 0.000 0.007 1.825
213 0 0.000 0.000 0.007 1.824
214 0 0.000 0.000 0.007 1.823
215 0 0.000 0.000 0.007 1.822
216 0 0.000 0.000 0.007 1.821
217 0 0.000 0.000 0.007 1.820
218 0 0.000 0.000 0.007 1.819
219 0 0.000 0.000 0.007 1.818
220 0 0.000 0.000 0.007 1.817
221 0 0.000 0.000 0.007 1.816
222 0 0.000 0.000 0.007 1.816
223 0 0.000 0.000 0.007 1.815
224 0 0.000 0.000 0.007 1.814
225 0 0.000 0.000 0.007 1.813
226 0 0.000 0.000 0.007 1.812
227 0 0.000 0.000 0.007 1.811
228 0 0.000 0.000 0.007 1.810
229 0 0.000 0.000 0.007 1.809
230 0 0.000 0.000 0.007 1.808
231 0 0.000 0.000 0.007 1.807
232 0 0.000 0.000 0.007 1.807
233 0 0.000 0.000 0.007 1.806
234 0 0.000 0.000 0.007 1.805

Cont.....

235 0 0.000 0.000 0.007 1.804
236 0 0.000 0.000 0.007 1.803
237 0 0.000 0.000 0.007 1.802
338 38 0.000 0.049 0.783 2.172 -0.253

The output for the α and β lines are given for illustration.

list of coordinates x and y and stress in x and y direction along the α-slip line

ib ia x y sxx syy sxy
200 0 0.200 0.000 0.009 0.005 0.003
200 10 0.190 0.005 0.220 0.132 0.076
200 20 0.180 0.011 0.430 0.258 0.149
200 30 0.170 0.016 0.641 0.384 0.222
200 40 0.159 0.022 0.851 0.511 0.295
200 50 0.149 0.027 1.062 0.637 0.368
200 60 0.139 0.033 1.272 0.763 0.441
200 70 0.129 0.038 1.483 0.890 0.514
200 80 0.119 0.044 1.694 1.016 0.587
200 90 0.109 0.049 1.904 1.142 0.660
200 100 0.098 0.055 2.115 1.269 0.733
200 110 0.088 0.060 2.325 1.395 0.806
200 120 0.078 0.066 2.536 1.522 0.878
200 130 0.068 0.071 2.746 1.648 0.951
200 140 0.058 0.077 2.957 1.774 1.024
200 150 0.048 0.082 3.168 1.901 1.097
200 160 0.038 0.088 3.378 2.027 1.170

list of coordinates x and y and stress in x and y direction along the β-slip line

ia ib x y sxx syy sxy
111 90 0.089 0.001 0.030 0.018 0.010
111 100 0.089 0.006 0.241 0.144 0.083
111 110 0.089 0.012 0.451 0.271 0.156
111 120 0.089 0.017 0.662 0.397 0.229

Cont.....

111 130 0.088 0.023 0.872 0.523 0.302
111 140 0.088 0.028 1.083 0.650 0.375
111 150 0.088 0.034 1.293 0.776 0.448
111 160 0.088 0.039 1.504 0.902 0.521
111 170 0.088 0.044 1.715 1.029 0.594
111 180 0.088 0.050 1.925 1.155 0.667
111 190 0.087 0.055 2.136 1.281 0.740
111 200 0.087 0.061 2.346 1.408 0.813
112 0 0.000 0.000 0.000 0.000 0.000
112 10 0.010 0.000 0.000 0.000 0.000
112 20 0.020 0.000 0.000 0.000 0.000
112 30 0.030 0.000 0.000 0.000 0.000
112 40 0.040 0.000 0.000 0.000 0.000

With minor modifications, the current program can be modified to the passive condition easily. The authors have noticed an interesting result by using the slip line program. For the passive pressure coefficient due to the self weight of soil, Euro code provides simplified formula for the normal component of the active and passive conditions as follows:

$$Ka = \frac{[1-\sin\phi\sin(2m_w-\phi)]}{1+\sin\phi\sin(2m_t-\phi)}e^{-2(m_t+\beta-m_w-\theta)\tan\phi} \quad \textbf{(A3.1)}$$

$$Kp = \frac{[1+\sin\phi\sin(2m_w+\phi)]}{1-\sin\phi\sin(2m_t+\phi)}e^{2(m_t+\beta-m_w-\theta)\tan\phi} \quad \textbf{(A3.2)}$$

$$\text{where } 2m_t = \cos^{-1}\left(\frac{-\sin\beta}{\pm\sin\phi}\right)\mp\phi-\beta \text{ and } 2m_w = \cos^{-1}\left(\frac{\sin\delta}{\sin\phi}\right)\mp\phi\mp\delta \text{ for passive and active} \quad \textbf{(A3.3)}$$

For a vertical retaining wall, using the parameters ϕ=35°, δ=30 or 35°, β=0, the normal passive pressure coefficients are given as 7.054 and 7.250 respectively. However, using the slip line method, the normal passive pressure coefficients are given as 8.114 and 8.605 respectively. Greater difference will be found when β is greater than 0. The authors have found that this is due to the use of a log-spiral curve which is actually distorted by the action of the self weight.

APPENDIX 3-2 - PROGRAM LEP FOR LIMIT ANALYSIS

This program runs under command prompt, while a version with graphics support can be obtained from the authors. The program accept wall and ground slope with any inclination and different friction angle and wall friction for the three failure mechanism as shown in Fig.(3.29). The operation of this program is interactive, and there is not any need to explain the operation of this program. The equations for the lateral earth pressure coefficients are given at the end of this chapter, and the results from limit analysis are usually close to that from slip line analysis. The active pressure coefficients and the passive pressure coefficients from limit analysis are slightly greater and smaller than those from slip line analysis, respectively. The small difference between the two results are generally negligible, but the authors prefer the use of slip line method in general, as it provides the stresses inside the failure mass, which is useful for many other uses. Limit analysis provides only the global results without the internal stresses. It should also be noted that based on the present program, the results are generally close to those by Chen (1975), and generally smaller Kp will be given by the present program as compared with that by Chen (1975).

In determining the critical solutions, the authors have adopted a highly inefficient pattern search method, with three levels of looping of the three controlling angles. The authors have developed another version using the gradient type optimization method, but have noticed that the solution may be trapped by the presence of local minima. To deal with this situation, the authors have generated several random seeds as the initial solution, and gradient type optimization is then applied. The most critical solution from all the different initial solution will then be taken as the global optimized solution. The structure of the program will then be more complex, but there is no major gain in the efficiency of computation. For such a simple problem, the authors view that a simple method of optimization may be the best method.

```
module methodcase

implicit none

integer :: case, file

end module methodcase


module reportdatatype
```

```
implicit none

real(kind=8) :: alpha, beta, phi, delta, h, cohesion, density, surcharge

real(kind=8), dimension(4) :: maxka, minkp, maxrol, maxneta, maxomega, maxsine

real(kind=8) :: dp

end module reportdatatype

module graphdatatype

implicit none

real(kind=4), dimension(6) :: x_0, y_0

real(kind=4),dimension(107)::x_1,y_1,x_2,y_2,x_3,y_3,x_4, y_4

real(kind=8), dimension(4) :: sl

real(kind=4), dimension(2) :: x_scale, y_scale

end module graphdatatype

program LEP

use methodcase

use reportdatatype

use graphdatatype

implicit none

real(kind=8) :: a, b, p, d

real(kind=8), parameter :: pi = 3.14159265358979323846_8

integer :: num, a_or_p

real(kind=8), dimension(8) :: chk

real(kind=8) :: ka1, ka2, ka3, ka4, ka5

real(kind=8) :: kp1, kp1a, kp2, kp3, kp4, kp5
```

```
real(kind=8), dimension(4) :: ka, kp
real(kind=8) :: r, n, o, s
integer(kind=4) :: rol_loop, neta_loop, omega_loop, sine_loop
real(kind=8) :: rol, neta, omega, sine
real(kind=8) :: dsine, r0, r1
integer(kind=4) :: i, j, x, y, z
character(1) :: continue

write(*,'(22x,a)') 'Lateral Earth Pressure Program'
write(*,'(22x,a)') '------------------------------'
write(*,*)
write(*,'(3x,a)') 'Select One Method : ( 1, 2 or 3 )'
write(*,'(3x,a)') '1. Two-Wedge Failure Method.'
write(*,'(3x,a)') '2. Log-Spiral Failure Method.'
write(*,'(3x,a)') '3. Arc-Sandwich Failure Method.'
do
write(*,'(3x,a)',advance='no') '( 1, 2 or 3 ) : '
read(*,*) num
if ((num > 0) .and. (num < 4)) exit
write(*,'(2x,a)') 'Invalid Input !'
end do
write(*,*)
write(*,'(3x,a)') 'Select Active Case or Passive Case : (1. Active Case 2.Passive Case)'
do
```

```
write(*,'(3x,a)',advance='no') '( 1 or 2 ) : '
read(*,*) a_or_p
if ((a_or_p > 0) .and. (a_or_p < 3)) exit
write(*,'(2x,a)') 'Invalid Input !'
end do
case = num*a_or_p
if ((num == 2) .and. (a_or_P == 1)) case = 3
if ((num == 3) .and. (a_or_P == 1)) case = 5
write(*,*)
do
write(*,'(3x,a)',advance='no') 'Angle of Wall (Alpha) [degree] : '
read (*,*) alpha
if ((alpha > 0) .and. (alpha < 180)) exit
write(*,'(2x,a)') 'Alpha should be larger than 0 and smaller than 180.'
end do
do
write(*,'(2x,a)',advance='no') 'Slope Angle (Beta) [degree] : '
read (*,*) beta
if ((beta > -90) .and. (beta < 90)) exit
write(*,'(2x,a)') 'Beta should be larger than -90 and smaller than 90.'
write(*,*)
write(*,'(a)',advance='no')
end do
do
```

```
write(*,'(2x,a)',advance='no') 'Shear Strength (Phi) [degree] : '
read (*,*) phi
if (((phi >= 45) .or. (phi <= 0)) .and. ((case == 5) .or. (case == 6))) then
write(*,'(2x,a)') 'In Arc-Sandwich Mechanism, Phi Should Line between 0 to 45 degrees.'
write(*,*)
write(*,'(a)',advance='no')
else
exit
end if
end do
write(*,'(2x,a)',advance='no') 'Wall Friction (Delta) [degree] : '
read (*,*) delta
do
write(*,'(2x,a)',advance='no') 'Height of Wall (H) [m] : '
read (*,*) h
if (h > 0) exit
write(*,'(2x,a)') 'Height of Wall Should be Larger Than 0.'
write(*,*)
write(*,'(a)',advance='no')
end do
write(*,'(2x,a)',advance='no') 'Cohesion of Soil (c) [kPa] : '
read (*,*) cohesion
write(*,'(2x,a)',advance='no') 'Density of Soil (Gamma) [kN/m sq.] : '
```

```
read (*,*) density

write(*,'(2x,a)',advance='no') 'Surcharge Loading (q) [kPa] : '

read (*,*) surcharge

dp=0.1

write(*,*)

write(*,'(3x,a)',advance='no') 'Press Enter to Continue, or Press N Button to Renew the Data : '

read(*,'(a)') continue


a = alpha*pi/180 ; b = beta*pi/180 ; p = phi*pi/180 ; d = delta*pi/180

if (alpha <=90) then

x_0(1) = 0 ; y_0(1) = real(h) ; x_0(2) = 0.5

y_0(2) = real(h) ; x_0(3) = 100

y_0(3) = y_0(2) + (x_0(3) - x_0(2))*dtan(b)

x_0(4) = 0.5 ; y_0(4) = real(h) ; y_0(5) = 0

x_0(5) = x_0(4) + (y_0(5) - y_0(4))/(-dtan(a))

x_0(6) = 0 ; y_0(6) = 0

else

x_0(6) = 0 ; y_0(6) = 0 ; x_0(5) = 0.5

y_0(5) = 0 ; y_0(4) = real(h) ; x_0(4) = x_0(5) + (y_0(4) - y_0(5))/(-dtan(a))

x_0(3) = 100 ; y_0(3) = y_0(4) + (x_0(3) - x_0(4))*(dtan(b))

x_0(2) = x_0(4) ; y_0(2) = y_0(4)

y_0(1) = real(h) ; x_0(1) = 0

end if
```

```
x_1(1) = x_0(5) ; y_1(1) = y_0(5) ; x_1(3) = x_0(2)
y_1(3) = y_0(2) ; x_2(1) = x_0(5) ; y_2(1) = y_0(5)
x_2(3) = x_0(2) ; y_2(3) = y_0(2) ; x_3(1) = x_0(5)
y_3(1) = y_0(5) ; x_3(3) = x_0(2) ; y_3(3) = y_0(2)
x_4(1) = x_0(5) ; y_4(1) = y_0(5) ; x_4(3) = x_0(2)
y_4(3) = y_0(2)
x_scale(1) = 0. ; x_scale(2) = real(h)*3 ; y_scale(1) = -real(h)/5
y_scale(2) = real(h)*9/5
do i = 1 , 4
maxka(i) = -99999999
minkp(i) = 99999999
end do
file = 1

SELECT CASE (case)
case (1)

rol_loop = (alpha+beta)/dp
do x = 1, int(rol_loop)+1
if (x-1 < rol_loop) then
rol = (x-1)*dp
else
rol = (alpha+beta)*dp
end if
R = rol*pi/180
```

```
neta_loop = (180-alpha+phi)/dp
do y = 1, int(neta_loop)+1
if (y-1 < neta_loop) then
neta = -phi + (y-1)*dp
else
neta = 180-alpha
end if
N = neta*pi/180
omega_loop = (90-neta)/dp
do z = 1, int(omega_loop)+1
if (z-1 < omega_loop) then
omega = neta + (z-1)*dp
else
omega = 90
end if
O = omega*pi/180
CHK(1) = Alpha - Delta
CHK(2) = Neta - Phi
CHK(3) = 180 - (Alpha - Delta - Phi + Neta)
CHK(4) = Omega + Neta
CHK(5) = Alpha + Neta - Rol - 2 * Phi
CHK(6) = 180 - (Alpha - Rol + Omega - 2 * Phi)
CHK(7) = Alpha - Rol + Neta
CHK(8) = Omega - Beta
```

```
i = 1

Do

if (i > 8) exit

If ((CHK(i) <= 0.00001) .Or. CHK(i) >= 180) Then

Go To 50

end if

i = i + 1

end do

KA1 = dsin(A - D) * dsin(A + N) * dsin(N - P) * dsin(R) / (dsin(N + A - D - P) * (dsin(A)) ** 2 * dsin(A + N - R))

KA2 = (dsin(A + N)) ** 2 * dsin(A - R + B) * dsin(A - R + O) * dsin(O - P) * dsin(A + N - R - 2 * P) * dsin(A - D)

KA3 = (dsin(A)) ** 2 * (dsin(A + N - R)) ** 2 * dsin(O - B) * dsin(A + N - D - P) * dsin(A - R + O - 2 * P)

KA4 = (dsin(A) - dtan(D) * dcos(A)) * dcos(D)

Ka(1) = (KA1 + KA2 / KA3) / KA4

KA1 = dsin(A - D) * dcos(P) * dsin(R) / (dsin(A + N - D - P) * dsin(A) * dsin(A + N - R))

KA2 = dsin(A - D) * dsin(O - N) * dcos(P) * dsin(A + N) / (dsin(A + N - D - P) * dsin(A - R + O - 2 * P) * dsin(A) * dsin(A + N - R))

KA3 = dsin(A-D)*dsin(A+N-R-2*P)*dcos(P)*dsin(A+N)*dsi-(A-R+B)/(dsin(A+N-D-P)*dsin(A-R+O-2*P)*dsin(A)*dsin(A+N-R)*dsin(O-B))

KA4 = (-dsin(A) + dtan(D) * dcos(A)) * dcos(D)

Ka(2) = (KA1 + KA2 + KA3) / KA4

KA1 = dsin(A - D) * dsin(A + N - R - 2 * P) * dsin(A - R + O) * dsin(A + N) * dsin(O - P)

KA2 = dsin(N + A - D - P) * dsin(A - R + O - 2 * P) * dsin(A) * dsin(A + N - R)
```

```
* dsin(O - B)
KA3 = (dsin(A) - dtan(D) * dcos(A)) * dcos(D)
Ka(3) = (KA1 / KA2 / KA3)
do i = 1, 3
If (MAXKA(i) < Ka(i)) Then
MAXKA(i) = Ka(i)
MAXRol(i) = Rol
MAXNeta(i) = Neta
MAXOmega(i) = Omega
End If
end do
Ka(4) = 0.5 * Ka(1) * Density * H ** 2 + Cohesion * Ka(2) * H + Ka(3) *
Surcharge * H
If (MAXKA(4) < Ka(4)) Then
MAXKA(4) = Ka(4)
MAXRol(4) = Rol
MAXNeta(4) = Neta
MAXOmega(4) = Omega
End If
50 end do
end do
end do
MAXKA(2) = (MAXKA(2) /(-2)) ** 2
do i = 1, 4
```

```
sl(4) = dtan(b) ; sl(1) = -dtan((alpha - maxrol(i))*pi/180)

sl(2) = dtan(maxneta(i)*pi/180) ; sl(3) = dtan(maxomega(i)*pi/180)

SELECT CASE (i)

CASE (1)

do j = 0, 1

x_1(3*j+2)= ((y_1(j+1)-x_1(j+1)*sl(j+2))-(y_1(3)-x_1(3)*sl(3*j+1)))/(sl(3*j-1)-sl(j+2))

y_1(3*j+2) = sl(3*j+1)*x_1(3*j+2)+(y_1(3)-x_1(3)*sl(3*j+1))

end do

x_1(4) = x_1(2) ; y_1(4) = y_1(2)

CASE (2)

do j = 0, 1

x_2(3*j+2)= ((y_2(j+1)-x_2(j+1)*sl(j+2))-(y_2(3)-x_2(3)*sl(3*j+1)))/(sl(3*j-1)-sl(j+2))

y_2(3*j+2) = sl(3*j+1)*x_2(3*j+2)+(y_2(3)-x_2(3)*sl(3*j+1))

end do

x_2(4) = x_2(2) ; y_2(4) = y_2(2)

CASE (3)

do j = 0, 1

x_3(3*j+2)= ((y_3(j+1)-x_3(j+1)*sl(j+2))-(y_3(3)-x_3(3)*sl(3*j+1)))/(sl(3*j-1)-sl(j+2))

y_3(3*j+2) = sl(3*j+1)*x_3(3*j+2)+(y_3(3)-x_3(3)*sl(3*j+1))

end do

x_3(4) = x_3(2) ; y_3(4) = y_3(2)

CASE (4)
```

```
do j = 0, 1

x_4(3*j+2)=   ((y_4(j+1)-x_4(j+1)*sl(j+2))-(y_4(3)-x_4(3)*sl(3*j+1)))/(sl(3*j-1)-sl(j+2))

y_4(3*j+2) = sl(3*j+1)*x_4(3*j+2)+(y_4(3)-x_4(3)*sl(3*j+1))

end do

x_4(4) = x_4(2) ; y_4(4) = y_4(2)

end select

end do

call report()

CASE (2)

rol_loop = (alpha+beta)/dp

do x = 1, int(rol_loop)+1

if (x-1 < rol_loop) then

rol = (x-1)*dp

else

rol = alpha+beta

end if

R = rol*pi/180

neta_loop = (180-alpha+phi)/dp

do y = 1, int(neta_loop)+1

if (y-1 < neta_loop) then

neta = -phi + (y-1)*dp

else
```

```
neta = 180-alpha
end if
N = neta*pi/180
omega_loop = (90-neta)/dp
do z = 1, int(omega_loop)+1
if (z-1 < omega_loop) then
omega = neta + (z-1)*dp
else
omega = 90
end if
O = omega*pi/180
CHK(1) = Delta + Alpha
CHK(2) = Neta + Phi
CHK(3) = 180 - (Alpha + Delta + Phi + Neta)
CHK(4) = Omega - Neta
CHK(5) = Alpha + Neta - Rol + 2 * Phi
CHK(6) = 180 - (Alpha - Rol + Omega + 2 * Phi)
CHK(7) = Alpha - Rol + Neta
CHK(8) = Omega - Beta
I = 1
Do
IF (I > 8) EXIT
If (CHK(I) <= 0.00001 .Or. CHK(I) >= 180) Then
Go To 60
```

```
END IF

I = I + 1

END DO

KP1 = dsin(A + D) * dsin(A + N) * dsin(N + P) * dsin(R)

KP1A = (dsin(N + A + D + P) * (dsin(A))**2 * dsin(A + N - R))

KP2 = (dsin(A + N))**2 * dsin(A - R + B) * dsin(A - R + O) * dsin(O + P) *
dsin(A + N - R + 2 * P) * dsin(A + D)

KP3 = (dsin(A))**2 * (dsin(A + N - R))**2 * dsin(O - B) * dsin(A + N + D + P)
* dsin(A - R + O + 2 * P)

KP4 = (dsin(A) + dtan(D) * dcos(A)) * dcos(D)

Kp(1) = (KP1 / KP1A + KP2 / KP3) / KP4

KP1 = dsin(A + D) * dcos(P) * dsin(R) / (dsin(A + N + D + P) * dsin(A) * dsin(A
+ N - R))

KP2 = dsin(A + D) * dsin(O - N) * dcos(P) * dsin(A + N) / (dsin(A + N + D + P)
* dsin(A - R + O + 2 * P) * dsin(A) * dsin(A + N - R))

KP3 = dsin(A+D)*dsin(A+N-R+2*P)*dcos(P)*dsin(A+N)*dsin(-
-R+B)/(dsin(A+N+D+P)*dsin(A-R+O+2*P)*dsin(A)*dsin(A+N-R)*dsin(O-B))

KP4 = (dsin(A) + dtan(D) * dcos(A)) * dcos(D)

Kp(2) = (KP1 + KP2 + KP3) / KP4

KP1 = dsin(A + D) * dsin(A + N - R + 2 * P) * dsin(A - R + O) * dsin(A + N) *
dsin(O + P)

KP2 = dsin(N + A + D + P) * dsin(A - R + O + 2 * P) * dsin(A) * dsin(A + N - R)
* dsin(O - B)

KP3 = (dsin(A) + dtan(D) * dcos(A)) * dcos(D)

Kp(3) = (KP1 / KP2 / KP3)

DO I = 1, 3

If (MINKP(I) > Kp(I)) Then
```

```
MINKP(I) = Kp(I) ; MAXRol(I) = Rol

MAXNeta(I) = Neta ; MAXOmega(I) = Omega

End If

END DO

Kp(4) = 0.5 * Kp(1) * Density * H**2 + Cohesion * Kp(2) * H + Kp(3) * Surcharge * H

If (MINKP(4) > Kp(4)) Then

MINKP(4) = Kp(4) ; MAXRol(4) = Rol

MAXNeta(4) = Neta ; MAXOmega(4) = Omega

End If

60 END DO

END DO

END DO

MINKP(2) = (MINKP(2) /(-2))**2

do i = 1, 4

sl(4) = dtan(b) ; sl(1) = -dtan((alpha - maxrol(i))*pi/180)

sl(2) = dtan(maxneta(i)*pi/180) ; sl(3) = dtan(maxomega(i)*pi/180)

SELECT CASE (i)

CASE (1)

do j = 0, 1

x_1(3*j+2)= ((y_1(j+1)-x_1(j+1)*sl(j+2))-(y_1(3)-x_1(3)*sl(3*j+1)))/(sl(3*j-1)-sl(j+2))

y_1(3*j+2) = sl(3*j+1)*x_1(3*j+2)+(y_1(3)-x_1(3)*sl(3*j+1))

end do

x_1(4) = x_1(2) ; y_1(4) = y_1(2)
```

```
CASE (2)

do j = 0, 1

x_2(3*j+2)=    ((y_2(j+1)-x_2(j+1)*sl(j+2))-(y_2(3)-x_2(3)*sl(3*j+1)))/(sl(3*j-
1)-sl(j+2))

y_2(3*j+2) = sl(3*j+1)*x_2(3*j+2)+(y_2(3)-x_2(3)*sl(3*j+1))

end do

x_2(4) = x_2(2) ; y_2(4) = y_2(2)

CASE (3)

do j = 0, 1

x_3(3*j+2)=    ((y_3(j+1)-x_3(j+1)*sl(j+2))-(y_3(3)-x_3(3)*sl(3*j+1)))/(sl(3*j-
1)-sl(j+2))

y_3(3*j+2) = sl(3*j+1)*x_3(3*j+2)+(y_3(3)-x_3(3)*sl(3*j+1))

end do

x_3(4) = x_3(2) ; y_3(4) = y_3(2)

CASE (4)

do j = 0, 1

x_4(3*j+2)=    ((y_4(j+1)-x_4(j+1)*sl(j+2))-(y_4(3)-x_4(3)*sl(3*j+1)))/(sl(3*j-
1)-sl(j+2))

y_4(3*j+2) = sl(3*j+1)*x_4(3*j+2)+(y_4(3)-x_4(3)*sl(3*j+1))

end do

x_4(4) = x_4(2) ; y_4(4) = y_4(2)

end select

end do

call report()

CASE (3)
```

```
rol_loop = (alpha+beta)/dp
do x = 1, int(rol_loop)+1
if (x-1 < rol_loop) then
rol = (x-1)*dp
else
rol = alpha+beta
end if
R = rol*pi/180 ; sine_loop = (alpha+beta-rol)/dp
do y = 1, int(sine_loop)+1
if (y-1 < sine_loop) then
sine = (y-1)*dp
else
sine = alpha+beta-rol
end if
s = sine*pi/180
CHK(1) = Alpha – Delta ; CHK(2) = 90 - (Alpha - Rol)
CHK(3) = 90 - (Rol - Delta) ; CHK(4) = 90 - (Alpha - Rol - Sine - Phi + Beta)
CHK(5) = Alpha + Beta - Rol – Sine ; CHK(6) = 90 - (Rol + Phi)
i = 1
do
if (i > 6) exit
If ((CHK(i) <= 0.00001) .Or. (CHK(i) >= 180)) then
go to 70
end if
```

```
i = i + 1

end do

KA1 = dsin(A - D) * dcos(R + P) / dcos(R - D) / (dsin(A)) ** 2 / dcos(P)

KA2 = dsin(R) * dcos(A - R)

KA3= dcos(R+P)*(dcos(A-R)*(3*dtan(P)+((-3)*dtan(P)*dcos(s)+dsin(s))*Exp-
(-3) &

*s*dtan(P)))+dsin(A-R)*(1+((-3)*dtan(P)*dsin(s)-dcos(s))*Ex-
((-3)*s*dtan(P))))/dcos(P)/(1+9*(dtan(P))**2)

KA4 = dcos(R + P) * dsin(A - R - s + B) * dcos(A - R - s) * Exp((-3) * s *
dtan(P)) / dcos(A - R - s + B - P)

KA5 = (dsin(A) - dtan(D) * dcos(A)) * dcos(D)

Ka(1) = KA1 * (KA2 + KA3 + KA4) / KA5

KA1 = dsin(A - D) * dsin(R) / dsin(A) / dcos(R - D)

KA2 = (dsin(A - D) * dcos(R + P) * dsin(A - R - s + B) * Exp((-2) * s * dtan(P)))
/ (dcos(R - D) * dsin(A) * dcos(A - R - s - P + B))

KA3 = dsin(A - D) * dcos(R + P) * (Exp((-2) * s * dtan(P)) - 1) / dcos(R - D) /
dsin(A) / dsin(P)

KA4 = (-dsin(A) + dtan(D) * dcos(A)) * dcos(D)

Ka(2) = (KA1 + KA2 - KA3) / KA4

KA1 = dsin(A - D) * dcos(R + P) * dcos(R + s - A) * Exp((-2) * s * dtan(P))

KA2 = dcos(R - D) * dsin(A) * dcos(A - R - s - P + B)

KA3 = (dsin(A) - dtan(D) * dcos(A)) * dcos(D)

Ka(3) = KA1 / KA2 / KA3

do i = 1, 3

If (MAXKA(i) < Ka(i)) Then

MAXKA(i) = Ka(i) ; MAXRol(i) = Rol ; MAXSine(i) = Sine
```

```
End If

end do

Ka(4) = 0.5 * Ka(1) * Density * H ** 2 + Cohesion * Ka(2) * H + Ka(3) *
Surcharge * H

If (MAXKA(4) < Ka(4)) Then

MAXKA(4) = Ka(4) ; MAXRol(4) = Rol

MAXSine(4) = Sine

End If

70 end do

end do

MAXKA(2) = (MAXKA(2) / (-2)) ** 2

sl(4) = dtan(B)

do i = 1, 4

sl(1) = -dtan((Alpha - MAXRol(i)) * pi / 180)

sl(2) = dtan((90 - Alpha + MAXRol(i) + Phi) * pi / 180)

sl(3) = dtan((90 - (Alpha - MAXRol(i) - MAXSine(i) - Phi)) * pi / 180)

select case (i)

case (1)

X_1(2) = ((Y_1(1) - X_1(1) * sl(2)) - (Y_1(3) - X_1(3) * sl(1))) / (sl(1) - sl(2))

Y_1(2) = sl(1) * X_1(2) + (Y_1(3) - X_1(3) * sl(1))

r0 = ((X_1(3) - X_1(2)) ** 2 + (Y_1(3) - Y_1(2)) ** 2) ** 0.5

x_1(4) = x_1(2) ; y_1(4) = y_1(2)

Dsine = MAXSine(i) / 100
```

```
do j = 5, 104
r1 = r0 * Exp((-Dsine) * pi / 180 * dtan(Phi * pi / 180))
X_1(j) = X_1(3) + r1 * dcos((Alpha - MAXRol(i) - Dsine) * pi / 180)
Y_1(j) = Y_1(3) - r1 * dsin((alpha - MAXRol(i) - Dsine) * pi / 180)
Dsine = Dsine + MAXSine(i) / 100
end do
x_1(105) = x_1(3) ; y_1(105) = y_1(3)
x_1(106) = x_1(104) ; y_1(106) = y_1(104)
X_1(107) = ((Y_1(106) - X_1(106) * sl(3)) - (Y_1(3) - X_1(3) * sl(4))) / (sl(4) -
sl(3))
Y_1(107) = sl(4) * X_1(107) + (Y_1(3) - X_1(3) * sl(4))
case (2)
x_2(2) = ((y_2(1) - x_2(1) * sl(2)) - (y_2(3) - x_2(3) * sl(1))) / (sl(1) - sl(2))
y_2(2) = sl(1) * x_2(2) + (y_2(3) - x_2(3) * sl(1))
r0 = ((x_2(3) - x_2(2)) ** 2 + (y_2(3) - y_2(2)) ** 2) ** 0.5
x_2(4) = x_2(2) ; y_2(4) = y_2(2)
Dsine = MAXSine(i) / 100
do j = 5, 104
r1 = r0 * Exp((-Dsine) * pi / 180 * dtan(Phi * pi / 180))
x_2(j) = x_2(3) + r1 * dcos((Alpha - MAXRol(i) - Dsine) * pi / 180)
y_2(j) = y_2(3) - r1 * dsin((alpha - MAXRol(i) - Dsine) * pi / 180)
Dsine = Dsine + MAXSine(i) / 100
end do
x_2(105) = x_2(3) ; y_2(105) = y_2(3)
```

```
x_2(106) = x_2(104) ; y_2(106) = y_2(104)

x_2(107) = ((y_2(106) - x_2(106) * sl(3)) - (y_2(3) - x_2(3) * sl(4))) / (sl(4) -
sl(3))

y_2(107) = sl(4) * x_2(107) + (y_2(3) - x_2(3) * sl(4))

case (3)

x_3(2) = ((y_3(1) - x_3(1) * sl(2)) - (y_3(3) - x_3(3) * sl(1))) / (sl(1) - sl(2))

y_3(2) = sl(1) * x_3(2) + (y_3(3) - x_3(3) * sl(1))

r0 = ((x_3(3) - x_3(2)) ** 2 + (y_3(3) - y_3(2)) ** 2) ** 0.5

x_3(4) = x_3(2) ; y_3(4) = y_3(2)

Dsine = MAXSine(i) / 100

do j = 5, 104

r1 = r0 * Exp((-Dsine) * pi / 180 * dtan(Phi * pi / 180))

x_3(j) = x_3(3) + r1 * dcos((Alpha - MAXRol(i) - Dsine) * pi / 180)

y_3(j) = y_3(3) - r1 * dsin((Alpha - MAXRol(i) - Dsine) * pi / 180)

Dsine = Dsine + MAXSine(i) / 100

end do

x_3(105) = x_3(3) ; y_3(105) = y_3(3)

x_3(106) = x_3(104) ; y_3(106) = y_3(104)

x_3(107) = ((y_3(106) - x_3(106) * sl(3)) - (y_3(3) - x_3(3) * sl(4))) / (sl(4) -
sl(3))

y_3(107) = sl(4) * x_3(107) + (y_3(3) - x_3(3) * sl(4))

case (4)

x_4(2) = ((y_4(1) - x_4(1) * sl(2)) - (y_4(3) - x_4(3) * sl(1))) / (sl(1) - sl(2))

y_4(2) = sl(1) * x_4(2) + (y_4(3) - x_4(3) * sl(1))

r0 = ((x_4(3) - x_4(2)) ** 2 + (y_4(3) - y_4(2)) ** 2) ** 0.5
```

```
x_4(4) = x_4(2) ; y_4(4) = y_4(2)

Dsine = MAXSine(i) / 100

do j = 5, 104

r1 = r0 * Exp((-Dsine) * pi / 180 * dtan(Phi * pi / 180))

x_4(j) = x_4(3) + r1 * dcos((Alpha - MAXRol(i) - Dsine) * pi / 180)

y_4(j) = y_4(3) - r1 * dsin((Alpha - MAXRol(i) - Dsine) * pi / 180)

Dsine = Dsine + MAXSine(i) / 100

end do

x_4(105) = x_4(3) ; y_4(105) = y_4(3)

x_4(106) = x_4(104) ; y_4(106) = y_4(104)

x_4(107) = ((y_4(106) - x_4(106) * sl(3)) - (y_4(3) - x_4(3) * sl(4))) / (sl(4) -
sl(3))

y_4(107) = sl(4) * x_4(107) + (y_4(3) - x_4(3) * sl(4))

end select

end do

call report()

CASE (4)

rol_loop = (alpha+beta)/dp

do x = 1, int(rol_loop)+1

if (x-1 < rol_loop) then

rol = (x-1)*dp

else

rol = alpha+beta
```

```
end if
R = rol*pi/180
sine_loop = (alpha+beta-rol)/dp
do y = 1, int(sine_loop)+1
if (y-1 < sine_loop) then
sine = (y-1)*dp
else
sine = alpha+beta-rol
end if
s = sine*pi/180
CHK(1) = Alpha + Delta ; CHK(2) = 90 - (Alpha - Rol)
CHK(3) = 90 - (Rol + Delta)
CHK(4) = 90 - (Alpha - Rol - Sine + Phi + Beta)
CHK(5) = Alpha + Beta - Rol – Sine ; CHK(6) = 90 - (Rol - Phi)
i = 1
do
if (i > 6) exit
If ((CHK(i) <= 0.00001) .Or. (CHK(i) >= 180)) Then
Go To 80
end if
i = i + 1
end do
KP1 = dsin(A + D) * dcos(R - P) / dcos(R + D) / (dsin(A)) ** 2 / dcos(P)
KP2 = dsin(R) * dcos(A - R)
```

```
KP3 = dcos(R-P)*(dcos(A-R)*((-3)*dtan(P)+(3*dtan(P)*dcos(s)+dsin(s))*Exp(3
*s*dtan(P)))+dsin(A-R)*(1+(3*dtan(P)*dsin(s-
-dcos(s))*Exp(3*s*dtan(P))))/dcos(P)/(1+9*(dtan(P))**2)

KP4 = dcos(R - P) * dsin(A - R - s + B) * dcos(A - R - s) * Exp(3 * s * dtan(P)) /
dcos(A - R - s + B + P)

KP5 = (dsin(A) + dtan(D) * dcos(A)) * dcos(D)

Kp(1) = KP1 * (KP2 + KP3 + KP4) / KP5

KP1 = dsin(A + D) * dsin(R) / dsin(A) / dcos(R + D)

KP2 = (dsin(A + D) * dcos(R - P) * dsin(A - R - s + B) * Exp(2 * s * dtan(P))) /
(dcos(R + D) * dsin(A) * dcos(A - R - s + P + B))

KP3 = dsin(A + D) * dcos(R - P) * (Exp(2 * s * dtan(P)) - 1) / dcos(R + D) /
dsin(A) / dsin(P)

KP4 = (dsin(A) + dtan(D) * dcos(A)) * dcos(D)

Kp(2) = (KP1 + KP2 + KP3) / KP4

KP1 = dsin(A + D) * dcos(R - P) * dcos(R + s - A) * Exp(2 * s * dtan(P))

KP2 = dcos(R + D) * dsin(A) * dcos(A - R - s + P + B)

KP3 = (dsin(A) + dtan(D) * dcos(A)) * dcos(D)

Kp(3) = KP1 / KP2 / KP3

do i = 1, 3

If (MINKP(i) > Kp(i)) Then

MINKP(i) = Kp(i) ; MAXRol(i) = Rol ; MAXSine(i) = Sine

End If

end do

Kp(4) = 0.5 * Kp(1) * Density * H ** 2 + Cohesion * Kp(2) * H + Kp(3) *
Surcharge * H

If (MINKP(4) > Kp(4)) Then
```

```
MINKP(4) = Kp(4) ; MAXRol(4) = Rol
MAXSine(4) = Sine
End If
80 end do
end do
MINKP(2) = (MINKP(2) / (-2)) ** 2
sl(4) = dtan(B)
do i = 1, 4
sl(1) = -dtan((Alpha - MAXRol(i)) * pi / 180)
sl(2) = dtan((90 - Alpha + MAXRol(i) - Phi) * pi / 180)
sl(3) = dtan((90 - (Alpha - MAXRol(i) - MAXSine(i) + Phi)) * pi / 180)
select case (i)
case (1)
X_1(2) = ((Y_1(1) - X_1(1) * sl(2)) - (Y_1(3) - X_1(3) * sl(1))) / (sl(1) - sl(2))
Y_1(2) = sl(1) * X_1(2) + (Y_1(3) - X_1(3) * sl(1))
r0 = ((X_1(3) - X_1(2)) ** 2 + (Y_1(3) - Y_1(2)) ** 2) ** 0.5
x_1(4) = x_1(2) ; y_1(4) = y_1(2)
Dsine = MAXSine(i) / 100
do j = 5, 104
r1 = r0 * Exp(Dsine * pi / 180 * dtan(Phi * pi / 180))
X_1(j) = X_1(3) + r1 * dcos((Alpha - MAXRol(i) - Dsine) * pi / 180)
Y_1(j) = Y_1(3) - r1 * dsin((Alpha - MAXRol(i) - Dsine) * pi / 180)
Dsine = Dsine + MAXSine(i) / 100
end do
```

```
x_1(105) = x_1(3) ; y_1(105) = y_1(3)

x_1(106) = x_1(104) ; y_1(106) = y_1(104)

X_1(107) = ((Y_1(106) - X_1(106) * sl(3)) - (Y_1(3) - X_1(3) * sl(4))) / (sl(4) -
sl(3))

Y_1(107) = sl(4) * X_1(107) + (Y_1(3) - X_1(3) * sl(4))

case (2)

x_2(2) = ((y_2(1) - x_2(1) * sl(2)) - (y_2(3) - x_2(3) * sl(1))) / (sl(1) - sl(2))

y_2(2) = sl(1) * x_2(2) + (y_2(3) - x_2(3) * sl(1))

r0 = ((x_2(3) - x_2(2)) ** 2 + (y_2(3) - y_2(2)) ** 2) ** 0.5

x_2(4) = x_2(2) ; y_2(4) = y_2(2)

Dsine = MAXSine(i) / 100

do j = 5, 104

r1 = r0 * Exp(Dsine * pi / 180 * dtan(Phi * pi / 180))

x_2(j) = x_2(3) + r1 * dcos((Alpha - MAXRol(i) - Dsine) * pi / 180)

y_2(j) = y_2(3) - r1 * dsin((Alpha - MAXRol(i) - Dsine) * pi / 180)

Dsine = Dsine + MAXSine(i) / 100

end do

x_2(105) = x_2(3) ; y_2(105) = y_2(3)

x_2(106) = x_2(104) ; y_2(106) = y_2(104)

x_2(107) = ((y_2(106) - x_2(106) * sl(3)) - (y_2(3) - x_2(3) * sl(4))) / (sl(4) -
sl(3))

y_2(107) = sl(4) * x_2(107) + (y_2(3) - x_2(3) * sl(4))

case (3)

x_3(2) = ((y_3(1) - x_3(1) * sl(2)) - (y_3(3) - x_3(3) * sl(1))) / (sl(1) - sl(2))
```

```
y_3(2) = sl(1) * x_3(2) + (y_3(3) - x_3(3) * sl(1))
r0 = ((x_3(3) - x_3(2)) ** 2 + (y_3(3) - y_3(2)) ** 2) ** 0.5
x_3(4) = x_3(2) ; y_3(4) = y_3(2)
Dsine = MAXSine(i) / 100
do j = 5, 104
r1 = r0 * Exp(Dsine * pi / 180 * dtan(Phi * pi / 180))
x_3(j) = x_3(3) + r1 * dcos((Alpha - MAXRol(i) - Dsine) * pi / 180)
y_3(j) = y_3(3) - r1 * dsin((Alpha - MAXRol(i) - Dsine) * pi / 180)
Dsine = Dsine + MAXSine(i) / 100
end do
x_3(105) = x_3(3) ; y_3(105) = y_3(3)
x_3(106) = x_3(104) ; y_3(106) = y_3(104)
x_3(107) = ((y_3(106) - x_3(106) * sl(3)) - (y_3(3) - x_3(3) * sl(4))) / (sl(4) - sl(3))
y_3(107) = sl(4) * x_3(107) + (y_3(3) - x_3(3) * sl(4))
case (4)
x_4(2) = ((y_4(1) - x_4(1) * sl(2)) - (y_4(3) - x_4(3) * sl(1))) / (sl(1) - sl(2))
y_4(2) = sl(1) * x_4(2) + (y_4(3) - x_4(3) * sl(1))
r0 = ((x_4(3) - x_4(2)) ** 2 + (y_4(3) - y_4(2)) ** 2) ** 0.5
x_4(4) = x_4(2) ; y_4(4) = y_4(2)
Dsine = MAXSine(i) / 100
do j = 5, 104
r1 = r0 * Exp(Dsine * pi / 180 * dtan(Phi * pi / 180))
x_4(j) = x_4(3) + r1 * dcos((Alpha - MAXRol(i) - Dsine) * pi / 180)
```

```
y_4(j) = y_4(3) - r1 * dsin((Alpha - MAXRol(i) - Dsine) * pi / 180)

Dsine = Dsine + MAXSine(i) / 100

end do

x_4(105) = x_4(3) ; y_4(105) = y_4(3)

x_4(106) = x_4(104) ; y_4(106) = y_4(104)

x_4(107) = ((y_4(106) - x_4(106) * sl(3)) - (y_4(3) - x_4(3) * sl(4))) / (sl(4) - sl(3))

y_4(107) = sl(4) * x_4(107) + (y_4(3) - x_4(3) * sl(4))

end select

end do

call report()

CASE (5)

rol_loop = (alpha+beta)/dp

do x = 1, int(rol_loop)+1

if (x-1 < rol_loop) then

rol = (x-1)*dp

else

rol = alpha+beta

end if

R = rol*pi/180

sine_loop = (alpha+beta-rol)/dp

do y = 1, int(sine_loop)+1

if (y-1 < sine_loop) then
```

```
sine = (y-1)*dp

else

sine = alpha+beta-rol

end if

s = sine*pi/180

CHK(1) = Alpha – Delta ; CHK(2) = 90 - (Alpha - Rol + Phi)

CHK(3) = 90 - (Rol - Delta - Phi) ; CHK(4) = Alpha - Rol - Sine + Beta

CHK(5) = 90 - (Alpha + Beta - Sine - Rol) ; CHK(6) = 90 - Rol

I = 1

Do

IF (I > 6) EXIT

If ((CHK(I) <= 0.00001) .Or. (CHK(I) >= 180)) then

go to 90

end if

I = I + 1

END DO

KA1 = dsin(A - D) / dcos(R - D - P)

KA2 = dcos(R) * dsin(R) * dcos(A - R + P)

KA3= (dcos(R))**2*(dcos(A-R+P)*(dtan(2*P)+(-dtan(2*P)*dcos(s)+dsin(s)))*&

Exp((-s)*dtan(2*P))+dsin(A-R+P)*(1+(-dtan(2*P)*dsin(s)-dcos(s)))*&

Exp((-s)*dtan(2*P)))/(1+(dtan(2*P))**2)

KA4 = Exp((-s) * dtan(2 * P)) * (dcos(R)) ** 2 * dcos(A - R - s + P) * dtan(A - R
- s + B)

KA5 = (dsin(A)) ** 2 * (dsin(A) - dtan(D) * dcos(A)) * dcos(D)
```

```
Ka(1) = KA1 * (KA2 + KA3 + KA4) / KA5

KA1 = dsin(A - D) * dcos(R) * dcos(P) / dcos(R - D - P) / dsin(A)

KA2 = dtan(R) + dtan(A - R - s + B) * Exp((-s) * dtan(2 * P)) - (Exp((-s) * dtan(2
* P)) - 1) * (1 + dcos(2 * P)) / dsin(2 * P)

KA3 = ((-dsin(A)) + dtan(D) * dcos(A)) * dcos(D)

Ka(2) = (KA1 * KA2) / KA3

KA1 = dcos(R) * dcos(A - R - s + P) * dsin(A - D) / (dsin(A) * dcos(A - R - s +
B) * dcos(R - D - P)) * Exp((-s) * dtan(2 * P))

KA2 = (dsin(A) - dtan(D) * dcos(A)) * dcos(D)

Ka(3) = KA1 / KA2

DO I = 1, 3

If (MAXKA(I) < Ka(I)) Then

MAXKA(I) = Ka(I) ; MAXRol(I) = Rol ; MAXSine(I) = Sine

End If

END DO

Ka(4) = 0.5 * Ka(1) * Density * H ** 2 + Cohesion * Ka(2) * H + Ka(3) *
Surcharge * H

If (MAXKA(4) < Ka(4)) Then

MAXKA(4) = Ka(4) ; MAXRol(4) = Rol ; MAXSine(4) = Sine

End If

90 END DO

END DO

MAXKA(2) = (MAXKA(2) / (-2)) ** 2

sl(4) = dtan(B)

do i = 1, 4
```

```
sl(1) = -dtan((Alpha - MAXRol(i)) * pi / 180)

sl(2) = dtan((90 - Alpha + MAXRol(i)) * pi / 180)

sl(3) = dtan((90 - (Alpha - MAXRol(i) - MAXSine(i))) * pi / 180)

select case (i)

case (1)

X_1(2) = ((Y_1(1) - X_1(1) * sl(2)) - (Y_1(3) - X_1(3) * sl(1))) / (sl(1) - sl(2))

Y_1(2) = sl(1) * X_1(2) + (Y_1(3) - X_1(3) * sl(1))

r0 = ((X_1(3) - X_1(2)) ** 2 + (Y_1(3) - Y_1(2)) ** 2) ** 0.5

x_1(4) = x_1(2) ; y_1(4) = y_1(2)

Dsine = MAXSine(i) / 100

do j = 5, 104

r1 = r0

X_1(j) = X_1(3) + r1 * dcos((Alpha - MAXRol(i) - Dsine) * pi / 180)

Y_1(j) = Y_1(3) - r1 * dsin((Alpha - MAXRol(i) - Dsine) * pi / 180)

Dsine = Dsine + MAXSine(i) / 100

end do

x_1(105) = x_1(3) ; y_1(105) = y_1(3)

x_1(106) = x_1(104) ; y_1(106) = y_1(104)

X_1(107) = ((Y_1(106) - X_1(106) * sl(3)) - (Y_1(3) - X_1(3) * sl(4))) / (sl(4) -
sl(3))

Y_1(107) = sl(4) * X_1(107) + (Y_1(3) - X_1(3) * sl(4))

case (2)

x_2(2) = ((y_2(1) - x_2(1) * sl(2)) - (y_2(3) - x_2(3) * sl(1))) / (sl(1) - sl(2))

y_2(2) = sl(1) * x_2(2) + (y_2(3) - x_2(3) * sl(1))
```

```
r0 = ((x_2(3) - x_2(2)) ** 2 + (y_2(3) - y_2(2)) ** 2) ** 0.5
x_2(4) = x_2(2) ; y_2(4) = y_2(2)

Dsine = MAXSine(i) / 100
do j = 5, 104
r1 = r0
x_2(j) = x_2(3) + r1 * dcos((Alpha - MAXRol(i) - Dsine) * pi / 180)
y_2(j) = y_2(3) - r1 * dsin((Alpha - MAXRol(i) - Dsine) * pi / 180)
Dsine = Dsine + MAXSine(i) / 100
end do
x_2(105) = x_2(3) ; y_2(105) = y_2(3)
x_2(106) = x_2(104) ; y_2(106) = y_2(104)
x_2(107) = ((y_2(106) - x_2(106) * sl(3)) - (y_2(3) - x_2(3) * sl(4))) / (sl(4) -
sl(3))
y_2(107) = sl(4) * x_2(107) + (y_2(3) - x_2(3) * sl(4))
case (3)
x_3(2) = ((y_3(1) - x_3(1) * sl(2)) - (y_3(3) - x_3(3) * sl(1))) / (sl(1) - sl(2))
y_3(2) = sl(1) * x_3(2) + (y_3(3) - x_3(3) * sl(1))
r0 = ((x_3(3) - x_3(2)) ** 2 + (y_3(3) - y_3(2)) ** 2) ** 0.5
x_3(4) = x_3(2) ; y_3(4) = y_3(2)

Dsine = MAXSine(i) / 100
do j = 5, 104
r1 = r0
x_3(j) = x_3(3) + r1 * dcos((Alpha - MAXRol(i) - Dsine) * pi / 180)
```

```
y_3(j) = y_3(3) - r1 * dsin((Alpha - MAXRol(i) - Dsine) * pi / 180)

Dsine = Dsine + MAXSine(i) / 100

end do

x_3(105) = x_3(3) ; y_3(105) = y_3(3)

x_3(106) = x_3(104) ; y_3(106) = y_3(104)

x_3(107) = ((y_3(106) - x_3(106) * sl(3)) - (y_3(3) - x_3(3) * sl(4))) / (sl(4) -
sl(3))

y_3(107) = sl(4) * x_3(107) + (y_3(3) - x_3(3) * sl(4))

case (4)

x_4(2) = ((y_4(1) - x_4(1) * sl(2)) - (y_4(3) - x_4(3) * sl(1))) / (sl(1) - sl(2))

y_4(2) = sl(1) * x_4(2) + (y_4(3) - x_4(3) * sl(1))

r0 = ((x_4(3) - x_4(2)) ** 2 + (y_4(3) - y_4(2)) ** 2) ** 0.5

x_4(4) = x_4(2) ; y_4(4) = y_4(2)


Dsine = MAXSine(i) / 100

do j = 5, 104

r1 = r0

x_4(j) = x_4(3) + r1 * dcos((Alpha - MAXRol(i) - Dsine) * pi / 180)

y_4(j) = y_4(3) - r1 * dsin((Alpha - MAXRol(i) - Dsine) * pi / 180)

Dsine = Dsine + MAXSine(i) / 100

end do

x_4(105) = x_4(3) ; y_4(105) = y_4(3)

x_4(106) = x_4(104) ; y_4(106) = y_4(104)

x_4(107) = ((y_4(106) - x_4(106) * sl(3)) - (y_4(3) - x_4(3) * sl(4))) / (sl(4) -
sl(3))
```

```
y_4(107) = sl(4) * x_4(107) + (y_4(3) - x_4(3) * sl(4))
end select
end do
CALL REPORT()
CASE (6)
rol_loop = (alpha+beta)/dp
do x = 1, int(rol_loop)+1
if (x-1 < rol_loop) then
rol = (x-1)*dp
else
rol = alpha+beta
end if
R = rol*pi/180
sine_loop = (alpha+beta-rol)/dp
do y = 1, int(sine_loop)+1
if (y-1 < sine_loop) then
sine = (y-1)*dp
else
sine = alpha+beta-rol
end if
s = sine*pi/180
CHK(1) = Alpha + Delta ; CHK(2) = 90 - (Alpha - Rol - Phi)
CHK(3) = 90 - (Rol + Delta + Phi) ; CHK(4) = Alpha - Rol - Sine + Beta
CHK(5) = 90 - (Alpha + Beta - Sine - Rol) ; CHK(6) = 90 - Rol
```

```
I = 1

Do

IF (I > 6) EXIT

If ((CHK(I) <= 0.00001) .Or. (CHK(I) >= 180)) Then

Go To 1000

END IF

I = I + 1

END DO

KP1 = dsin(A + D) / dcos(R + D + P)

KP2 = dcos(R) * dsin(R) * dcos(A - R - P)

KP3 = (dcos(R))**2*(dcos(A-R-P)*((-dtan(2*P))+(dtan(2*P)*dcos(s)+dsin(s))*&

Exp(s*dtan(2*P)))+dsin(A-R-P)*(1+(dtan(2*P)*dsin(s)-dcos(s))*&

Exp(s*dtan(2*P))))/(1+(dtan(2*P))**2)

KP4 = Exp(s * dtan(2 * P)) * (dcos(R)) ** 2 * dcos(A - R - s - P) * dtan(A - R - s
+ B)

KP5 = (dsin(A)) ** 2 * (dsin(A) + dtan(D) * dcos(A)) * dcos(D)

Kp(1) = KP1 * (KP2 + KP3 + KP4) / KP5

KP1 = dsin(A + D) * dcos(R) * dcos(P) / dcos(R + D + P) / dsin(A)

KP2 = dtan(R) + dtan(A - R - s + B) * Exp(s * dtan(2 * P)) + (Exp(s * dtan(2 *
P)) - 1) * (1 + dcos(2 * P)) / dsin(2 * P)

KP3 = (dsin(A) + dtan(D) * dcos(A)) * dcos(D)

Kp(2) = (KP1 * KP2) / KP3

KP1 = dcos(R) * dcos(A - R - s - P) * dsin(A + D) / (dsin(A) * dcos(A - R - s + B)
* dcos(R + D + P)) * Exp(s * dtan(2 * P))

KP2 = (dsin(A) + dtan(D) * dcos(A)) * dcos(D)
```

```
Kp(3) = KP1 / KP2

DO I = 1, 3

If (MINKP(I) > Kp(I)) Then

MINKP(I) = Kp(I) ; MAXRol(I) = Rol ; MAXSine(I) = Sine

End If

END DO

Kp(4) = 0.5 * Kp(1) * Density * H ** 2 + Cohesion * Kp(2) * H + Kp(3) *
Surcharge * H

If (MINKP(4) > Kp(4)) Then

MINKP(4) = Kp(4) ; MAXRol(4) = Rol ; MAXNeta(4) = Neta

MAXSine(4) = Sine

End If

1000 END DO

END DO

MINKP(2) = (MINKP(2) / (-2)) ** 2

sl(4) = dtan(B)

do i = 1, 4

sl(1) = -dtan((Alpha - MAXRol(i)) * pi / 180)

sl(2) = dtan((90 - Alpha + MAXRol(i)) * pi / 180)

sl(3) = dtan((90 - (Alpha - MAXRol(i) - MAXSine(i))) * pi / 180)

select case (i)

case (1)

X_1(2) = ((Y_1(1) - X_1(1) * sl(2)) - (Y_1(3) - X_1(3) * sl(1))) / (sl(1) - sl(2))
```

```
Y_1(2) = sl(1) * X_1(2) + (Y_1(3) - X_1(3) * sl(1))
r0 = ((X_1(3) - X_1(2)) ** 2 + (Y_1(3) - Y_1(2)) ** 2) ** 0.5
x_1(4) = x_1(2) ; y_1(4) = y_1(2)
Dsine = MAXSine(i) / 100
do j = 5, 104
r1 = r0
X_1(j) = X_1(3) + r1 * dcos((Alpha - MAXRol(i) - Dsine) * pi / 180)
Y_1(j) = Y_1(3) - r1 * dsin((Alpha - MAXRol(i) - Dsine) * pi / 180)
Dsine = Dsine + MAXSine(i) / 100
end do
x_1(105) = x_1(3) ; y_1(105) = y_1(3)
x_1(106) = x_1(104) ; y_1(106) = y_1(104)
X_1(107) = ((Y_1(106) - X_1(106) * sl(3)) - (Y_1(3) - X_1(3) * sl(4))) / (sl(4) -
sl(3))
Y_1(107) = sl(4) * X_1(107) + (Y_1(3) - X_1(3) * sl(4))
case (2)
x_2(2) = ((y_2(1) - x_2(1) * sl(2)) - (y_2(3) - x_2(3) * sl(1))) / (sl(1) - sl(2))
y_2(2) = sl(1) * x_2(2) + (y_2(3) - x_2(3) * sl(1))
r0 = ((x_2(3) - x_2(2)) ** 2 + (y_2(3) - y_2(2)) ** 2) ** 0.5
x_2(4) = x_2(2) ; y_2(4) = y_2(2)
Dsine = MAXSine(i) / 100
do j = 5, 104
r1 = r0
x_2(j) = x_2(3) + r1 * dcos((Alpha - MAXRol(i) - Dsine) * pi / 180)
```

```
y_2(j) = y_2(3) - r1 * dsin((Alpha - MAXRol(i) - Dsine) * pi / 180)

Dsine = Dsine + MAXSine(i) / 100

end do

x_2(105) = x_2(3) ; y_2(105) = y_2(3)

x_2(106) = x_2(104) ; y_2(106) = y_2(104)

x_2(107) = ((y_2(106) - x_2(106) * sl(3)) - (y_2(3) - x_2(3) * sl(4))) / (sl(4) -
sl(3))

y_2(107) = sl(4) * x_2(107) + (y_2(3) - x_2(3) * sl(4))

case (3)

x_3(2) = ((y_3(1) - x_3(1) * sl(2)) - (y_3(3) - x_3(3) * sl(1))) / (sl(1) - sl(2))

y_3(2) = sl(1) * x_3(2) + (y_3(3) - x_3(3) * sl(1))

r0 = ((x_3(3) - x_3(2)) ** 2 + (y_3(3) - y_3(2)) ** 2) ** 0.5

x_3(4) = x_3(2) ; y_3(4) = y_3(2)

Dsine = MAXSine(i) / 100

do j = 5, 104

r1 = r0

x_3(j) = x_3(3) + r1 * dcos((Alpha - MAXRol(i) - Dsine) * pi / 180)

y_3(j) = y_3(3) - r1 * dsin((Alpha - MAXRol(i) - Dsine) * pi / 180)

Dsine = Dsine + MAXSine(i) / 100

end do

x_3(105) = x_3(3) ; y_3(105) = y_3(3)

x_3(106) = x_3(104) ; y_3(106) = y_3(104)

x_3(107) = ((y_3(106) - x_3(106) * sl(3)) - (y_3(3) - x_3(3) * sl(4))) / (sl(4) -
sl(3))

y_3(107) = sl(4) * x_3(107) + (y_3(3) - x_3(3) * sl(4))
```

```
case (4)
x_4(2) = ((y_4(1) - x_4(1) * sl(2)) - (y_4(3) - x_4(3) * sl(1))) / (sl(1) - sl(2))
y_4(2) = sl(1) * x_4(2) + (y_4(3) - x_4(3) * sl(1))
r0 = ((x_4(3) - x_4(2)) ** 2 + (y_4(3) - y_4(2)) ** 2) ** 0.5
x_4(4) = x_4(2) ; y_4(4) = y_4(2)
Dsine = MAXSine(i) / 100
do j = 5, 104
r1 = r0
x_4(j) = x_4(3) + r1 * dcos((Alpha - MAXRol(i) - Dsine) * pi / 180)
y_4(j) = y_4(3) - r1 * dsin((Alpha - MAXRol(i) - Dsine) * pi / 180)
Dsine = Dsine + MAXSine(i) / 100
end do
x_4(105) = x_4(3) ; y_4(105) = y_4(3)
x_4(106) = x_4(104) ; y_4(106) = y_4(104)
x_4(107) = ((y_4(106) - x_4(106) * sl(3)) - (y_4(3) - x_4(3) * sl(4))) / (sl(4) - sl(3))
y_4(107) = sl(4) * x_4(107) + (y_4(3) - x_4(3) * sl(4))
end select
end do

CALL REPORT()
END SELECT

stop
end program kk
```

```
subroutine report()
use methodcase
use reportdatatype
implicit none
integer :: i

select case (case)
case (1)
write (*,*) 'Two-Wedge Failure Method : Active Case'
case (2)
write (*,*) 'Two-Wedge Failure Method : Passive Case'
case (3)
write (*,*) 'Log-Spiral Failure Method : Active Case'
case (4)
write (*,*) 'Log-Spiral Failure Method : Passive Case'
case (5)
write (*,*) 'Arc-Sandwich Failure Method : Active Case'
case (6)
write (*,*) 'Arc-Sandwich Failure Method : Passive Case'
end select
write (*,*) 'Result of Limit Analysis'

do i = 1, 10
if (dp-1 == 0) then
dp = -1
```

```
exit
end if
if (dp*10**i >= 1) then
dp = i
exit
end if
end do

if ((case == 1) .or. (case == 3) .or. (case == 5)) then
write (*,*) ' Kar = ',maxka(1)
write (*,*) ' Kac = ',maxka(2)
write (*,*) ' Kaq = ',maxka(3)
else
write (*,*) ' Kpr = ',minkp(1)
write (*,*) ' Kpc = ',minkp(2)
write (*,*) ' Kpq = ',minkp(3)
end if
return
end subroutine report
```

APPENDIX 3-3-AXI-SYMMETRIC LATERAL EARTH PRESSURE SLIP LINE PROGRAM

This program is based on the theory as discussed in 3.2.2. In the first step, the boundary condition on the ground surface outside the well (which may not be horizontal) is determined. The computer code follows the equations as given in

3.2.2, and the readers should find it easy to modify the code, if necessary.

```
! This program is used to evaluate active pressure coefficient on
! the axial retaining wall with the inclination of surcharge q and
! show the points in boundary slipline
!
!
! INPUT:
!
! Qload(a1) : uniformly distributed load q ... ...(kPa)
! Weigh(a2) : unit weight of soil (gama) ... ...(kN/m3)
! Asoil(a3) : angle of friction of soil ... ...(degree)
! Afiwa(a4) : angle of friction of retaining wall ... ...(degree)
! Ahlwa(a5) : angle between horizontal line and the
! retaining wall ... ...(degree)
! Mdivz(a6) : no. of division in transition zone
! Ahlfi(a7) : angle of inclination of backfill to the
! horizontal line ... ...(degree)
! Mbofi(c1) : no. of division on the boundary line
! of backfill
! Dstep(c2) : separation of division
! cohe : cohesion ... ...(kN/m2)
!
!
```

```
! OUTPUT:
!
! ARRAY OF X,Y,Sigma,Theta and Pa
!
! x and y are used to show the position of each point
! sigma(m) is the stress measuring from centre of Mohr's Circle to
! the point that tangent of Mohr's circle intersect with
! the axis of abscissae
! Theta(n) is angle between the major principal stress to the X axis
! Pa(p) is the actual active pressure acting on the retaining wall
!
PROGRAM ACTIVE
implicit none
integer :: Mbofi,Mdivz,Malli,Mbeli
integer :: Nalli,Nbeli,Mcoun
integer :: ialli,ibeli,kcoun,jalli,jbeli,Nbozo
INTEGER :: ierror
real (kind=8) :: Dstep,Qload,Weigh,Asoil,Afiwa,Ahlwa,Ahlfi,Rinit,pi
real (kind=8) :: Temp1,Temp2,Temp3,Temp4,Temp5,Temp6,temp7, &
& ztem1,ztem1,rtem2,ztem2,Stem1,Ttem1,Stem2,Ttem2,rtem3,ztem3,Stem3, &
& Ttem3,rempi,cohe,Nstre,Sstre,Rdlta,judg2
real (kind=8) :: Rsoil,Rfiwa,Rhlwa,Rhlfi,Rmu,judg1,beta,beta1,Rbeta
real (kind=8), allocatable :: r(:,:),z(:,:),Sigma(:,:),Theta(:,:),Pa(:,:)
pi=3.1415926535
```

```
write (*,*) " Axi-symmetric Active earth pressure determination by solution of
write (*,*) " two partial axi-symmetric slip line equations"

open (UNIT=8 , FILE='parameter.txt' , STATUS='OLD' , ACTION='READ' ,
IOSTAT=ierror)
read (8,*) Asoil , rinit , Afiwa , Ahlfi , beta
close (UNIT=8)
!
Weigh=18. ; Ahlwa=90. ; Mdivz=40 ; Mbofi=80 ; Dstep=0.1d0 !
cohe = 0. ; Mcoun=16 ; Qload=1.0d0

!!!! cohe, Qload, weigh can be set to any value
!!!! Mdivz, Mbofi and Dstep can be set to any value
!
! Output file name : a_rwall, a_bline
!
Mbeli = Mbofi ; Malli=Mdivz+Mbeli+Mbofi
Rsoil=pi*Asoil/180. ; Rfiwa=pi*Afiwa/180.
Rhlwa=pi*Ahlwa/180. ; Rmu=pi/4.-Rsoil/2.
Rhlfi=pi*Ahlfi/180.
!
! Temp1=(dsin(Rhlfi)+beta*dcos(Rhlfi))/(dcos(Rhlfi)-beta*dsin(Rhlfi))
! Rbeta=datan(Temp1)
!
! Malli(k1) : max of alpha slipline
```

```
! Mbeli(k2) : max of beta slipline
! Rsoil(b1) : angle of friction of soil ... ... radian
! Rfiwa(b2) : angle of friction of retaining wall ... ... radian
! Rhlwa(b3) : angle between horizontal line and the
! retaining wall ... ... radian
! Rmu(b4):angle between slipline and major principal stress ...... radian
! Rhlfi(b5) : angle of inclination of backfill to
! the horizontal line ... ... radian
! Rbeta force angle with the consideration of earthquake ........radian
!
allocate (r(0:Malli,0:Mbeli),z(0:Malli,0:Mbeli),Sigma(0:Malli,0:Mbeli), &
& Theta(0:Malli,0:Mbeli),Pa(0:Mbeli,0:Malli))
Nalli=Malli; Nbeli=Mbeli
!
! Nalli and Nbeli are used for Max of counting alpha and beta slipline
! ialli and ibeli are used for couting alpha and beta slipline
!
do 490 ibeli = 0, Nbeli
do 480 ialli = 0, Nalli
if (ialli > Mbofi) go to 310
if (ialli < Mbofi-ibeli) go to 480
if (ialli == Mbofi-ibeli) go to 300
go to 340
```

```
300 r(ialli,ibeli)=(Mbofi-ialli)*Dstep*dcos(Rhlfi)+rinit
z(ialli,ibeli)=-(Mbofi-ialli)*Dstep*dsin(Rhlfi)
if (cohe==0) then
Temp1=0
Nstre=Qload*dcos(Rhlfi)*dcos(Rhlfi)
else
Temp1=cohe/dtan(Rsoil)
Nstre=Qload*dcos(Rhlfi)*dcos(Rhlfi)+Temp1
endif
Sstre=Qload*dsin(Rhlfi)*dcos(Rhlfi)
Judg1=((Nstre*dsin(Rsoil))**2.)-((Sstre*dcos(Rsoil))**2.)
if (judg1 < 0.) then
Sigma(ialli,ibeli)=Nstre/(dcos(Rsoil)**2.)
Temp2=Nstre-Sigma(ialli,ibeli)-Temp1
Theta(ialli,ibeli)=(datan(Sstre/Temp2))/2.+pi/2.-Rhlfi
else
Sigma(ialli,ibeli)=(Nstre-(judg1**0.5))/(dcos(Rsoil)**2.)
Temp2=Nstre-Sigma(ialli,ibeli)-Temp1
Theta(ialli,ibeli)=(datan(Sstre/Temp2))/2.+pi/2.-Rhlfi
ENDif
go to 480
!
! 300 ... ... calculate all the points on boundary line of backfill
!
```

```
! Nstre : normal stress acting on the plane of backfill due to q

! Sstre : shear stress acting on the plane of backfill due to q

!

310 Nbozo=Mbofi+Mdivz

if (ialli > Nbozo) go to 320

if (ibeli > 0) go to 340

if (Asoil==Afiwa) then

Rdlta=pi/2.

else

Temp1=dsin(Rfiwa)/dsin(Rsoil)

Rdlta=dasin(Temp1)

end if

Temp2=Rhlwa+(Rdlta-Rfiwa)/2.-Theta(Mbofi,0)

Theta(ialli,ibeli)=Theta(Mbofi,0)+(ialli-Mbofi)*Temp2/Mdivz

Temp3=Theta(Mbofi,0)-Theta(ialli,ibeli)

Sigma(ialli,ibeli)=Sigma(Mbofi,0)*dexp(2*Temp3*dtan(Rsoil))

r(ialli,ibeli)=rinit; z(ialli,ibeli)=0.

go to 480

!

! 310 ... ... calculate Sigma & Theta of point O (0,0) in each division in transition
zone

! Temp2 is angle between active I division and II divsion

!

320 if (ialli > Nbozo+ibeli) go to 480
```

```
if (ialli == Nbozo+ibeli) go to 330

go to 340

330 Theta(ialli,ibeli)=Rhlwa+(Rdlta-Rfiwa)/2.

rtem2=r(ialli-1,ibeli) ; ztem2=z(ialli-1,ibeli)

Stem2=Sigma(ialli-1,ibeli) ; Ttem2=Theta(ialli-1,ibeli)

Ttem3=(Theta(ialli,ibeli)+Ttem2)/2.

if (dabs(Rhlwa-pi/2.) < 0.001) then

r(ialli,ibeli)=rinit

z(ialli,ibeli)=ztem2+(rinit-rtem2)*dtan(Ttem3+Rmu)

else

Temp1=ztem2+rinit*dtan(Rhlwa)-rtem2*dtan(Ttem3+Rmu)

r(ialli,ibeli)=Temp1/(dtan(Rhlwa)-dtan(Ttem3+Rmu))

z(ialli,ibeli)=(r(ialli,ibeli)-rinit)*dtan(Rhlwa)

z(ialli,ibeli)=z(ialli,ibeli)/2.+(ztem2+(r(ialli,ibeli)-rtem2) &

*dtan(Ttem3+Rmu))/2.

endif    Temp1=Weigh/dcos(Rsoil)*(dsin(Rsoil)-beta*dcos(Rsoil))*(r(ialli,ibeli-
-rtem2)
Temp2=Weigh/dcos(Rsoil)*(dcos(Rsoil)+beta*dsin(Rsoil))*(z(ialli,ibeli)-ztem2)
Temp3=(dsin(Rsoil)-1)*(r(ialli,ibeli)-rtem2)+dcos(Rsoil)*(z(ialli,ibeli)-ztem2)

Temp4=Temp3*dtan(pi/4+Rsoil/2)*dtan(Rsoil)/(r(ialli,ibeli)+rtem2)
Temp5=Temp4+(Theta(ialli,ibeli)-Ttem2)*dtan(Rsoil)

Sigma(ialli,ibeli)=(Temp1+Temp2-Stem2*(Temp5-1))/(Temp5+1)

go to 480

!

! 340 ... ... calculate all points on the retaining wall

!
```

```
340 jbeli=ibeli-1

rtem1=r(ialli,jbeli); ztem1=z(ialli,jbeli)

Stem1=Sigma(ialli,jbeli); Ttem1=Theta(ialli,jbeli)

jalli=ialli-1

rtem2=r(jalli,ibeli); ztem2=z(jalli,ibeli)

Stem2=Sigma(jalli,ibeli); Ttem2=Theta(jalli,ibeli)

kcoun=0

rtem3=(rtem1+rtem2)/2 ; ztem3=(ztem1+ztem2)/2

Stem3=(Stem1+Stem2)/2 ; Ttem3=(Ttem1+Ttem2)/2

!

! calculate the r and z

!

31                                    Temp1=rtem1*dtan((Ttem1+Ttem3)/2-Rmu-
-rtem2*dtan((Ttem2+Ttem3)/2+Rmu)  &

+ztem2-ztem1

Temp2=dtan((Ttem1+Ttem3)/2-Rmu)-dtan((Ttem2+Ttem3)/2+Rmu)

r(ialli,ibeli)=Temp1/Temp2

temp3=(r(ialli,ibeli)-rtem1)*dtan((Ttem1+Ttem3)/2-Rmu)+ztem1
z(ialli,ibeli)=(temp3+(r(ialli,ibeli)-rtem2)*dtan((Ttem2+Ttem3)/2+Rmu) &

+ztem2)/2

!

! calculate the stress sigma(::) and angle Theta(::)

!

Temp1=Weigh/dcos(Rsoil)*(beta*dcos(Rsoil)+dsin(Rsoil))*(r(ialli,ibeli)-rtem1)
Temp1=Temp1+Weigh/dcos(Rsoil)*(beta*dsin(Rsoil)-
dcos(Rsoil))*(z(ialli,ibeli)-ztem1)                    Temp2=Weigh/dcos(Rsoil)*(-
```

```
beta*dcos(Rsoil)+dsin(Rsoil))*(r(ialli,ibeli)-rtem2)
Temp2=Temp2+Weigh/dcos(Rsoil)*(beta*dsin(Rsoil)+dcos(Rsoil))*(z(ialli,ibeli)
-ztem2)

Temp6=(dsin(Rsoil)+1)*dtan(Rsoil)/dcos(Rsoil)          Temp3=(dsin(Rsoil)-
1)*(r(ialli,ibeli)-rtem1)-dcos(Rsoil)*(z(ialli,ibeli)-ztem1)   Temp4=(dsin(Rsoil)-
1)*(r(ialli,ibeli)-rtem2)+dcos(Rsoil)*(z(ialli,ibeli)-ztem2)

Temp3=-Temp3*Temp6/(r(ialli,ibeli)+rtem1)

Temp4=Temp4*Temp6/(r(ialli,ibeli)+rtem2)          Temp7=(Temp4-
Temp3)+(Theta(ialli,jbeli)-Theta(jalli,ibeli))*dtan(Rsoil)
Temp5=(Temp7*(Sigma(ialli,jbeli)+Sigma(jalli,ibeli))+Temp1-Temp2)/(Temp7+
2)

Temp6=Temp1*Sigma(jalli,ibeli)-Temp2*Sigma(ialli,jbeli)
Temp6=(Temp6+(Temp7-2)*Sigma(jalli,ibeli)*Sigma(ialli,jbeli))/(Temp7+2)

judg1=Temp5*Temp5-4*temp6

if ( judg1 < 0. ) then

write (*,103) ialli, ibeli

103 format (1x,'ialli=', I5,'ibeli=', I5)

write (*,*) " press value have not real value, do not calculate press in line 215"

go to 450

else

Sigma(ialli,ibeli)=(-temp5+(judg1**0.5))/2.

Temp5=(Sigma(ialli,ibeli)+Sigma(jalli,ibeli))*Theta(jalli,ibeli)          &
+(Sigma(ialli,ibeli)+Sigma(ialli,jbeli))*Theta(ialli,jbeli)
Temp5=Temp5*dtan(Rsoil)+Temp2+Temp1+Sigma(jalli,ibeli)-Sigma(ialli,jbeli)

Temp5=Temp5-Temp3*(Sigma(ialli,jbeli)+Sigma(ialli,ibeli)) &

-Temp4*(Sigma(jalli,ibeli)+Sigma(ialli,ibeli))

Temp6=(Sigma(ialli,jbeli)+Sigma(jalli,ibeli)+2*Sigma(ialli,ibeli)) & *dtan(Rsoil)

Theta(ialli,ibeli)=Temp5/Temp6
```

```
endif
!
! judge and iteration
!
judg2=(r(ialli,ibeli)-rtem3)**2+(z(ialli,ibeli)-ztem3)**2
judg2=judg2+(Sigma(ialli,ibeli)-Stem3)**2+(Theta(ialli,ibeli)-Ttem3)**2
judg2=judg2**0.5

! write (*,104) ialli,ibeli,kcoun,judg2
! 104 FORMAT('1x,element i,j=', I5,I5,'iterative time=',I5,F12.7)

if (judg2<0.001) go to 480
kcoun=kcoun+1
if (kcoun > Mcoun) then
write (*,100) ialli,ibeli
100 FORMAT ('the programe do not divergence, element i,j=',I5,I5)
go to 480
endif
rtem3=r(ialli,ibeli) ; ztem3=z(ialli,ibeli)
Stem3=Sigma(ialli,ibeli) ; Ttem3=Theta(ialli,ibeli)
go to 31

! Temp5=-Sigma(ialli,jbeli)+Sigma(jalli,ibeli)
!
Temp5=Temp3+2.*dtan(Rsoil)*(Stem1*Theta(ialli,jbeli)+Stem2*Theta(jalli,ibeli
))      Temp4=2*(dsin(Rsoil)-1)*((r(ialli,ibeli)-rtem1)+dcos(Rsoil)*(z(ialli,ibel-
)-ztem1))
Temp4=Stem1*Temp4*dtan(Pi/4.+Rsoil/2.)*dtan(Rsoil)/(rtem1+r(ialli,ibeli))
```

```
Temp5=2*(dsin(Rsoil)-1)*((r(ialli,ibeli)-rtem2)+dcos(Rsoil)*(z(ialli,ibel-
)-ztem2))
Temp5=Temp5*Stem2*dtan(Pi/4.+Rsoil/2.)*dtan(Rsoil)/(rtem2+r(ialli,ibeli))

! 420 ... ... calculate all points within boundary lines

! kcoun : circular calculating times

! Mcoun : total circular calculating times

!


480 continue

490 continue

!

! Remark: x, y axis are rotated with the plane of backfill in the above calculation

!

! beta1=dsqrt(1+beta*beta)

!

open (UNIT=27 , FILE='Paexcel.dat' , STATUS='new' , ACTION='write' ,
IOSTAT=ierror)

do ibeli=1,Nbeli

ialli=Mdivz+Mbofi+ibeli

if (Rfiwa < 0.001) then

Temp1=Sigma(ialli,ibeli)*(1-dsin(Rsoil))

else

Temp1=Sigma(ialli,ibeli)*dsin(Rdlta-Rfiwa)/dsin(Rdlta)

end if

if (cohe==0) then

Temp2=Temp1*dcos(Rfiwa)
```

```
else
Temp2=Temp1*dcos(Rfiwa)-cohe/dtan(Rsoil)
END if
Temp3=Temp1*dsin(Rfiwa)
Pa(ialli,ibeli)=((Temp2**2.)+(Temp3**2.))**(0.5)
Temp4=Pa((Mdivz+Mbofi+ibeli),ibeli)/z((Mdivz+Mbofi+ibeli),ibeli)/Weigh
write (27,117) z(ialli,ibeli) , Pa(ialli,ibeli) , Temp4
117 FORMAT(1x, 3F10.3)
end do
close (UNIT=27)
!
! Print out all the points on retaining wall in output file A_RWALL
!
do 471 ibeli = 0, Nbeli
do 461 ialli = 0, Nalli
if (ialli < Mbofi-ibeli) then
jalli=Mbofi-ibeli
r(ialli,ibeli)=r(jalli,ibeli) ; z(ialli,ibeli)=z(jalli,ibeli)
endif
if (ialli > Nbozo+ibeli) then
jalli=nbozo+ibeli
r(ialli,ibeli)=r(jalli,ibeli) ; z(ialli,ibeli)=z(jalli,ibeli)
endif
461 continue
```

```
471 continue

open (UNIT=26 , FILE='brzexcel.dat' , STATUS='new' , ACTION='write' , IOSTAT=ierror)

do 472 ialli = 0, Nalli

write (26,*) (r(ialli,ibeli) , z(ialli,ibeli) , ibeli=0, Nbeli,10)

472 continue

close (UNIT=26)

do 475 ibeli = 0, Nbeli

do 465 ialli = 0, Nalli

if (ialli < Mbofi-ibeli) then

jbeli=Mbofi-ialli

r(ialli,ibeli)=r(ialli,jbeli) ; z(ialli,ibeli)=z(ialli,jbeli)

endif

if (ialli > Nbozo+ibeli) then

jbeli=ialli-Nbozo

r(ialli,ibeli)=r(ialli,jbeli) ; z(ialli,ibeli)=z(ialli,jbeli)

endif

465 continue

475 continue

open (UNIT=27 , FILE='arzexcel.dat' , STATUS='new' , ACTION='write' , IOSTAT=ierror)

do 473 ibeli = Nbeli, 0, -1

write (27,*) (r(ialli,ibeli) , z(ialli,ibeli) , ialli=0, Nalli,10)
```

```
473 continue

close (UNIT=27)

open (UNIT=25 , FILE='rzoutput.dat' , STATUS='replace' , ACTION='write' ,
IOSTAT=ierror)

write (25,*) " list of cylindrical coordinates r and z "

write (25,*) " ialli ibeli r z sigma Theta"

do 470 ibeli = 0, Nbeli

do 460 ialli = 0, Nalli

if (ialli < Mbofi-ibeli) go to 460

if (ialli > Nbozo+ibeli) go to 460

write (25,110) ialli,ibeli,r(ialli,ibeli),z(ialli,ibeli),Sigma(ialli,ibeli), &

Theta(ialli,ibeli)

110 FORMAT (1x,I6,',',1x,I6,',',F11.3,',',F11.3,',',F11.3,',',F11.3)

460 continue

WRITE(25,115)

115 format (' ')

470 continue

close (UNIT=25)

!

open (UNIT=30 , FILE='zsoutput.dat' , STATUS='replace' , ACTION='write' ,
IOSTAT=ierror)

write (30,*) " list of cylindrical coordinates z and sigma "

write (30,*) " ialli ibeli r z sigma Theta"

do 440 ibeli = 0, Nbeli

do 430 ialli = 0, Nalli
```

```
if (ialli == Nbozo+ibeli) then
write (30,116) ialli,ibeli,r(ialli,ibeli),z(ialli,ibeli),Sigma(ialli,ibeli), &
Theta(ialli,ibeli)
116 FORMAT (1x,I6,',',1x,I6,',',F11.3,',',F11.3,',',F11.3,',',F11.3)
ENDif
430 continue
440 continue
close (UNIT=30)
450 continue
END PROGRAM ACTIVE
```

Some sample output from this program is given below.

List of cylindrical coordinates r and z

ialli ibeli r z sigma Theta
80, 0, 10.000, -0.000, 3.347, 1.381
81, 0, 10.000, 0.000, 3.317, 1.388
82, 0, 10.000, 0.000, 3.287, 1.394
83, 0, 10.000, 0.000, 3.258, 1.400
84, 0, 10.000, 0.000, 3.229, 1.407
85, 0, 10.000, 0.000, 3.200, 1.413
86, 0, 10.000, 0.000, 3.171, 1.420
87, 0, 10.000, 0.000, 3.143, 1.426
88, 0, 10.000, 0.000, 3.115, 1.432
89, 0, 10.000, 0.000, 3.087, 1.439
90, 0, 10.000, 0.000, 3.060, 1.445
91, 0, 10.000, 0.000, 3.032, 1.452
92, 0, 10.000, 0.000, 3.005, 1.458
93, 0, 10.000, 0.000, 2.978, 1.464

Cont.....

94, 0, 10.000, 0.000, 2.952, 1.471
95, 0, 10.000, 0.000, 2.925, 1.477
96, 0, 10.000, 0.000, 2.899, 1.484

List of cylindrical coordinates z and sigma

ialli ibeli r z sigma Theta
120, 0, 10.000, 0.000, 2.338, 1.637
121, 1, 10.000, 0.210, 2.268, 1.637
122, 2, 10.000, 0.426, 2.199, 1.637
123, 3, 10.000, 0.648, 2.132, 1.637
124, 4, 10.000, 0.876, 2.067, 1.637
125, 5, 10.000, 1.109, 2.003, 1.637
126, 6, 10.000, 1.349, 1.940, 1.637
127, 7, 10.000, 1.594, 1.878, 1.637
128, 8, 10.000, 1.846, 1.818, 1.637
129, 9, 10.000, 2.105, 1.760, 1.637
130, 10, 10.000, 2.369, 1.702, 1.637
131, 11, 10.000, 2.641, 1.646, 1.637
132, 12, 10.000, 2.919, 1.592, 1.637
133, 13, 10.000, 3.204, 1.538, 1.637
134, 14, 10.000, 3.496, 1.486, 1.637
135, 15, 10.000, 3.795, 1.435, 1.637
136, 16, 10.000, 4.102, 1.386, 1.637
137, 17, 10.000, 4.415, 1.337, 1.637
138, 18, 10.000, 4.737, 1.290, 1.637
139, 19, 10.000, 5.066, 1.244, 1.637
140, 20, 10.000, 5.402, 1.199, 1.637
141, 21, 10.000, 5.747, 1.155, 1.637
142, 22, 10.000, 6.100, 1.113, 1.637
143, 23, 10.000, 6.460, 1.071, 1.637
144, 24, 10.000, 6.830, 1.031, 1.637
145, 25, 10.000, 7.207, 0.992, 1.637
146, 26, 10.000, 7.594, 0.953, 1.637
147, 27, 10.000, 7.989, 0.916, 1.637

Cont.....

148, 28, 10.000, 8.393, 0.880, 1.637
149, 29, 10.000, 8.806, 0.845, 1.637
150, 30, 10.000, 9.229, 0.812, 1.637

References

Adami, S. (2014). Modeling and Simulation of Multiphase Phenomena with Smoothed Particle Hydrodynamics.Doktor-Ingenieurs, Technischen Universität München.

Adoko, A.C., Jiao, Y.Y., Wu, L., Wang, H., Wang, Z.H. (2013). Predicting tunnel convergence using multivariate adaptive regression spline and artificial neural network. *Tunn. Undergr. Space Technol, 38*, 368-376.
[http://dx.doi.org/10.1016/j.tust.2013.07.023]

Al-Defae, A.H., Knappett, J.A. (2015). Newmark sliding block model for pile reinforced slopes under earthquake loading. *Soil. Dyn. Earthquake Eng, 75*, 265-278.
[http://dx.doi.org/10.1016/j.soildyn.2015.04.013]

Al-Bittar, T., Soubra, A.H. (2013). Probabilistic Analysis of Strip Footings Resting on Spatially Varying Soils and Subjected to Vertical or Inclined Loads. *J. Geotech. Geoenviron. Eng.*
[http://dx.doi.org/10.1061/(ASCE)GT.1943-5606.0001046]

Alonso, E. (1976). Risk analysis of slopes and its application to slopes in Canadian sensitive clays. *Geotechnique, 26*(3), 453-472.
[http://dx.doi.org/10.1680/geot.1976.26.3.453]

Aluru, N. (2000). A point collocation method based on reproducing kernel approximation. *Int. J. Numer. Methods Eng, 47*(6), 1083-1121.
[http://dx.doi.org/10.1002/(SICI)1097-0207(20000228)47:6<1083::AID-NME816>3.0.CO;2-N]

Anderheggen, E., Knopfel, H. (1972). Finite element limit analysis using linear programming. *Int. J. Solids Struct, 8*(12), 1413-1431.
[http://dx.doi.org/10.1016/0020-7683(72)90088-1]

Ang, A.H., Tang, W. (1984). *Probability concepts in engineering planning and design* (Vol. 2). New York: John Wiley and Sons Inc.

Arai, K., Tagyo, K. (1985). Determination of noncircular slip surfaces giving the minimum factor of safety in slope stability analysis. *Soil Found, 25*, 43-51.
[http://dx.doi.org/10.3208/sandf1972.25.43]

Arai, K., Tagyo, K., Barber, J.R. (2010). Determination of noncircular slip surfaces giving the minimum factor of safety in slope stability analysis, Soils and Foundations. *Elasticity* (Vol. 25, pp. 43-51). London: Springer.
[http://dx.doi.org/10.1007/978-90-481-3809-8]

Au, S.K., Beck, J.L. (2001). Estimation of small failure probabilities in high dimensions by subset simulation. *Probab. Eng. Mech, 16*(4), 263-277.
[http://dx.doi.org/10.1016/S0266-8920(01)00019-4]

Au, S.K., Beck, J.L. (2003). Subset simulation and its application to seismic risk based on dynamic analysis. *J. Eng. Mech, 129*(8), 901-917.
[http://dx.doi.org/10.1061/(ASCE)0733-9399(2003)129:8(901)]

Au, S.K., Ching, J., Beck, J.L. (2007). Application of subset simulation methods to reliability benchmark problems. *Struct. Saf, 29*(3), 183-193.

[http://dx.doi.org/10.1016/j.strusafe.2006.07.008]

Au, S.K., Cao, Z.J., Wang, Y. (2010). Implementing advanced Monte Carlo simulation under spreadsheet environment. *Struct. Saf, 32*(5), 281-292.
[http://dx.doi.org/10.1016/j.strusafe.2010.03.004]

Au, S.K., Wang, Y. (2014). *Engineering risk assessment with subset simulation.* Singapore: John Wiley & Sons.
[http://dx.doi.org/10.1002/9781118398050]

Baars, S.V. (2018). *100 years of Prandtl's wedge.* IOS Press.

Babuska, I., Melenk, J.M. (1997). The partition of unity method. *Int. J. Numer. Methods Eng, 40*, 727-758.
[http://dx.doi.org/10.1002/(SICI)1097-0207(19970228)40:4<727::AID-NME86>3.0.CO;2-N]

Baker, R., Garber, M. (1978). Theoretical analysis of the stability of slopes. *Geotechnique, 28*, 395-411.
[http://dx.doi.org/10.1680/geot.1978.28.4.395]

Baker, R. (1980). Determination of the critical slip surface in slope stability computations. *Int. J. Numer. Anal. Methods Geomech, 4*, 333-359.
[http://dx.doi.org/10.1002/nag.1610040405]

Baker, R. (2003). Sufficient conditions for existence of physically significant solutions in limiting equilibrium slope stability analysis. *Int. J. Solids Struct, 40*(13-14), 3717-3735.
[http://dx.doi.org/10.1016/S0020-7683(03)00075-1]

Bathe, K.J. (1985). A Four node plate bending element based on Mindlin/Ressiner plate theory and mixed interpolation. *Int. J. Numer. Methods Eng, 21*, 367-383.
[http://dx.doi.org/10.1002/nme.1620210213]

Bathe, K.J. (2014). *Finite element Procedures.* K.J. Bathe.

Bathurst, R., Rothenburg, L. (1989). Investigation of micromechanical features of idealized granular assemblies using DEM *Proceeding of the 1st US Conference on Discrete Element Methods,* Golden, Colo12.

Bauer, S., Lackner, R. (2015). Gradient-based adaptive discontinuity layout optimization for the prediction of strength properties in matrix-inclusion materials. *Int. J. Solids Struct, 63*, 82-98.
[http://dx.doi.org/10.1016/j.ijsolstr.2015.02.042]

Baecher, G.B., Ingra, T.S. (1981). Stochastic FEM in settlement predictions. *J. Geotech. Eng. Div, 107*(4), 449-463.

Baecher, G.B. (1987). *Statistical analysis of geotechnical data.* Vicksburg, Mississippi: USACE Waterways Experiment Station.

Baecher, G.B., Christian, J.T. (2003). *Reliability and statistics in geotechnical engineering.* John Wiley.

Belytschko, T., Liu, W.K., Moran, B., Elkhodary, K.I. (2014). *Nonlinear finite elements for continua and structures.* John Wiley.

Belytschko, T., Lu, Y.Y., Gu, L. (1994). Element-free Galerkin methods. *Int. J. Numer. Methods Eng, 37*, 229-256.
[http://dx.doi.org/10.1002/nme.1620370205]

Berezantzev, V.G. (1958). Earth pressure on the cylindrical retaining wall. *Proc., Brussels Conf. on Earth*

Pressure Problems, 2, 21-27.

Bhattacharya, G., Jana, D., Ojha, S., Chakraborty, S. (2003). Direct search for minimum reliability index of earth slopes. *Comput. Geotech, 30*, 445-462.
[http://dx.doi.org/10.1016/S0266-352X(03)00059-4]

Birolini, A. (2014). *Reliability engineering* Springer.
[http://dx.doi.org/10.1007/978-3-642-39535-2]

Bishop, A.W. (1955). The use of the slip circle in the stability analysis of earth slopes. *Geotechnique, 5*(1), 7-17.
[http://dx.doi.org/10.1680/geot.1955.5.1.7]

Bolton, H.P.J., Heymann, G., Groenwold, A. (2003). Global search for critical failure surface in slope stability analysis. *Eng. Optim, 35*(1), 51-65.
[http://dx.doi.org/10.1080/0305215031000064749]

Blum, C., Raidl, G.R. (2016). *Hybrid Metaheuristics Powerful Tools for Optimization.* Springer.

Booker, J.R., Zheng, X. (2000). *Application of the theory of classical plasticity to the analysis of the stress distribution in wedges of a perfectly frictional material, Modelling in Geomechanics.* New York: John Wiley.

Bottero, A., Negre, R., Pastor, J., Turgeman, S. (1980). Finite element method and limit analysis theory for soil mechanics problems. *Comput. Methods Appl. Mech. Eng, 22*, 131-149.
[http://dx.doi.org/10.1016/0045-7825(80)90055-9]

Breitung, K. (1984). Asympotic approximations for multinormal integrals. *J. Eng. Mech, 110*(3), 357-366.
[http://dx.doi.org/10.1061/(ASCE)0733-9399(1984)110:3(357)]

Bucher, C.G., Bourgund, U. (1990). A fast and efficient response surface approach for structural reliability problems. *Struct. Saf, 7*(1), 57-66.
[http://dx.doi.org/10.1016/0167-4730(90)90012-E]

Bui, H.H. (2007).

Bui, H.H., Fukagawa, R., Sako, K., Ohno, S. (2008). Lagrangian meshfree particles method (SPH) for large deformation and failure flows of geomaterial using elastic-plastic soil constitutive model. *Int. J. Numer. Anal. Methods Geomech, 32*, 1537-1570.
[http://dx.doi.org/10.1002/nag.688]

Busby, D. (2009). Hierarchical adaptive experimental design for Gaussian process emulators. *Reliab. Eng. Syst. Saf, 94*(7), 1183-1193.
[http://dx.doi.org/10.1016/j.ress.2008.07.007]

Cai, Y., Liang, G.P., Shi, G.H., Cook, N.G.W. (1996). Studying an impact problem by using LDDA method.

Cao, W.G., Shu, B.Y. (2001). A study on techniques of automatically forming of cover system of numerical manifold method. *Chinese Journal of Geotechnical Engineering, 23*(2), 187-190.

Castilo, E., Luceno, A. (1980). Evaluation of variational methods in slope analysis. *Proceedings of International Symposium on Landslides,* New-Delhi, India*1*, 255-258.

Castilo, E., Luceno, A. (1982). A critical analysis of some variational methods in slope stability analysis. *Int. J. Numer. Anal. Methods Geomech, 6*, 195-209.

[http://dx.doi.org/10.1002/nag.1610060206]

Celestino, T.B., Duncan, J.M. (1981). Simplified Search for Non-Circular Slip Surface *Proceedings of 10th International Conference on Soil Mechanics and Foundation Engineering,* Stockholm, Sweden391-394.

Chang, W.A. (2006). *Advances in Evolutionary Algorithms Theory, Design and Practice.* Springer.

Chapman, S.J. (2007). *Fortran 95/2003 for Scientists & Engineers* McGraw Hill.

Chen, Z. (2003). *Stability analysis of soil slopes: theories, methods and programs.* Beijing: Water Power Press. (in Chinese)

Chen, H., Lee, C.F. (2000). Numerical simulation of debris flows. *Can. Geotech. J, 37*(1), 146-160. [http://dx.doi.org/10.1139/t99-089]

Chen, W.F., Giger, M.W. (1971). Limit analysis of stability of slopes. Journal of the Soil Mechanics and Foundations Division. *Proc. Am. Soc. Civ. Eng, 97*, 19-26.

Chen, W.F. (1975). *Limit Analysis and Soil Plasticity.* US: Elsevier.

Chen, W.F., Liu, X.L. (1990). *Limit analysis in soil mechanics.* New York: Elsevier.

Chen, Z., Morgenstern, N.R. (1983). Extensions to generalized method of slices for stability analysis. *Can. Geotech. J, 20*(1), 104-109.
[http://dx.doi.org/10.1139/t83-010]

Chen, Z., Shao, C. (1983). Evaluation of minimum factor of safety in slope stability analysis. *Can. Geotech. J, 25*(4), 735-748.
[http://dx.doi.org/10.1139/t88-084]

Chen, Z. (1998). On Pan's principles of rock and soil stability analysis. *Journal of Tsinghua University (Sci & Tech), 38*, 1-4.

Chen, Y.P., Lee, J.D., Eskandarian, A. (2006). *Meshless Methods in Solid Mechanics.* Springer.

Chen, G.Q., Ohnishi, Y., Ito, T. (1997). Development of high order manifold method. *Proceedings of the Second International Conference on Analysis of Discontinuous Deformation,* Kyoto, Japan132-154.

Chen, G.Q., Ohnishi, Y., Ito, T. (1998). Development of High-order Manifold Method. *Int. J. Numer. Methods Eng, V43*, 685-712.
[http://dx.doi.org/10.1002/(SICI)1097-0207(19981030)43:4<685::AID-NME442>3.0.CO;2-7]

Chen, G.Q., Miki, S., Ohnishi, Y. (1996). Practical improvements on DDA.

Chen, G.Q., Ohnishi, Y., Zhen, L., Sasaki, T. (2013). Frontiers of Discontinuous Numerical Methods and Practical Simulations in Engineering and Disaster Prevention *Proceedings of the 11th International Conference on analysis of Discontinuous Deformation, ICADD11.* Fukuoka, Japan27-29 August
[http://dx.doi.org/10.1201/b15791]

Chen, G.Q., Ohnishi, Y. (1999). A non-liner model for discontinuous in DDA. *The Third International Conference on Analysis of Discontinuous Deformation from Theory to Practice,* Vail, Colorado, USA57-64.

Chen, W., Qiu, T. (2011). Numerical Simulations of Granular Materials Using Smoothed Particle Hydrodynamics Method. *Int. J. Geomech, 12*(2), 157-164.

Cheng, Y.P., Nakata, Y., Bolton, M.D. (2003). Distinct element simulation of crushable soil. *Geotechnique, 53*(7), 633-641.
[http://dx.doi.org/10.1680/geot.2003.53.7.633]

Cheng, Y.M. (1989). An Efficient and Flexible Out-of-Core Equation Solver. *Microcomputers in Civil Engineering, 4*, 297-306.
[http://dx.doi.org/10.1111/j.1467-8667.1989.tb00031.x]

Cheng, Y.M., Tsui, Y. (1993). A Simple and Flexible Finite Element Mesh Graphics Program. *Comput. Struc, 48*, 555-574.

Cheng, Y.M. (1993). An Effective Storage Minimizer Program. *J. Struct. Eng, 20*, 103-110.

Cheng, Y.M. (1996). Iterative Solution using Sparse Matrix Storage in Finite Element Analysis *International Conference on Computing & Information Technology for Architecture, Engineering,* Construction, Singapore16-17 May93-96.

Cheng, Y.M., Tsui, Y. (1996).

Cheng, Y.M., Tsui, Y. (1997). Use of Discrete Element Analysis in Geotechnical Engineering *The Second International Symposium on Structures and Foundations in Civil Engineering,* Jan. 7-10153-156.

Cheng, Y.M. (1997). Application of Discrete Element Method/Discontinuous Deformation Analysis, The International Symposium on Rock Mechanics and Environmental Geotechnology, April, Chongqing, p. 214-218.

Cheng, Y.M. (1998). Advancement and Improvements in Discontinuous Deformation Analysis. *Comput. Geotech, 22*(2), 153-163.
[http://dx.doi.org/10.1016/S0266-352X(98)00002-0]

Cheng, Y.M., Zhang, Y.H. (1998). The Extension and Application of DDA Method. *Chinese Journal of Geotechnical Engineering, 20*(3), 109-111. a

Cheng, Y.M., Zhang, Y.H. (1998). Block Rotation in DDA and Its Application to Rolling Stone. *Chinese Journal of Rock Mechanics and Engineering, 17*, 834-839. b

Cheng, Y.M., Zhang, Y.H. (2000). Rigid Body Rotation and Internal Block Discretization in DDA Analysis. *Int. J. Numer. Anal. Methods Geomech, 24*, 567-578.
[http://dx.doi.org/10.1002/(SICI)1096-9853(200005)24:6<567::AID-NAG83>3.0.CO;2-N]

Cheng, Y.M., Zhang, Y.H., Wang, K.J. (2000). Coupling of Fem and DDA methods in engineering. *J. Geotech. Eng, 22*, 727-730.

Cheng, Y.M. (2002). Slip Line Solution and Limit Analysis for Lateral Earth Pressure Problem, the Ninth Conference on Computing in Civil and Building Engineering, April 3-5, Taipei, Taiwan, p.311-314.

Cheng, Y.M. (2003). Seismic lateral earth pressure coefficients by slip line method. *Comput. Geotech, 30*(8), 661-670.
[http://dx.doi.org/10.1016/j.compgeo.2003.07.003]

Cheng, Y.M., Zhang, Y.H., Chen, W.S. (2002). Wilson non-conforming Element in Numerical Manifold Method. *Commun. Numer. Methods Eng, 18*, 877-884.
[http://dx.doi.org/10.1002/cnm.545]

Cheng, Y.M., Chen, W.S., Guo, X.R. (2002). Distinct Element Analysis of Ground Deformation arising from Underground Mining. *Chinese Journal of Rock Mechanics and Engineering, 21*(8), 1130-1135.

Cheng, Y.M., Zhang, Y.H. (2002). Coupling of FEM and DDA Methods. *Int. J. Geomech, 2*(4), 503-517. [http://dx.doi.org/10.1061/(ASCE)1532-3641(2002)2:4(503)]

Procedure to the detect of three-dimensional blocks using penetration edges method Cheng, Y.M., Chen, W.S., Guo, X.R. (2002). Geotechnical Special PublicationASCE.

Cheng, Y.M. (2003). Locations of Critical Failure Surface and some Further Studies on Slope Stability Analysis. *Comput. Geotech, 30*, 255-267.
[http://dx.doi.org/10.1016/S0266-352X(03)00012-0]

Cheng, Y.M., Zhang, Y.H. (2003). Application of NMM in Underground excavation *Proceedings of GeoEng2003,* Beijing1159-1164.

Cheng, Y.M., Zhang, Y.H. (2004). Application and discussion of three dimensional numerical manifold method based on hexahedron element. *Chinese Journal of Rock Mechanics and Engineering,* 1745-1754.

Cheng, Y.M., Chen, W.S., Zheng, H., Zhang, Y.H. (2004). Detection of 3D block contacts by penetration edges. *Chinese Journal of Rock mechanics and engineering, 23*(4), 565-571.

Cheng, Y.M., Zhu, L.J. (2004). Unified Formulation for Two Dimensional Slope Stability Analysis and Limitations in Factor of Safety Determination. *Soils and Foundations, 44*(6), 121-128.

Cheng, Y.M., Chen, W.S., Zhang, Y.H. (2006). A robust method for the detection of contacts for three-dimensional blocks. *Int. J. Geomech, 6*(5), 303-310.
[http://dx.doi.org/10.1061/(ASCE)1532-3641(2006)6:5(303)]

Cheng, Y.M., Au, S.K. (2005). Slip line solution of bearing capacity problems with inclined ground. *Can. Geotech. J, 42*, 1232-1241.
[http://dx.doi.org/10.1139/t05-037]

Cheng, Y.M., Liu, H.T., Wei, W.B., Au, S.K. (2005). Location of critical three- dimensional non-spherical failure surface by NURBS functions and ellipsoid with applications to highway slopes. *Comput. Geotech, 32*(6), 387-399.
[http://dx.doi.org/10.1016/j.compgeo.2005.07.004]

Cheng, Y.M., Yip, C.J. (2007). Three-dimensional asymmetrical slope stability analysis – Extension of Bishop's, Janbu's, and Morgenstern–Price's techniques. *J. Geotech. Geoenviron. Eng, 133*(12), 1544-1555.
[http://dx.doi.org/10.1061/(ASCE)1090-0241(2007)133:12(1544)]

Cheng, Y.M., Hu, Y.Y., Wei, W.B. (2007). General Axi-symmetric active earth pressure by Methods of Characteristics – Theory and Numerical Formulation. *International Journal on Geomechanics, ASCE, 7*(1), 1-15.

Cheng, Y.M., Li, L. (2007). Particle swarm optimization algorithm for non-circular failure surface in two dimensional lope stability analysis. *Comput. Geotech, 34*(2), 92-103.
[http://dx.doi.org/10.1016/j.compgeo.2006.10.012]

Cheng, Y.M. (2007). Global optimization analysis of slope stability by simulated annealing with dynamic bounds and Dirac function. *Eng. Optim, 39*(1), 17-32.
[http://dx.doi.org/10.1080/03052150600916294]

Cheng, Y.M., Li, L., Chi, S.C. (2007). Studies on six heuristic global optimization methods in the location of critical slip surface for soil slopes. *Comput. Geotech, 34*, 462-484.
[http://dx.doi.org/10.1016/j.compgeo.2007.01.004]

Cheng, Y.M., Lansivaara, T., Wei, W.B. (2007). Two-dimensional Slope Stability Analysis by Limit Equilibrium and Strength Reduction Methods. *Comput. Geotech, 34*, 137-150.
[http://dx.doi.org/10.1016/j.compgeo.2006.10.011]

Cheng, Y.M., Zhang, Y.H. (2008). Three-dimensional Numerical Manifold Method - Tetrahedron and Hexahedron Mesh. *International for Rock Mechanics and Rock Engineering, 41*(4), 601-628.
[http://dx.doi.org/10.1007/s00603-006-0120-9]

Cheng, Y.M., Hu, Y.Y., Au, S.K., Wei, W.B. (2008). Active pressure for circular cut with Berezantzev's and Prater's Theories. *Soil Found, 48*(5), 621-632.
[http://dx.doi.org/10.3208/sandf.48.621]

Cheng, Y.M., Lansivaara, T., Siu, J. (2008). Impact of Convergence on Slope Stability Analysis and Design. *Comput. Geotech, 35*(1), 105-115.
[http://dx.doi.org/10.1016/j.compgeo.2007.02.011]

Cheng, Y.M., Li, L., Chi, S.C., Wei, W.B. (2008). Determination of critical slip surface using artificial fish swarms algorithm. *J. Geotech. Geoenviron. Eng, 134*(2), 244-251.
[http://dx.doi.org/10.1061/(ASCE)1090-0241(2008)134:2(244)]

Cheng, Y.M., Li, L., Lansivaara, T., Chi, S.C., Sun, Y.J. (2008). Minimization of factor of safety using different slip surface generation methods and an improved harmony search minimization algorithm. *Eng. Optim, 40*(2), 95-115.
[http://dx.doi.org/10.1080/03052150701618153]

Cheng, Y.M., Law, C.W. (2008). Development of a New and Efficient Thick Plate Element. *Struct. Eng. Mech, 29*(3), 327-354.
[http://dx.doi.org/10.12989/sem.2008.29.3.327]

Cheng, Y.M., Liu, Z.N., Song, W.D., Au, S.K. (2009). Laboratory test and Particle Flow Simulation of Silos problem with nonhomogeneous materials. *J. Geotech. Geoenviron. Eng, 135*, 1754-1761.
[http://dx.doi.org/10.1061/(ASCE)GT.1943-5606.0000125]

Cheng, Y.M., Chau, K.T., Xiao, L.J., Li, N. (2010). Flow pattern for silo with two layers of materials with single and double openings. *Journal of Geotechnical and Geoenvironmental Engineering, ASCE, 136*(9), 1278-1286.

Cheng, Y.M., Zhao, Z.H., Sun, Y.J. (2010). Evaluation of interslice force function and discussion on convergence in slope stability analysis by the lower bound method. *J. Geotech. Geoenviron. Eng, 136*(8), 1103-1113.
[http://dx.doi.org/10.1061/(ASCE)GT.1943-5606.0000317]

Cheng, Y.M., Li, D.Z., Li, L., Sun, Y.J., Baker, R., Yang, Y. (2011). Limit equilibrium method based on approximate lower bound method with variable factor of safety that can consider residual strength. *Comput. Geotech, 38*, 628-637.
[http://dx.doi.org/10.1016/j.compgeo.2011.02.010]

Cheng, Y.M., Li, L., Sun, Y.J., Au, S.K. (2012). A coupled particle swarm and harmony search optimization algorithm for difficult geotechnical problems. *Struct. Multidiscipl. Optim, 45*, 489-501.

[http://dx.doi.org/10.1007/s00158-011-0694-z]

Cheng, Y.M., Lansivaara, T., Baker, R., Li, N. (2013). The use of internal and external variables and extremum principle in limit equilibrium formulations with application to bearing capacity and slope stability problems. *Soil Found, 53*(1), 130-143.
[http://dx.doi.org/10.1016/j.sandf.2012.12.009]

Cheng, Y.M., Li, D.Z., Li, N., Li, Y.Y., Au, S.K. (2013). Solution of some engineering partial differential equations governed by the minimal of a functional by global optimization method. *J. Mech, 29*(3), 493-506.

Cheng, Y.M., Lau, C.K. (2014). *Soil Slope stability analysis and stabilization - new methods and insights* Spon Press.
[http://dx.doi.org/10.1201/b17015]

Cheng, Y.M., Li, L., Liu, L.L. (2015). Simplified Approach for Locating the Critical Probabilistic Slip Surface in Limit Equilibrium Analysis. *Nat. Hazards Earth Syst. Sci, 15*, 2241-2256.
[http://dx.doi.org/10.5194/nhess-15-2241-2015]

Cheng, Y.M., Li, N. (2017). Equivalence between Bearing Capacity, Lateral Earth Pressure and Slope Stability Problems by Slip-line and Extremum Limit Equilibrium Methods. *International Journal of Geomechanics, 17*(12) ASCE, 04017113.

Cheng, Y.M., Li, N., Fung, W.H., Li, L. (2019). Laboratory and Field Test and Distinct Element Analysis of Debris Flow. *Nat. Hazards Earth Syst. Sci, 19*, 181-199.
[http://dx.doi.org/10.5194/nhess-19-181-2019]

Cheng, Y.M., Au, S.K., Wong, H. (2019). *Fracture Grouting and Geonails for Soft Soil Tunnelling.* Geomechanics and Geoengineering.
[http://dx.doi.org/10.1080/17486025.2019.1573321]

Ching, J.Y., Phoon, K.K., Hu, Y.G. (2009). Efficient evaluation of reliability for slopes with circular slip surfaces using importance sampling. *J. Geotech. Geoenviron. Eng, 135*(6), 768-777.
[http://dx.doi.org/10.1061/(ASCE)GT.1943-5606.0000035]

Ching, J.Y., Phoon, K.K. (2013). Effect of element sizes in random field finite element simulations of soil shear strength. *Comput. Struc, 126*, 120-134.
[http://dx.doi.org/10.1016/j.compstruc.2012.11.008]

Chiou, Y.J., Tsay, R.J., Chuang, W.L. (1997). Crack propagation using manifold method. *Proceedings of the Second International Conference on Analysis of Discontinuous Deformation,* Kyoto, Japan298-308.

Chiou, Y.J., Tsay, R.J. (2000). Nonlinear analysis of low-rise reinforced concrete shear walls by manifold method. *Zhongguo Gongcheng Xuekan, 23*(6), 721-729.
[http://dx.doi.org/10.1080/02533839.2000.9670593]

Chiou, Y.J., Lee, Y.M., Tsay, R.J. (2002). Mixed mode fracture propagation by manifold method. *Int. J. Fract, 114*(4), 327-347.
[http://dx.doi.org/10.1023/A:1015713428989]

Cho, S.E. (2009). Probabilistic stability analyses of slopes using the ANN-based response surface. *Comput. Geotech, 36*, 787-797.
[http://dx.doi.org/10.1016/j.compgeo.2009.01.003]

Cho, S.E. (2010). Probabilistic assessment of slope stability that considers the spatial variability of soil properties. *J. Geotech. Geoenviron. Eng, 136*(7), 10.

[http://dx.doi.org/10.1061/(ASCE)GT.1943-5606.0000309]

Cho, S.E. (2013). First-order reliability analysis of slope considering multiple failure Modes. *Eng. Geol, 154*, 98-105.
[http://dx.doi.org/10.1016/j.enggeo.2012.12.014]

Chowdhury, R.N., Xu, D.W. (1995). Geotechnical system reliability of slopes. *Reliab. Eng. Syst. Saf, 47*, 141-151.
[http://dx.doi.org/10.1016/0951-8320(94)00063-T]

Christian, J.T., Ladd, C.C., Bolcher, G.B. (1992). Reliability and probability in stability analysis. *Stability and Performance of Slopes and Embankments - II* Geotechnical Special PublicationASCE.

Christian, J.T., Ladd, C., Baecher, G.B. (1994). Reliability applied to slope stability analysis. *J. Geotech. Eng, 120*(12), 2180-2207.
[http://dx.doi.org/10.1061/(ASCE)0733-9410(1994)120:12(2180)]

Clarke, S.D., Smith, C.C., Gilbert, M. (2013). Modelling discrete soil reinforcement in numerical limit analysis. *Can. Geotech. J, 50*, 705-715.
[http://dx.doi.org/10.1139/cgj-2012-0387]

Cornell, C.A. (1969). A probability based structural code. *J. Am. Concr. Inst, 66*(12), 974-985.

Coley, D.A. (1999). *An Introduction to Genetic Algorithms for Scientists and Engineers.* World Scientific.
[http://dx.doi.org/10.1142/3904]

Collins, I.F. (1974). A note on the interpretation of Coulomb's analysis of the thrust on a rough retaining wall in terms of the limit theorems of limit plasticity. *Geotechnique, 24*, 106-108.
[http://dx.doi.org/10.1680/geot.1974.24.1.106]

Cornell, C.A. (1969). A probability-based structural code. *J. Am. Concr. Inst, 66*(12), 974-985.

Cressie, N. (1993). *Statistics for spatial data.* John Wiley & Sons.

Crouch, S.L., Starfield, A.M. (1983). *Boundary element methods in solid mechanics, S. L. Crouch and A. M. Starfield.* London: George Allen & Unwin.

Cuevas, E., Zaldívar, D., Cisneros, M.P. (2018). *Advances in Metaheuristics Algorithms: Methods and Applications.* Springer.
[http://dx.doi.org/10.1007/978-3-319-89309-9]

Cundall, P.A. (1971). A computer model for simulating progressive, large-scale movements in blocky rock systems.

Cundall, P.A. (1978). *Ball - A computer program to model granular medium using the distinct element method, Technical note TN-LN-13* (pp. 129-163). London: Advanced Technology Group, Dames and Moore.

Cundall, P.A., Strack, O.D.L. (1979). A discrete model for granular assemblies. *Geotechnique, 29*(1), 47-65.
[http://dx.doi.org/10.1680/geot.1979.29.1.47]

Cundall, P.A., Strack, O.D.L. (1979). A discrete model for granular assemblies. *Geotechnique, 29*(1), 47-65.

Cundall, P.A., Hart, R.D. (1985). *Development of generalized 2-D and 3-D distinct element programs for modeling jointed rock, Misc. Paper SL-85-1.* US Army Corps of Engineers.

Cundall, P.A. (1987). Distinct Element Models of Rock and Soil structure. *Analytical and Computational Methods in Engineering Rock Mechanics* (pp. 129-163). London: George Allen and Unwin.

Cundall, P.A. (1988). Formulation of a 3-D distinct element model-Part I. A scheme to detect and represent contacts in system composed of many polyhedral blocks. *Int. J. Rock Mech. Min. Sci. Geomech. Abstr, 25*(3), 107-116.
[http://dx.doi.org/10.1016/0148-9062(88)92293-0]

Cundall, P.A., Hart, D.H. (1992). Numerical Modelling of Discontinua. *Eng. Comput, 9*, 101-111.
[http://dx.doi.org/10.1108/eb023851]

Dasaka, S.M., Zhang, L. (2012). Spatial variability of in situ weathered soil. *Geotechnique, 49*(6), 835-840.
[http://dx.doi.org/10.1680/geot.8.P.151.3786]

Davis, R.O., Selvadurai, A.P.S. *Cambridge University Press.*

De Jong, D.J.G. (1957). Graphical method for the determination of slip-line fields in soil mechanics. *Ingenior, 69*, 61-65.

De Jong, D.J.G. (1980). Application of calculus of variation to the vertical cut-off in cohesive frictionless soil. *Geotechnique, 30*(1), 1-16.
[http://dx.doi.org/10.1680/geot.1980.30.1.1]

De Jong, D.J.G. (1981). Variational fallacy. *Geotechnique, 31*(4), 289-290.
[http://dx.doi.org/10.1680/geot.1981.31.2.289]

DeGroot, D.J., Baecher, G.B. (1993). Estimating autocovariance of insitu soil properties. *J. Geotech. Eng, 119*(1), 147-166.
[http://dx.doi.org/10.1061/(ASCE)0733-9410(1993)119:1(147)]

Der Kiureghian, A., Ke, J.B. (1988). The stochastic finite element method in structural reliability. *Probab. Eng. Mech, 3*(2), 83-91.
[http://dx.doi.org/10.1016/0266-8920(88)90019-7]

Der Kiureghian, A., Dakessian, T. (1998). Multiple design points in first and second order reliability. *Struct. Saf, 20*(1), 37-49.
[http://dx.doi.org/10.1016/S0167-4730(97)00026-X]

Desai, C.S., Zaman, M.M., Lightner, J.G., Siriwardane, H.J. (1984). Thin-layer element for interfaces and joints. *Int. J. Numer. Anal. Methods Geomech.*
[http://dx.doi.org/10.1002/nag.1610080103]

Diaby, M., Karwan, M.H. (2016). *Advances in combinatorial optimization.* World Scientific Publishing.
[http://dx.doi.org/10.1142/9725]

Dinis, L.M.J.S., Jorge, R.M.N., Belinha, J. (2007). Analysis of 3D solids using the natural neighbor radial point interpolation method. *Comput. Methods Appl. Mech. Eng, 196*(13-16), 2009-2028.
[http://dx.doi.org/10.1016/j.cma.2006.11.002]

Dolinski, K. (1983). First order second moment approximation in reliability of structural systems -critical review and alternative approach. *Struct. Saf, 1*(3), 211-231.
[http://dx.doi.org/10.1016/0167-4730(82)90027-3]

Donald, I.B., Chen, Z. (1997). Slope stability analysis by the upper bound approach: fundamentals and

methods. *Can. Geotech. J, 34*, 853-862.
[http://dx.doi.org/10.1139/t97-061]

Doss, L.J.T., Arathi, P. (2016). A constructive bandwidth reduction algorithm-A variant of GPS algorithm. *AKCE International Journal of Graphs and Combinatorics, 13*, 241-254.
[http://dx.doi.org/10.1016/j.akcej.2016.06.014]

Ditlevsen, O. (1979). Generalized second moment reliability index. *Journal of Structural Mechanics, 7*(4), 435-451.
[http://dx.doi.org/10.1080/03601217908905328]

Dow, J.O. (2015). *A concise overview of the finite element method.* Momentum Press.

Drescher, A. (1983). Limit plasticity approach to piping in bins. *ASME Trans. J. Appl. Mech, 50*, 549-553.
[http://dx.doi.org/10.1115/1.3167089]

Drescher, A. (1986). Kinematics of axisymmetric vertical slopes at collapse. *Int. J. Numer. Anal. Methods Geomech, 10*, 431-444.
[http://dx.doi.org/10.1002/nag.1610100407]

Drescher, A., Detournay, E. (1993). Limit load in translational failure mechanisms for associative and non-associative materials. *Geotechnique, 43*(3), 443-456.
[http://dx.doi.org/10.1680/geot.1993.43.3.443]

Drucker, D.C., Prager, W. (1952). Soil mechanics and plastic analysis or limit design. *Q. Appl. Math, 10*, 157-165.
[http://dx.doi.org/10.1090/qam/48291]

Drucker, D.C., Greenberg, W., Prager, W. (1951). The safety factor of an elastic plastic body in plane strain, Transactions of the ASME. *J. Appl. Mech, 73*, 371.

Duarte, C.A., Oden, J.T. (1996). h-p clouds - an h-p meshless method. *Numer. Methods Partial Differ. Equ, 12*(6), 673-705.
[http://dx.doi.org/10.1002/(SICI)1098-2426(199611)12:6<673::AID-NUM3>3.0.CO;2-P]

Duncan, J.M. (1996). State of the art: Limit equilibrium and finite element analysis of slopes. *J. Geotech. Eng, 122*(7), 577-596.
[http://dx.doi.org/10.1061/(ASCE)0733-9410(1996)122:7(577)]

Duncan, J.M. (2000). Factors of safety and reliability in geotechnical engineering. *J. Geotech. Geoenviron. Eng, 126*(4), 307-316.
[http://dx.doi.org/10.1061/(ASCE)1090-0241(2000)126:4(307)]

Ellingwood, B.R., Galambos, T.V., MacGregor, J.G., Cornell, C.A. (1980). Development of probability-based load criterion for American National Standard A58. *Special Publication 577, National Bureau of Standards, Washington.*
[http://dx.doi.org/10.6028/NBS.SP.577]

Fan, K., Fredlund, D.G., Wilson, G.W. (1986). An interslice force function for limit equilibrium slope stability analysis. *Can. Geotech. J, 23*(3), 287-296.
[http://dx.doi.org/10.1139/t86-042]

Fellin, W., Lessmann, H., Oberguggenberger, M., Vieider, R. (2005). *Analyzing Uncertainty in Civil Engineering.* Springer.

[http://dx.doi.org/10.1007/b138177]

Fenton, G.A. (1999). Random field modeling of CPT data. *J. Geotech. Eng, 125*(6), 486-498. [http://dx.doi.org/10.1061/(ASCE)1090-0241(1999)125:6(486)]

Fenton, G.A., Vanmarcke, E.H. (1998). Spatial variation in liquefaction risk. *Geotechnique, 48*(6), 819-831. [http://dx.doi.org/10.1680/geot.1998.48.6.819]

Fenton, G.A., Griffiths, D.V. (2002). Probabilistic foundation settlement on spatially random soil. *J. Geotech. Geoenviron. Eng, 128*(5), 381-390. [http://dx.doi.org/10.1061/(ASCE)1090-0241(2002)128:5(381)]

Fenton, G.A., Griffiths, D.V. (2003). Bearing-capacity prediction of spatially random c-phi soils. *Can. Geotech. J, 40*(1), 54-65. [http://dx.doi.org/10.1139/t02-086]

Fenton, G.A., Griffiths, D.V. (2005). Three-dimensional probabilistic foundation settlement. *J. Geotech. Geoenviron. Eng, 131*(2), 232-239. [http://dx.doi.org/10.1061/(ASCE)1090-0241(2005)131:2(232)]

Fenton, G.A. (2007). Data analysis/geostatistics. *Probabilistic methods in geotechnical engineering* (pp. 51-73). New York: Springer.

Fiessler, B., Neumann, H.J., Rackwitz, R. (1979). Quadratic limit states in structural reliability. *J. Eng. Mech, 105*(4), 661-676.

Fleming, K., Weltman, A., Randolph, M., Elson, K. (2009). *Piling engineering* Taylors and Francis. Flibs.http://flibs.sourceforge.net/

Floudas, C.A., Pardalos, P.M. (2009). *Encyclopedia of Optimization.* Springer. [http://dx.doi.org/10.1007/978-0-387-74759-0]

Fredlund, D.G., Krahn, J. (1984). Analytical methods for slope analysis *International Symposium on Landslides,* 229-250.

Freudenthal, A.M. (1947). Safety of Structures. *Trans. Am. Soc. Civ. Eng, 112*, 125-159.

Frey, P.L., George, P.J. (2000). *Mesh generation* John Wiley.

Friedman, J.H. (1991). Multivariate adaptive regression splines. *Ann. Stat, 19*(1), 1-67. [http://dx.doi.org/10.1214/aos/1176347963]

Ganesan, S., Tobiska, L. (2017). *Finite elements theory and algorithms.* Cambridge University Press. [http://dx.doi.org/10.1017/9781108235013]

Ghanem, R., Saad, G., Doostan, A. (2007). Efficient solution of stochastic systems: application to the embankment dam problem. *Struct. Saf, 29*(3), 238-251. [http://dx.doi.org/10.1016/j.strusafe.2006.07.015]

Gharti, H.N., Komatitsch, D., Oye, V., Martin, R., Tromp, J. (2012). Application of an elastoplastic spectral-element method to 3D slope stability analysis. *Int. J. Numer. Methods Eng, 91*(1), 1-26. [http://dx.doi.org/10.1002/nme.3374]

Gharti, H.N., Tromp, J. (2017).

Gilbert, M., Smith, C., Pritchard, T. (2010). Masonry arch analysis using discontinuity layout optimisation. *Proc. Instn Civ. Engrs - Engng. Comput. Mech, 163*(3), 155-166.

Gingold, R.A., Monaghan, J.J. (1977). Smoothed particle hydrodynamics: theory and application to non-spherical stars. *Mon. Not. R. Astron. Soc, 181*, 375-389. [http://dx.doi.org/10.1093/mnras/181.3.375]

Glover, F., Laguna, M. (1996). *Tabu Search.* Dordrecht, Netherlands: Kluwer.

Goh, A.T.C. ((1984)). Genetic algorithm search for critical slip surface in multiple-wedge stability analysis. In: Graham , J., (Ed.), *Canadian Geotechnical Journal, 36(2), 382–391. Graham J* Slope Instability, John Wiley & Sons.36(2),

Goldreich, O.P. (2010). *NP, and NP-Completeness.* Cambridge University Press. [http://dx.doi.org/10.1017/CBO9780511761355]

Gong, W.P., Wang, L., Khoshnevisan, S., Juang, C.H., Huang, H.W., Zhang, J. (2015). Robust geotechnical design of earth slopes using fuzzy sets. *J. Geotech. Geoenviron. Eng, 141*(1), 1-9. [http://dx.doi.org/10.1061/(ASCE)GT.1943-5606.0001196]

Goodman, R.E. (1976). *Methods of geological engineering in discontinuous rocks.* San Francisco, CA, USA: West Publishing Company.

Greco, V.R. (1996). Efficient Monte Carlo technique for locating critical slip surface. *J. Geotech. Eng, 122*, 517-525. [http://dx.doi.org/10.1061/(ASCE)0733-9410(1996)122:7(517)]

Gregory, B. B., John, T. C. (2005). Reliability and Statistics in Geotechnical Engineering. .Wiley Press, Baecher Christian.

Griffiths, D.V., Fenton, G.A. (2000). Influence of soil strength spatial variability on the stability of an undrained clay slope by finite element, Slope stability 2000. *GSP, 101*, 184-193.

Griffiths, D.V., Fenton, G.A. (2001). Bearing capacity of spatially random soil:The undrained clay Prandtl problem revisited. *Geotechnique, 51*(4), 351-359. [http://dx.doi.org/10.1680/geot.2001.51.4.351]

Griffiths, D.V., Fenton, G.A. (2004). Probabilistic slope stability analysis by Finite element methods. *J. Geotech. Geoenviron. Eng, 130*(5), 507-518. [http://dx.doi.org/10.1061/(ASCE)1090-0241(2004)130:5(507)]

Griffiths, D.V., Fenton, G.A. (2007). Probabilistic slope stability analysis by Finite element methods. *J. Geotech. Geoenviron. Eng, 130*(5), 507-518. [http://dx.doi.org/10.1061/(ASCE)1090-0241(2004)130:5(507)]

Griffiths, D.V., Fenton, G.A. (2009). Probabilistic Settlement Analysis by Stochastic and Random Finite element methods. *J. Geotech. Geoenviron. Eng, 135*(11), 1629-1637. [http://dx.doi.org/10.1061/(ASCE)GT.1943-5606.0000126]

Griffiths, D.V., Huang, J., Fenton, G.A. (2011). Probabilistic infinite slope analysis. *Comput. Geotech, 38*, 577-584. [http://dx.doi.org/10.1016/j.compgeo.2011.03.006]

Guenin, B., Konemann, J., Tuncel, L. (2014). *A Gentle Introduction to Optimization.* Cambridge University

Press.
[http://dx.doi.org/10.1017/CBO9781107282094]

Haddad, O.B. (2018). *Advanced Optimization by Nature-Inspired Algorithms.* Springer.

Haldar, A., Mahadevan, S. (2000). *Reliability assessment using stochastic finite Reliability assessment using stochastic finite Reliability assessment using stochastic finite analysis.* John Wiley.

Haldar, A., Mahadevan, S. (2000). *Probability, reliability and statistical methods in engineering design.* John Wiley. a

Haldar, A., Mahadevan, S. (2000). *Reliability assessment using stochastic finite Reliability assessment using stochastic finite Reliability assessment using stochastic finite analysis.* John Wiley. b

Halsey, T., Mehta, A. (2002). *Challenges in Granular Physics.* World Scientific.

Hansen, J.B. (1970). A revised and extended formula for bearing capacity, Danish Geotechnical Institute. *Bulletin,* 28. [Copenhagen.].

Harr, M.E. (1977). *Mechanics of particulate media: A probabilistic approach.* New York: McGraw Hill.

Hartmann, F., Katz, C. (2004). *Structural Analysis with Finite Elements.* Springer.
[http://dx.doi.org/10.1007/978-3-662-05423-9]

Hasofer, A.M., Lind, N.C. (1974). Exact and invariant second-moment code format. *J. Eng. Mech. Div, 100*(EM1), 111-121.

Hassan, A.M., Wolff, T.F. (1999). Search algorithm for minimum reliability index of earth slopes. *J. Geotech. Geoenviron. Eng, 125*(4), 301-308.
[http://dx.doi.org/10.1061/(ASCE)1090-0241(1999)125:4(301)]

Henderson, C.R. (1975). Best linear unbiased estimation and prediction under a selection model. *Biometrics, 31*(2), 423-447.
[http://dx.doi.org/10.2307/2529430]

Hertz, H. (1881). Ueber die Berührung fester elastischer Körper. *J. Reine Angew. Math, 92*, 156-171.

Hicks, M.A., Spencer, W. (2010). Influence of heterogeneity on the reliability and failure of a long 3D slopes. *Comput. Geotech, 37*(7-8), 948-955.
[http://dx.doi.org/10.1016/j.compgeo.2010.08.001]

Hilbert, L.B., Yi, W., Cook, N.G.W., Cai, Y., Liang, G.P. (1994). A new discontinuous finite element method for interaction of many deformable bodies in geomechanics.

Hill, R. (1950). *Mathematical theory of plasticity.* Oxford University Press.

Hill, J.M., Cox, G.M. (2000). Cylindrical cavities and classical rat-hole theory occurring in bulk materials. *Int. J. Numer. Anal. Methods Geomech, 24*, 971-990.
[http://dx.doi.org/10.1002/1096-9853(200010)24:12<971::AID-NAG107>3.0.CO;2-G]

Hinton, E., Owen, D.R.J. (1977). *Programming the finite element method.* Academic Press.

Homayouni, S.M., Tang, S.H., Motlagh, O. (2014). A genetic algorithm for optimization of integrated scheduling of cranes, vehicles, and storage platforms at automated container terminals. *J. Comput. Appl. Math, 270*, 545-556.

[http://dx.doi.org/10.1016/j.cam.2013.11.021]

Hohenbichler, M., Rackwitz, R. (1988). First order concept in system reliability. *Struct. Saf, 1*(3), 177-188. [http://dx.doi.org/10.1016/0167-4730(82)90024-8]

Hong, H.P. (1999). Simple approximations for improving second-order reliability estimates. *J. Eng. Mech, 125*(5), 592-595.
[http://dx.doi.org/10.1061/(ASCE)0733-9399(1999)125:5(592)]

Hong, H.P., Roh, G. (2008). Reliability evaluation of earth slopes. *J. Geotech. Geoenviron. Eng, 134*(12), 1700-1705.
[http://dx.doi.org/10.1061/(ASCE)1090-0241(2008)134:12(1700)]

Honjo, Y., Kuroda, K.A. (1991). New look at fluctuating geotechnical data for reliability design. *Soil Found, 31*(01), 110-120.
[http://dx.doi.org/10.3208/sandf1972.31.110]

Houlsby, G.T., Wroth, C.P. (1982). Direct solution of plasticity problems in soils by the method of characteristics *Proc., 4th Int. Conf. on Numerical Methods in Geomechancis,* Edmonton, Alta., Canada*3*, 1059-1071.

Hovland, H.J. (1977). Three-Dimensional Slope Stability analysis method. *J. Geotech. Eng. Div, 103*, 971-986.

Huang, C.C., Tsai, C.C. (2000). New method for 3D and asymmetrical slope stability analysis. *J. Geotech. Geoenviron. Eng, 126*(10), 917-927.
[http://dx.doi.org/10.1061/(ASCE)1090-0241(2000)126:10(917)]

Huang, C.C., Tsai, C.C., Chen, Y.H. (2002). Generalized method for three- dimensional slope stability analysis. *J. Geotech. Geoenviron. Eng, 128*(10), 836-848.
[http://dx.doi.org/10.1061/(ASCE)1090-0241(2002)128:10(836)]

Huang, J.S., Griffiths, D.V., Fenton, G.A. (2010). Probabilistic analysis of coupled soil consolidation. *J. Geotech. Geoenviron. Eng, 136*(3), 417-430. a
[http://dx.doi.org/10.1061/(ASCE)GT.1943-5606.0000238]

Huang, J.S., Griffiths, D.V., Fenton, G.A. (2010). System reliability of slopes by RFEM. *Soil Found, 50*(3), 345-355. b
[http://dx.doi.org/10.3208/sandf.50.343]

Hughes, T.J.R., Hinton, E. (1986). *Finite Element Methods for Plates and Shell Structures* Pineridge Press.

Ibrahimbegovic, A. (1993). Quadrilateral Finite Elements for Analysis of Thick and Thin Plates. *Comput. Methods Appl. Mech. Eng, 10*, 195-209.
[http://dx.doi.org/10.1016/0045-7825(93)90160-Y]

Irfan, T.Y. (1986). *Mode and mechanism of a complex failure at Tin Wan Hill, Special Projects Report 2/86.* Hong Kong: Geotechnical Control Office.

Irfan, T.Y. (1998). Structurally controlled landslides in saprolitic soils in Hong Kong. *Geotech. Geol. Eng, 16*, 215-238.
[http://dx.doi.org/10.1023/A:1008805827178]

Isukapalli, S.S., Roy, A., Georgopoulos, P.G. (1998). Stochastic response surface methods for uncertainty propagation: application to environmental and biological systems. *Risk Anal, 18*(3), 351-363.

[http://dx.doi.org/10.1111/j.1539-6924.1998.tb01301.x]

(1995). PFC3D 1.0, User Guide.

(2004). PFC3D 3.1, User Guide.

Iwashita, K., Oda, M. (1998). Rolling resistance at contacts in simulation of shear band development by DEM. *J. Eng. Mech, 124*(3), 285-292.
[http://dx.doi.org/10.1061/(ASCE)0733-9399(1998)124:3(285)]

Izbicki, R.J. (1981). Limit plasticity approach to slope stability problems, Journal of the Geotechnical Engineering Division. *Proc. Am. Soc. Civ. Eng, 107*, 228-233.

Janbu, N. (1957). Earth pressure and bearing capacity by generalized procedure of slices *Proc. 4th International Conference on Soil Mechanics,* 207-212.

Janbu, N. (1973). Slope stability computations. *Embankment-Dam Engineering* John Wiley.

Jaksa, M.B., Kaggwa, W.S., Brooker, P.I. (1999). Experimental evaluation of the scale of fluctuation of a stiff clay

Jaksa, M.B., Fenton, G.A. (2000). Random field modeling of CPT data. *J. Geotech. Geoenviron. Eng, 126*(12), 1212-1216.
[http://dx.doi.org/10.1061/(ASCE)1090-0241(2000)126:12(1212)]

Jebahi, M., André, D., Terreros, I., Iordanoff, I. (2015). *Discrete Element Method to Model 3D Continuous Materials.* John Wiley.
[http://dx.doi.org/10.1002/9781119103042]

Jenike, A.J., Yen, B.C. (1962). Slope stability in axial symmetry *Proc., 5th Symp. on Rock Mechanics,* Univ. of Minnesota, Pergamon.New York: 689-711.

Jha, S.K., Ching, J.Y. (2013). Simplified reliability method for spatially variable undrained engineered slopes. *Soil Found, 53*(5), 708-719.
[http://dx.doi.org/10.1016/j.sandf.2013.08.008]

Ji, J., Low, B.K. (2012). Stratified response surfaces for system probabilistic evaluation 781 of slopes. *J. Geotech. Geoenviron. Eng, 138*(11), 1398-1406.
[http://dx.doi.org/10.1061/(ASCE)GT.1943-5606.0000711]

Ji, J., Liao, H.J., Low, B.K. (2012). Modeling 2-D spatial variation in slope reliability analysis using interpolated autocorrelations. *Comput. Geotech, 40*, 135-146.
[http://dx.doi.org/10.1016/j.compgeo.2011.11.002]

Jing, L.R., Stephansson, O., Erling, N. (1993). Study of rock joints under cyclic loading conditions. *Rock Mech. Rock Eng, 26*(3), 215-232.
[http://dx.doi.org/10.1007/BF01040116]

Jiang, Q.H. (2000). *Research on three dimensional discontinuous deformation analysis method.*

Jiang, M.J., Yu, H.S., Harris, D. (2005). A novel discrete model for granular material incorporating rolling resistance. *Comput. Geotech, 32*, 340-357.
[http://dx.doi.org/10.1016/j.compgeo.2005.05.001]

Jiao, Y.Y. (1998). *Three dimension DDA and its application. Ph.D. Thesisi.* Wuhan, China: Wuhan Institute of Rock & Soil Mechanics.

Jing, R.J. (2007). *Fundamentals of Discrete Element Methods for Rock Engineering: Theory and Applications.* Elsevier.

Jiang, S.H., Huang, J.S. (2016). Efficient slope reliability analysis at low-probability levels in spatially variable soils. *Comput. Geotech, 75*, 18-27.
[http://dx.doi.org/10.1016/j.compgeo.2016.01.016]

Jiang, S.H., Li, D.Q., Zhang, L.M., Zhou, C.B. (2014). Slope reliability analysis considering spatially variable shear strength parameters using a non-intrusive stochastic finite element method. *Eng. Geol, 168*, 120-128.
[http://dx.doi.org/10.1016/j.enggeo.2013.11.006]

Jiang, S.H., Qi, X.H., Cao, Z.J., Li, D.Q. (2015). System reliability analysis of slope with stochastic response surface method. *Chin J Rock Soil Mech, 36*(3), 809-818.

Jirousek, J., Wroblewski, A., Szybinski, B. (1995). A new 12 d.o.f. Quadrilateral element for analysis of thick and thin plates. *Int. J. Numer. Methods Eng, 38*, 2619-2638.
[http://dx.doi.org/10.1002/nme.1620381508]

Juang, C.H., Rosowsky, D.V., Tang, W.H. (1999). Reliability-based method for assessing liquefaction potential of soils. *J. Geotech. Geoenviron. Eng, 128*(8), 684-689.
[http://dx.doi.org/10.1061/(ASCE)1090-0241(1999)125:8(684)]

Kang, F., Han, S.X., Salgado, R., Li, J.J. (2015). System probabilistic stability analysis of soil slopes using Gaussian process regression with Latin hypercube sampling. *Computers and Geotechnics, 63:13:25.*

Kang, F., Li, J. (2015). Artificial bee colony algorithm optimized support vector regression for system reliability analysis of slopes. *J. Comput. Civ. Eng.*04015040

Karel, K. (1977). Application of energy method, Journal of the Geotechnical Engineering Division. *Proc. Am. Soc. Civ. Eng, 103*, 381-397. a

Karel, K. (1977). Energy method for soil stability analyses, Journal of the Geotechnical Engineering Division. *Proc. Am. Soc. Civ. Eng, 103*, 431-445. b

Kaymaz, I. (2005). Application of kriging method to structural reliability problems. *Struct. Saf, 27*(2), 133-151.
[http://dx.doi.org/10.1016/j.strusafe.2004.09.001]

Ke, T.C. (1995). Modification of DDA with respect to rigid-body rotation. *Proceedings of the First International Conference on Analysis of Discontinuous Deformation,* Chungli, Taiwan260-273.

Ke, T.C., Thapa, B., Goodman, R.E. (1994). Stability analysis of a penstock slope. *Proceedings of the First North American Rock Mechanics Symposium,* Austin, TX, USA1109-1116.

Ke, T.C., Tang, J.H. (1996). Modeling of solid-fluid Interactions using the manifold method. *Proceedings of the Second North American Rock Mechanics Symposium,* Montreal, Quebec, Canada1815-1822.

Kennedy, J., Eberhart, R. (1995). Particle Swarm Optimization *International Conference on Neural Networks,* Perth, Australia1942-1948.

Kerisel, J., Absi, E. (1990). *Active and passive earth pressure tables.* U.K.: Taylor and Francis.

Khoei, A.R. (2015). *Extended finite element method.* John Wiley.

Kim, J.M., Sitar, N. (2013). Reliability approach to slope stability analysis with spatially correlated soilproperties. *Soil Found,* *53*(1), 1-10.
[http://dx.doi.org/10.1016/j.sandf.2012.12.001]

Kirk, P.A., Campbell, S.D.G., Fletcher, C.J.N., Merriman, R.J. (1997). The significance of primary volcanic fabrics and clay distributions in landslides in Hong Kong. *J. Geol. Soc. London,* *154*, 1009-1019.
[http://dx.doi.org/10.1144/gsjgs.154.6.1009]

Knill, J. (2006). *Report on Shum Wan Road Landslide of 13 August 1995.* Geotechnical Engineering Office, Hong Kong SAR Government.

Knill and GEO (2006). Report on the Fei Tsui road landslide of 13 August 1995, Hong Kong SAR Government.

Krige, D.G. (1994). A statistical approach to some basic mine valuation problems on the witwatersrand. *J. S. Afr. Inst. Min. Metall,* *94*(3), 95-112.

Komzsik l. (2009). *Applied Calculus of Variations for Engineers.* CRC Press.

Koo, C.Y., Chern, J.C., Chen, S. (1995). The development of the second-order displacement function for discontinuous deformation analysis. *Proceedings of the Sixth Conference on Geotechnical Engineering,* Taiwan665-674.

Koo, C.Y., Chern, J.C. (1996). *The development of DDA with third displacement function. DDA and Simulations of Discontinuous Media* TSI press.

Konietzky, H. (2004). Numerical Modelling of Discrete Materials in Geotechnical Engineering, Civil Engineering and Earth Sciences: Proceedings of the First International UDEC/3DEC Symposium, Bochum, Germany, 29 September - 1 October 2004, CRC Press.

Korte, B., Vygen, J. (2018). *Combinatorial Optimization Theory and Algorithms6th edition.* Springer.
[http://dx.doi.org/10.1007/978-3-662-56039-6]

Kosinski, W. (2008). *Advances in Evolutionary algorithms, AvE4EvA.*

Kotter, F. (1903). *Die Bestimmung des Druckes an gekrmmten Gleitflchen, eine Aufgabe aus der Lehre vom Erddruck, Berlin Akad* (pp. 229-233). Berlin: Wiss.

Koutromanos, I. (2018). *Fundamentals of Finite Element Analysis, Linear Finite Element Analysis.* John Wiley.

Koyluoglu, H.U., Nielsen, S.R.K. (1994). New approximations for SORM integrals. *Struct. Saf,* *13*(4), 235-246.
[http://dx.doi.org/10.1016/0167-4730(94)90031-0]

Krahn, J., Fredlund, D.G. (1997). Evaluation of the University of Saskatchewan slope stability program." R.T.A.C Forum, Roads and Transportation Association of Canada, Quebec City, Canada, 69–76.

Kulhawy, F.H., Mayne, P.W. (1990). *Manual on estimating soil properties for foundation design.* Ithaca, NY, USA: Cornell Univ.

Kumar, J. (2001). Seismic passive earth pressure coefficients for sands. *Can. Geotech. J,* *38*, 876-881.
[http://dx.doi.org/10.1139/t01-004]

Kumar, J., Chitikela, S. (2002). Seismic passive earth pressure coefficients using the method of characteristics. *Can. Geotech. J, 39*, 463-471.
[http://dx.doi.org/10.1139/t01-103]

Lasdon, L.S., Warren, A.D., Jain, A., Ratner, M. (1978). Design and testing of a generalized reduced gradient code for nonlinear programming. *ACM Trans.Math, 4*(1), 34-50.
[http://dx.doi.org/10.1145/355769.355773]

Lam, L., Fredlund, D.G., Barbour, S.L. (1987). Transient Seepage Model for Saturated-Unsaturated Soil Systems: a Geotechnical Engineering Approach. *Can. Geotech. J, 24*, 565-580.
[http://dx.doi.org/10.1139/t87-071]

Lam, L., Fredlund, D.G. (1993). A general limit equilibrium model for three-dimensional slope stability analysis. *Can. Geotech. J, 30*(6), 905-919.
[http://dx.doi.org/10.1139/t93-089]

Lawrence, C.T., Tits, A.L. (1996). Nonlinear equality constraints in feasible sequential quadratic programming. *Optim. Methods Softw, 6*, 265-282.
[http://dx.doi.org/10.1080/10556789608805638]

Lawrence, S. (1984). *Resource constrained project scheduling: An experimental investigation of heuristic scheduling techniques, Pittsburgh: School of industrial Administration, Carnegie.* USA: Mellon University.

Lemos, J.V. (1983). *A Hybrid Distinct Element-Boundary Element Computational Model for the Half-Plane.*

Leshchinsky, B. (2015). Bearing Capacity of Footings Placed Adjacent to c′-ϕ′ Slopes. *J. Geotech. Geoenviron. Eng, 141*(6)04015022
[http://dx.doi.org/10.1061/(ASCE)GT.1943-5606.0001306]

Leshchinsky, B., Ambauen, S. (2015). Limit Equilibrium and Limit Analysis: Comparison of Benchmark Slope Stability Problems. *J. Geotech. Geoenviron. Eng, 141*(10)04015043
[http://dx.doi.org/10.1061/(ASCE)GT.1943-5606.0001347]

Li, K.S., Lee, I.K. (1991). The assessment of geotechnical safety. *Selected Topics in Geotechnical Engineering, The Lumb Volume* (pp. 195-229). Canberra: University College, University of New South Wales.

Li, K.S., Lumb, P. (1987). Probabilistic design of slopes. *Can. Geotech. J, 24*(4), 520-535.
[http://dx.doi.org/10.1139/t87-068]

Li, L., Chu, X.S. (2015). Multiple response surfaces for slope reliability analysis. *Int. J. Numer. Anal. Methods Geomech, 39*(2), 175-192.
[http://dx.doi.org/10.1002/nag.2304]

Li, D.Q., Jiang, S.H., Cao, Z.J., Zhou, W., Zhou, C.B., Zhang, L.M. (2015). A multiple response surface method for slope reliability analysis considering spatial variability of soil properties. *Eng. Geol, 187*, 60-72.
[http://dx.doi.org/10.1016/j.enggeo.2014.12.003]

Li, S.F., Liu, W.K. (2007). *Meshfree Particle Methods.* Springer.

Li, X.K., Feng, Y.T., Mustoe, G. (2017). *Proceedings of the 7th International Conference on Discrete Element Methods*Springer.

Li, D.Z., Cheng, Y.M. (2012). Lower Bound Limit Analysis Using Nonlinear Failure Criteria, 2012

International Conference on Structural Computation and Geotechnical Mechanics. Procedia Earth and Planetary Science City: Kun Ming China, pp. 170-174.

Li, D.Z., Cheng, Y.M., Wang, J.A., Yang, Y., Li, N. (2013). Application of finite-element-based limit analysis with mesh adaptation in geotechnical engineering, Chinese. *J. Geotech. Eng, 35*(5), 922-929.

Li, N., Cheng, Y.M. (2015). Laboratory and 3D-Distinct Element Analysis of Failure Mechanism of Slope under External Surcharge. *Nat. Hazards Earth Syst. Sci, 15*, 35-43.
[http://dx.doi.org/10.5194/nhess-15-35-2015]

Li, N., Cheng, Y.M., Fung, W.H. (2018). Transition Zone in Bearing Capacity Problem from Plasticity Method, Discrete Element Method and Laboratory Tests. *Adv. Geosci, 2*(1), 1-21.

Li, H., Mulay, S.S. (2013). *Meshless methods and their numerical properties.* CRC Press.
[http://dx.doi.org/10.1201/b14492]

Li, L., Chu, X.S. (2016). Locating the Multiple Failure Surfaces for Slope Stability Using Monte Carlo Technique. *Geotech. Geol. Eng, 34*(5), 1475-1486.
[http://dx.doi.org/10.1007/s10706-016-0055-1]

Li, X.Y., Zhang, L.M., Gao, L., Zhu, H. (2016). Simplified slope reliability analysis considering spatial soil variability. *Eng. Geol.*
[http://dx.doi.org/10.1016/j.enggeo.2016.11.013]

Libersky, L.D., Petschek, A.G., Carney, T.C., Hipp, J.R., Allahdadi, F.A. (1993). High strain lagrangian hydrodynamics: A three-dimensional SPH code for dynamic material response. *J. Comput. Phys, 109*(1), 67-75.
[http://dx.doi.org/10.1006/jcph.1993.1199]

Lind, N.C. (1971). Consistent partial safety factors. *J. Struct. Div, 97*(ST6), 1651-1669.

Lin, J.S., Lee, D.H. (1996). Manifold method using polynomial basis function of any order.

Lin, D.Z., Mo, H.H. (1994). *Manifold method of material analysis.* The University of Oklahoma.

Lind, N.C. (1983). Modelling uncertainty in discrete dynamical systems. *Appl. Math. Model, 7*(3), 146-152.
[http://dx.doi.org/10.1016/0307-904X(83)90001-X]

Liu, H.P. (2009). *Taylor kriging metamodeling for simulation interpolation, sensitivity analysis and optimization.*

Liu, H.P., Shi, J., Erdem, E. (2010). Prediction of wind speed time series using modified taylor kriging method. *Energy, 35*(12), 4870-4879.
[http://dx.doi.org/10.1016/j.energy.2010.09.001]

Liu, G.R. (2010). *Meshfree methods Moving Beyond the Finite Element Method.* CRC Press.

Liu, M.B., Liu, G.R. (2016). *Particle methods for multi-scale and multi-physics.* World Scientific Publishing.
[http://dx.doi.org/10.1142/9017]

Liu, L.L., Cheng, Y.M. (2016). Efficient system reliability analysis of soil slopes using multivariate adaptive regression splines-based Monte Carlo simulation. *Comput. Geotech, 79*, 41-54.
[http://dx.doi.org/10.1016/j.compgeo.2016.05.001]

Liu, L.L., Cheng, Y.M., Wang, X.M. (2017). Genetic algorithm optimized Taylor Kriging surrogate model

for system reliability analysis of soil slopes. *Journal of Landslides, 14*, 535-546.
[http://dx.doi.org/10.1007/s10346-016-0736-0]

Liu, L.L., Cheng, Y.M. (2017). Conditional random field reliability analysis of a cohesion-frictional slope. *Comput. Geotech, 82*, 173-186.
[http://dx.doi.org/10.1016/j.compgeo.2016.10.014]

Liu, L.L., Cheng, Y.M., Wang, X.M., Zhang, S.H., Wu, Z.H. (2017). System reliability analysis and risk assessment of a layered slope in spatially variable soils considering stratigraphic boundary uncertainty. *Comput. Geotech, 89*, 213-225.
[http://dx.doi.org/10.1016/j.compgeo.2017.05.014]

Liu, L.L., Deng, Z., Zhang, S.H., Cheng, Y.M. (2018). Simplified framework for system reliability analysis of slopes in spatially variable soils. *Eng. Geol, 239*, 330-343.
[http://dx.doi.org/10.1016/j.enggeo.2018.04.009]

Liu, L.L., Cheng, Y.M., Zhang, S.H, Z., Deng (2018). Simplified framework for system reliability analysis of slopes in spatially variable soils. *Engineering Geology, 239*, 330-343.

Liu, L.L., Zhang, S.H., Cheng, Y.M. (2018). Advanced reliability analysis of slopes in spatially variable soils using multivariate adaptive regression splines. *Geoscience Frontiers*.
[http://dx.doi.org/10.1016/j.gsf.2018.03.013]

Liu, W.K., Jun, S., Zhang, Y. (1995). Reproducing kernel particle methods. *Int. J. Numer. Methods Eng, 20*, 1081-1106.
[http://dx.doi.org/10.1002/fld.1650200824]

Liu, W.K., Li, S., Belytschko, T. (1997). Moving least square reproducing kernel methods part i: Methodology and convergence. *Comput. Methods Appl. Mech. Eng, 143*(1-2), 113-154.
[http://dx.doi.org/10.1016/S0045-7825(96)01132-2]

Liu, Y.X., Zheng, Y.R. (2019). *Plastic Mechanics of Geomaterial.* Science Press and Springer.
[http://dx.doi.org/10.1007/978-981-13-3753-6]

Liu, G.R., Liu, M.B. (2003). *Smoothed Particle Hydrodynamics: A Meshfree Particle Method.* Singapore: World Scientific Publishing.
[http://dx.doi.org/10.1142/5340]

Livesley, R.K., Sabin, M.A. (1991). Algorithms for numbering the nodes of finite-element meshes. *Comput. Syst. Eng, 2*(1), 103-114.
[http://dx.doi.org/10.1016/0956-0521(91)90042-4]

Lloret-Cabot, M., Fenton, G.A., Hicks, M.A. (2014). On the estimation of scale of fluctuation in geostatistics. Georisk. *Assessment and Management of Risk for Engineered Systems and Geohazards, 8*(2), 129-140.
[http://dx.doi.org/10.1080/17499518.2013.871189]

Lo, S.H. (2015). *Finite element mesh generation.* CRC Press.

Lophaven, S.N., Nielsen, H.B., Søndergaard, J. (2002). *Dace, a matlab kriging toolbox.*

Lorig, L.J. (1984). *A Hybrid Computational Model for Excavation and Support Design in Jointed Media.*

Low, B.K., Tang, W.H. (1997). Reliability analysis of reinforced embankments on soft ground. *Can. Geotech. J, 34*, 672-685. a

[http://dx.doi.org/10.1139/t97-032]

Low, B.K., Tang, W.H. (1997). Probabilistic slope analysis using Janbu's generalized procedure of slices. *Comput. Geotech, 21*(2), 121-142. b
[http://dx.doi.org/10.1016/S0266-352X(97)00019-0]

Low, B.K. (2003). Practical probabilistic slope stability analysis.

Low, B.K., Lacasse, S., Nadim, F. (2007). Slope reliability analysis accounting for low spatial variation. *Georisk, 1*(4), 177-189.

Low, B.K., Zhang, J., Tang, W.H. (2011). Efficient system reliability analysis illustrated for a retaining wall and a soil slope. *Comput. Geotech, 38*(2), 196-204.
[http://dx.doi.org/10.1016/j.compgeo.2010.11.005]

Lowe, J., Karafiath, L. (1960). Stability of Earth Dams Upon Drawdown *Proceedings of the 1st Pan-American Conference on Soil Mechanics and Foundation Engineering,* Mexico2, 537-552.

Lu, Z.M., Zhang, D.X. (2007). Stochastic simulations for flow in nonstationary randomly heterogeneous porous media using a KL-based moment-equation approach. *Multiscale Model. Simul, 6*(1), 228-245.
[http://dx.doi.org/10.1137/060665282]

Lucy, L.B. (1977).)A numerical approach to the testing of the fission hypothesis. *The Astronomical Journal, 82*(1977), 1024.

Lumb, P., Lee, C.F. (1975). Clay mineralogy of the Hong Kong soils. *Proceedings 4th South East Asian Conference on Soil Engineering,* Kuala Lumpur, Malaysia1-41.

Lumb, P. (1970). Safety factors and the probability distribution of soil strength. *Can. Geotech. J, 7*(3), 225-242.
[http://dx.doi.org/10.1139/t70-032]

Lumb, P., Lee, C.F. (1975). Clay mineralogy of the Hong Kong soils. Proceedings 4th South East Asian Conference on Soil Engineering.(pp. 1-41). Kuala Lumpur, Malaysia,.

Luo, S.M., Zhang, X.W., Cai, Y.C. (2000). The variational principle and application of nonlinear numerical manifold method. *Chinese Journal of Applied Mathematics and Mechanics, 21*(12), 1265-1270.

Luo, S.M., Zhang, X.W., Cai, Y.C. (2001). The variational principle and application of numerical manifold method. *Chinese Journal of Applied Mathematics and Mechanics, 22*(6), 587-592.

Luo, X.F., Li, X., Zhou, J., Cheng, T. (2012). A Kriging-based hybrid optimization algorithm for slope reliability analysis. *Struct. Saf, 34*(1), 401-406.
[http://dx.doi.org/10.1016/j.strusafe.2011.09.004]

Luo, X.F., Li, X., Zhou, J., Cheng, T. (2012). A kriging-based hybrid optimization algorithm for slope reliability analysis. *Struct. Saf, 34*(1), 401-406. b
[http://dx.doi.org/10.1016/j.strusafe.2011.09.004]

Lyamin, A.V., Sloan, S.W. (1997). A comparison of linear and nonlinear programming formulations for lower bound limit analysis In: Pietruszczak, S., Pande, G.N., (Eds.), *Proceedings of the 6th International Symposium on Numerical Models in Geomechanics,* 367-373.Balkema.Rotterdam:

Lysmer, J. (1970). Limit analysis of plane problems in soil mechanics. *J. Soil Mech. Found. Div, 96*(SM4), 1311-1334.

Ma, S., Zhang, X., Qiu, X.M. (2009). Comparison study of MPM and SPH in modeling hypervelocity impact problems. *Int. J. Impact Eng, 36*(2), 272-282.
[http://dx.doi.org/10.1016/j.ijimpeng.2008.07.001]

Maclaughlin, M.M., Sitar, N. (1996). Rigid-body rotations in DDA Proceedings of the First International Forum on Discontinuous Analysis(DDA) and Simulations of Discontinuous Media, TSI Press, Berkeley, California, USA, p. 620-636.

Maday, Y., Patera, A.T. (1989). Spectral element methods for the incompressible Navier-Stokes equations. *State-of-the-Art surveys on computational mechanics.* New York: ASME.

Malkawi, A.I.H., Hassan, W.F., Sarma, S.K. (2001). Global search method for locating general slip surface using Monte Carlo techniques. *J. Geotech. Geoenviron. Eng, 127*, 688-698.
[http://dx.doi.org/10.1061/(ASCE)1090-0241(2001)127:8(688)]

Mao, Z., Liu, G.R., Dong, X. (2017). A comprehensive study on the parameters setting in smoothed particle hydrodynamics (SPH) method applied to hydrodynamics problems. *Comput. Geotech, 92*, 77-95.
[http://dx.doi.org/10.1016/j.compgeo.2017.07.024]

Martin, C.M. (2004). *User Guide for ABC - Analysis of Bearing Capacity (v1.0).* Department of Engineering Science, University of Oxford.

Matsuo, M., Kuroda, K. (1974). Probabilistic approach to design of embankments. *Soil Found, 14*(2), 1-17.
[http://dx.doi.org/10.3208/sandf1972.14.2_1]

Matuttis, H.G., Chen, J. (2014). *Understanding the discrete element method.* John Wiley.
[http://dx.doi.org/10.1002/9781118567210]

Mayne, D.Q., Polak, E. (1976). Feasible directions algorithms for optimization problems with equality and inequality constraints. *Math. Program, 11*, 67-80.
[http://dx.doi.org/10.1007/BF01580371]

Mester, D., Ronin, D., Frenkel, M., Korol, A., Braysy, Z., Dullaert, O., Raa, W. (2010). *Discrete optimization for TSP-Like genome mapping problems.* Nova Science Publishers.

Meyerhof, G.G. (1963). Some recent research on the bearing capacity of foundations. *Can. Geotech. J, 1*, 16-26.
[http://dx.doi.org/10.1139/t63-003]

Michalowski, R.L. (1989). Three-dimensional analysis of locally loaded slopes. *Geotechnique, 39*(1), 27-38.
[http://dx.doi.org/10.1680/geot.1989.39.1.27]

Michalowski, R.L. (1995). Slope stability analysis: A kinematical approach. *Geotechnique, 45*(2), 283-293.
[http://dx.doi.org/10.1680/geot.1995.45.2.283]

Michalowski, R.L. (1997). An estimate of the influence of soil weight on bearing capacity using limit analysis. *Soil Found, 37*(4), 57-64.
[http://dx.doi.org/10.3208/sandf.37.4_57]

Mohammadi, S. (2003). *Discontinuous mechanics Using finite and discrete elements.* WIT Press.

Monaghan, J.J., Lattanzio, J.C. (1985). A refined particle method for astrophysical problems. *Astron. Astrophys, 149*, 135-143.

Mononobe, N., Matsuo, H. (1929). On the determination of earth pressure during earthquakes *Proceeding of the World Engineering Congress,* Tokyo*Vol.9*, 179-187.

Morgenstern, N.R., Price, V.E. (1965). The analysis of stability of general slip surface. *Geotechnique, 15*(1), 79-93.
[http://dx.doi.org/10.1680/geot.1965.15.1.79]

The Evaluation of Slope Stability - A 25-Year Perspective, Stability and Performance of Slopes and Embankments - II Morgenstern, N.R. (1992). Geotechnical Special PublicationASCE.

Morgenstern, N.R. (1994). *Report on Kwun Lung Lau Landslide of 23 July, 1994; Causes of the Landslide and Adequacy of Slope Safety Practice in Hong Kong.* Hong Kong: Civil Engineering Department.

Morgenstern, N.R. (1995). Managing risk in geotechnical engineering. *Proceedings 10th Pan-American Conference on Soil Mechanics and Foundation Engineering,* Guadalajara *4*, 102-126.

Morrison, E.E., Jr, Ebeling, R.M. (1995). Limit equilibrium computation of dynamic passive earth pressure. *Can. Geotech. J, 32*, 481-487.
[http://dx.doi.org/10.1139/t95-050]

Mroz, Z., Drescher, A. (1969). Limit plasticity approach to some cases of flow of bulk solids, Journal of Engineering for Industry. *Trans. Am. Soc. Mech. Eng, 51*, 357-364.

Munjiza, A.A., Knight, E.E., Rougier, E. (2012). *Computational mechanics of discontinua.* John Wiley.

Munjiza, A. (2004). *The Combined Finite-Discrete Element Method.* John Wiley.
[http://dx.doi.org/10.1002/0470020180]

Nash, D. (1987). A comparative review of limit equilibrium methods of stability analysis. *Slope Stability* (pp. 11-75). New York: John Wiley & Sons.

Nayroles, B., Touzot, G., Villon, P. (1992). Generalizing the finite element method: diffuse approximation and diffuse elements. *Comput. Mech, 10*, 307-318.
[http://dx.doi.org/10.1007/BF00364252]

Nayyar, A., Le, D.N., Nguyen, N.G. (2019). *Advances in Swarm Intelligence for Optimizing Problems in Computer Science.* CRC Press.

Nedjah, N., Alba, E. (2006). Parallel Evolutionary Computations, Springer.

Nedderman, R.M. (1992). Statistics and kinematics of granular materials, Cambridge University Press.

Nguyen, V.U. (1985). Determination of critical slope failure surfaces. *J. Geotech. Eng, 111*(2), 238-250.
[http://dx.doi.org/10.1061/(ASCE)0733-9410(1985)111:2(238)]

Nithiarasu, P., Lewis, R.W., Seetharamu, K.N. (2016). *Fundamentals of the Finite Element Method for Heat and Mass Transfer* John Wiley.

Nocedal, J., Wright, S.J. (2006). *Numerical Optimization.* Springer.

Norouzi, H.R., Zarghami, R., Sotudeh-Gharebagh, R., Mostoufi, N. (2016). *Coupled CFD-DEM Modeling.* John Wiley.
[http://dx.doi.org/10.1002/9781119005315]

Novotortsev, V.I. (1938). Experience with the application of the theory of plasticity to problems of determination of the bearing capacity of foundation beds of structures, Izv. Nauchno-Issled. *Inst. Anzh. Geol, 22*, 115-128.

Oka, Y., Wu, T.H. (1990). System reliability of slope stability, Journal of Geotechnical Engineering, ASCE. *Geotech. Eng, 116*(8), 1185-1189.
[http://dx.doi.org/10.1061/(ASCE)0733-9410(1990)116:8(1185)]

Ohnishi, Y., Tanaka, M., Sasaki, T. (1995). Modification of DDA for elasto-plastic analysis with illustrative generic problems. *Proceedings of the 35th U.S. Symposium On Rock Mechanics,* Lake Tahoe, California45-50.

Ohnishi, Y., Miki, S. (1996). Development of circular and elliptic disc elements for DDA.

Ohtsubo, H., Suzuki, K., Terada, K., Nakanishi, K. (1997). Utilization of finite covers in the manifold method for accuracy control. *Proceedings of the Second International Conference on Analysis of Discontinuous Deformation,* Kyoto, Japan317-322.

Ohnishi, Y., Tanaka, R., Koyama, T. (1999). Manifold method in saturated-unsaturated groundwater flow analysis.

Okereke, M., Keates, S. (2018). *Finite Element Applications A Practical Guide to the FEM Process.* Springer.
[http://dx.doi.org/10.1007/978-3-319-67125-3]

Oñate, E., Idelsohn, S. (1998). A mesh-free finite point method for advective-diffusive transport and fluid flow problems. *Comput. Mech, 21*(4-5), 283-292.

Oñate, E., Owen, D.R.J. (2011). *Particle-Based Methods.* Springer. a
[http://dx.doi.org/10.1007/978-94-007-0735-1]

Oñate, E., Owen, D.R.J. (2011). *Particle-based Methods II Fundamentals and Applications, International Center for Numerical Methods in Engineering.* CIMNE. b

Onwubolu, G., Davendra, D. (2009). *Differential Evolution: A Handbook for Global Permutation-Based Combinatorial Optimization.* Springer.
[http://dx.doi.org/10.1007/978-3-540-92151-6]

O'Sullivan, C. (2002). *The application of discrete element modelling to finite deformation problems in geomechanics.*

O□Sullivan, C. (2011). *Particulate Discrete Element Modelling.* Spon Press.

Ourique, C.O., Biscaia, E.C., Pinto, J.C. (2002). The use of particle swarm optimization for dynamic analysis in chemical processes. *Comput. Chem. Eng, 26*, 1783-1793.
[http://dx.doi.org/10.1016/S0098-1354(02)00153-9]

Owen, D.R.J., Hinton, E. (1980). *Finite element in Plasticity.* Pineridge Press Limited.

Paice, G.M., Griffiths, D.V., Fenton, A. (1996). Finite element modeling of settlements on spatially random soil. *J. Geotech. Eng, 122*(9), 777-779.
[http://dx.doi.org/10.1061/(ASCE)0733-9410(1996)122:9(777)]

Pan, J. (1980). *Analysis of Stability and Landslide for Structures* (pp. 25-28). Beijing: Hydraulic Press. (in

Chinese)

Paradalos, P.M. (1993). *Complexity in numerical optimization.* Utopia Press.
[http://dx.doi.org/10.1142/2041]

Pardalos, P.M., Du, D.Z., Graham, R.L. (2005). *Handbook of Combinatorial Optimization.* Springer.

Patera, A. T. (1984).), A spectral element method for fluid dynamics: laminar flow in a channel expansion. *J. Comput. Phys,* 468-488.

Pearce, C., Bicanic, N., Thavalingham, A., Liao, Z.H. (1999). On the DDA framework for modelling concrete fracture.

Peck, R.B. (1973). Influence of non-technical factors on the quality of embankment dams. *Embankment Dam Engineering, Casagrande Volume* John Wiley.

Pepper, D.W., Heinrich, J.C. (2006). *The Finite Element Method Basic Concepts and Applications* CRC Press.

Phoon, K., Kulhawy, F.H. (1999). Characterization of geotechnical variability. *Can. Geotech. J, 36*(4), 612-624. a
[http://dx.doi.org/10.1139/t99-038]

Phoon, K., Kulhawy, F.H. (1999). Evaluation of geotechnical property variability. *Can. Geotech. J, 36*(4), 625-639. b
[http://dx.doi.org/10.1139/t99-039]

Phoon, K.K., Ching, J. (2015). *Risk and reliability in geotechnical engineering.* CRC Press.

Pillo, G.D., Roma, M. (2006). *Large scale nonlinear optimization.* Springer.
[http://dx.doi.org/10.1007/0-387-30065-1]

Plewa, T., Linde, T., Weirs, V.G. (2003). Adaptive mesh refinement - theory and applications *Proceedings of the Chicago Workshop on Adaptive Mesh Refinement Methods.*Sept. 3-5, 2003

Polidori, D.C., Beck, J.L., Papadimitriou, C. (1999). New approximations for reliability integrals. *J. Eng. Mech, 125*(4), 466-475.
[http://dx.doi.org/10.1061/(ASCE)0733-9399(1999)125:4(466)]

Pozrikidis, C. (2014). *Introduction to Finite and Spectral Element Methods Using MATLAB* Chapman and Hall.

Prandtl, L. (1920). Uber die Harte plasticher Korper, Nachrichten von der Koniglichen Gesellschaften, Gottingen, Math.-phys. Klasse, p.74-85.

Prater, E.G. (1977). An examination of some theories of earth pressure on shaft lining. *Can. Geotech. J, 14*(1), 91-106.
[http://dx.doi.org/10.1139/t77-007]

Press, W.H., Teukolsky, S.A., Vetterling, W.T., Flannery, B.P. Numerical recipes in Fortran 90. .Cambridge University Press.

Pugsley, A. (1955). Report on structural safety. *Struct. Eng, 33*(5), 141-149.

Qiu, X.J. (1996). Manifold method without use of penalty springs.

Rackwitz, R., Fiessler, B. (1978). Structural reliability under combined random load sequences. *Comp. & Struct, 9*, 484-494.
[http://dx.doi.org/10.1016/0045-7949(78)90046-9]

Rackwitz, R. (2001). Reliability analysis: A review and some perspectives. *Struct. Saf, 23*(4), 365-395.
[http://dx.doi.org/10.1016/S0167-4730(02)00009-7]

Rao, S.S. (2018). *The finite element method in engineering* (6th ed.). London: Butterworth Heinemann.

Rao, S.S. (2009). *Engineering Optimization.* Wiley.
[http://dx.doi.org/10.1002/9780470549124]

Rassam, D.W., Williams, D.J. (1999). 3-dimensional effects on slope stability of high waste rock dumps. *International Journal of Surface Mining, 13*, 19-24.
[http://dx.doi.org/10.1080/09208119908944197]

Ravindra, M.K., Galambos, T.V. (1978). Load and resistance factor design for steel. *J. Struct. Div, 104*(ST9), 1427-1441.

Reale, C., Xue, J.F., Pan, Z.M., Gavin, K. (2015). Deterministic and probabilistic multi-modal analysis of slope stability. *Comput. Geotech, 66*, 172-179.
[http://dx.doi.org/10.1016/j.compgeo.2015.01.017]

Reddy, J.N. (2000). *Theory and Analysis of Elastic Plates.* Taylor and Francis.

Reddy, J.N. (2015). *An Introduction to Nonlinear Finite Element Analysis.* Oxford University Press.

Rees, D.W.A. (2006). *Basic engineering plasticity.* Elsevier.

Reissner, H. (1924). Zum Erddruckproblem In: Biezeno, C.B., Burgers, J.M., (Eds.), *Proc. 1st Int. Congress for Applied Mechanics,* Delft295-311.

Reissner, H. (1924). Zum Erddruckproblem In: Biezeno, C.B., Burgers, J.M., (Eds.), *Proc. 1st Int. Congress for Applied Mechanics,* Delft295-311.

Revilla, J., Castillo, E. (1977). The calculus of variations applied to stability of slopes. *Geotechnique, 27*, 1-11.
[http://dx.doi.org/10.1680/geot.1977.27.1.1]

Rodriguez, R.J., Sitar, N., Chacon, J. (2006). System reliability approach to rock slope stability. *Int. J. Rock Mech. Min. Sci.*
[http://dx.doi.org/10.1016/j.ijrmms.2005.11.011]

Rosenblueth, E. (1981). Two-point estimates in probabilities. *Appl. Math. Model, 5*(2), 329-335.
[http://dx.doi.org/10.1016/S0307-904X(81)80054-6]

Rourke, J. (1998). *Computational Geometry in C* Cambridge University Press.
[http://dx.doi.org/10.1017/CBO9780511804120]

Sarhosis, V., Bagi, K., Lemos, J.V., Milani, G. (2016). *Computational Modeling of Masonry Structures Using the Discrete Element Method.* Engineering Science Reference.
[http://dx.doi.org/10.4018/978-1-5225-0231-9]

Sarma, S.K., Tan, D. (2006). Determination of critical slip surface in slope analysis. *Geotechnique, 56*(8),

539-550.
[http://dx.doi.org/10.1680/geot.2006.56.8.539]

Schultze, E. (1975). Some aspects concerning the application of statistics and probability to foundation structures *Proceedings of the 2nd International Conference on Applications of Statistics and Probability in Soil and Structural Engineering,* Aachen457-494.

Senders, J.W., Moray, N.P. (1991). *Human Error: Cause, Prediction and Reduction.* New Jersey: Lawrence Erlbaum Associates.

Sherali, H.D., Terlaky, T., Ye, Y. (2016). *Advances in Stochastic and Deterministic Global Optimization.* Springer.

Shi, G.H., Goodman, R.E. (1985). Two dimensional discontinuous deformation analysis. *Int. J. Numer. Anal. Methods Geomech, 39,* 541-556.
[http://dx.doi.org/10.1002/nag.1610090604]

Shi, G.H. (1988). *DDA-a new numerical model for the static and dynamics of block system.*

Sh, G.H. (1993). Block system modelling by discontinuous deformation analysis. *Comput. Mech,* 11.

Shinozuka, M., Deodatis, G. (1988). Response variability of stochastic finite-element systems. *J. Eng. Mech, 114*(3), 499-519.
[http://dx.doi.org/10.1061/(ASCE)0733-9399(1988)114:3(499)]

Shooman, M. (1968). *Probabilistic Reliability an Engineering Approach.* McGraw Hill.

Shyu, K., Salami, M.R. (1995). Manifold with four-node isoparametric finite element method. *Work Forum on Manifold Method of Material, California, USA, V1,* 165-182.

Silvestrini, R.T., Montgomery, D.C., Jones, B. (2013). Comparing computer experiments for the Gaussian process model using integrated prediction variance. *Qual. Eng, 25*(2), 164-174.
[http://dx.doi.org/10.1080/08982112.2012.758284]

Sloan, S.W. (1988). Lower bound limit analysis using finite elements and linear programming. *Int. J. Numer. Anal. Methods Geomech, 12*(1), 61-77.
[http://dx.doi.org/10.1002/nag.1610120105]

Sloan, S.W. (1989). Upper bound limit analysis using finite-elements and linear-programming. *Int. J. Numer. Anal. Methods Geomech, 13*(3), 263-282.
[http://dx.doi.org/10.1002/nag.1610130304]

Sloan, S.W., Randolph, M.F. (1983). Automatic element reordering for finite element analysis with frontal solution schemes. *Int. J. Numer. Methods Eng, 19,* 1153-1181.
[http://dx.doi.org/10.1002/nme.1620190805]

Sloan, S.W. (1989). A Fortran program for profile and wavefront reduction. *Int. J. Numer. Methods Eng, 28,* 2651-2679.
[http://dx.doi.org/10.1002/nme.1620281111]

Smith, I.M., Griffiths, D.V., Margetts, L. (2014). *Programming the Finite Element Method* John Wiley.

Smith, C., Gilbert, M. (2007). Application of discontinuity layout optimization to plane plasticity problems. *Proc. R. Soc. A, 463*(2086), 2461-2484.

[http://dx.doi.org/10.1098/rspa.2006.1788]

Stuedlein, A., Kramer, S., Arduino, P., Holtz, R. (2012). Geotechnical characterization and random field modeling of desiccated clay. *J. Geotech. Geoenviron. Eng, 138*(11), 1301-1313.
[http://dx.doi.org/10.1061/(ASCE)GT.1943-5606.0000723]

Sokolovskii, V.V. (1965). *Statics of Granular Media.* Pergamon Press.

Solnon, C. (2010). *Ant Colony Optimization and Constraint Programming.* ISTE Ltd.

Soubra, A.H. (2000). Static and seismic passive earth pressure coefficients on rigid retaining structures. *Can. Geotech. J, 37*, 463-478.
[http://dx.doi.org/10.1139/t99-117]

Sowers, G.F. (1991). *The human factor in failures.* Civil Engineering, ASCE, March, p. 72-73.

Spanos, P.D., Ghanem, R. (1989). Stochastic finite element expansion for random medium. *J. Eng. Mech, 115*(5), 1035-1053.
[http://dx.doi.org/10.1061/(ASCE)0733-9399(1989)115:5(1035)]

Spencer, E. (1967). A method of analysis of the stability of embankments assuming parallel inter-slice forces. *Geotechnique, 17*, 11-26.
[http://dx.doi.org/10.1680/geot.1967.17.1.11]

Strack, O.D.L., Cundall, P.A. (1984). *Fundamental studies of fabrics in granular materials, Interim report to National Science Foundation, NSF CEE-8310729.* Department of Civil and Mineral Engineering, University of Minnesota.

Suchomel, R., Masin, D. (2010). Comparison of different probabilistic methods for predicting stability of a slope in spatially variable c- soil. *Comput. Geotech, 37*(1), 132-140.
[http://dx.doi.org/10.1016/j.compgeo.2009.08.005]

Sudret, B., Der Kiureghian, A. (2002). Comparison of finite element reliability methods. *Probab. Eng. Mech, 17*, 337-348.
[http://dx.doi.org/10.1016/S0266-8920(02)00031-0]

Sukumar, N., Moran, B. (2001). Natural neighbour Galerkin Methods. *Int. J. Numer. Methods Eng, 50*, 1-27.
[http://dx.doi.org/10.1002/1097-0207(20010110)50:1<1::AID-NME14>3.0.CO;2-P]

Sulsky, D., Chen, Z., Schreyer, H. (1994). A particle method for history-dependent materials. *Comput. Methods Appl. Mech. Eng, 118*(1-2), 179-196.
[http://dx.doi.org/10.1016/0045-7825(94)90112-0]

Sze, K.Y. (2002). Three-dimensional continuum finite element models for plate/shell analysis. *Prog. Struct. Eng. Mater, 4*, 400-407.
[http://dx.doi.org/10.1002/pse.133]

Szilard, R. (2004). *Theory and Applications of Plates Analysis of Plates.* John Wiley.
[http://dx.doi.org/10.1002/9780470172872]

Talbi, E.G., Nakib, A. (2019). *Bioinspired Heuristics for Optimization.* Springer.
[http://dx.doi.org/10.1007/978-3-319-95104-1]

(1985). Some considerations of an acceptable level of risk in the Netherlands. Report by TAW workgroup 10, "Probabilistic methods".

Temme, Y., Goh, C.K. (2010). *Computational Intelligence in Expensive Optimization Problems.* Springer.

Thapa, B. (1996). Joint shear displacement-dilation analysis using in-situ opposing profiles. *Proceedings of 35th U.S. Symposium on Rock Mechanics,* Lake Tahoe, CA, USA44-51.

Thornton, C. (2015). *Granular Dynamics, Contact Mechanics and Particle System Simulations.* Springer. [http://dx.doi.org/10.1007/978-3-319-18711-2]

Thornton, C., Randall, C.W. (1988). *Applications of theoretical contact mechanics to solid particle system simulations, Mechanics of granular materials.* Elsevier Science Publishing.

Timoshenko, S., Krieger, W. (1959). *Theory of Plates and Shells.* McGraw-Hill.

Ting, J.M., Corkum, B.T., Kauffman, C.R. (1989). Discrete element model for soil mechanics. *J. Geotech. Eng, 115*(3), 379-398.
[http://dx.doi.org/10.1061/(ASCE)0733-9410(1989)115:3(379)]

Tiwari, R.C., Bhandary, N.P., Yatabe, R. (2014). Spectral element analysis to evaluate the stability of long and steep slopes. *Acta Geotech, 2014*(9), 753-770.
[http://dx.doi.org/10.1007/s11440-013-0292-x]

Tiwari, R.C., Bhandary, N.P., Yatabe, R. (2015). 3-D elasto-plastic spectral element application to evaluate the stability of large-scale landslides. *Geomechanics and Geoengineering, 10*(4), 271-289.
[http://dx.doi.org/10.1080/17486025.2014.985337]

Tiwari, R.C. (2015). *3D Slope Stability Modeling: Numerical Methods and Applications: Slope stability, Prog. Failure, Finite-element method, Spectral-element method, Vegetation, Stability charts, Excavation.* LAP LAMBERT Academic Publishing.

Tsay, R.J., Chiou, Y.J., Chuang, W.L. (1999). Crack growth prediction by manifold method. *J. Eng. Mech, 125,* 884-890.
[http://dx.doi.org/10.1061/(ASCE)0733-9399(1999)125:8(884)]

Tuy, H. (1998). *Convex analysis and global optimization.* Kluwer Academic Publishers.
[http://dx.doi.org/10.1007/978-1-4757-2809-5]

Tvedt, L. (1990). Distribution of quadratic forms in normal space: Application to structural reliability. *J. Eng. Mech, 116*(6), 1183-1197.
[http://dx.doi.org/10.1061/(ASCE)0733-9399(1990)116:6(1183)]

Ugai, K. (1988). 3-D Slope Stability Analysis by Slice Methods *Proceedings of the 6th International Conference on Numerical Methods in Geomechanics,* Innsbruck, Austria1369-1374.

Ugai, K., Leshchinsky, D. (1995). Three-dimensional limit equilibrium and finite element analysis: a comparison of results. *Soil Found, 35*(4), 1-7.
[http://dx.doi.org/10.3208/sandf.35.4_1]

Urugal, A.C. (1999). *Stress in Plates and Shells* McGraw-Hill.

Vahedifard, F., Leshchinsky, B., Sehat, S., Leshchinsky, D. (2014). Impact of Cohesion on Seismic Design of Geosynthetic-Reinforced Earth Structures. *J. Geotech. Geoenviron. Eng, 140*(6)04014016
[http://dx.doi.org/10.1061/(ASCE)GT.1943-5606.0001099]

Vanmarcke, E.H. (1977). Probabilistic modeling of soil profiles. *J. Geotech. Eng, 103*(11), 1227-1246. a

Vanmarcke, E.H. (1977). Reliability of earth slopes. *J. Geotech. Eng, 103*(11), 1247-1265. b

Vanmarcke, E., Grigoriu, M. (1983). Stochastic finite element analysis of simple beams. *J. Eng. Mech, 109*(5), 1203-1214.
[http://dx.doi.org/10.1061/(ASCE)0733-9399(1983)109:5(1203)]

Vanmarcke, E.H. (2010). *Random Fields: Analysis and Synthesis (revised and expanded).* New ed. World Scientific Publishing Co. Pte. Ltd.Singapore.:

Vesic, A.S. (1973). Analysis of ultimate loads for shallow foundations. *Journal of Soil Mechanics and Foundations Division, ACSE, 99*(SM1), 45-73.

Verlet, L. (1967). Computer "Experiments" on Classical Fluids. I. Thermodynamical Properties of Lennard-Jones Molecules'. *Phys. Rev, 159*, 98-103.
[http://dx.doi.org/10.1103/PhysRev.159.98]

Vick, S.G. (1995). *Geotechnical risk and reliability - from theory to practice in dam safety.* In The Earth, Engineers and Education - Whitman Symposium, Massachusetts Institute of Technology, p. 45-58.

Vrijling, J.K., van Hengel, W., Houben, R.J. (1998). *Acceptable risk as a basic for design. Journal of Reliability Engineering and System Safety* Elsevier.

Walton, O.R. (1982). Explicit particle dynamics model for granular materials, The 4th International Conference on Numerical Methods in Geomechanics, p.1261-1268.

Wang, S.F. (2000). *Research on numerical manifold method of discontinuous and nonlinear deformation analysis for rockmass.*

Wang, Y. (2012). Uncertain parameter sensitivity in Monte Carlo simulation by sample Reassembling. *Comput. Geotech, 46*, 39-47.
[http://dx.doi.org/10.1016/j.compgeo.2012.05.014]

Wang, Y. (2013). Mcs-based probabilistic design of embedded sheet pile walls. *Georisk, 7*(3), 151-162.
[http://dx.doi.org/10.1080/17499518.2013.765286]

Wang, Y., Cao, Z.J., Au, S.K. (2010). Efficient Monte Carlo simulation of parameter sensitivity in probabilistic slope stability analysis. *Comput. Geotech, 37*(7-8), 1015-1022.
[http://dx.doi.org/10.1016/j.compgeo.2010.08.010]

Wang, Y., Cao, Z.J., Au, S.K. (2011). Practical reliability analysis of slope stability by advanced Monte Carlo simulations in a spreadsheet. *Can. Geotech. J, 48*, 162-172.
[http://dx.doi.org/10.1139/T10-044]

Wang, C.Y., Sheng, J., Chen, M.H., Chuang, C.C. (1995). Approximation theories for the manifold method. *Work Forum on Manifold Method of Material, California, USA, 1*, 61-86.

Wang, Q., Shi, X.W. (2009). An improved algorithm for matrix bandwidth and profile reduction in finite element analysis. *Progress In Electromagnetics Research Letters, 9*, 29-38.
[http://dx.doi.org/10.2528/PIERL09042305]

Wang, S.L., Ge, X.R., Zhang, G. (1999). Manifold method with four physical covers forming an element and its application. *The Third International Conference on Analysis of Discontinuous Deformation from Theory to Practice,* Vail, Colorado, USA193-201.

Wang, S.F., Zhu, W.S., Li, S.C., Qiu, X.B. (2001). Numerical manifold method considering lateral effect and its application. *Chinese Journal of Rock Mechanics and Engineering, 20*(3), 297-300.

Wang, S.F., Zhu, W.S., Li, S.C., Chen, S.H. (2002). Numerical manifold method of elastoplastic analysis for rockmass. *Chinese Journal of Rock Mechanics and Engineering, 21*(6), 900-904. a

Wang, S.F., Zhu, W.S., Li, S.C., Chen, S.H. (2002). Numerical manifold method of deformation analysis for bolt supported rockmass. *Chinese Journal of Rock Mechanics and Engineering, 21*(8), 1120-1123. b

Wang, Z.Y., Wang, S.J., Yang, Z.F. (1997). Manifold method in analysis of large deformation for rock. *Chinese Journal of Rock Mechanics and Engineering, 16*(5), 513-516.

Wang, C.M., Reddy, J.N., Lee, K.H. (2000). *Shear deformable beams and plates.* Elsevier.

Wang, L., Liu, B. (2008). *Particle swarm optimization and scheduling algorithms.* China: Tsinghua University Press.

Wei, W.B., Cheng, Y.M., Li, L. (2009). Three-dimensional slope failure by strength reduction and limit equilibrium methods. *Comput. Geotech, 36*, 70-80.
[http://dx.doi.org/10.1016/j.compgeo.2008.03.003]

Wei, W.B., Cheng, Y.M. (2009). Soil nailed slope by strength reduction and limit equilibrium methods. *Comput. Geotech, 37*, 602-618.
[http://dx.doi.org/10.1016/j.compgeo.2010.03.008]

Wei, W.B., Cheng, Y.M. (2009). Strength reduction analysis for slope reinforced with one row of piles. *Comput. Geotech, 36*, 1176-1185.
[http://dx.doi.org/10.1016/j.compgeo.2009.05.004]

Wei, W.B., Cheng, Y.M. (2010). Stability analysis of slope with water flow by strength reduction method. *Soil Found, 50*(1), 83-92.
[http://dx.doi.org/10.3208/sandf.50.83]

Wilson, E.L., Taylor, R.L., Doherty, W., Ghaboussi, J. (1973). *Incompatible Displacement Models, Numerical and Computer Methods in Structural Mechanics.* Academic press.

Wolff, T.F. (1985).

Wolff, T.F. (1991). Embankment reliability versus factor of safety - Before and after slide repair. *Int. J. Numer. Anal. Methods Geomech, 15*(1), 41-50.
[http://dx.doi.org/10.1002/nag.1610150104]

Wriggers, P., Panagiotopoulos, P. (1999). *New developments in contact problems.* Springer.
[http://dx.doi.org/10.1007/978-3-7091-2496-3]

Wu, H.Y. (2012). *Discrete Element Modelling of Particulate Media.* RSC Publishing.
[http://dx.doi.org/10.1039/9781849735032]

Wu, T.H. (2008). *Reliability analysis of slopes, Reliability based design in geotechnical engineering: Computations and Applications* (pp. 413-447). London: Taylor and Francis.

Wu, L.Y., Tsai, Y.F. (2004). Analysis of Earth Pressure for Retaining Wall and Ultimate Bearing Capacity for Shallow Foundation by Variational method. *J. Mech, 20*, 45-55.

Wu, LY, Tsai, YF (2005). Variational Stability Analysis of Cohesive Slope by Applying Boundary integral Equation Method. *Journal of Mechanics, 21*, 187-195.

Wu, Z.J., Wang, S.L., Ge, X.R. (2009). Slope reliability analysis by random FEM under constraint random field. *Chincese Journal of Rock and Soil Mechanics, 30*(10), 3086-3092.

Xing, Y.F., Liu, B.O., Liu, G. (2010). A differential quadrature finite element method. *Int. J. Appl. Mech, 02*, 207.
[http://dx.doi.org/10.1142/S1758825110000470]

Xu, B., Low, B.K. (2006). Probabilistic stability analyses of embankments based on finite-element method. *J. Geotech. Geoenviron. Eng, 132*(11), 1444-1454.
[http://dx.doi.org/10.1061/(ASCE)1090-0241(2006)132:11(1444)]

Xue, J.F., Gavin, K. (2007). Simultaneous Determination of Critical Slip Surface and Reliability Index for Slopes. *J. Geotech. Geoenviron. Eng, 133*(7), 878-886.
[http://dx.doi.org/10.1061/(ASCE)1090-0241(2007)133:7(878)]

Yang, Y., Huang, R.Q., Cheng, Y.M., Wang, J.F. (2014). Investigation of the Deformable Behavior of Loose and Dense Sand through DEM. *Adv. Mat. Res, 871*, 124-128.

Yang, Y., Cheng, Y.M. (2014). The fractal features of contact force chains for crushable granular materials under confined compression. *Proceedings of ASCE Engineering Mechanics Institute International Conference,* Hong Kong7-9.

Yang, Y., Cheng, Y.M. (2015). A fractal model of contact force distribution and the unified coordination distribution for crushable granular materials under confined compression. *Powder Technol, 279*, 1-9.
[http://dx.doi.org/10.1016/j.powtec.2015.03.006]

Yang, Y., Cheng, Y.M. (2016). Quantified evaluation of particle shape effects from micro-to-macro scales for non-convex grains. *Particuology, 25*, 23-35.
[http://dx.doi.org/10.1016/j.partic.2015.01.008]

Yang, Y., Cheng, Y.M. (2016). Exploring the contact types within mixtures of different shapes at the steady state by DEM. *Powder Technol, 301*, 440-448.
[http://dx.doi.org/10.1016/j.powtec.2016.06.016]

Yang, Y., Cheng, Y.M. (2017). The effects of rolling resistance and non-convex particle on the mechanics of the undrained granular assembles in 2D. *Powder Technol, 318*, 528-542.
[http://dx.doi.org/10.1016/j.powtec.2017.06.027]

Yang, Y., Cheng, Y.M., Sun, Q.C. (2018). Revisiting the confined comminution of granular materials with the consideration of the initial particle size distributions and repetitive loadings. *Powder Technol, 329*, 149-157.
[http://dx.doi.org/10.1016/j.powtec.2018.01.045]

Yang, Y.S. (2010). *Engineering Optimization An Introduction with Metaheuristic Applications.* Wiley.
[http://dx.doi.org/10.1002/9780470640425]

Yeung, M., Klein, S.J., Max, M. (1994). Application of the discontinuous deformation analysis to the evaluation of rock reinforcement for tunnel stabilization. *Proceedings of the First North American Rock Mechanics Symposium,* Austin, TX607-614.

Yi, P., Wei, K.T., Kong, X.J., Zhu, Z. (2015). Cumulative PSO-Kriging model for slope reliability analysis.

Probab. Eng. Mech, 39, 39-45.
[http://dx.doi.org/10.1016/j.probengmech.2014.12.001]

Yim, S.C. (2007). *3-D Wave-Structure Interaction with Coastal Sediments-A Multi-Physics/Multi-Solutin-Techniques Approach.* Corvallis: Defense Technical Information Center.

Yin, P.Y. (2004). A discrete particle swarm algorithm for optimal polygonal approximation of digital curves. *J. Vis. Commun. Image Represent, 15*, 241-260.
[http://dx.doi.org/10.1016/j.jvcir.2003.12.001]

Yu, H.S., Salgado, R., Sloan, S.W., Kim, J.M. (1998). Limit analysis versus limit equilibrium for slope stability. *J. Geotech. Geoenviron. Eng, 124*(1), 1-11.
[http://dx.doi.org/10.1061/(ASCE)1090-0241(1998)124:1(1)]

Yu, H.S. (2006). *Plasticity and geotechnics.* Springer.

Yu, X.J., Gen, M.S. (2010). *Introduction to Evolutionary Algorithms.* Springer.
[http://dx.doi.org/10.1007/978-1-84996-129-5]

Yuan, X., Lu, Z., Zhou, C., Yue, Z. (2013). A novel adaptive importance sampling algorithm based on markov chain and low-discrepancy sequence. *Aerosp. Sci. Technol, 29*(1), 253-261.
[http://dx.doi.org/10.1016/j.ast.2013.03.008]

Zelinka, I., Snasel, V., Abraham, A. (2013). *Handbook of Optimization From Classical to Modern Approach.* Springer.
[http://dx.doi.org/10.1007/978-3-642-30504-7]

Zhang, X. (1988). Three-dimensional stability analysis of concave slopes in plan view. *J. Geotech. Eng, 114*(6), 658-671.
[http://dx.doi.org/10.1061/(ASCE)0733-9410(1988)114:6(658)]

Zhang, X. (1999). Slope stability analysis based on the rigid finite element method. *Geotechnique, 49*(5), 585-593.
[http://dx.doi.org/10.1680/geot.1999.49.5.585]

Zhang, Y.H., Zhu, W.S., Qiu, X.B., Li, S.C. (1998). The application of DDA method on the excavation of the ground factory of Xiledu hydropower station. *Chinese Journal of Rock Mechanics and Engineering, 18*, 945-947.

Zhang, X.W., Cai, Y.C., Liao, L.C. (2000). Auto mesh algorithm of the finite cover system in the numerical manifold method. *Chinese Journal of Chongqing University, 23*(1), 28-31. [Natural Science Edition].

Zhang, X., Sanderson, D.J. (2002). *Numerical Modelling and Analysis of Fluid Flow and Deformation of Fractured Rock Masses.* Elsevier.

Zhang, G.X., Sigiura, Y., Hasegawa, H. (1997). Crack propagation and thermal fracture analysis by manifold method. *Proceedings of the Second International Conference on Analysis of Discontinuous Deformation,* Kyoto, Japan282-297.

Zhang, G.X., Sugiura, Y., Saito, K. (1999). Application of manifold method to jointed dam foundation. *The Third International Conference on Analysis of Discontinuous Deformation from Theory to Practice,* Vail, Colorado, USA211-220.

Zhang, J., Huang, H.W., Phoon, K.K. (2013). Application of the Kriging-Based Response Surface Method to the System Reliability of Soil Slopes. *J. Geotech. Geoenviron. Eng, 139*(4), 651-655.

[http://dx.doi.org/10.1061/(ASCE)GT.1943-5606.0000801]

Zhang, X., Chen, Z., Liu, Y. (2017). *The Material Point Method: A Continuum-Based Particle Method for Extreme Loading Cases.* Tsinghua University Press Computational Mechanics Series. [http://dx.doi.org/10.1016/B978-0-12-407716-4.00003-X]

Zhang, J., Zhang, L.M., Tang, W.H. (2011). New methods for system reliability analysis of soil slopes. *Can. Geotech. J, 48*(7), 1138-1148. [http://dx.doi.org/10.1139/t11-009]

Zhang, J., Huang, H.W., Juang, C.H., Li, D.Q. (2013). Extension of Hassan and Wolff method for system reliability analysis of soil slopes. *Eng. Geol, 60*, 81-88. [http://dx.doi.org/10.1016/j.enggeo.2013.03.029]

Zhang, W.G., Goh, A.T.C. (2013). Multivariate adaptive regression splines for analysis of geotechnical engineering systems. *Comput. Geotech, 48*, 82-95. [http://dx.doi.org/10.1016/j.compgeo.2012.09.016]

Zhao, J., Ohnishi, Y., Zhao, G.F., Sasaki, T. (2011). Advances in Discontinuous Numerical Methods and Applications in Geomechanics and Geoengineering *Proceedings of the 10th International Conference on advances in Discontinuous numerical methods and applications in geomechanics and geoengineering, ICADD 10.*Honululu, Hawaii6□8 Dec.

Zhao, H.B. (2008). Slope reliability analysis using a support vector machine. *Comput. Geotech, 35*(3), 459-467. [http://dx.doi.org/10.1016/j.compgeo.2007.08.002]

Zhao, W., Wang, W., Dai, H., Xue, G. (2010). Structural reliability analysis based on the cokriging technique. *IOP Conf. Series Mater. Sci. Eng, 10*012204 [http://dx.doi.org/10.1088/1757-899X/10/1/012204]

Zhao, Y.G., Ono, T. (1999). New approximations for SORM: Part 1. *J. Eng. Mech, 125*(1), 79-85. a [http://dx.doi.org/10.1061/(ASCE)0733-9399(1999)125:1(79)]

Zhao, W., Wang, W. (2011). Application of cokriging technique to structural reliability analysis. *2011 International Conference,* 170-174. [http://dx.doi.org/10.1109/ICFCSA.2011.46]

Zhao, Y.G., Ono, T. (1999). New approximations for SORM: Part 2. *J. Eng. Mech, 125*(1), 86-93. b [http://dx.doi.org/10.1061/(ASCE)0733-9399(1999)125:1(86)]

Zheng, H., Zhou, C.B., Liu, D.F. (2009). A robust solution procedure for the rigorous methods of slices. *Soil Found, 49*(4), 537-544. [http://dx.doi.org/10.3208/sandf.49.537]

Zhou, J.L., Tits, A.L. (1992). *User's Guide for FSQP Version 3.0c: A FORTRAN Code for Solving Constrained Nonlinear (Minimax) Optimization Problems, Generating Iterates Satisfying All Inequality and Linear Constraints, Technical research report.* University of Maryland.

Zhou, W.Y., Yang, Q., Kou, X.D. (1997). Development of high order manifold method. *Proceedings of the Second International Conference on Analysis of Discontinuous Deformation,* Kyoto, Japan274-281.

Zhou, W.Y., Kou, X.D., Yang, R.Q. (1999). Crack propagation by manifold method coupled with element free method. *The Third International Conference on Analysis of Discontinuous Deformation from Theory to Practice,* Vail, Colorado, USA283-289.

Zhu, Y.W. (1999). Augment manifold method in big deformation of rock. *Chinese Journal of Rock Mechanics and Engineering, 18*(1), 1-5.

Zhu, B.F. (2018). *The Finite Element Method, Fundamentals and Applications in Civil, Hydraulic, Mechanical and Aeronautical Engineering.* Tsinghua University Press.

Zhu, D.Y., Lee, C.F., Qian, Q.H., Zou, Z.S., Sun, F. (2001). A new procedure for computing the factor of safety using the Morgenstern–Price's method. *Can. Geotech. J, 38*(4), 882-888. [http://dx.doi.org/10.1139/t01-002]

Zhu, D.Y., Lee, C.F., Jiang, H.D. (2003). Generalised framework of limit equilibrium methods for slope stability analysis. *Geotechnique, 53*(4), 377-395. [http://dx.doi.org/10.1680/geot.2003.53.4.377]

Zhu, T., Atluri, S.N. (1998). A modified collocation method and a penalty formulation for enforcing the essential boundary conditions in the element free Galerkin method. *Comput. Mech, 21*(3), 211-222. [http://dx.doi.org/10.1007/s004660050296]

Zolfaghari, A.R., Heath, A.C., McCombie, P.F. (2005). Simple genetic algorithm search for critical non-circular failure surface in slope stability analysis. *Comput. Geotech, 32*(3), 139-152. [http://dx.doi.org/10.1016/j.compgeo.2005.02.001]

Zoyama, A.Y. (2006). *Handbook of nature-inspired and innovative computing.* Springer. [http://dx.doi.org/10.1007/0-387-27705-6]

Zienkiewicz, O.C., Taylor, R.L., Zhu, J.Z. (2011). *The finite element method: Its basis and fundamentals* (6th ed.). London: Butterworth Heinemann.

SUBJECT INDEX

A

B

Y.M. Cheng, J. H. Wang, L. Liang, W. H. Fung Ivan (Eds.)

E

F

G

H

M

N

O

P

Q

R

S

T

U

V

W

Y

Z

www.ingramcontent.com/pod-product-compliance
Lightning Source LLC
LaVergne TN
LVHW070942210926
848482LV00002B/75
9789811437380